D1647200

WATER AT THE SURFACE OF THE EARTH

An Introduction to Ecosystem Hydrodynamics

Student Edition

This is Volume 21 in
INTERNATIONAL GEOPHYSICS SERIES
A series of monographs and textbooks

A complete list of the books in this series appears at the end of this volume.

WATER AT THE SURFACE OF THE EARTH

An Introduction to Ecosystem Hydrodynamics

Student Edition

DAVID H. MILLER

Department of Geological Sciences
University of Wisconsin-Milwaukee
Milwaukee, Wisconsin

1977

ACADEMIC PRESS
A Subsidiary of Harcourt Brace Jovanovich, Publishers

New York London
Paris San Diego San Francisco São Paulo Sydney Tokyo Toronto

COPYRIGHT © 1977, BY ACADEMIC PRESS, INC.
ALL RIGHTS RESERVED.
NO PART OF THIS PUBLICATION MAY BE REPRODUCED OR TRANSMITTED IN ANY FORM OR BY ANY MEANS, ELECTRONIC OR MECHANICAL, INCLUDING PHOTOCOPY, RECORDING, OR ANY INFORMATION STORAGE AND RETRIEVAL SYSTEM, WITHOUT PERMISSION IN WRITING FROM THE PUBLISHER.

ACADEMIC PRESS, INC.
111 Fifth Avenue, New York, New York 10003

United Kingdom Edition published by
ACADEMIC PRESS, INC. (LONDON) LTD.
24/28 Oval Road, London NW1 7DX

LIBRARY OF CONGRESS CATALOG CARD NUMBER: 76-13947

ISBN 0-12-496752-3

PRINTED IN THE UNITED STATES OF AMERICA

82 83 84 85 9 8 7 6 5 4 3 2 1

To Fritz

CONTENTS

Chapter X
Evaporation from Wet Surfaces

Chapter XI
Evaporation from Well-Watered Ecosystems

Chapter XII
Evaporation from Drying Ecosystems

Chapter XIII
Water in the Local Air

Chapter XIV
Percolation from Ecosystems

PREFACE

I have tried to express in this book some of the ways that biological, physical, cultural, and urban systems at the surface of the earth operate. Of the many different forms of mass and energy these systems receive and transform, this book deals primarily with water seen in association with other forms of matter, including pollutants, and with several forms of energy; in other words, with the hydrodynamics of ecosystems.

Since it concentrates on the reception, processing, and transformation of water by ecosystems at the earth/air interface, the book is not a conventional hydrology or hydrometeorology text. It considers off-site flow, for instance, not from the viewpoint of channel hydrology, but as ecosystem yield, which is a counterpoint to input in these systems and a consequence of the modes of transformation.

The book approaches the dynamics of water in terrestrial systems through the budgets of water in each zone or environment of a system, e.g., the canopy, the ground surface, the soil, and so on. These zones extend the overall water budget in hydrology, which, expressed in numerous models and prediction procedures, long ago proved its worth, and which I met in flood engineering in 1941. Shortly thereafter I saw its association with the energy budget at the earth's surface in the generation of snow-melt floods. My view of these interface budgets from experience in engineering groups was later expanded as I worked with land managers, principally foresters, and with meteorologists. Each of these groups—engineers, meteorologists, and foresters—has from both pragmatic and fundamental standpoints contributed much to the study of hydrology in the United States and indeed throughout the world. In particular, I owe much to such people as Cleve Milligan, S. E. Rantz, the late Bill Bottorf, Henry

Anderson, and Frank Snyder, hydrologists in the Corps of Engineers, Forest Service, and Geological Survey.

In a different perspective, I have tried to blend the contributions of these hydrologists with those of such research workers, geographers, and meteorologists as John Leighly, Gilbert White, the late C. W. Thornthwaite, J. R. Mather, K. R. Knoerr, IUrii L'vovich Rauner, R. A. Muller, Gene Wilken, and Canute VanderMeer, working with rainfall, floodplain management, applications of the water budget in many areas, forest hydrology, and techniques of water management.

Many of the ideas in this book were first expressed in a manuscript that I wrote under the twin stimuli of viewing Australian water problems during a Fulbright year and the thoughts of Alan Tweedie and Alec Costin. Expansion of the material during a second Fulbright year and a term at Hawaii reflects the encouragement of James Auchmuty at Newcastle and Jen-hu Chang at Hawaii. I have also profited from comments and questions from my students in hydrology, meteorology, and climatology classes and seminars at Clark, Georgia, Berkeley, Newcastle, Macquarie, Hawaii, and, of course, Milwaukee. I am grateful to the Fulbright-Hays program and its Australian counterpart, the Australian-American Educational Foundation, for the years in Australia, and to the encouragement of my wife Enid for the whole book.

The drawings have been prepared under the able supervision of James J. Flannery, and photographs not otherwise credited were made by Enid Miller.

Chapter I

INTRODUCTION

Water at the surface of the earth represents a convergence of two objects of highest human interest: water, and the outer active surface of our planet.

Water is a unique molecule, present in three physical states and in bulk quantities on the earth. The outer active surface of our planet is the place where intense physical and chemical changes and almost all biological and cultural phenomena are concentrated.

This essential substance, water, is of obvious practical importance in ecosystems at and near the earth's surface. Its manifestations in these systems also present problems of intellectual significance, which have been studied by many fields of science —hydrology and oceanography, climatology and micrometeorology, soil science, geology and geophysics, ecology, and geography. Other problems have been examined (although sometimes incompletely) by practitioner disciplines—civil engineering, forestry, agronomy, resource management, city planning, and sanitary or environmental engineering. Each field, having its own focus elsewhere, fails to a degree to follow through on the coupling stated above: water as it is manifested in systems at and near the surface of the earth.

JUST WHAT IS THE EARTH'S SURFACE?

We will focus here on the complex outer skin of the earth and its ecosystems through which the pulses of water delivered by rainstorms make their ways. It is the earth's surface and its mantling ecosystems that are emphasized. We are less concerned about water in channels or captive in the hands of man than with water at the surface.

Such an examination of on-site processes identifies a series of water storages at different levels in ecosystems. These storages are seen on the leaf canopy, on the litter and ground, and in the soil of the ecosystems that cover the lands. The term "ecosystem" in the subtitle of the book encompasses the set of environments through which water moves. The term "hydrodynamics" indicates the movements themselves. Associated with the biological realm of the ecosystem are two zones that are connected with it; both the local air and the rock formations underlying the soil hold, take in, and give out water.

The storages of water in ecosystems are connected by water fluxes. Rain and snow are impacted from the atmosphere onto foliage; water blows or drips from leaves to the ground; water is infiltrated into the soil from which it percolates deeper into the ground or is extracted by transpiring plants. Movement of water off the site to which the atmosphere delivered it finally takes place in the form of vapor, surface or near-surface runoff, or groundwater movement.

The approach we are taking to water at the earth's surface does not follow the so-called hydrology "cycle," that often-cited, little used relic of the 17th century. Instead, we seek to follow the sequence of events occurring as water moves through ecosystems—a sequence that provides a chronological framework in which we can pursue the successive storage of water in the different levels and the fluxes that connect them. This framework of alternating storages and fluxes demonstrates the now rapid, now halting, progress of water through the ecosystems of the interfacial zone between the atmosphere and the bedrock of our planet.

THE BUDGET IDEA

Before we can discuss water in each of these ecosystem environments, we must understand that all the mass and energy of an ecosystem are as carefully counted as a miser's hoard. For every input there are equivalent outgoes; for a credit there is a balancing debit. Ecosystems operate on a budget of water as well as of nitrogen, carbon, or other forms of matter. Everything has a price and everything has to be paid for.

So in each environment at and near the earth's surface we will try to strike an account of inflows and outflows. We speak, in general, of "the water budget," but in actuality we make a budget for each environment within an ecosystem. For example, we make quantitative statements about the snow mantle on a meadow by measuring the input to it from snowstorms and the outputs from it by evaporation, off-site drifting, and downward movement of meltwater. Following this, we can strike another budget for the underlying soil, totting up inputs and outputs. We can use the same procedure for the deeper groundwater. In each environment, the water budget simply states the law of the conservation of matter. Its value is limited only by the accuracy with which we measure each flux or storage of water. In fact, casting a budget often warns us to look for unreliable measuring instruments or procedures.

Concomitant Budgets

The movement of water through the sequence of storages and fluxes is accompanied by the movement of waterborne materials of many kinds: dissolved gases and salts, nutrients, eroded soil particles, and even man-made molecules of the new biocides. These flows are nearly ubiquitous companions of the water flows. For instance, the salt that is spread on the roads inconsiderately moves into the groundwater body. Such mass budgets are a useful means of analyzing problems of environmental pollution.

We also recognize that no form of matter moves unless energy in some form is being expended. Therefore in each environment where we construct a water budget, say for the snow intercepted by trees in a winter storm, we can also construct an energy budget. The movement of intercepted snow out of the tree crowns is powered by applications of energy and does not take place otherwise. Evaporation of intercepted snow, for instance, is not as common as was once thought, because the large energy supply required is usually just not available.

Energy takes many forms. Some of those associated with water budgets are radiation, both short wave (solar energy with wavelengths shorter than 3 μm) and long wave (emitted by clouds, surfaces, and some atmospheric gases at wavelengths greater than 3 μm), and the sensible and latent forms of heat. Sensible heat is perceived as warmth of air or soil.

Latent heat, a form connected in many ways with water, represents the heat added to water when it changes physical state, as from liquid to gas. This is the heat of vaporization, 2500 kJ kg^{-1} of water. It is also, in the reverse process, the heat of condensation, released in clouds when vapor condenses into droplets.

The first law of thermodynamics states that the energy inputs and outputs to a system, plus the change of stored energy, add to zero at any instant. The amount of energy used in vaporizing 1 kg of water is equaled by the energy inputs, e.g., from solar radiation (or reduced heat storage in the water), received in the same period of time. If inputs fall short, evaporation does also. Analogies to the water budget are plain.

Joint analysis of energy and water budgets provides a double-barreled attack on many hydrologic problems, with a better chance of success because both fundamental continuity equations—of mass and of energy—have to be satisfied. A proposed water budget that does not check out in terms of inputs and outputs of energy is telling us that some of our measurements are wrong and need to be checked and improved.

Patterns of the Water Budget in Time

Solution of the water budget in a specific environment for long-term conditions, let us say over an entire year, must be compatible with its solution minute by minute. The law of conservation of mass applies as much to short periods of time as to long ones. Especially in short periods, the absence of steady-state conditions is compensated for by fluctuating amounts of water held in storage in the local environment.

Local storages in the environmental sequence through which water progresses sometimes tend to smooth the initial fluctuations fed into terrestrial systems by the episodic deliveries of water from rainstorms. The soil holds water from sporadic rain, feeding it out more gradually as vegetation transpires during succeeding days. On the other hand, some local storages generate their own fluctuations; snow builds on a fir branch during a storm, then suddenly slides to the ground.

The water budget helps us characterize the regular regimes of the day and the year insofar as they emerge in the various water fluxes. Seasonality over the span of the year is evidenced in many of the interactions between water and ecosystem processes, and is succinctly expressed in budget terms.

Similarly, the effects brought about by climatic change or by man over time can be examined by constructing water budgets for conditions before and after the change. This means of assessing the consequences of man's impact on the environment has been applied where logging, severe grazing, prescribed use of fire, clearing a forest for cultivation, urbanization, or other alterations of ecosystems have occurred. The budget is a powerful tool for analyzing these impacts on the environment.

Spatial Patterns of the Water Budget

Patterns in the landscape can be made concrete if we examine the spatial distribution of components of the water budget. We can depict the areal pattern of snowfall in a mountain valley, or radiant energy and other forms of heat supplied to the melting snow cover, and therefore the pattern of meltwater formation and the generation of off-site flow. While the budgets in each ecosystem in the valley are in balance, the mix of components will vary from place to place. We have a quantitative means of comparing ecosystems on north and south slopes, on ridges, in valleys, on granite or andesitic agglomerate, and in forest and cleared land, and we can see how they differ. On a medium spatial scale we can then construct a single water budget for the whole mosaic of ecosystems in the valley and its drainage basin. This areally averaged budget can then be compared with those characterizing other drainage basins to explain why one yields more streamflow than another, or sends it out sooner in the spring.

Similarly, we can strike a water budget for a large region, such as the snow zone of the California Sierra Nevada. On a still larger scale, we can make one for all of eastern North America, for a whole continent, or a whole ocean. For the entire earth, the budget is simple; an annual precipitation input to the surface of 1000 kg m^{-2} approximates the annual output by evaporation from the surface—the budget idea again!

WATER IN SYSTEMS

Although the earth's surface and its lower atmosphere taken together form a virtually closed system for water, the surface alone represents an open system, as does any sector of it or any ecosystem. Water moves in and out of each of these systems. Inputs and outputs can also be defined for levels or environments within each ecosystem, a set of environments that provides a logical sequence of water budgets, which feed one another. The outflow from the forest canopy becomes the inflow into the water system at the forest floor, infiltration through the forest floor becomes the input into the soil, and so on.

WATER SUPPLIED BY THE ATMOSPHERE TO THE EARTH'S SURFACE

Rain and Snow

The principal input of water to ecosystems on the earth's surface is rain and snow extracted by systems of vertical motion from vapor in the moving currents of the atmosphere. Although it is an areal phenomenon, rainfall is mostly known to us from measurements at specific points where rain gages are located. At these points the rain pattern has the dimensions of duration, depth during a storm, and intensity. From point data we can reconstruct the individual rain area or hydrologic storm. Over a period these provide a picture of seasonal and yearly rainfall to a whole region and its ecosystems.

We begin by describing storms in the atmosphere. These are systems that convert inflows of water vapor into outflows of raindrops and snowflakes that are precipitated to the underlying surface. Their budgets, involving the rates of inflow and outflow, are the fundamental idea in the next chapter. The chapter sequence in this book follows this downward progress of water from the lower atmosphere, through ecosystems at the earth's surface, through the soil and mantle rock, to the "waters under the earth." Four chapters describe how water is delivered from the atmosphere to surface ecosystems; four describe water budgets at the surface and in the soil; three describe evaporation from these systems back to the atmosphere; the following three discuss water in the local air and rocks, zones associated with ecosystems; and the last two chapters describe horizontal movement of water transformed by ecosystems where the preceding storages and fluxes were located. The book begins with input of water to ecosystems, then describes how it is processed in these systems, and ends with the liquid water yield from them.

Chapter II

ATMOSPHERIC VAPOR FLOWS AND ATMOSPHERIC STORMS

WATER VAPOR AND ITS MOVEMENT OVER THE EARTH'S SURFACE

Water vapor emanating from water bodies and vegetation systems covering the earth's surface is mixed upward into the earth's atmosphere. Sometimes it enters a storm cell in the same day and is precipitated back to the earth in the same region of the world. More often it gets caught in one of the great airstreams that move restlessly over the globe and is carried a great distance. Sooner or later, however, it is pulled into a cell of vertical motion, lifted, and cooled by expansion to the temperature of condensation. When amalgamated into snowflakes or raindrops the condensed water falls to the ground, perhaps a thousand kilometers from where it became airborne.*

Atmospheric Water Vapor

One characteristic of the amount of vapor in the atmosphere is its small mass. The areal average over the conterminous United States is 17 kg m^{-2}; Table I shows how it varies throughout the year. In the cold season it is about one-third of what it is in the warm season, an

* A great 17th century work by John Ray, analyzed by Tuan (1968, p. 104), distinguishes between "the rain that moistens the soil, thus making it productive, and the rain that causes rivers to flood." The former, beneficial, type comes from "vapours that are exhaled out of dry land; the latter is caused by the condensation of 'surplus' vapours which the Winds bring over the land from the great oceans." While we still try to distinguish local and remote sources of vapor, we do not associate either with beneficial or harmful hydrologic effects.

TABLE I

Atmospheric Content of Water Vapor over the Conterminous United States from the Surface to the 300-mb Level[a]

Month:	D	J	F	M	A	M	J	J	A	S	O	N	D	Mean
Water vapor:	11	10	9	10	14	19	25	30	29	23	17	12	11	17

[a] Unit: kg m^{-2}. Source: Reitan (1960).

expression of the relation between vapor pressure and temperature. The warm-season increase is particularly large in the interior of the continent because here in winter the air was dry at all levels.

Regions of small vapor content are those like the Arctic Archipelago of Canada in winter (only 2 kg m^{-2}). Here the atmosphere receives little vapor from the cold underlying surface or by transport from distant warm and moist surfaces (Hay, 1971). Even here the content increases in summer to 16 kg m^{-2} as the underlying surface of the whole continent becomes warm and wet.*

Over the northern hemisphere, the yearly average vapor content varies with latitude approximately as follows (kg m^{-2}):

North Pole	5
60N	10
45N	20
30N	30
Equator	45

The annual average over the entire earth is about 25 kg m^{-2}. This represents a total mass of about 13×10^{15} kg. When this number is compared with those characterizing masses of water in other locations, it is seen to be minute (see Table II).

In mass, atmospheric water exceeds only the water in river channels. Both these locations, however, are important beyond what is suggested by the momentary mass of water in them because this water is in rapid motion. Its turnover is short.

A small amount of water is airborne in liquid or solid state. Clouds contain liquid droplets and solid crystals of water, mostly in forms so

* In spite of the dominant long-distance movement of vapor molecules, it is clear that a large regional deviation in evaporation from the underlying surface is important. For example, slow evaporation from the cold surface of the Great Lakes in summer is evidenced in Hay's map as a depression of 2–3 kg m^{-2} in atmospheric vapor content; that is, about 0.1.

finely divided that they fall slowly or not at all. In many latitudinal zones "more than 30% of the lower troposphere is filled with clouds" (Junge, 1963), indicating a substantial amount of airborne water not in vapor form. The global mean liquid-water content of the atmosphere (about 0.9 kg m^{-2}), however, is considerably smaller than the mean water-vapor content of about 25 kg m^{-2}.

The total vapor content integrated vertically over an atmospheric column is called "precipitable water," for historical, rather than physical reasons (it can never all be precipitated out, even in the greatest storms). Its world pattern is strongly zonal, deriving from the pattern of vapor flux from the underlying surface. Table III presents latitudinal means, which display a maximum in the equatorial latitudes. In the Northern Hemisphere winter, there is a rapid poleward decrease of vapor content, reflecting the dryness over the large northern continents mentioned earlier for Canada. Mid-latitude and high-latitude places therefore experience a large variation from winter to summer. At Milwaukee this range is from about 9 kg m^{-2} in winter to 20 or more in summer, contrasting with the limited variation above the Caribbean Sea from 40 in winter to 45 in summer.

At particular places and times the vapor content of the whole atmospheric column is not necessarily closely associated with the vapor content near the earth's surface; advection at high levels of dry air above a surface layer that remains moist, or vice versa, can account for a discrepancy between low-level conditions and those aloft. The advection of vapor at middle and upper levels is an important member in the global balance of the water budget.

TABLE II

Water Substance in Different Domains of the Globe[a]

Location	Mass (10^{15} kg)
Water in the atmosphere	13
Water in the world ocean	1,320,000
Water on the continents:	
in ice caps and glaciers	29,200
in the ground to a depth of 4 km	8350
in lakes (fresh and salt)	229
in soil	67
in rivers	1.2

[a] Source: Nace (1967).

TABLE III

Atmospheric Content of Water Vapor over the Globe[a]

Latitude	January	July
90N	2	11
60	4	21
40	10	29
30	17	34
20	26	41
10	37	45
Equator	42	43
10S	44	38
20	39	26
30	29	18
40	21	14
60	(11)[b]	(9)[b]

[a] Unit: kg m^{-2}. Source: Kessler (1968, p. 19).
[b] Parentheses indicate estimated values.

This mobility can be seen if we isolate a segment of the atmosphere for closer examination, say the air space of Milwaukee County. A west wind of 20 km hr^{-1} moving across its boundary, if transporting a moisture charge of 30 kg m^{-2} in summer, is conveying a mass of 0.6×10^6 kg hr^{-1} of water across each meter of the county boundary. If the county is 15 km wide, such an inflow would cover it 40 mm deep within 1 hr. It is obvious that most, if not all, the water that crosses the western county line keeps right on going across the county and out over its east side. Any precipitating process is working only a few percent of the time, and even then only a small fraction is precipitated out of the large advective current of vapor carried in the winds.

Vapor Transport

The average time that a molecule of water is resident in the atmosphere, as vapor or in a droplet of a nonprecipitating cloud, is 10–12 days, a period sufficient for a long journey. The processes of evaporation from the earth's surface and delivery back to it are likely to occur far distant from each other, separated by many atmospheric events that we can only note here. It is most important to us that the mobility of water vapor while airborne is one cause of an enormous variability in the way water returns to the surface of the planet.

Transport of Other Forms of Matter Besides water vapor, other forms of matter also move over great distances in the atmosphere. For example, sulfur in acid rainfall that has reduced the productivity of forest ecosystems and soils in Scandinavia has been traced back to the industrial districts of England and Germany (Førland, 1973).

While part of the lead emitted into the atmosphere over Los Angeles from industrial sources and car exhausts (5 tons day^{-1}) is returned to the source by local rain, much of it is widely dispersed. Adding the San Francisco contribution of lead, and assuming uniform dispersal over an area of diameter 2000 km, the deposition rate should be 0.7 mg m^{-2} yr^{-1} (Hirao and Patterson, 1974). In fact, the deposition in a remote valley in the Sierra Nevada is 0.85 mg m^{-2} yr^{-1}, as determined from isotope measurements of the lead content of the snow cover, ecosystems, animals, and streams.

Transport over North America Over a large area of land, the longer-traveling molecule is likely to encounter a precipitating system and be halted. The flux of vapor across the North American continent, for example, can be studied by looking at the amounts crossing its boundaries. This was done by Benton and Estoque (1954) twice daily during the year 1949, using upper-air observations at stations around the periphery of the continent north of Mexico. Table IV shows the flux across each segment of the boundary, averaged over the summer.

TABLE IV

Net Inflow (+) or Outflow (−) of Water Vapor across Boundaries of Anglo-America in Summer 1949[a]

Segment of boundary	Approximate length (10^5 m)	Vapor flux
Coast of Gulf of Mexico	22	+17
Border between U.S. and Mexico	16	+3
Coast of Pacific Ocean and Bering Sea	60	+22
Arctic coast	55	−2
Labrador	16	−17
Atlantic coast	38	−31
South boundary	38	+20
West boundary	60	+22
North boundary	55	−2
East boundary	54	−48
Boundaries with net inflow (south and west)	98	+42
Boundaries with net outflow (north and east)	109	−50

[a] Units: 10^7 kg sec^{-1} (from Benton and Estoque, 1954).

Over any given segment, say the coast of the Gulf of Mexico, vapor flows now inward, now seaward, depending on wind fluctuations as storms pass, but the net flow over the summer averages 17×10^7 kg sec^{-1} inward (shown as +). Three things stand out in this table:

(1) There is a difference in intensity of flow per unit length of boundary. The flow is far stronger across the coast of the Gulf of Mexico, where it averages 80 kg sec^{-1} m^{-1} of boundary, than it is across the Pacific Coast, where it averages 36 kg sec^{-1} m^{-1}. The small net flow across the Mexican border and the shores of the Arctic reflects a drier atmosphere as well as wind direction alternations such that inflow and outflow come nearly into equilibrium.

(2) Vapor moves from the Gulf and the Pacific, across the continent, and off again in the direction of Europe. This motion accords with the general westerly airflow of middle latitudes fed by southerly flow around the end of the Bermuda anticyclone. Figure II-1, a cross section along the 30th parallel, shows how close to the ground the strong vapor flow occurs.

A specific water molecule following this hypothetical path, coming ashore over Galveston and leaving the continent over Megalopolis, would travel about 2500 km. Considering that the mean residence time of vapor in the atmosphere is 12 days, this molecule has a good chance of making this traverse without being precipitated to earth. Many a

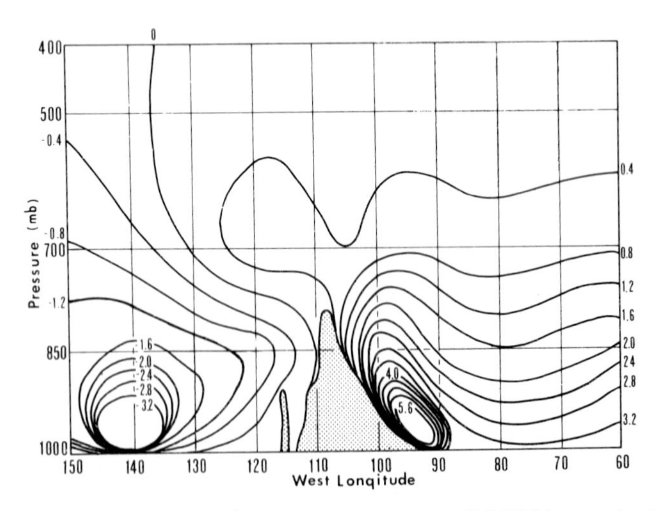

Fig. II-1. Meridional transport of water vapor across parallel 30N (approximately at the coastline of the Gulf of Mexico), averaged over the summer of 1949. Values are in g cm^{-1} mb^{-1} sec^{-J} (from Benton and Estoque, 1954).

molecule goes all the way across the continent without encountering a zone of vertical motion that might condense it and precipitate it out of the airstream.

(3) In the season illustrated in Table IV, net inflow of water is not as large as net outflow. During this summer the airstreams moving across North America, although suffering attrition as rain was precipitated from them, nevertheless left the continent with more moisture than they carried when they entered. In crossing the continent, they picked up from the underlying surface as much water as they precipitated to it, plus an additional 8×10^7 kg sec^{-1}. What accounts for this growth?

Vapor-Flux Divergence and Convergence

In summer North America is a vast transpiring surface as forests and corn fields energetically pump water from the soil into the atmosphere.* An airstream moving across the eastern part of the continent, especially if it comes from the west or northwest, takes up vast quantities of water, as Holzman showed long ago (1937), and carries it off. He pointed out that for this reason, providing more evaporation opportunities—trees in Nebraska, ponds in Oklahoma—would result in little, if any, additional precipitation in the same county or even the same state. Over the continent in 1949 this net outflow (called "flux divergence") was 8×10^7 kg sec^{-1}, equivalent to a total of 62×10^{13} kg over the summer. When this mass is averaged over the area of the continent, the unit-area amount of water is 80 kg m^{-2}. In terms of water depth at the surface this is 80 mm.

The transpiring surface of the continent thus gave out 80 mm of water more than it received as precipitation, which in the summer of 1949 was about 170 mm. Total evapotranspiration from the continent during the three months thus was 80 + 170 = 250 mm. This value, calculated by casting a water budget for the atmosphere of eastern North America, is independent of surface measurements of evapotraspiration, and serves as a check on them.†

* In energy terms, this pumping is powered by transformation of solar energy at a rate exceeding 10^5 million kW (1 million kW represents a large electric power station's output).

† Wide use of this method, unfortunately, is made difficult by our poorly defined knowledge of upper-air flows, a result of the lack of sufficient observing stations. Calculated flux divergences are reasonably accurate over such large areas as those discussed here, but suffer from increasing error in smaller drainage areas. For large regions, however, daily inventory calculations of vapor-flux divergence would be most useful.

This behavior of the continent during the summer of 1949 is unexpected if one customarily thought of the summer water situation chiefly in terms of heavy inputs to the land surface—general rains, frequent showers, squall lines, thunderstorms that dump 50–100 mm of water. We think of summer, in fact, in terms of the likelihood of getting wet at the 5 o'clock rush hour. We find, however, in working out the water budget, that while the surface was receiving 170 mm of water it was giving off 250. Evapotranspiration, an invisible process, receives less notice than rainfall, yet bulks larger in the budget.

Let us now look at the winter season. Figure II-2 shows strong vapor transport across the Gulf Coast. A later study of atmospheric vapor transport (Rasmusson, 1968) presents flux-convergence calculations for the winter of 1962 to 1963. In this period the atmosphere brought more vapor into the eastern North American airspace than it took out. In terms of an equivalent layer of water over the area from the Rockies to the Atlantic, the excess was 60 mm.

This depletion of the atmospheric vapor streams that cross the continent represents the difference between moisture picked up along the way and moisture precipitated out of the airstreams. Rasmusson estimates that on a 5-yr average 60 mm moved into the atmosphere each winter from the underlying surface, and 130 mm was precipitated to the surface.

The surplus of water in the budget at the earth's surface in winter is more visible than is the deficit of summer. Water precipitated to the

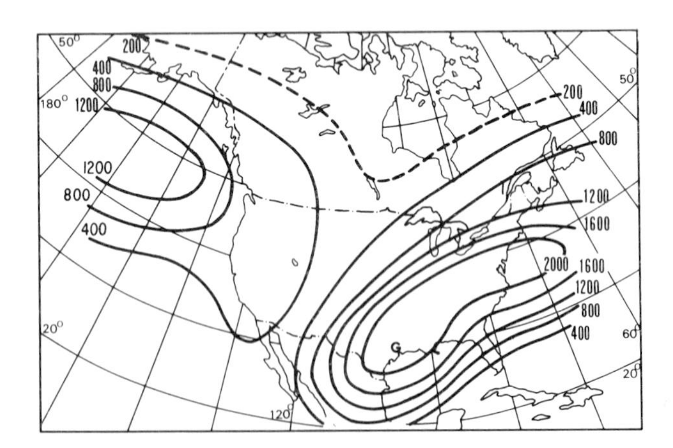

Fig. II-2. Transport of water vapor, integrated from surface to 400 mb and averaged over the winter of 1949. Peak inflow—to the continent near Galveston—averages more than 2000 g cm^{-1} sec^{-1} (Benton and Estoque, 1954).

surface in excess of what evaporates from it piles up. Some of it accumulates to the point that it moves to the stream networks, down the rivers, and then off the continent. The rest accumulates in and on the ground, in the familiar forms of moist soil, saturated subsoil, high groundwater table, and rising levels of swamps and lakes. There is mud in the South and ice and snow in the North. The excess of incoming atmospheric water over outgoing is evident by the end of winter; we live in a soggy or ice-encrusted landscape.

This example of vapor-flux convergence in the atmosphere and the resulting accumulation of liquid and solid water on the earth's surface applies to a whole winter and a large area of land. It sums up an uncounted number of individual convergences in separate atmospheric motion systems or storms, which we will now discuss.

ATMOSPHERIC STORMS

"For many years I was self-appointed inspector of snow-storms and rain-storms, and did my duty faithfully" (Thoreau, 1854). We look now at the processes by which water is extracted from the mobile atmosphere and precipitated to the earth. This series of phenomena, in spite of being commonplace on our planet, is not necessarily inevitable. It is not inconceivable that a planetary atmosphere might remain humid, even cloud filled, without releasing any water that could fall to the surface of the planet. On earth this situation occurs between storms, much more than half the time.

Even when clouds form, they are, in 99% of the cases, nonprecipitating. Some classes of clouds never produce snow or rain. The initial step of transferring water from atmospheric vapor to the underlying surface has taken place, i.e., vapor has condensed into ice crystals, snowflakes, or water droplets, but the next process has not followed. These crystals and droplets must be transformed into heavier particles if they are to fall through the encumbering air fast enough to get to earth before they evaporate.

Gilman (1964) notes four conditions necessary for production of rain: "(1) a mechanism to produce cooling of the air, (2) a mechanism to produce condensation, (3) a mechanism to produce growth of cloud droplets, and (4) a mechanism to produce accumulation of moisture. . . ." To continue with the inflow concept used earlier in this chapter, let us first look at his fourth point.

Vapor Flux into Storms

Gilman's condition (4) is that moisture must flow into a storm system. The vapor flow into a storm system may be compared with that into a continent, considering that the storm is smaller, may be moving, and is a vigorous system that crams a high rate of vapor inflow and vapor-flux convergence into its short life.

In the water budget of a storm, the storage term, i.e., the moisture content of the atmospheric column at the site, is seldom greater than 20–40 kg m^{-2}. This amount cannot support long-continued precipitation. For instance, a convective cloud as a small system looks solid and we visualize it as a possible source of cloudburst, yet its actual content of liquid water is unimpressive. If the cloud is 1 km deep and has a liquid-water content of 1 g m^{-3}, the mass of liquid water totals only 1 kg in a column of 1-m^2 cross-sectional area or 1-mm depth of water. Twenty times as much mass is present as uncondensed vapor in the cloud.

In large storms rainfall may exceed the initial water content of the atmosphere by ten times. No storm of any size can exist without strong vapor influx. Since vapor carries latent heat, the influx plays a role in the energy budget of a storm that we cannot take time to consider here except to note that convective cells often move not with the wind but at an angle to it so as to maximize the vapor inflow that is fueling them (Newton and Fankhauser, 1964). Cyclonic storms crossing North America often begin rapid development in the Midwest, where they start to pull in moisture originating in the Gulf of Mexico (Petterssen, 1956).

Influx of vapor into a storm is a concentrating process. An isolated thunderstorm, for example, organizes and concentrates water vapor (and air) from 1000 km^2 around it. In the process it inhibits the growth of other systems. The many small clouds of late morning are replaced by one giant system in the afternoon that has starved out its neighbors.

Water Budgets of Storms The mass budgets of water and air in a large squall-line thunderstorm (Table V) are illustrated by Newton (Palmén and Newton, 1969, p. 416). Fluxes of water vapor and air, the two forms of matter being processed through the storm system, are shown in Fig. II-3. Note that a considerable tonnage of water vapor is carried out of the storm. Of the 9.5×10^6 kg sec^{-1} entering the system, half flows out still in the vapor state. About half leaves it as liquid water—some as cloud droplets in the anvil, most as raindrops precipitated to the underlying surface.

TABLE V

Mass Budgets of Thunderstorm of 21 May 1961 over Oklahoma City[a]

Realm of storm	Flow	Air	Water	
Low level	Inflow	+700	+8.8	
	Downdraft	−400	−4.3	
Middle level	Inflow	+400	+0.7	
Anvil	Expansion	−700	−0.6	as vapor
			−0.6	as liquid or solid water particles
Underlying surface	Precipitation	0	−4.0	
Sum		±0	±0	

[a] Unit: 10^6 kg sec^{-1}. Data from Palmén and Newton (1969, p. 416).

Budgets of total storm quantities of water and beta radioactivity were cast for another midwestern storm (Gatz, 1967):

Total inflow of vapor in the layer between the surface and 650 mb	$+74 \times 10^9$ kg
Water precipitated to the underlying surface	-54×10^9
Outflow of vapor and cloud droplets from the storm system (largely from its upper levels) into other parts of the atmosphere	-20×10^9
Sum	0

A budget for beta radioactivity also showed that low-level inflow was the important source of substances of this nature that the storm deposited.

The idea of vapor flowing into a storm cell from a broader area implies that cells must be widely spaced. In a region of convective rain Sharon (1974) found cells at a preferred spacing of 40–60 km. These cells are "competing in the scavenging of the water vapor present in a uniform air-mass. The spacing of precipitating cells corresponds to the area required by a single cell, or cell-group, for the supply of water vapor to be condensed." Here the supply area is about three orders of magnitude greater than the area in which vapor is actively being condensed. The cell's water budget involves a large area of the atmosphere.

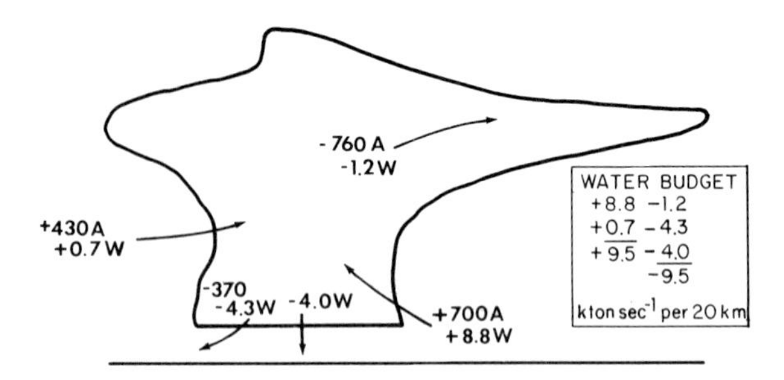

Fig. II-3. Flows in the budgets of air (A) and water (W) in a thunderstorm, expressed as ktons sec^{-1} for a 20-km length of the squall line. Flows of water vapor are: +8.8 and +0.7 into the lower part of the storm, −4.3 out in the downdraft, and −0.6 out in the anvil. Flows of droplets and ice crystals in the anvil also are −0.6. Rain precipitated to the underlying surface amounts to −4 (from Palmén and Newton, 1969).

In one hurricane (Palmén and Riehl, 1957), vapor inflow was calculated to be 200×10^6 kg sec^{-1}. This amount is 25 times as large as the 20-km-long line storm analyzed by Newton, and of the same order as the average rate of vapor inflow into the whole North American continent. While not all of this input was precipitated out of the hurricane, the rate of removal was still very large—150 kg m^{-2} daily over a central area covering 125,000 km^2. The area from which vapor was contributed, i.e., the total oceanic airspace organized by the hurricane, was 7×10^6 km^2, about the size of the eastern North America area studied by Benton and Estoque and noted earlier. Such a storm is a giant mechanism that collects water from a large sector of the earth and concentrates it in a smaller area of intense activity. In this central zone, which is still very large, changes in physical state occur and large amounts of water are precipitated to the underlying surface.

Altitude Effects on Vapor Flux While it is not feasible here to go into the vertical structure of the atmosphere, we should note two characteristics that are important in storms as water-converting systems. One is the decrease of temperature with altitude, which will be mentioned later in this chapter; the other is the change in vapor concentration and wind speed with height, which will be discussed here.

The vertical distribution of vapor and the vapor flux has an important influence on precipitation. For years it was believed that in mountains a zone of maximum precipitation existed at some intermediate level, above which precipitation decreased with altitude. The reasoning was plausible, since little water vapor is generally found at high altitudes. Hann generalized the decrease of vapor with height in the atmosphere as

$$\log_{10} e_h = \log_{10} e_0 - h/6.5,$$

in which e represents vapor pressure, 0 and h represent sea-level and any other altitude h, expressed in kilometers (Hann, 1897, Vol. 1, p. 279). This expression says that at 6.5-km altitude the vapor pressure is 0.1 of its sea-level value, and derives from the circumstance that vapor in the atmosphere has its source at the underlying surface only. In the atmosphere it can undergo nothing but depletion.

In the upglide parts of mid-latitude frontal storms, however, airstreams of high moisture content that initially were at low levels are forced into the middle troposphere. Weischet (1965) shows that this situation during storms, rather than the situation during clear weather, is important when we try to explain the precipitation that reaches highland surfaces.

Improved measurements of mountain precipitation suggest that in many ranges in the middle latitudes there is no zone of maximum precipitation below the summit. An apparent decrease in precipitation near the summit turned out to be an error in measurement. In the Sierra Nevada of California, for example, a weather station near the crest was long reported to have a mean annual precipitation of 1150 mm, which is somewhat less than reported at stations lower and west of it. Careful hydrometeorological studies in the crest region indicated, however, that the true precipitation is of the order of 1600 mm. This value, which is consistent with measurements of the snow cover and of streamflow in the region, is larger than precipitation readings at lower altitudes.

Furthermore, in middle-latitude storms, the normal upward increase in wind speed increases the vapor flux at high altitudes, as seen in vertical profiles in the Alps (Havlik, 1969). Vapor inflow into mountain storms therefore can be very rapid.

In low latitudes, on the other hand, where no systematic increase of wind speed with height occurs, a decrease of rainfall at high altitudes is the usual case. In fact, Weischet (1965) distinguishes lowlands from highlands in these latitudes by their distinct rainfall systems.

Storm Mechanisms

> All the fountains of the great deep burst apart,
> And the flood-gates of the sky broke open.
>
> *Genesis 7*

Lifting Convergence of the atmospheric gases represents an excess of mass in the lower layers that forces them upward. This ascent is one mechanism for cooling water vapor (the other constituents of air are cooled too, but not as far as their condensation temperatures). A storm requires massive lifting of air, usually in convergence, and a large inflow of vapor, often accompanying such convergence.

Convergence of airstreams is the mechanism most capable of causing rapid condensation of vapor, over a sustained period (part (2) of Gilman's conditions listed earlier). Without convergence, the straight-line glide of a moist airstream up a warm front or a mountain slope cools vapor relatively slowly; precipitation is prolonged but seldom heavy. The conventional model of a frontal cyclone is, by itself, inadequate to explain the actual patterns of precipitation (Gilman, 1964) although it qualitatively explains many types of rain in the middle latitudes.

Prolonged heavy precipitation occurs when mass convergence, alone or in conjunction with other factors, generates strong vertical movement and cooling. For example, heavy rain poleward of the center of an extratropical cyclone usually results from a combination of low-level convergence above the warm front, and convective instability in the warm air that is triggered by upslope movement. The action of mountain slopes in setting off convective instability, which is discussed in meteorology textbooks, is probably more significant than the pure orographic lifting itself; the mountain lift does not extract vapor from the air as efficiently as do convective or frontal systems (Gilman, 1964, p. 24).

Orographic effects provide a plausible explanation for rainfall, enhanced by the pictorial appeal of bulging rain clouds running into mountain ranges. However, in a classic monograph on frontal cyclones analyzed by the kinds of rain they produce, Bjerknes and Solberg (1921) assess the lifting processes and come to a more realistic evaluation. They state that even in the most favorable parts of western Norway, certainly a mountainous country, orographic rainfall seldom exceeds $\frac{1}{5}$ mm hr^{-1} and generally is far less.

General convergence or inflow usually extends over a large area, promoting either actual upward motion of the atmosphere or an environment that favors local spots of concentrated upward motion. Accordingly, most precipitation has been found to be organized over large areas. For example, in the upper Colorado River basin the greater part of the precipitation in both summer and winter occurs in organized systems that affect most of an area of 200,000 km^2 (Marlatt and Riehl, 1963), even though the lifting mechanisms are different in the two seasons.

It was once thought that thunderstorms in the midwestern United States were isolated motion systems touched off in random fashion by surface heating. It now appears, however, that most thunderstorms are associated with frontal or nonfrontal large-scale convergence. They are found in organized populations, as is seen in radar and satellite photos.

Physical Processes in Clouds Some effects of attempted modification of clouds and supercooled fogs with respect to Gilman's mechanisms (2) and (3) are obvious, but, as Neiburger and Chin (1969) point out, we cannot yet predict whether the modification of precipitation processes will be upward or downward. A hydrologist who wants to augment the deliveries of atmospheric water to solve a regional water problem would be well advised to look at the decades of unfulfilled promises and to study realistic assessments by meteorologists like Houghton (1968).

One might consider other points also. Are there economic and ethical aspects involved in attempting to alter such a common as the atmosphere? It is not the exclusive property of one landowner or of one professional field. One might consider possible crop and insect changes in a region if its weather were modified. Ecologists tell us they do not understand atmosphere–plant relationships well enough to be able to predict what directions changes might take (Sargent, 1967).

Inadvertent modification of mechanisms (2) and (3) by the injection of waste material from industries and internal-combustion engines might occur. Nuclei, heat, and vapor are introduced into the air by most machines, including airplanes; cities and industrial regions change the patterns of precipitation within them and in some cases for a distance downwind. The lead from urban traffic might have nucleating properties. We need a better understanding of these mechanisms within clouds, along the lines of the METROMEX study of rain cells downwind from St. Louis (Huff and Schickedanz, 1974; and others).

Condensed water in clouds is in equilibrium with CO_2 in the air, so that cloud droplets are slightly acid, with a pH of 5.7, more or less (Carroll, 1962). At lower values of pH rainwater "contains gases or acid such as SO_2, H_2SO_4, or HCl." Some of these are of local origin, others have been transported a long distance, and are removed from the atmosphere as a concomitant to the process of water-vapor condensation in a precipitating system.

From the only comprehensive surveys of precipitation chemistry ever made in the United States, in 1955–1956 and 1965–1966, Cogbill and Likens (1974) mapped acidity. In the period between the two surveys, the 5.6 pH isarithm expanded greatly, and at both dates the 4.5 pH isarithm covered Pennsylvania, New York State, and New England outside of Maine. This pattern apparently represents a massive transport from the Midwest in the southwesterly and westerly flow that we saw earlier in this chapter as transporting vapor across the eastern part of the continent.

Sulfates often occur as particles, sometimes as a result of coagulation and other processes within the atmosphere. Many such particles are eventually incorporated in cloud droplets during the condensation processes in clouds. Gaseous sulfur compounds dissolve in cloud droplets and raindrops, adding to the wet-fallout phase of the cycle. Sulfur damage to forests in Canada is credited with a direct loss of as much as $3 million annually, and seems to be a major agent in producing changes in the species composition of forests throughout the eastern United States.

Storm Temperature

Along with the basic conditions of vapor inflow and vertical motion in the atmosphere, a storm system also is characterized by its vertical profile of temperature. In the middle latitudes, liquid-water droplets seldom grow big enough to fall, but snowflakes do. If, as they fall, they pass from below-freezing air to above-freezing air, they melt and arrive at the underlying surface as rain.

If, on the other hand, the whole atmospheric profile is colder than 0°C, the snowflakes arrive at the ground without change. Once landed in contact with earlier-arrived brethren, their chances of survival in solid form are reasonably good. The critical question is whether they melt on the way down. For this reason storm temperature, often expressed as the height of the freezing level above the surface, is an important parameter.

The high latitudes, the interiors of the middle-latitude continents, mountains of middle latitudes, and very high mountains even at low latitudes are locations where cold air hugs the earth's surface. The cold air of underrunning polar air in an extratropical cyclone, the coldness of high altitudes in the atmosphere, or of high latitudes, all make it possible for snowflakes to get to the earth's surface.

Lowland and Mountain Snowfall. In some of these regions, like the middle latitudes, snowfall is a winter phenomenon, usually connected with frontal storms. The cold air, especially in interior lowlands, forces the warm moist air into an upglide that provides the cooling mechanism for precipitation; it also provides a cold layer that preserves the falling particles until they reach the surface.

In middle-latitude mountains the solid earth itself provides the lift and intercepts the falling snowflakes at elevations high enough to be below the freezing point during the storm. In a three-day storm in California, for example, maritime air came ashore at a sea-level temperature of +12°C. All precipitation falling on the Coast Ranges, the Central Valley, and on lower Sierra slopes at altitudes less than 1.5 km had melted on the way down and arrived as rain. At higher altitudes the air near the surface averaged near or below the freezing level over the period of the storm, and most of the precipitation arrived unmelted (Table VI).

The winter snows of these mid-latitude mountains differ in genesis from those of the mid-latitude lowlands. Furthermore, the mountains, by anchoring the zones of vertical atmospheric motion to one locality in storm after storm, localize the accumulations of many snowstorms. By the end of winter the total accumulation is far greater in this area than in the equally stormy lowlands, where snow is spread over much more extensive areas. For example, in California the narrow crest region of the Sierra Nevada receives 1000–1500 kg m^{-2} of water as snow each winter; in contrast, winter snowfall in the upper Midwest

TABLE VI

Temperatures and Snowfall along a Transect across the Sierra Nevada, 13–15 November 1950[a]

Station	Altitude (m)	Mean temperature (°C)	Depth of snowfall (cm)
Fort Ross	30	12	0
Sacramento Airport	5	11	0
Auburn	375	9	0
Colfax	740	7	0
Blue Canyon	1600	−1	39
Soda Springs	2055	−2	75
Central Sierra Snow Laboratory	2100	−3	94
Truckee Ranger Station	1825	0	39
Boca	1690	0	13
Reno	1340	+4	trace

[a] Source: U. S. Weather Bureau Climatological Data, California, 54, No. 11: 373–413 (1950).

(except in the immediate lee of the Great Lakes) is spread over a large area and is thin—of the order of 100–200 kg m^{-2}.

Freezing Level This is the level in the atmosphere above which falling snowflakes do not melt. Where it intersects a mountain range the snowflakes reach the ground upslope from it without melting, while those downslope arrive as raindrops. The location of the freezing level during a storm is an important part of the forecast. If it is low, little precipitation is delivered as rain and the flood-producing potential of the storm is negligible. If the freezing level is high—and it may be as high as 2 or even 3 km in a Sierra storm—the rain area is large and the flood hazard very serious.

In the lowlands the freezing level intersects the earth's surface at a low angle, such that a small shift in storm temperature will cause it to move north or south 100 km. South of this line precipitation falls as rain, freely running off through the storm drainage network of a city; north of it precipitation comes as snow, and is hardly an unmixed blessing to a lowland city. To a person coming to a midwestern city from the mountainous west, the contrast in lowland and mountain snowfall is striking. In the western mountains, snow forms a welcome play area and a water reserve that alleviates concerns about summer irrigation, whereas in lowlands winter snowfall is a burden that has to be bodily removed from roads and streets, the degree of disruption being increased by one or two classes in a 10-class ranking if storm temperature is low (Freitas, 1975).

Urban Snowfall Snowfall on a city in amounts greater than 10–15-cm depth must be plowed and shoveled off streets and parking lots. The cost of this operation in Milwaukee, for instance, is not less than \$20,000 cm^{-1}; the public expense of removing this gift of nature is about 2¢ per capita for each centimeter of snowfall, and at least as much more is expended privately.

The exact figure varies from city to city. In some cities the implications of snowfall are ignored, and little public money is spent in preparation for snowfall (Rooney, 1967). Instead, when a snowstorm comes a period of confusion ensues. Individuals bear a heavy cost in struggling with the snow and face a long period of disruption of their activities. This contrast suggests that the inherent variability of precipitation events is differently perceived, even by people in the same North American culture.*

* Milwaukee's usual state of preparedness for snowfall apparently derives from a bad experience at the end of January 1947; the city was tied up from 2100 on the 29th until

Precipitation arriving as snow in the mountains has future hydrologic value but brings immediate concern only on the transmountain roads. Below the freezing level, heavy rain can cause a flood. The contrasting effects of the same atmospheric storm system in California is illustrated in the novel, "Storm," by George R. Stewart. Rain in the lowlands caused local flooding and highway accidents; rain in the hills produced large floods in the major rivers; snow in the mountains created problems for communications people—the highway and telephone crews—but had no immediate hydrologic effects. Storm temperature, affecting the physical state in which precipitation is delivered to the earth, is an important characteristic of a storm.

SIZES AND MOVEMENT OF ATMOSPHERIC STORMS

Maximum Storms

What do we find if we assign very large values to the storm parameters of vapor influx and vertical motion? This question is of philosophical interest: What are Nature's limits? Models of the dynamics of cloud and precipitation systems (Kessler, 1969) bear on this question. The question also has immediate importance to a hydrologist designing a dam on a river of uncertain behavior, who needs to delimit the capabilities of the atmosphere to deliver rain to the surface of the drainage basin that he plans to obstruct.

Historical and Physical Methods One approach is historical, that is, to see what rates of vapor inflow (determined from upper-air winds

---

the 31st, with 50 cm on the ground on the 31st (U.S. Weather Bureau, 1947). The disruption is still vivid in memory and apparently led to a decision: Never again! Yet the mass of water that made this strong impression on the whole city was not large, about 50 kg m^{-2}.

At this time upper-air flow changed abruptly from two months of mild winter weather to very cold (Namias, 1947), a sudden shift in the general circulation with which was associated a large storm that moved east from Colorado. This storm came over Wisconsin as a double system, one type of snow-precipitating shield being replaced by another and generating heavy snow for a duration of 48 hr in Milwaukee. In an interlude between these periods, a change in storm temperature structure produced freezing rain (George Blandino, National Weather Service, personal communication). As a result, the second-story drifts piled up by the gale-force winds were internally reinforced by thick layers of ice, making removal difficult with the war-worn equipment the city then possessed. The storm produced "the longest, worst, and costliest tie-up in Milwaukee's history" (Anon., 1947) and total direct and indirect costs were estimated as $75 million, which in today's dollars would be several hundred dollars per capita. No Milwaukeean wants to repeat this experience, hence the present planning and preparation for whatever the atmospheric systems of winter can bring.

and moisture content) generated observed rainfalls in the past. The lifting mechanisms in a large number of storms can be generalized as they must have operated on a given moisture inflow to produce the rainfalls actually observed.

This method is illustrated by Snyder (1964) for the basin of the Black Warrior River, Alabama; the area is about 11,600 km^2. The isohyetal pattern of a very large storm that occurred elsewhere in Alabama in March 1929 (Fig. II-4) was transposed to the most critical location on the Black Warrior basin and rotated 20°. It indicates the lifting mechanism.

The effect of vapor inflow in the hypothetical storm was maximized by factors derived from climatological studies of the maximum humidity over periods equal to the duration of the storm (84 hr). The largest such factor, 1.19, was for May. The resulting mean depth of rainfall over the Black Warrior basin was estimated as 560 kg m^{-2}. It was three times the size of the greatest storm that had occurred over that particular basin during the period in which rainfall has been recorded there, and represents the potential of atmospheric storms that will

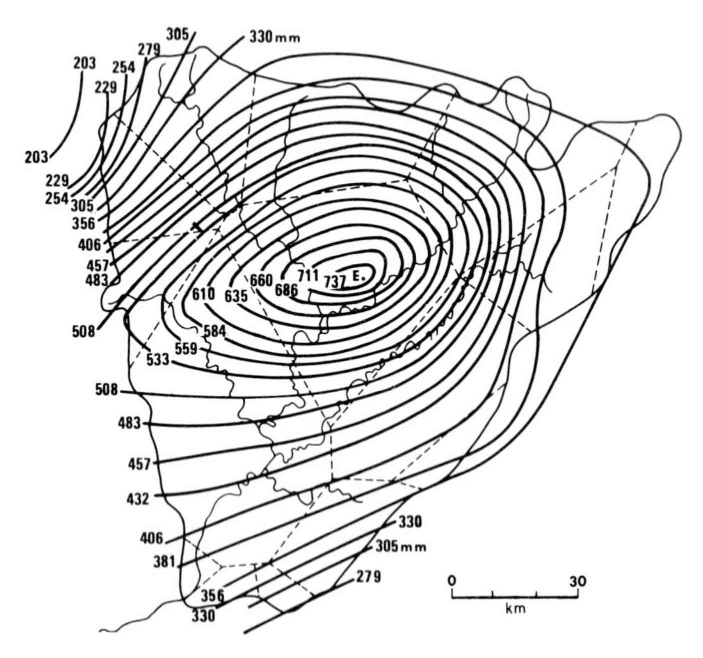

Fig. II-4. Transposition of Elba, Alabama storm of 11–16 March 1929 to the basin of the Black Warrior River above Holt Dam site (drainage area about 11,600 km^2); peak rainfall 750 mm (Snyder, 1964).

some day occur—knowledge necessary for conservative hydrologic design of water-control works.

The model of a probable maximum storm for Oahu is a slowly moving zone of strong convergence in the lower atmosphere that is "only remotely related to topography" (Schwarz, 1963, p. 9). From a very humid atmosphere, in which the vapor content approaches 75 kg m^{-2}, this model generates rainfall of 1000 kg m^{-2} in 24 hr in the saddle between the two mountain ranges of the island. Its mechanism is derived largely from storm physics of vapor inflow and convergent lifting, but the physics framework is based on the historical facts that at times severe storms have occurred in southerly flow (Kona storms) and that inflows of very high moisture content also have occurred—a contrast with the usual trade-wind air.

Like other climatological maximizing studies carried out to obtain information to guide the design engineer, this model does not look to a specific date in the future, because the designer has little interest in such a date. He does not care exactly when a big storm will come, as long as it comes within the useful life of his project; but he does care how big it will be. What can the fickle atmosphere dump on his project?

Camille One tends to think of the hypothetical probable maximum rainstorm as some sort of remote, unapproachable fiction. It is a shock when a real event comes close to this ideal; yet such a super-storm happens every once in a while in one part of the country or another.

The aftermath of hurricane Camille in 1969 was such an event. As will be seen later, its rainfall came within 0.80–0.85 of the probable maximum storm for central Virginia. While some minor orographic effect was involved, the condition that revitalized the remnants of a storm that had traveled 1500 km over land was the sudden access to a large body of very moist air over central North Carolina. Atmospheric water vapor here amounted to 75 kg m^{-2} * (Schwarz, 1970), which is nearly three times the U.S. mean in August that was shown in the tabulation on p. 8.

When the circulation system of Camille's remnants tapped this rich source of vapor, a high rate of vapor inflow began. Its effects were augmented by an apparent coupling of low-level convergence and high-level divergence (Schwarz, 1970). Moreover, the storm made good "use of the high moisture in the form of persistent, efficient thundershowers," which were continuous at Charlottesville, for in-

* The surface value of vapor pressure was 32 mb.

stance, for 8 hr. In this storm an effective lifting mechanism thus operated on a strong vapor inflow.

Less-Than-Maximum Storms

A person concerned with an engineering work that might be permitted occasional failure, like a bridge on a county road, or a levee protecting a corn field, finds that designing his structure to the probable maximum storm is prohibitively expensive. He is more interested in less-than-maximum storms, which are more likely to affect his project.

His situation exposes an area of ignorance about storms. Small storms can be watched from the ground, but larger ones cover the whole sky, and the observer loses perspective. Large cyclonic storms and hurricanes can be watched as they are portrayed on the synoptic weather map, but the coarse mesh of stations does not delineate some storms accurately, and lets smaller ones slip unseen through its mesh.

The descending shafts of raindrops and the windblown trails of snowflakes coming down from the clouds represent deterministic events of influx of vapor, updraft, and lifting that are so diverse that we treat the results principally as random events. In some ways we are not much closer to intellectual mastery of the processes of generation and fall of rain than our forefathers, to whom it represented an uncertain gift from the gods.

We cannot predict the falling of 230 kg m^{-2} of rain in a single day, as happened from a nonfrontal system over Sydney on a day in April 1966. We even find it difficult to explain, after this enormous mass of water had fallen and the wreckage was being cleared away, why 250 million tons fell out of the sky at this particular time and place. Such an amount of water, enough to cave in streets and wash away cars, was nearly two orders of magnitude greater than the average daily rainfall received in the Sydney Observatory gage over the last century. Few other events in our environment depart so far from their habitual behavior; rainfall is unique in its variability.

Broad-scale low-level convergence of airflow that sets the stage for vertical motion and condensation does not necessarily imply uniform uplift over a large area. Rather, it indicates a general propensity toward vertical motion, which might be concentrated in a few favored places—cells or lines of convergence, or places where upward motion has already been initiated, as above mountains, or where instability results from surface heating, as over a warm lake surface in winter, or simply a random site where local convection has gotten an earlier start than over adjacent sites. General convergence sets the scene for vertical motion that is likely to be centered in a few places, and to take one or several physical shapes.

Yields from Lifting Mechanisms Precipitation falling on Chicago during the year was classified by the mechanism of vertical motion (Table VII). Cold- and warm-front mechanisms were dominant in April, for example, where they contributed three-quarters of the total amount of precipitation. In July they contributed only one-quarter; squall-line thunderstorms and nonfrontal showers and thunderstorms were then more important. The amount of precipitation delivered by each type varies from 2 to 11 mm per event, as might be expected from their relative access to moisture and scale of vertical motion.

In other climates different atmospheric systems are the rainproducers. In China, for example, cold fronts are the most important (Chang, 1972, p. 294), as in Illinois; but cold vortices (pools of cold air aloft) and monsoon heat lows, which are lacking in Illinois, are important. Warm fronts are productive of precipitation in both places, but hurricane (or typhoon) rainfall is important in China and nearly absent in Illinois. If you travel southeastward from Illinois, however, you soon come into an area where hurricane rainfall makes up 0.10 of the autumnal total. A little farther it reaches 0.25 (Fig. II-5) (Cry, 1967), where these vast oceanic storms have not yet been weakened by friction and reduced moisture input over land surfaces.

TABLE VII

Kinds of Atmospheric Disturbance Associated with Precipitation at Chicago[a]

Atmospheric situation	Total precip.	Fraction of annual total	Precipitation per event	Months
Cold fronts	285	0.33	7	No seasonal concentration
Warm fronts	265	0.31	9	Few in J A S O
Squall lines	105	0.12	11	Mostly M J J A
Stationary fronts	100	0.11	8	Little concentration
Nonfrontal in warm air	80	0.10	6	Mostly in M J J A S
Post cold front	30	0.03	2	Few in J J A
		1.00		

[a] Unit: mm. Source: Hiser (1956).

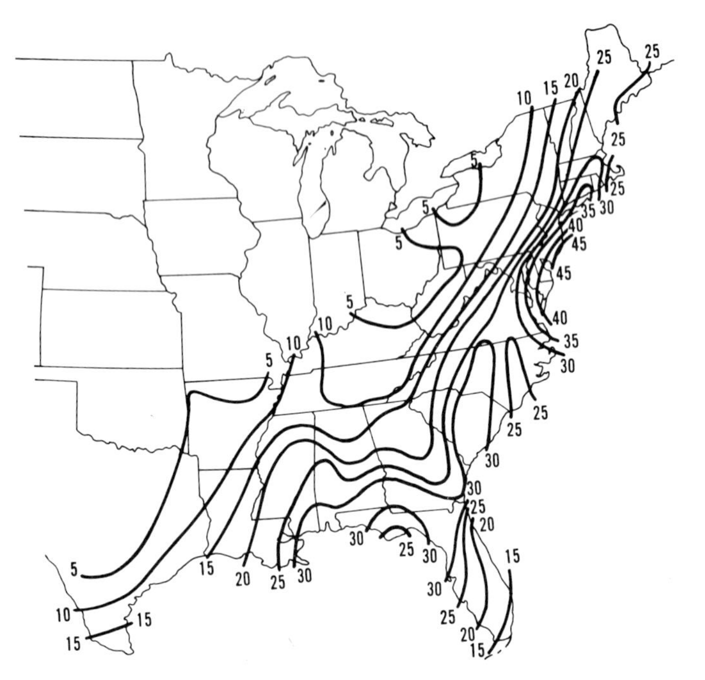

Fig. II-5. Percentage of September rainfall that is delivered by hurricanes (mean of 1931 to 1960) (Cry, 1967).

Yields from Size Classes of Storms The widening of our vision by satellite and radar photographs shows us many ways in which vertical motion is organized in the atmosphere. The spectacular spirals of extratropical and tropical cyclones represent rain-generating systems in which the areas of precipitation have appreciable widths as well as great length.

Radar has been used for many years to watch particularly intense rain-generating cells as they move across the country, affording a sampling of systems of less than synoptic size. Photographs of the radar screen every few minutes therefore form a series of basic census maps of precipitating systems. Analyses of such data by Pauline Austin and collaborators (1972) at MIT suggest that in New England at least four scales of size can be identified (Table VIII).

The associations among systems of these four scales are complicated; cells apparently are found in all small mesoscale areas. Water was precipitated from large mesoscale areas at rates from 10 to 100 $\times$ 10^6 kg sec^{-1} (cf. rates earlier mentioned for a large thunderstorm, 4 $\times$ 10^6 kg sec^{-1}). Of these amounts approximately 0.2 came from cells,

0.5 from regions of small mesoscale areas excluding cells, and 0.3 from large mesoscale areas excluding small mesoscale areas and cells.

There is an indication that much of the condensate generated by cells does not fall out immediately, but later gives rise to what is classified as "rain from regions of mesoscale areas outside of cells." The contribution of the cells, systems of a few minutes life and 10-km^2 area, is probably basic, whether or not the cells are "imbedded in more widespread precipitation" (Austin *et al.*, 1972).

As a result of their painstaking cartographic analysis of these atmospheric systems, the authors conclude that "the precipitation patterns, which initially appeared very dissimilar, turned out to be composed of subsynoptic-scale precipitation areas with rather clearly definable characteristics and behavior." It might therefore be possible eventually "to describe the distribution of precipitation in any storm in a parameterized manner by giving the number, intensities, height, and horizontal extent of precipitation areas of each scale" (Austin and Houze, 1972).

Yields in Relation to Storm Area Another way of analyzing radar data yields information on the vertically integrated content of liquid water in the atmospheric column (Greene and Clark, 1972) (Fig. II-6). Good resolution of the mean spatial pattern of this quantity is feasible, as shown in Fig. II-6, in which grid intersections are 8 km apart. The center at the lower left of the figure, about 25 km^2 in area, represents liquid content exceeding 20 kg m^{-2}, mostly in the form of raindrops. (Cloud droplets are poor radar targets, and in the whole atmospheric column hardly total more than 1 kg m^{-2}.) By averaging the liquid-water content in areas of different size, we can obtain an idea of the

TABLE VIII

Yields of Rain Areas of Different Sizes over Boston[a]

Class	Size (km^2)	Approximate duration	Average rates of rainfall (mm hr^{-1})
1. Synoptic rain areas	>10,000	Several days	$\frac{1}{2}$–1
2. Large mesoscale areas	1000–10,000	Several hours	1–5
3. Small mesoscale areas	100–400	An hour	up to 20
4. Cells	About 10	A few minutes	up to 100

[a] Source: Austin and Houze (1972).

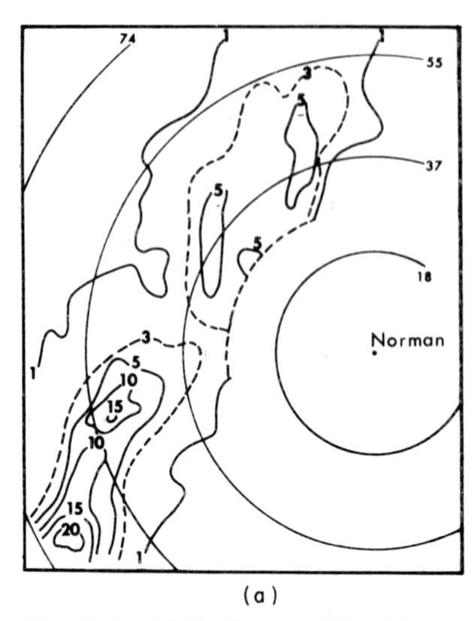

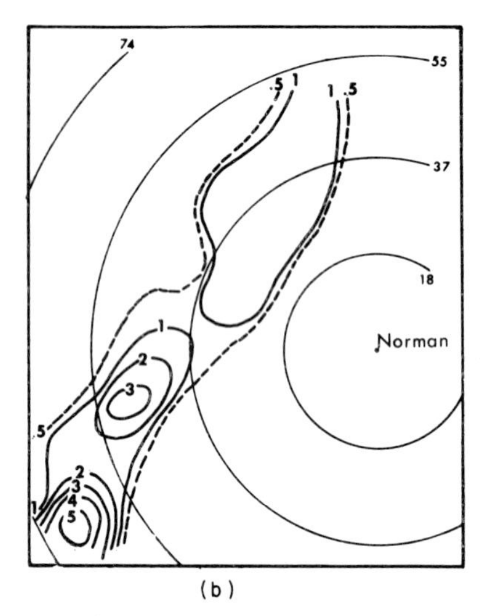

Fig. II-6. (a) Total mass of liquid water in vertical column through lower atmosphere at 1640 to 1647 CST on 26 April 1969 near Norman, Oklahoma (unit: kg m^{-2}). (b) Rainfall rate (in. hr^{-1}) from 1640 to 1645 CST on 26 April 1969, near Norman, Oklahoma. The peak rate exceeds 125 mm hr^{-1} (from Greene and Clark, 1972).

spatial concentration of water in the atmosphere over this area. In a small area, convergence of vapor and condensation processes are highly concentrated, producing a large liquid-water content; as larger areas come into consideration, the intensity of the convergence and condensation processes averages decreasing amounts, as might be expected. This can be measured in these terms:

Area (km^2)	Water content (kg m^{-2})
25	22
250	13
2500	3

When we apply the same measuring process to the map of rainfall intensity at about the same time as the previous map, we obtain a rough set of numbers such as:

Area (km^2)	Rainfall intensity (kg m^{-2} hr^{-1})
25	150
250	70
2500	20

This relation shows a similar decrease in water flux with increasing area. It will be discussed further in a later chapter.

The spatial pattern of the two maps is similar. The cell at the lower left delivered about 10 kg m^{-2} of water in 5 min; considering that more water probably fell after this period, the total delivery might approximate 15–20 kg m^{-2}, in good correspondence with the amount of airborne water shown in the first map. This comparison illustrates the coupling of an atmospheric system that converts an inflow of vapor into liquid water with the arrival of this water at the underlying surface.

Systems in Motion

Relative to the face of the earth, water-extracting atmospheric systems are generally in motion, steered by the upper winds, tending to drift in the direction from which their moisture is coming. They leave trails of water behind them, wetting the underlying surface. Cells leave heavy but short streaks; the mesoscale rain areas leave a wider, longer band, striated by heavier rain from their interior cells. Single synoptic-scale storms leave a wide swatch that often exhibits a complicated internal pattern of rainfall.

One way to stop the blurring motion of precipitation systems and to understand their rainfall pattern is to average them on an artificial set of coordinates applied to each storm. Such an abstraction from synoptic maps of many winter storms is shown in Fig. II-7 (Jorgensen, 1963)—a composite storm expressed in terms of the probability of more than 0.2-mm precipitation falling in the ensuing 6 hr. The storm center is at point 0. Note that measurable precipitation is most probable north of the center, although the direction of movement of the storm is toward the northeast, and the direction of winds within it forms a circulation in a counterclockwise sense. The swath left by such a storm is elongated from southwest to northeast and is several hundred kilometers wide.

Frontal storms moving in families leave trails on top of the trails of earlier storms. Sometimes a large zone of convergence will attain a balanced state, moving neither north or south, but experiencing the inflow of air from both directions; waves moving through this zone generate deluge after deluge of rain onto a surface that eventually cannot accommodate any more. Such a case is an extraordinary concurrence of storm tracks in January 1937. Bergeron (1960) calls to our attention the fact that this pattern is not orographic. It is aligned along the Ohio River, not the Appalachian Mountains, which in fact experienced only moderate rain. He points out that this gigantic

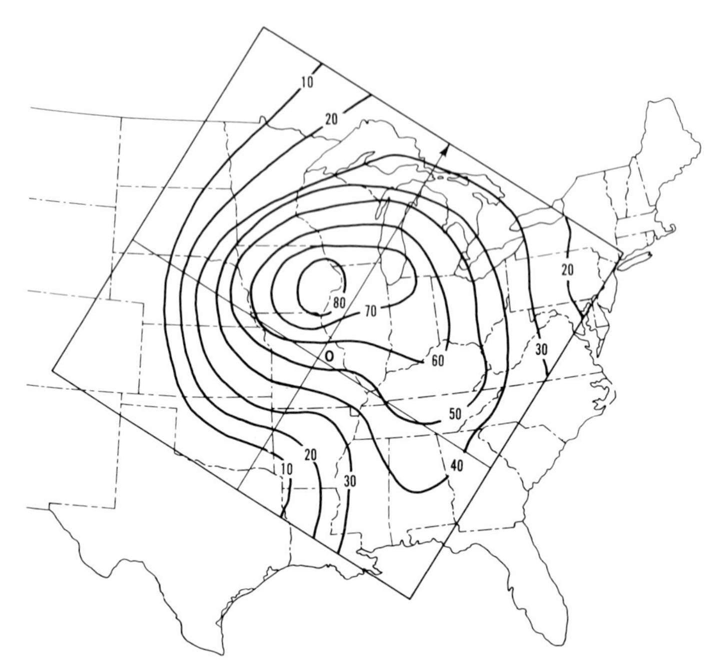

Fig. II-7. Frequency of occurrence of precipitation exceeding rates of 0.2 mm in 6 hr following the time at which the storm center was located at point 0. (The arrow parallels the lines of mean thickness.) (Jorgensen, 1963.)

precipitating mechanism was not orographic, not frontal, but something much more complex.

ATMOSPHERIC STORMS: CAUSES OF VARIABILITY IN RAINFALL

Limited space dictates that we look only briefly at such atmospheric processes as condensation and growth of precipitation particles, but longer at the larger-scale phenomena of moisture inflow into precipitating storms, lifting mechanisms in them, and spatial patterns. Lifting mechanisms are not well understood in quantitative terms, but we did examine numerical data on vapor inflows into storms and thereby saw them as systems governed by their inflows and outflows of water. The vertical flux of water precipitated to the underlying surface is, of course, of greatest interest in this book, and will be the subject of following chapters.

Atmospheric conditions that cause the precipitation process to yield a solid or liquid product, in small areas or large, are highly complicated. Their pattern in space and time is irregular, and can best be summed up in the one word, variability. Variability of the water input to ecosystems is one of the most important characteristics of their water budgets, and is often frustrating to man's activities also. We have seen in this chapter how diverse in size and intensity, in spacing and in movement are the rain-producing cells. Yet the systems at the underlying surface are wholly dependent for their water supply upon these erratically moving, growing, then dying systems of vertical motion floating in the mobile atmosphere above the earth.

From week to week these atmospheric systems bring, or fail to bring, the water needed by the ecosystems at the underlying surface. Where these ecosystems are fields of corn or wheat, the variable size and spacing of individual atmospheric systems determines the progress and steadiness of crop growth during the summer. On a broader time scale, the tracks these systems follow in different summers determine the stability of growing-season moisture input to a crop region. Variation in rainfall depth and timing from one growing season to the next is one of the most difficult conditions facing agriculture. In a period when the world grain-storage cushion has shrunk, this uncertainty becomes a driving trend in our national economy as important as energy costs.

In the following chapters we will examine the variability of rainfall that is produced by these atmospheric systems—the fluctuations of storm rain at a given place, the spatial differences, and the long-term patterns of precipitation. Later chapters will indicate how this variability affects the hydrologic processes at and near the earth's surface.

REFERENCES

Anon. (1947). Severe local storms for [sic] January 1947. *Mon. Weather Rev.* **75,** 13.

Austin, P. M., and Houze, R. A., Jr. (1972). Analysis of the structure of precipitation patterns in New England, *J. Appl. Meteorol.* **11,** 926–935.

Benton, G. S., and Estoque, M. A. (1954). Water-vapor transfer over the North American continent, *J. Meteorol.* **11,** 462–477.

Bergeron, T. (1960). Problems and methods of rainfall investigation, Am. Geophys. Union, *Geophys. Monog.* **5,** 5–25. Disc. 25–30.

Bjerknes, J., and Solberg, H. (1921). Meteorological Conditions for the Formation of Rain, *Geofys. Publikasjoner* (Oslo) **2,** No. 3. 60 pp.

Carroll, D. (1962). Rainwater as a chemical agent of geologic processes—a review, U. S. Geolog. Surv., Water-Supply Paper 1535-G, 18 pp.

Chang, J.-H. (1972). "Atmospheric Circulation Systems and Climates." Oriental Publ., Honolulu, 328 pp.

Cogbill, C. V., and Likens, G. E. (1974). Acid precipitation in the northeastern United States, *Water Resources Res.* **10,** 1133–1137. Discussion, *ibid.* **12,** 569–571.
Cry, G. W. (1967). Effects of tropical cyclone rainfall on the distribution of precipitation over the eastern and southern United States. U. S. Dept. Commerce Environ. Sci. Serv. Admin., ESSA Prof. Paper 1, 67 pp.
Førland, E. J. (1973). A study of the acidity in southwestern Norway, *Tellus* **25,** 291–299.
Freitas, C. R. de (1975). Estimation of the disruptive impact of snowfalls in urban areas, *J. Appl. Meteorol.* **14,** 1166–1173.
Gatz, D. F. (1967). Low-altitude input of artificial radioactivity to a severe convective storm—comparison with deposition, *J. Appl. Meteorol.* **6,** 530–535.
Gilman, C. S. (1964). Rainfall. *In* "Handbook of Applied Hydrology" (V. T. Chow, ed.), Chapter 9, pp. 1–68. McGraw-Hill, New York.
Greene, D. R., and Clark, R. A. (1972). Vertically integrated liquid water—a new analysis tool, *Mon. Weather Rev.* **100,** 548–552.
Hann, J. (1897). "Handbuch der Klimatologie," 2te Aufl., I Band: Allgemeine Klimatologie. Engelhorn, Stuttgart, 404 pp.
Havlik, D. (1969). Die Höhenstufe maximaler Niederschlagssummen in den Westalpen, Nachweis und dynamische Begründung, *Freiburger Geog. Heft.* No. 7, 66 pp.
Hay, J. E. (1971). Precipitable water over Canada: II Distribution, *Atmosphere* **9,** 101–111.
Hirao, Y., and Patterson, C. C. (1974). Lead aerosol pollution in the High Sierra overrides natural mechanisms which exclude lead from a food chain, *Science* **184,** 989–992.
Hiser, H. W. (1956). Type distributions of precipitation at selected stations in Illinois, *Am. Geophys. Un. Trans.* **37,** 421–424.
Holzman, B. (1937). Sources of moisture for precipitation in the United States, U. S. Dept. of Agriculture, Tech. Bull. 589, 42 pp.
Houghton, H. G. (1968). On precipitation mechanisms and their artificial modification, *J. Appl. Meteorol.* **7,** 851–859.
Huff, F. A., and Schickedanz, P. T. (1974). METROMEX: Rainfall analyses, *Bull. Am. Meteorol. Soc.* **55,** 90–92.
Jorgensen, D. L. (1963). A computer derived synoptic climatology of precipitation from winter storms, *J. Appl. Meteorol.* **2,** 226–234.
Junge, C. E. (1963). "Air Chemistry and Radioactivity." Academic Press, New York, 382 pp.
Kessler, A. (1968). Globalbilanzen von Klimaelementen. Ein Beitrag zur allgemeinen Klimatologie der Erde. Hannover, Hochschule Inst. Meteorologie Klimatologie, *Ber.* **3,** 141 pp.
Kessler, E. (1969). On the distribution and continuity of water substance in atmospheric circulation, *Meteorol. Monogr.* **10,** No. 32, 84 pp.
Marlatt, W., and Riehl, H. (1963). Precipitation regimes over the upper Colorado River, *J. Geophys. Res.* **68,** 6447–6458.
Nace, R. L. (1967). Water resources: a global problem with local roots, *Environ. Sci. Technol.* **1,** 550–560.
Namias, J. (1947). Characteristics of the general circulation over the Northern Hemisphere during the abnormal winter 1946–1947, *Mon. Weather Rev.* **75,** 145–152.
Neiburger, M., and Chin, H.-C. (1969). The meteorological factors associated with the precipitation effects of the Swiss hail suppression project, *J. Appl. Meteorol.* **8,** 264–273.
Newton, C. W. (1966). Circulation in large sheared cumulonimbus, *Tellus* **18,** 699–713.

Newton, C. W., and Fankhauser, J. C. (1964). On the movements of convective storms, with emphasis on size discrimination in relation to water-budget requirements, *J. Appl. Meteorol.* **3,** 651–668.
Palmén, E., and Riehl, H. (1957). The budget of angular momentum and energy in tropical cyclones, *J. Meteorol.* **14,** 150–159.
Palmén, E. H., and Newton, C. W. (1969). "Atmospheric Circulation Systems: Their Structure and Physical Interpretation." Academic Press, New York, 603 pp.
Petterssen, S. (1956). "Weather Analysis and Forecasting," Vol. II, Weather and Weather Systems. McGraw-Hill, New York, 266 pp.
Rasmusson, E. M. (1968). Atmospheric water vapor transport and the water balance of North America. II. Large-scale balance investigations, *Mon. Weather Rev.* **96,** 720–734.
Reitan, C. H. (1960). Distribution of precipitable water vapor over the continental United States, *Bull. Am. Meteorol. Soc.* **44,** 79–87.
Rooney, J. F. Jr. (1967). The urban snow hazard in the United States; an appraisal of disruption, *Geogr. Rev.* **57,** 538–559.
Sargent, F. II (1967). A dangerous game: taming the weather. *Bull. Am. Meteorol. Soc.* **48,** 452–458.
Schwarz, F. K. (1963). Probable maximum precipitation in the Hawaiian Islands. U. S. Weather Bureau and Corps of Engineers, *Hydrometeorol. Rep.* **39,** 98 pp.
Schwarz, F. K. (1970). The unprecedented rains in Virginia associated with the remnants of hurricane Camille, *Mon. Weather Rev.* **98,** 851–859.
Sharon, D. (1974). The spatial pattern of convective rainfall in Sukumuland, Tanzania—a statistical analysis, *Archiv. Meteorol. Geophys. Bioklimat.,* **B 22,** 55–65.
Snyder, F. F. (1964). Hydrology of spillway design: Large structures—adequate data, *Proc. Am. Soc. Civil Engrs. J. Hydr. Div.* **90** (HY-3), 239–259.
Thoreau, Henry David (1854). Walden. I, Economy.
Tuan, Yi-fu (1968). "The Hydrologic Cycle and the Wisdom of God: A Theme in Geoteleology." Univ. Toronto Press, Toronto, 160 pp.
U. S. Weather Bureau (1947). Climatological Data for Wisconsin, January.
U. S. Weather Bureau (1950). Climatological Data for California, November, Vol. 54, pp. 373–413.
Weischet, W. (1965). Der tropisch-konvektive und der aussertropisch-advektive type der vertikalen Niederschlagverteilung, *Erdkunde* **19,** 6–14.

Chapter III

POINT RAINFALL—THE DELIVERY OF WATER TO AN ECOSYSTEM

In Chapter II we looked at the atmospheric storm systems that concentrate water vapor into small cells of vertical motion, condense it, and precipitate it to earth. In the following chapter we shall look at hydrologic storms, i.e., the swaths of water left behind storms as they move through the atmosphere above the ecosystems at the surface of the earth. Here we will look at how much rain a particular ecosystem that is fixed at a point on the surface receives and how often.

Point rainfall refers to a rain gage such as you might have in your backyard, or in a cornfield, or in a parking lot, or to a small basin draining into a culvert. For many practical purposes, point rainfall is one of the most important of all water fluxes.

MEASURING RAIN AND SNOW

Like some other climatological and hydrologic instruments bequeathed to us by the 18th century, the gage for measuring precipitation at a place on the earth's surface leaves much to be desired. It is nothing more than a can set up to catch vertically falling snowflakes and raindrops. Often its catch is amplified by a funnel to an inside can to make reading easier; sometimes it is set on a scale or provided with some other means of recording the changing catch over short periods of time. Basically it remains a cylinder sitting in the airstreams that blow the snowflakes and raindrops along as they approach the ground.

Problems in Measuring

No real solution has been found to the aerodynamic problem of a bluff object in the wind. Palliatives like windshields are sometimes attached to reduce updrafts at the mouth of the can, but an exhaustive study (Weiss and Wilson, 1958) concluded that "attempts to secure 'true catch' by adjoining a windshield to the gage are vain. The importance of adequate site and environment characteristics is now almost universally recognized."

For measuring water delivered to croplands of low vegetation by storms that are not too windy, the site exposure problems generate errors up to 20–30%. A report to the International Association of Scientific Hydrology states that "no other records are as readily accepted as being correct when in fact they are not" (Rodda, 1971). "Should a gauge be devised that will provide a measure of the amount of water reaching the ground to a known degree of accuracy, then the solution of most of the other problems should be made simpler" (Rodda, 1971). The problems of inherent variability of precipitation falling from atmospheric storms, discussed in the preceding chapter, is made more difficult by the fact that our primitive mode of measurement of rain and snow does not cope with the wind energy that is a part of a storm. (See Plate 1.) A report on a scientific symposium (Rodda, 1972) says that "there was agreement that the effect of wind on the gauge was the chief problem, but no agreement on what to do about the wind."

Landscapes Poorly Sampled for Precipitation In types of ground cover other than low vegetation, measurement becomes more difficult. We have, for example, no clear indication of the amount of water delivered to a point—any point—in a forest. We *hope* that cans located in small clearings countersunk into the forest body receive the same amounts of water as are being impacted on the canopy. If the aerodynamic properties of the countersunk hole happen to be such as to cause the can to catch 20 mm in the same storm that brings 20 mm to the forest canopy, then the can is getting a good sample. In another storm, however, when winds are stronger or from a different direction, the aerodynamics of the hole are different and the can sample is incorrect. We have a set of measurements that in occasional storms are true measurements of precipitation—but do not know which storms these are.

Similar reservations attach to catches in all wind-swept sites: wide plains, mountain ridges, coastal lands. The can on top of Mount

Plate 1. This battery of rain gages, encountered by accident in a remote part of Tasmania, illustrates some of the ways that hydrometeorologists have tried to improve the accuracy of rainfall measurement. Note several types of wind shields at various heights above the ground, and just to the right of the thermometer shelter, a grating around a buried gage, its orifice flush with the ground surface. This gage battery is observed by the Hydroelectric Commission, which operates Lake St. Clair in the background for storage.

Hamilton, California, for example, often contained more pebbles than water after a storm. Stations on the coasts where oceanic storms come ashore suffer from the defects of the rain gage, as do those on the plains of the Arctic where very low vegetation hardly slows the wind at the gage mouth.

Measurements are seldom attempted at sea, or on the ice caps. We have no direct indication of the inputs to the budgets of these massive bodies of water substance.

Size of Errors The Soviet Hydrometeorological Service, an organization concerned with both the atmospheric and the surface manifestations of water, has studied the methods of measuring precipitation in order to develop correction procedures. Pertinent factors in corrective adjustments include the fraction of rainfall arriving at rates less than 2 mm hr^{-1}, the wind speed, and the temperature of snowstorms (Struzer *et al.*, 1965).

These correction procedures are not minor. For example, in a wind of 3 m sec^{-1} they range up to 0.1 for rain and from 0.3 to 0.8 for snowfall. In windier exposures they are far greater (Fig. III-1).

The growing employment of formal modeling in hydrology creates a demand for accurate inputs into such models as the National Weather Service River Forecast System with its 33 parameters (Larson and Peck, 1974). It appears that the long tolerance of hydrologists for relative or so-called "index" values of precipitation, snow-cover mass, and other quantities has finally, under the pressure of modeling, given way to an insistence on real values, which can be obtained only by correcting the readings of admittedly inaccurate precipitation gages. Larson and Peck's model for meltwater yield in the Pemigewasset River basin in New Hampshire, for example, reconstitutes observed yields best if snowfall measurements are increased by a factor of 0.3, thus approaching a true water-budget hydrology.

Problems Due to Storm Movement

Except in mountains, where storm systems tend to find a temporary anchor, precipitation-yielding systems in the atmosphere are continually in motion relative to the face of the earth. Even if we obtained perfect measurements of rain and snow at a precipitation station, we would continue to be confronted with the question of what the measurements mean.

Because the sizes of atmospheric precipitation areas, as well as their life spans, range greatly in size, a precipitation can at a point fixed in the earth's surface receives variable amounts of water. Its catch over a period of time represents a mix of contributions from different kinds

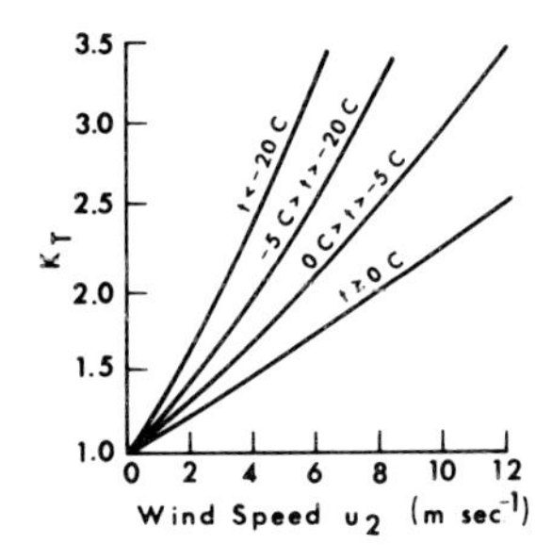

Fig. III-1. Relation of correction factor for gage catch K_T to wind speed u_2 at different levels of storm temperature (Struzer, 1965).

and sizes of precipitating cells, yielding water at different stages in their life histories. The shorter the period of time, however, the less mixed is the origin of the rain falling into the gage and the more diagnostic value is attached to its record.

Atmospheric conditions are clearly responsible for much uncertainty about the rainfall that ecosystems receive, in view of the disturbed, turbulent nature of the local air during a rainstorm. A conference of hydrologists and meteorologists was summarized thus: "The meteorologists were anxious to convince their audience that these [meteorological] approaches are of value to the hydrologist while the hydrologists were equally anxious to be convinced. However, it is difficult to be certain whether this really happened. Perhaps the difference in the scale of operation is still the main problem . . ." (Rodda, 1972). This scale difference between systems in the atmosphere that produce rain and ecosystems at the earth's surface that receive it is indeed something that we must keep in mind as we look at point rainfall in this chapter in view of what we learned in the preceding one.

THE DIMENSIONS OF POINT RAINFALL

Both vapor inflow and vertical motion in the atmosphere tend to be concentrated in areas that usually are in motion relative to the location of the measuring station. As a result, the variability of point rainfall may differ from the variability of the atmospheric systems. Point rainfall, that is, the catch of a specific precipitation can in an unchanging (we hope) local exposure over the years, integrates the experience of many atmospheric motion systems that have passed overhead. It is possible to learn something from such records by analyzing the dimensions of point rainfall—its duration, intensity, depth, and frequency over a period of years.

From a gage read daily, we find out how many days were rainy and how much rain fell. From the continuously recording gage we find out how bursts of rain fall in short periods of time (Plate 2): How much, how often, how long does it rain, and how hard does it rain? These questions can be answered through an analysis of rain-can data, but only in empirical terms, until we have better information about the processes in the rain-generating systems in the atmosphere. Without reliable statistical descriptions of the variations of rainfall in time and space, we cannot yet put together a general theory that will account for the way water accumulates in any one rain gage as time passes.

Plate 2. Float and siphon gage for rainfall intensity (Hellman type) at observation plot of the Central Asia Hydrometeorological Research Institute in Tashkent. The siphon empties the float chamber (brass tube) and allows the pen to retraverse the chart (on drum at upper left), thereby providing a precise recording of intensity through a large depth of rainfall. Note the irrigated vegetation of the Tashkent urban oasis in the background; the plot itself remains in its natural state.

Duration

We do not always think in terms of the mass of water substance. Sometimes we simply want to know (a) how much of the time is it raining or snowing, i.e., how much of the time (e.g., number of hours) would you go out the door and encounter falling precipitation?

(b) How long would you have to wait for it to cease? (c) How often does rain or snow occur, i.e., how many days are marked by an appreciable amount of rain or snow?

From measurements at the climatological station of the Argonne National Laboratory (Moses and Bogner, 1967), we can answer these questions as follows from the Chicago area:

(a) In December it is snowing (or raining) 0.12 of the time. Over all seasons of the year, the fraction of hours when measurable precipitation was falling is 0.09. Your chances of going out the door and encountering falling rain or snow are thus 1 in 11.

(b) To supplement this information, we may want to know how long rain or snow continues. From the Argonne measurements we have the following:

Durations	of 1, 2, and 3 hr account for	0.19 of all hours with precipitation;
	of 4, 5, and 6 hr	0.18;
	of 7, 8, and 9 hr	0.15;
	of 10, 11, and 12 hr	0.12;
	of 13–18 hr	0.16;
	of 19–24 hr	0.12;
	of greater than 24 hr	0.08.

The median duration is 9 hr. Half of all the hours with precipitation come in sequences shorter than 9 hr, half in sequences longer than 9 hr. In winter this number is larger than in summer, since in winter atmospheric systems are likely to be large and take a long time to pass over a place.

(c) For the question about how often rain or snow occurs, let us look at the Argonne data on daily depths greater than 2.5 mm.* Such days occur more frequently in June, when there are six on the average (or a frequency equal to 0.2), than in December, when there are only three. The count has never exceeded ten such days in June or five in December, which in the Midwest is usually a dry month.

Durations of rain events used in the design of engineering works depend on the area involved and the cost of failure of the structure. Drainage of a parking lot might utilize data on the depth of rain in the heaviest 5 or 10 min of a storm; for an urban storm-sewer network, a 1-hr duration is appropriate (Jens and McPherson, 1964); for drainage of ordinary lowland farm land a 48-hr duration is common.

* The value 2.5 mm is of the order of the daily needs of plants for water in the growing season. In the city 2.5 mm is enough to tell a driver or pedestrian that rain or snow had fallen, was falling, or was about to fall, and induce him or her to drive with more caution or to put on a coat.

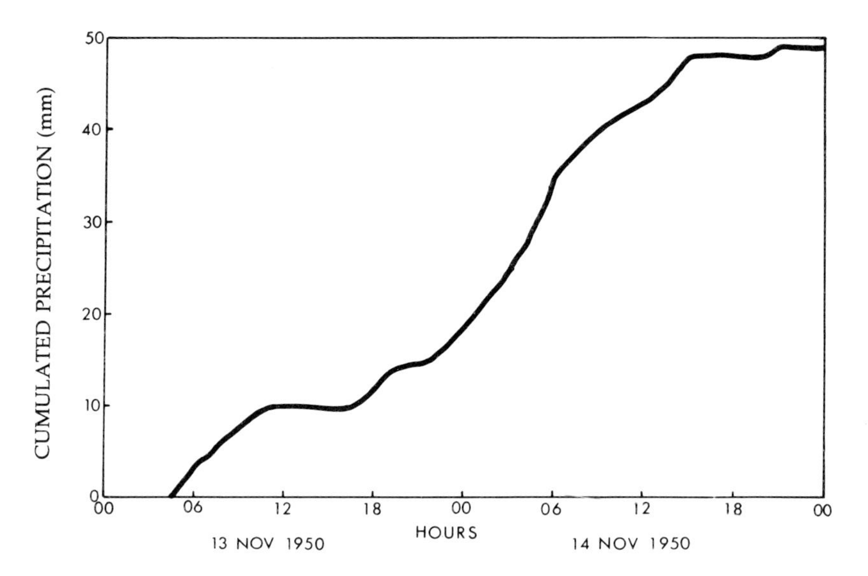

Fig. III-2. Mass curve of the accumulated depth of precipitation at Central Sierra Snow Laboratory, 13–14 November 1950 (mm) (U.S. Weather Bureau, 1950).

Depth of Precipitation

The foregoing example took us from mere duration of precipitation into another dimension: how much water arrived, i.e., the depth as if measured on a level surface, in millimeters.*

Episodes bringing less than about 0.2 mm of water are called *traces,* being too small to measure in most gages. From a mass-budget standpoint, this depth is negligible, but in some conditions it is highly important—say if it comes as freezing rain that forms a slippery layer not easily dislodged from sidewalks. In summer such small depths of rain chiefly have a transient effect, producing a little steam from the hot pavements or cooling plant leaves.

Mean storm types at Chicago, discussed in the preceding chapter, deliver amounts that range from 2 mm per event in post-cold-front showers to 11 mm per event in squall lines. These may be taken as typical of the usual range of depth of precipitation in average storms.

Heavy storms register depths to 100 mm or more, and extreme storms in regions of strong moisture flow approach 1000 mm. The total variation in storm depth is therefore from less than 1 mm to about three orders greater.

* In terms of mass per unit area, 1 kg m^{-2} is equal to 1-mm depth.

Intensity

Intensities, or rates of precipitation, are derived from gages fitted with recording devices that produce a chart or digital tape from which hourly amounts of snow or rain can be read.

A Snowstorm in the Mountains Table I and Figure III-2 report the catch in a gage in a mountain valley in California hour by hour through a small snowstorm in November 1950 in terms of mass, or water equivalent. Snowfall was caught in the gage at varying rates during these two days. The largest intensities came near sunrise and sunset on 13 November and through 14 November up to midafternoon. In more precise terms, we can say that in the heaviest hour snow was caught at a rate of 5 mm hr^{-1}; during the heaviest 3 hr 11 mm of snow was caught, for a mean intensity of almost 4 mm hr^{-1}. Such facts can be arrayed as shown in the accompanying tabulation. They plot in a curve showing that average intensities are less over long periods than over short, as would be expected (Fig. III-3).

Period	Duration (hr)	Depth (mm)[a]	Intensity (mm hr^{-1})
Heaviest clock hour (0500–0600, 14 Nov)	1	5	5
Heaviest 3-hr period (0300–0600, 14 Nov)	3	11	4−
Heaviest 6-hr period (0100–0700, 14 Nov)	6	17	3−
Heaviest 12-hr period (2100, 13 Nov–0900, 14 Nov)	12	26	2+
Heaviest 24-hr period (1500, 13 Nov–1500, 14 Nov)	24	38	1.6
Total storm duration	42	49	1.2

[a] Or kg m^{-2}.

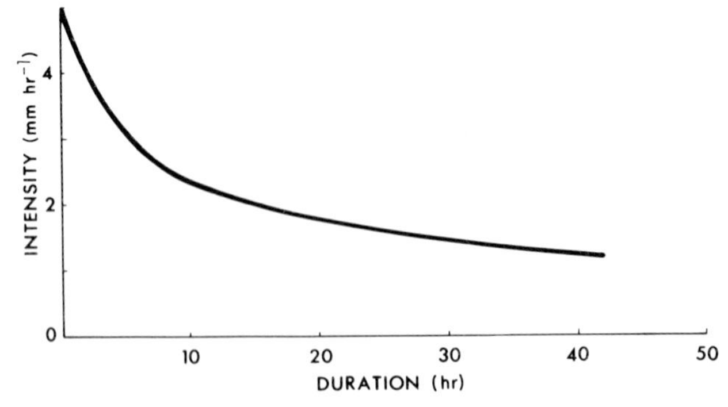

Fig. III-3. Duration–intensity curve of snowstorm at Central Sierra Snow Laboratory, 13–14 November 1950.

TABLE I

Hourly Precipitation, Central Sierra Snow Laboratory Headquarters Gage, 13–14 November 1950[a]

Date	Hour ending																							
	01	02	03	04	05	06	07	08	09	10	11	12	13	14	15	16	17	18	19	20	21	22	23	24
13 Nov	0	0	0	Tr	1	2	1	2	1	1	1	1	Tr	Tr	Tr	Tr	Tr	2	1	1	Tr	1	1	2
14 Nov	2	2	2	3	3	5	2	1	2	1	1	1	1	2	2	Tr	0	Tr	Tr	Tr	1	0	0	0

[a] Units: mm hr^{-1} or kg m^{-2} hr^{-1}. Source: U. S. Weather Bureau. Climatological Data, California, Vol. 54, No. 11, pp. 373–413 (November 1950) (units converted).

TABLE II

Hourly Precipitation, Honolulu, Hawaii (Central Business District) Federal Building Gage, 5–6 March 1958[a]

	Hour ending																							
Date	01	02	03	04	05	06	07	08	09	10	11	12	13	14	15	16	17	18	19	20	21	22	23	24
5 Mar	0	1	6	14	9	11	35	65	27	38	24	8	Tr	Tr	6	5	37	28	13	4	4	14	30	11
6 Mar	22	19	12	4	2	3	2	Tr	0	0	Tr	0	0	0	0	0	0	0	0	0	0	0	Tr	Tr

From these data, we find:

Period	Duration (hr)	Depth (mm)	Intensity (mm hr^{-1})
Heaviest clock hour	1	65	65
3 hr	3	127	42
6 hr	6	206	34
12 hr	12	273	23
24 hr	24	432	18
Total storm duration	30	450	15

[a] Units: mm hr^{-1}. Source: U. S. Weather Bureau, Local Climatological Data, Honolulu, Hawaii (March 1958).

TABLE III

Duration, Depth, and Intensity Values in a Rainstorm on Kauai in January 1956[a]

Duration (hr)	Depth (mm)	Intensity (mm hr^{-1})
0.5	150	300
1	280	280
24	965	40
48	1105	23

[a] Source: Jennings and Price (1956).

As a mass flux, these intensities are not large. As unmelted snow, these rates denote increases in the depth on the ground by 1–5 cm hr^{-1}. The 3-hr amount alone (approximately 10 cm) is enough to immobilize many wheeled vehicles. It corresponds to the depth at which many cities call out the snow plows.

Two Heavy Rains in Hawaii This rather mild snowstorm in the Sierra may be compared with a heavy rain from a convergence line in very moist equatorial air flowing northward into Oahu in March 1958. Clouds in this line reached heights of 13–16 km (Blumenstock, 1961) indicating extreme lifting along a zone that remained in the same position over Honolulu for about 30 hr (Table II).

These intensities lie in the top one percentile of all rainfalls everywhere, although they do not closely approach the recorded extremes. Compared with the intensities of snowfall at the Sierra Snow Laboratory, they indicate both a more vigorous lifting mechanism acting through a great depth of atmosphere (15 km!) and a much more rapid inflow of water vapor, the concentration of which in the equatorial airstream might well have been nearly an order greater than that in the maritime air above the Sierra. As can be seen, the Hawaiian intensities were about an order greater also than those in the Sierra storm.

Duration–intensity relations from a similar kind of storm, also in a lowland part of one of the islands of Hawaii, show the potential of atmospheric systems (Table III). This rain fell on Kilauea Plantation (Plate 3) near the northeast coast of Kauai and some distance from the great wet plateau that forms the center of the island.* Direct oro-

* The highest annual totals in the United States accumulate here from an almost continuous milking of trade-wind clouds. This different atmospheric system generates an average intensity of 2–3 mm hr^{-1} that continues over extremely long durations.

Plate 3. Recording and nonrecording rain gages at the headquarters of Kilauea Plantation, Kauai, Hawaii, looking toward the direction from which the moisture flow came in the great rain of 1956. Also shown is a thermometer shelter, and beside it a solar radiation recorder.

graphic effects were unimportant in this storm, although indirect ones might have made a contribution.

For comparison, the statistically estimated 0.01-probability storm (once a century) would bring only half as much rain in the 24-hr period as this storm did. The probable maximum storm's 24-hr depth is calculated as approximately 1200 mm, somewhat greater than the 965 mm measured in the 1956 Kauai storm; if not, it would be back to the drawing board for the probable max!

Camille Another storm that came near the probable maximum for its region was an unexpected resurgence of the dying phases of hurricane Camille in 1969, described in Chapter II. Sudden introduction of water vapor from a new source resulted in an extremely heavy rain over about 5 hr one night, well-described by an eyewitness:

> I stayed in bed until two and tried to sleep, but it wasn't really possible. The rain struck the roof like a giant flail being whopped onto the metal. Then, at intervals, there would come an entirely different kind of noise. Whomp! Whomp! Whomp! It was like the roof was under a waterfall. Not single blobs of rain hitting it but

> like a cascade, falling in jets and spurts. Then that kind of rain would stop for a while.
>
> (E. Kinkead, Big rain. *The New Yorker*, 31 July 1971, pp. 66–74).

Fortunately, the frequency of rainfalls of such catastrophic intensity is very small, otherwise the earth's surface would look quite different.

Significance of Precipitation Intensity The impact of raindrops, which have a falling speed of 6–8 m sec^{-1} (compared with, at most, 1 m sec^{-1} for snowflakes), releases a large amount of kinetic energy (see Plate 4). Its effects are seen in the pounding action of heavy rain on exposed soil, and the splash in which soil particles are broken and blasted out of the surface. The deepening film of water then helps float the detached particles downslope. Erosion—a combination of detachment and removal of soil particles—is in progress. Laboratory measurements on three soils that separate the effect of raindrop impact from that of flowing water found that impact is the primary cause for detachment of soil particles (Young and Wiersma, 1973).

Plate 4. Detention film of water accumulating in an unusually heavy intensity of summer rainfall in Wisconsin on 30 July 1970. Note the size of the raindrop splashes as indicators of the energy release.

Studies of soil erosion have shown that two dimensions of point rainfall are important: its 30-min intensity and the production of energy by the whole storm. This last figure is determined from data on the mass of water falling in different intensity classes (Wischmeier and Smith, 1958).

Snow falling at high intensities can effectively tie up a city as it accumulates on the streets; rain at high intensities causes local flooding of streets and underpasses, but more briefly than snow because natural and man-made drainage systems remove it. Even while falling through the air, however, snow at high intensities reduces visibility and slows urban functions that depend on long fields of vision, like airport operations. Interruptions due to rain are rare, but they do occur and we are usually unprepared. For example, a short rain at a rate of 90 mm hr^{-1} in Milwaukee one August caused three multiple-car accidents with numerous injuries and one death resulting. All three occurred on the freeways; no such serious accidents were reported on the much larger network of city streets. One wonders if the freeway as a road plus driving-pattern system is efficient in the Wisconsin environment.

Such short-term intensities, frequently in periods shorter than 30 min, are often heavy, although not all recording gages can be read to this degree of fineness. The limiting case is found in very brief periods when a column of raindrops held up in a strong updraft during a convective storm is suddenly left unsupported and falls to earth at a rate, calculated in a study of convection dynamics (Kessler, 1969), of 5 mm min^{-1} (equivalent to 300 mm hr^{-1}). Such shafts of descending water are small, so that this astronomically large rate of water input is not likely to be experienced for long at any point on the underlying surface.

THE FREQUENCY OF PRECIPITATION-INTENSITY EVENTS

Length of the Precipitation Record

Events That Occurred in One Year The relations between depth and intensity of precipitation over relatively short durations can be illustrated from the record of rainfall and snowfall at Milwaukee in the year 1973 (U.S. Environmental Data Service, 1974, p. 39). In Fig. III-4a, the raw depths that accumulated in each duration are plotted for the maximum day in June and in December. As durations increase the accumulated total depth must also increase (except that the rain in June lasted only 100 min). The December storm, in contrast (an early,

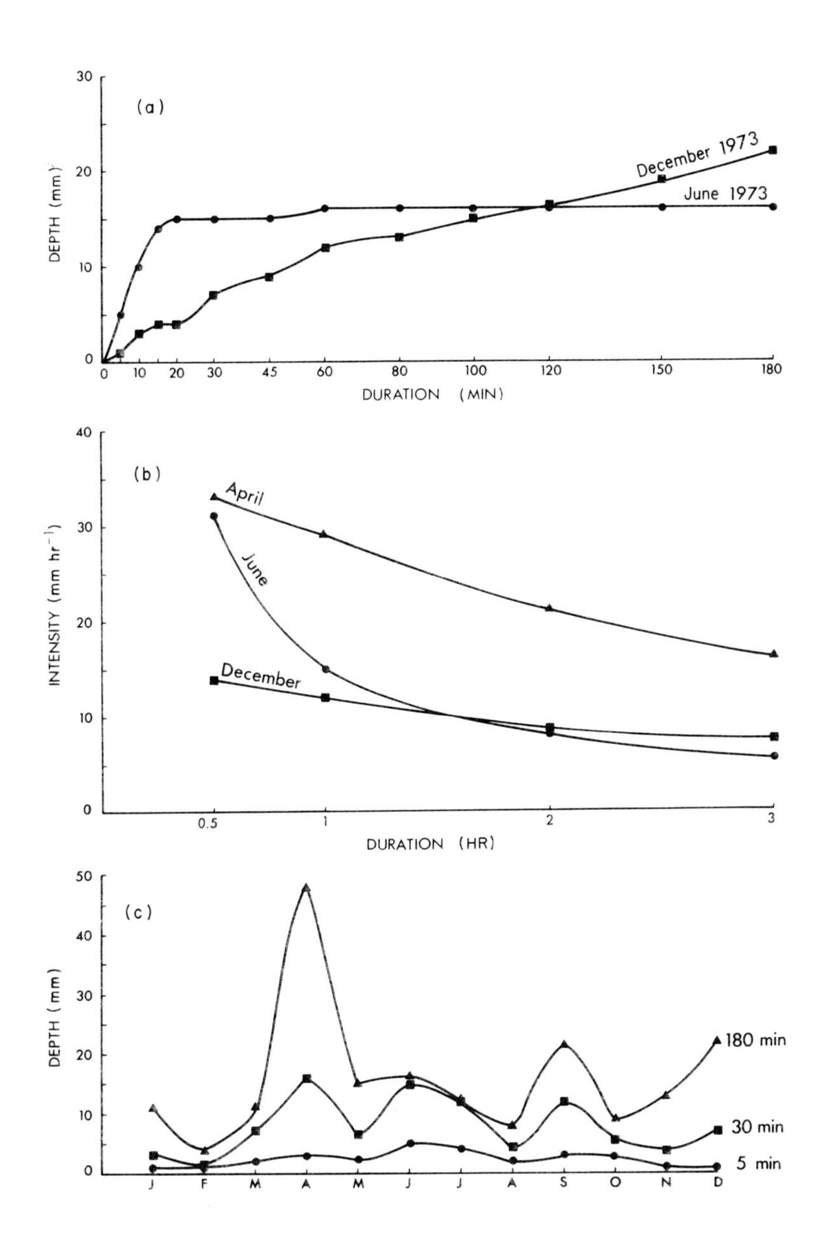

Fig. III-4. Relations of depth and intensity of precipitation over durations from 5 min to 3 hr at Milwaukee in 1973 (from U. S. Environmental Data Service, 1974, p. 39).

heavy snow), continued at the steady rate typical of cyclonic disturbances of great areal extent in the atmosphere; this contrasts with the convective cells of summer, which are deeper (hence give higher rates of rain generation) but horizontally smaller (shorter in duration).

In Fig. III-4b, this same relation is reinforced from a different viewpoint. Here the mean intensities, in units of millimeters per hour, are derived from the depth–duration information of part (a). This presentation brings out the steady rates of delivery in the winter storm and the high but brief intensities (54 mm hr^{-1} at the shortest time interval tabulated, undoubtedly heavier over shorter durations) of the storm in June.

Figure III-4c shows maximum depths in each month of 1973 and emphasizes that while basic differences are due to storm type, these types do not neatly divide the year between them. The April storm, for instance, combines the high rates of convective rain cells with the persistence of a general cyclonic system.

Events That Occurred over a Long Period of Record After making an examination of the volume of data that has accumulated from regular readings of one rain gage over an 82-yr period at Darwin, Australia, we can make such statements as the following:

(1) 25 mm hr^{-1} or more rain fell in 1‰ of all the hours in January.

(2) Measurable amounts of rain fell in 11% of all the hours (Table IV from Gibbs, 1964).

(3) Going to a longer period—the day—we find that rain equal to 25 mm or more fell on 15% of all the January days. Going a step further, to the whole month, we can say that more than 25 mm has fallen in every January in the 82 yr the gage has been read at this place.

These statements indicate that the frequency of occurrence is increased by lengthening the period of time over which we measure. Rain fell on 0.11 of the hours, but on 0.65 of the days. The average rainy day has six rainy hours.

The variation can be looked at in terms of a fixed frequency level; for example, take 1%. On 1% of all hours in January, rainfall exceeded 15 mm; on 1% of all days it exceeded 100 mm; and on 1% of all months it exceeded 750 mm.

Frequencies for Designers

Information of this kind has been compiled for many durations in each season of the year, at thousands of places over the earth. Results

TABLE IV

Rainfall in Darwin, Australia for January[a]

Intensity	Frequency
Hourly periods:	
Exceeding 25 mm hr^{-1}	0.001 of all hours
Measurable rates	0.11
Days:	
Exceeding 100 mm day^{-1}	0.01 of all days
Exceeding 25 mm day^{-1}	0.15
Measurable daily amounts	0.65
Months:	
Exceeding 750 mm $month^{-1}$	0.01 of all Januaries
Exceeding 100 mm $month^{-1}$	0.97
Exceeding 25 mm $month^{-1}$	1.00

[a] Source: Gibbs (1964).

are plotted on thousands of graphs drawn on various scales, usually distorted so that the distribution will appear less nonlinear. These provide frequency data on rain events of various sizes in the past. If the same distribution of intensities and sizes of rain events should hold true into the future, these tables and graphs can be accepted as the probable manner in which water fluxes will be delivered to the surface.

They say nothing about whether this distribution will occur in any *particular* year or month, nor how far apart the rain episodes will come. The fact that they do not tell us which date in January—the 25th, the 1st, or some other—will be the date that witnesses 75 mm or more, does not interfere with the use of the information for design and long-term planning. If the designer of a culvert expects to replace it after a 100-mm storm hits the small catchment above the road, the particular date of loss of the culvert is not as important in his calculations as is the anticipated cost of replacing it 10 times in a century. He can compare this cost with the higher initial cost of a larger culvert, which might have to be replaced only twice in a century.

Frequencies common in hydrologic design work are 0.2 ("5-yr storm") in urban storm-water sewer systems and often in airport and expressway drainage systems (Jens and McPherson, 1964). In these situations water in excess of the capacity to remove it causes inconvenience but little structural damage. A more severe condition (p = 0.1) is used in designing land terraces and levees protecting farm land, in which overtopping means destruction of the earthwork as well as flooding (Ogrosky, 1964). Larger storms (from $p = 0.04$ up to half of the probable maximum rainstorm) are used in design of levees and floodwalls protecting urban land. The frequency chosen depends on the probable damage when the wall is overtopped.

Spillways, the most critical of man's works put in the paths of water, require the most conservative design criteria. An inadequate spillway can result in overtopping and failure of the dam, which immediately converts a bad flood into a catastrophe. For spillways in farm ponds, where downstream damage is not likely to be great, storms of frequencies of 0.04–0.01 are recommended. For small floodwater-retarding structures, which would typically hold more water than a farm dam, a storm from one-quarter to one-half the size of the probable maximum storm is taken. For spillways in dams above settled areas, the probable maximum storm itself is used, along with conservative assumptions about the water-shedding condition of the basin (Jens and McPherson, 1964; Ogrosky, 1964).

There usually is no guarantee that two events even of low probability will not occur one right after the other. In fact, the chances are better than random that this will happen since the large-scale atmospheric conditions that generated the first episode could favor a succession of rainstorms; in this respect, the term "recurrence interval," which is sometimes used in speaking of probabilities, is misleading.

For example, the 24-hr rainfall in Washington, D.C. is calculated to reach 100 mm at a frequency of 0.2 yr^{-1}. This does not, however, mean once every 5 yr; many years went by from 1906 to 1921 without a single such rainfall experience (Hershfield and Wilson, 1958), while on the other hand amounts larger than 100 mm occurred in each of the consecutive years 1933, 1934, and 1935.

Intensity and Frequency over Different Durations

At Seattle Table V shows the rates of rainfall that occur in Seattle with various frequencies from once in 2 yr ($p = 0.5$)* to once in 50 yr ($p = 0.02$) (approximately the length of this record). Durations in the table range from those measured in minutes (significant in soil erosion

* This level of frequency is common in hydrologic work because it more or less corresponds to the frequency at which streamflow reaches the bankfull stage.

TABLE V

Rainfall Intensities at Seattle, 1903–1951[a]

Duration of rain	Frequency of occurrence			
	0.5	0.1	0.05	0.02
10 min	21	40	49	60
30 min	12	21	24	30
1 hr	8	13	15	18
6 hr	4	5	6	8
24 hr	2	3	4	5
48 hr	1	2	2	3
96 hr	1	1	2	2

[a] Units: mm hr^{-1}. Source: U. S. Weather Bureau (1961, 1964).

and in the runoff from such small basins as urban storm-sewer networks or parking-lot drains, or from rural catchments above road culverts), to those measured in hours (important in regulating river flow in small basins), to days (for larger river basins).

At a frequency of 0.5 there is a decline of intensity from 12 mm hr^{-1} during the heaviest $\frac{1}{2}$-hr period to 2 mm hr^{-1} over 24 hr and 1 mm hr^{-1} over 96 hr. This decline illustrates the fact that cells of strong vertical motion that produce heavy downpours are usually short lived, or are carried away downwind from the gage. The gage does not receive heavy rain continuously over a long period. While one cell may later be followed by another, the gap between them gives a respite that lowers the mean intensity. Even when bursts of heavy rain are embedded in a band of continuous general rainfall, periods of relatively lighter rain between the bursts reduce the average intensity.

At Other Places The manner in which intensity decreases as duration increases is not the same at different places, or even in different storms. Local relations between intensity and duration at a standard frequency, such as once in two years, show great variation from place to place, quite apart from differences in level of intensity (Fig. III-5).

At Santa Fe, where heavy convectional rains are common but brief, the curves of intensity versus duration have a steeper slope than do those for Seattle. For example, at the 50% frequency ($p = 0.5$) the intensity over 30 min is 21 times as large as the intensity over the day (Table VI).

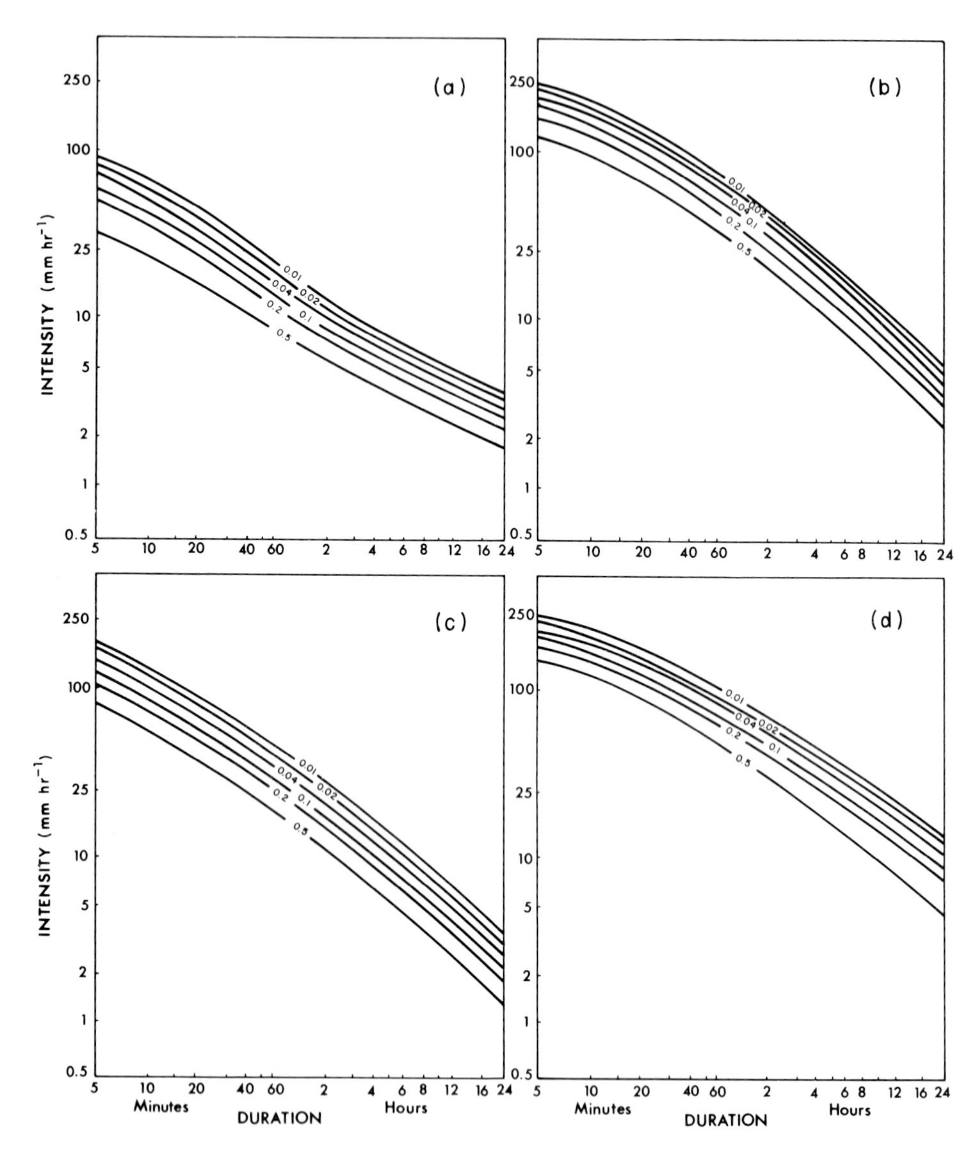

Fig. III-5. Duration–intensity curves of rainfall duration less than 24 hr at four cities, at frequencies from 0.5 to 0.01. (a) Seattle; (b) Chicago; (c) Santa Fe; (d) New Orleans. Note that a log scale is used for intensity (mm hr^{-1}). At the 0.5 frequency of occurrence, the intensity of a 1-hr rain at Santa Fe is twice that at Seattle; at Chicago and New Orleans intensities are still higher. All families of curves show the decline of intensity as storm duration lengthens; however, this decline tends to be reduced in the longer rainstorms typical of Seattle (from U. S. Weather Bureau, in Gilman, 1964).

TABLE VI

Rainfall Intensities at Frequency = 0.5 over Two Durations of Time[a]

Duration	Santa Fe	Seattle	New Orleans	Chicago
30 min	27	12	90	55
24 hr	1.3	1.8	5.9	2.8
Ratio of 30-min to 24-hr intensity	21:1	7:1	15:1	19:1

[a] Units: mm hr^{-1}. Source: U. S. Weather Bureau (1961).

At Seattle long frontal rains are usual and the 30-min intensity is relatively gentle, being only seven times as large as the 24-hr mean intensity. After several years in Seattle, experiencing prolonged rains that hardly wet a person in normal outdoor clothing, it is astonishing to be caught out in a California rain, which, while brief, wets one to the skin.

Intermediate relations between short-term and 24-hr intensities are found at New Orleans and Chicago. Due in part to the greater vapor content of Gulf air at New Orleans than at Chicago, intensities at New Orleans are higher over both short and long durations.

Both intensity and duration are significant parameters of rainfall. The drainage network of a city might be overwhelmed by a 30-min storm, with streets flooded and cars drowned out, while the level of a river that drains a large basin would scarcely reflect the occurrence of a brief storm and respond only to a long one.

Areal Distributions of Intensity–Frequency Information

The data that can be derived from compiling the intensities of rains of varying durations at different frequencies of occurrence in the past have been published for many places in voluminous tables and graphs, and somewhat condensed by means of empirical generalizations. From these data it is possible to map information that might be desired by a designer (WMO, 1969).

The map in Fig. III-6, rainfall intensities over durations of 30-min at frequency 0.5 shows the greatest depth, exceeding 50 mm, at the coast of the Gulf of Mexico. (The intensity would be 100 mm hr^{-1}.) Half-hour depths exceed 30 mm over most of the Mississippi and lower Missouri valleys and the southeastern United States. They are less than 10 mm in the Pacific Northwest with its gentle, long rains.

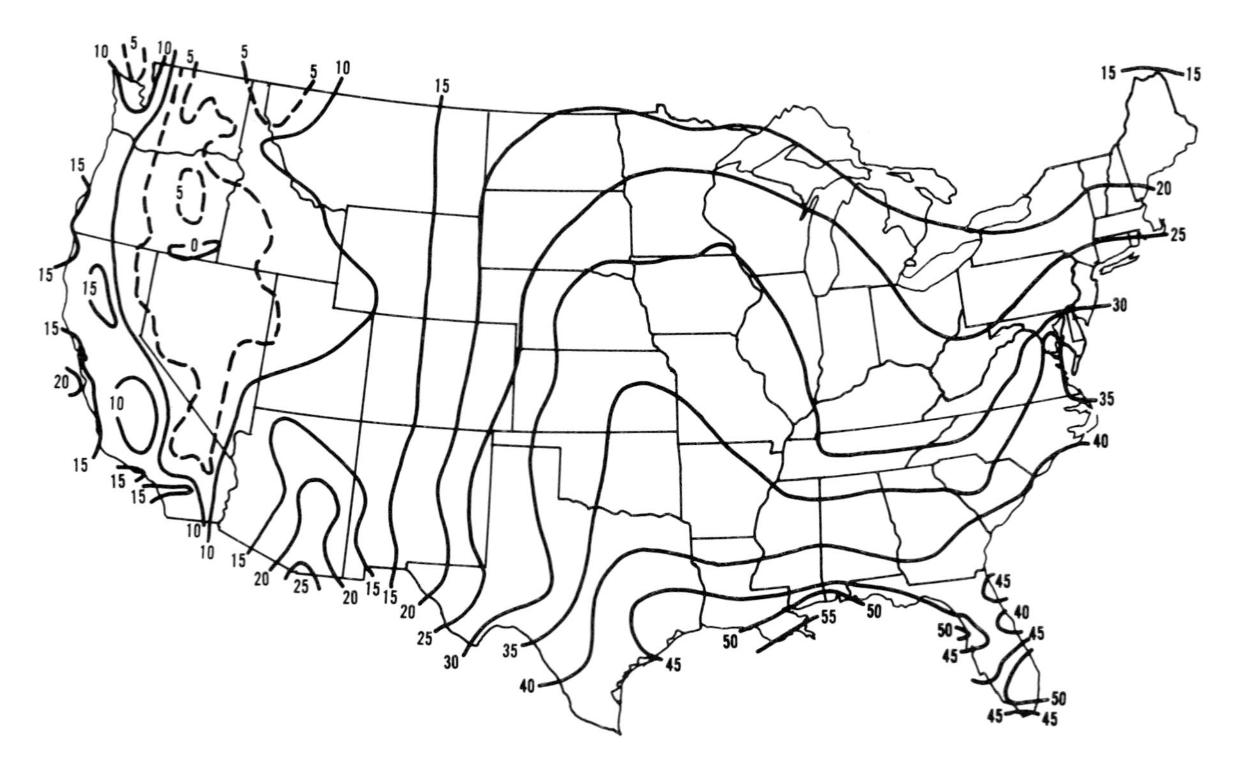

Fig. III-6. Spatial distribution of depth of rainfall (mm) in 30 min at 0.5 frequency of occurrence. (Intensities are double the figures shown.) The tongue of large depths extending north lies some distance west of the line of the Mississippi River, indicating convective potential (U. S. Weather Bureau, 1961).

Implications of these intensities for erosion and dissection of the landscape, and for such small drainage networks as those of cities, are obvious. The seemingly contradictory effects of rainfall, which on the one hand fosters the vegetation cover that protects the ground and on the other hand supplies energy and a flotation medium for detaching and transporting soil particles, are explained by considering the time scale of delivery. Abundant rain over a long period aids vegetation; heavy rain in 30 min or shorter periods ordinarily benefits neither vegetation nor soil.

The Chemical Dimension

We once thought of rainwater and fresh snow as nearly as pure as distilled water. Never quite true, this equivalence has been completely lost in the present era when raindrops and snowflakes bring down all kinds of pollutants from the air. In fact, it is becoming apparent that many ecosystems are being injured by the acidity and sulfur content of precipitation, changing their species composition and declining in productivity.

The mass budget of a storm cell described in an earlier chapter was paralleled by a budget of radioactive particles drawn into the storm and precipitated out along with the raindrops. A budget of radioactivity could be cast to parallel the budget of water.

From a long record of radioactive fallout in rain, Hicks (1972) identifies an annual cycle with a maximum in spring that is about 2.5 times as great as the minimum. The amounts of fallout are related to residence times of the particles in the stratosphere, and show a dependence on the square root of rainfall amounts.

Salt and many other kinds of particles become entrained in the rising air columns of precipitation cells, some of them serving as nuclei of condensation or sublimation. Some dissolve in water droplets or are adsorbed on snow crystals, or may simply be swept out of the air below the cloud level by falling drops and especially by snowflakes, which appear to be very efficient at this scavenging activity. As a result, rain and snow bring many substances to ecosystems at the earth's surface much faster than they would settle out or be impacted on vegetation in dry weather.

At least as much sulfur comes back to the earth in rain as through impaction on vegetation, fallout, or gaseous diffusion. In present conditions, this annual return in rain amounts to about 1.8 g m^{-2} (Kellogg *et al.*, 1972) and makes up about 0.9 of the total movement from atmosphere to earth. While sulfur is an essential element in some physiological processes in ecosystems, this downward flux, especially in the form of sulfates, is both too great and too acid for many ecosystems. It produces a steady acidifying pressure that substantially reduces productivity.

The acidity of even unpolluted rainwater has a soil-forming effect by selective removal of cations (Carroll, 1962). The intricacies of clay mineralogy are involved, and as a result "the relations between rainwater, the soil solution, and the removal or retention of cations in any soil or weathering rock are extremely complex."

WATER DELIVERY TO ECOSYSTEMS

In the preceding pages we have described the variation in the vertical flux of water from the atmosphere to the surface of the earth as received at a rain gage. At present we have no overall theory to unify this descriptive material. Each grouping of the inputs of water over different periods of time seems to bring out a different aspect of the protean nature of rainfall and snowfall. These diverse aspects include what can be called the "Noah–Joseph" property: Extreme precipitation tends to be *very* extreme (Noah), and long dry (or wet) periods tend to be *very* long (Mandelbrot and Wallis, 1968).

For these reasons point rainfall is collected in many durations or time classes for different purposes. A sampling in the preceding pages of these groupings shows some of the inherent characteristics and applications of each kind. It is apparent that an adequate rainfall climatology of even one place would fill pages of tables and graphs with an empirical representation of facts. When the day comes in which we have a general theory of the distribution of precipitation through time, these many pages can be replaced by a single paragraph setting forth the fundamental ideas. That day is not yet here.

Our ignorance of the time variations of rainfall is, however, somewhat lessened by another way of analyzing rainfall data, that is with respect to its spatial organization at the surface of the earth. This method is also a key to the significance of water inputs in terrestrial systems, which is the subject of the next chapter.

REFERENCES

Blumenstock, D. I. (1961). Climate of Hawaii, U. S. Weather Bureau, Climatography of the United States, No. 60-51 (Climates of the States: Hawaii). GPO, Washington, 20 pp.

Carroll, D. (1962). Rainwater as a chemical agent of geologic processes—a review. U. S. Geol. Surv., Water-Supply Paper 1535-G, 18 pp.

Gibbs, W. J. (1964). Space–time variation of rainfall in Australia, *In* "Water Resources Use and Management," pp. 71–79. Melbourne Univ. Press, Melbourne.

Gilman, C. S. (1964). Rainfall, *In* "Handbook of Applied Hydrology" (V. T. Chow, ed.), Chapter 9, pp. 1–68. McGraw-Hill, New York.

Hershfield, D. M., and Wilson, W. T. (1958). Generalizing of rainfall–intensity–frequency data, *Int. Assoc. Sci. Hydrol. Publ.* **43,** 499–506.

Hicks, B. B. (1972). Radioactive fallout in rain at Melbourne, 1958 through 1970, *Tellus* **24,** 277–281.

Jennings, A. H., and Price, S. (1956). The unprecedented Kauai rainfall of January 1956 (abstract), *Proc. Hawaiian Acad. Sci., 31st Meeting.*

Jens, S. W., and McPherson, M. B. (1964). Hydrology of urban areas, *In* "Handbook of Applied Hydrology" (V. T. Chow, ed.), Chapter 20. McGraw-Hill, New York, 45 pp.

Kellogg, W. W., Cadle, R. D., Allen, E. R., Lazrus, A. L., and Martell, E. A. (1972). The sulfur cycle, *Science* **175,** 587–596.

Kessler, E. (1969). On the distribution and continuity of water substance in atmospheric circulation, *Meteorol. Monogr.* **10,** No. 32, 84 pp.

Kinkead, E. (1971). Big rain. *The New Yorker*, 31 July 1971, pp. 66–74.

Larson, L. W., and Peck, E. L. (1974). Accuracy of precipitation measurements for hydrologic modeling, *Water Resources Res.* **10,** 857–863.

Mandelbrot, B. B., and Wallis, J. R. (1968). Noah, Joseph, and operational hydrology, *Water Resources Res.* **4,** 909–918.

Moses, H., and Bogner, M. A. (1967). Fifteen-Year Climatological Summary, January 1, 1950–December 31, 1964. Argonne Nat. Lab., DuPage County, Argonne, Illinois, 671 pp. (ANL-7084).

Ogrosky, H. O., and Mockus, V. (1964). Hydrology of agricultural lands, *In* "Handbook of Applied Hydrology" (V. T. Chow, ed.), Chapter 21. McGraw-Hill, New York, 97 pp.

Rodda, J. C. (Chairman) (1971). Report on precipitation, *Int. Assoc. Sci. Hydrol. Bull.* **16,** No. 4, 37–47.

Rodda, J. C. (1972). A review of the symposium on distribution of precipitation in mountainous areas, *Int. Assoc. Hydrol. Sci. Bull.* **17,** 401–403.

Struzer, L. R., Nechayev, I. N., and Bogdanova, E. G. (1965). Systematic errors of measurements of atmospheric precipitation, *Sov. Hydrol.* 500–504; *Met. Gidrol.* No. 10, 50–54.

U. S. Environmental Data Service (1974). Climatological Data National Summary, Annual.

U. S. Weather Bureau (1950). Climatological Data for California, November. (Vol. 54, pp. 373–413).

U. S. Weather Bureau (1958). Local Climatological Data, Honolulu, Hawaii, March 1958.

U. S. Weather Bureau (1961). Rainfall frequency atlas of the United States, for durations from 30 minutes to 24 hours and return periods from 1 to 100 years. D. M. Hershfield, comp. Tech. Rep. 40.

U. S. Weather Bureau (1964). Two- to ten-day precipitation for return periods of 2 to 100 years in the contiguous United States. Tech. Paper 49, 29 pp.

Weiss, L. L., and Wilson, W. T. (1958). Precipitation gage shields, *Int. Assoc. Sci. Hydrol. Publ.* **43,** 462–484.

Wischmeier, W. H., and Smith, D. D. (1958). Rainfall energy and its relationship to soil loss, *Trans. Am. Geog. Un.* **39,** 285–291.

WMO (1969). Manual for depth–area–duration analysis of storm precipitation, WMO No. 237, TP 129, 114 pp.

Young, R. A., and Wiersma, J. L. (1973). The role of rainfall impact in soil detachment and transport, *Water Resources Res.* **9,** 1629–1636.

Chapter IV

HYDROLOGIC STORMS

Usually a sizable area is wetted when a precipitating system moves over the earth's surface. The snowstorm that went on for 42 hr in the Sierra valley, which was discussed earlier, was almost certainly delivering snow to the whole crest region, and had brought rain to a large area of the foothills. Storms moving across lowlands typically travel 1000 km during their most active precipitation-generating period, and leave a wetted area of 100 km or more in width.

The wetted area of the earth's surface differs from the atmospheric water-generating system not only because of the motion of the atmospheric storm but also because of wind shear and continued growth of falling precipitation particles. This difference is seen in a comparison of the rainfall flux in a radar beam, say at 2 km height (typical for a distance of about 100 km from the radar set) and the rain delivered to the surface. In a study near Lake Ontario, overlapping radar beams registered much smaller rates of rainfall from hurricane Agnes in 1972 than were recorded in gages at the underlying surface (Wilson and Pollock, 1974). The discrepancy is due in part to continued growth of raindrops below the beam. Also, the pattern of falling drops is redistributed as a result of wind shear and forms a new spatial pattern.

We can now consider the horizontal dimensions of the underlying surface that is wetted by a storm, a concept we call the "rain area" or the "hydrologic storm." To the dimensions of duration, depth, and intensity employed in Chapter III we now add the wetted area of the earth's surface.

Areal rainfall is less easily measured than is point rainfall. About 1945, wartime radar was being reconverted to meteorological employment, particularly for watching moving rain-generating cells. It has since been much improved for this purpose (as well as for monitoring severe local storms), and a montage of photographs or a radar plan-position indicator scope taken every 6 min gives an idea of the size of the hydrologic storm. In spite of great efforts over three decades to develop quantitative measures of rainfall, radar data, however, remain pictures more than numbers. The hope that was raised in 1945, that one radar set located on a ridge near the center of a mountainous drainage basin would read out the integrated input of atmospheric water into any subbasin or into the whole basin itself, is slow in coming. Measurement of hydrologic storms still begins with readings of point rainfall by gages of one kind or another; the readings are integrated, with all their errors and uncertainties, by procedures to be described later.

THE AREA OF HYDROLOGIC STORMS

The concept of a storm as a hydrologic entity combines the factor of the area of the atmospheric system itself with its movement during its life span. Let us look first at the areal dimension of a small storm that moved relatively little during its short life.

Table I presents information on the areas receiving rainfall over several periods of time during a heavy storm over Sydney. These data purposely are limited to figures announced to the general public by radio and newspaper, in order to illustrate how even such fragmentary data can be made more meaningful when it is put into a framework of duration, depth, and wetted area of the storm unit. The 1-hr rainfall intensity over the 25-km^2 area is not unusual, corresponding to an approximately once-a-year frequency. Depths at the longer durations and over areas exceeding 25 km^2 are more rare. Their frequencies of occurrence are 0.05 or less. The storm area was small, but centered

TABLE I

Rainfall at Sydney, N.S.W., April 1966[a]

Duration (hr)	Area (km^2)	
	25	250
1	28	—
6	97	—
24	230	170

[a] Units: mm.

over the city's congested central business district, where water was least needed. It brought little rain to dry lawns in the suburbs, and none to the depleted reservoirs or dry catchments of the urban water-supply lands.

Shape Factors

The total storm isohyetal map portrays the spatial pattern of depth measurements. A typical pattern of ellipses is shown in Fig. IV-1, which is a model for a standard project storm of a 96-hr duration (Dalrymple, 1964, p. 25–30). Such patterns have led to varying expressions of an empirical kind. Court (1961) feels that a simple Gaussian distribution is as good as any other, and at least as well based in logic. In this expression the exponential decrease of depth outward from the center follows the "familiar bell-shaped curve." The depth z at a point x, y measured from the point of maximum depth m as the origin of coordinates is given by

$$z_{x,y} = m \exp(-a^2x^2 - b^2y^2),$$

where a and b are coefficients of ellipticity of the isohyets.

With respect to thunderstorms, a diversity of ideas exists. Some postulate a rapid initial decrease of depth as area increases outward from the storm center, and a slower decrease farther out; however, Arizona storms seem to fit a simple linear decrease of depth plotted against the relative area inside the isohyets (Smith, 1974).

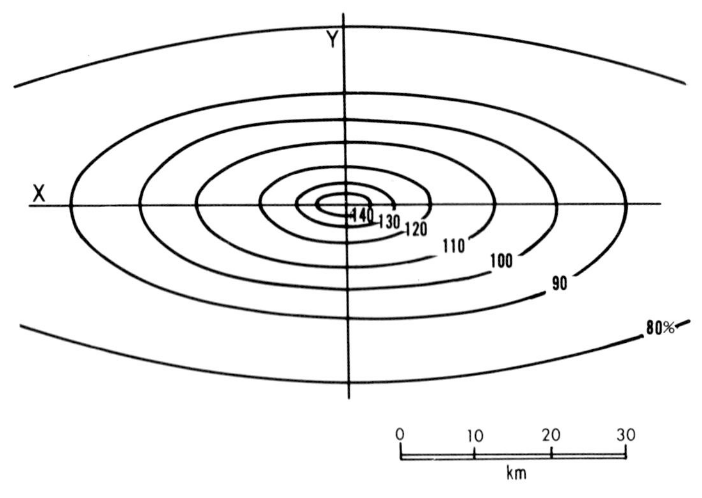

Fig. IV-1. Elliptical isohyets of a standard project storm, in percentages of index rainfall (U. S. Corps of Engineers, as presented in Dalrymple, 1964, pp. 25–30).

As in most sample situations, several measurements are better than one. Some of the irregularity in our information about delivery of rain at a single point is removed when we combine the reports from several gages to detect the shape and size of the wetted area. Rain gages do not travel along with the storm, but receive only samples from each passing shaft of descending drops or flakes. In other words, measurements at situations in a network represent the Eulerian point of view in which the frame of reference is geographically fixed on the earth's surface.*

Variation within the Area of a Hydrologic Storm

At places where topography would not be expected to disturb significantly the pattern of falling rain sweeping across the face of the earth, networks of rain gages have been installed at unusually close spacings in order to detect small-scale patterns in water delivery at the surface. These gages are usually located 1–5 km apart, a density of 100 per 1000 km^2. This spacing is in contrast with the density of gages in national networks—*one* gage for 1000 km^2 or more!

These dense networks depict the fine structure of rainfall. One of the first, Thornthwaite's (1937) network, yielded information on the "life history" of small storms. Networks of the postwar Thunderstorm project in Ohio and Florida provided detailed areal information about these energetic rain-producing systems. The networks have a close mesh, both to define the areal patterns of rainfall and also just to touch small storms. Small storms can slip through the usual 30-km mesh of stations, and, without the use of radar, might not be detected, especially if in their brief lives of a few hours they move less than 100 km. Tornadoes, for example, seldom happen to touch a weather station; their tracks are short and their swaths narrow.

Hail storms, likewise, deliver an areally restricted swath of water and often miss rain gages. Areal surveys of hail swaths by their coldness relative to surrounding areas show them to be relatively small; one large hailstorm left a patch of hail 4 × 14 km in area on an elliptical area 7 × 17 km (Roads, 1973). The usual coarse network of stations detected only five hail days per year in a part of South Africa, where denser sampling in a close-meshed sensing network registered 80 hail days per year (Carte, 1967). The change from isolated point

* The contrasting, or Lagrangian, method has a frame of reference that follows the trajectory of the precipitating cell. It can be illustrated by the models of Newton and Jorgensen described in Chapter II.

measurements to a kind of area-wide surveillance resulted in a radical increase in the number of small systems counted.

In the case of larger storm systems, the observational gain from close-meshed networks lies in having better knowledge of their characteristic internal structure. This has meteorologic interest and also shows the area distribution of the water flux at the surface, which is our interest here. As is true with many natural events, a closer look raises questions of how accurately we sample the event when we make only a few observations.

The Question of a Representative Sample Differences in samples caught in gages a few kilometers apart are often large. On a specific day, 10 July 1958, rainfall in 50 gages in an Illinois network over 1000 km^2 of very flat country varied from a minimum of 40 to a maximum of 101 mm. The range is of the same order of magnitude as the mean catch. It may be that the one gage in such a network that got the least catch, or the one that got the most, or any other, could *happen* to be the location selected for the one gage that is set apart by being included in the national precipitation network; yet the findings from these special studies with dense nets demonstrate that no single point measurement can represent the rainfall in the surrounding territory.

The effect of distance can be studied by correlating the catch in some selected central gage with the others in the dense network, at different distances from the central one. In another Illinois study, correlation coefficients were found to be highest for rain durations between 6 and 12 hr (Table II), for at shorter durations the corresponding rain cells are small and rain patterns are not stable over distance. At longer durations the tracks of rain-producing cells are likely to shift sideways, delivering water to distant gages out of proportion to that delivered to the central gage.

One-minute rainfall intensities (Huff, 1970) yield even lower correlations than the 10-min intensities shown in Table II. The rapid "correlation decay with distance" means that for short periods in a storm it may be virtually impossible to obtain an accurate areal average, say of an intense rain on the area draining into a storm-sewer network.

Number of Gages Needed Suppose that we select a certain level of correlation coefficient, say 0.90, as a criterion for accepting a calculated estimate of the reading at a point rather than going to the expense of placing a gage there and measuring it. From Table II it is clear that we cannot use calculated values at a distance as far as 16 km from the central gage, but must install another gage. Huff and Shipp (1969) show, as in the accompanying tabulation, the distances at which new

TABLE II

Correlation Coefficients of Precipitation Catch at Different Distances from a Central Station—Illinois Gage Networks[a]

Duration	Distance (km)		
	3	6	16
Single storms			
10 min	0.61	0.44	—
3 hr or less	0.91	0.82	0.65
3.1–6.0 hr	0.95	0.90	0.76
6.1–12.0 hr	0.96	0.93	0.87
12.1–24 hr	0.95	0.82	0.66
More than one storm			
Monthly sums[b]	0.96	0.92	0.84
Seasonal sums[b]	0.95	0.92	0.86

[a] Gage densities between 36 and 200 per 1000 km^2. Source: Huff and Shipp (1969, Tables 1 and 6).

[b] Since several storms that followed different paths are taken together in these sums, the correlations may be expected to improve.

gages will be found necessary if we apply this criterion, in various kinds of warm-season and cold-season storms. Closer spacing of gages is needed in summer than in winter rainstorms, as might be expected from the smaller areal sizes of most summer storms. Snowfall obviously presents a problem in sampling, due in part to the fact that "measurement accuracy is very poor" (Huff and Shipp, 1969). Random error is introduced by the faulty method of measuring snowfall; in addition, as noted earlier, this measurement method has systematic as well as random error, being biased on the low side.

Storm type	Distance between gages (km)	Number of gages per 1000 km^2 [a]
May–September		
Showers, thunderstorms	7	20
Continuous rain	>32	1
October–April		
Rain showers, thunderstorms	16	4
Continuous rain	>32	1
Rain and snow mixed	13	6
Snow	3	>100

[a] Assuming uniform placement of gages in a square grid. Calculated from Huff and Shipp (1969).

At Coshocton, Ohio, in more hilly country than the Illinois networks, records in a 60-gage network were employed to determine the area represented by each gage. (The 0.90 correlation coefficient was used as criterion.) A spacing of 2–3 km seems to be required (Hershfield, 1965), which is equivalent to 100–200 gages over an area of 1000 km^2.

Spacing distances calculated in this way at 15 gage networks over the United States were found to be associated with two measures of rainfall at the 0.5 frequency level—depths of rain over the 1-hr and the 24-hr durations (At Coshocton these depths are 30 mm over 1 hr and 65 mm over 24 hr). These depths seem to be associated with the uniformity of precipitation in hydrologic storms and indicate the gage density necessary for adequate sampling of water from the atmosphere. The reader will recall their use in the discussion of rainfall intensity in Chapter III.

The close spacing of precipitation measurements that these studies show to be necessary in order to define the areal distribution of the incoming water flux at the earth's surface is usually found only in special locations. Actual gage densities of some state and national networks are shown in the accompanying tabulation. It is apparent that many rain areas slip through the mesh of most networks. Even in such regions as Hawaii, where great effort and public support* have gone into establishing a dense rain-gage network, many important areas, according to local hydrologists, are not adequately gaged—in particular the high-rainfall windward mountains.

Location	Gages per 1000 km^2 [a]
Hawaii[b]	19
Connecticut	7
Nevada	1
Conterminous U.S.	2
United Kingdom	22
Italy	13
Germany	9
Japan	12
Taiwan	17
Israel	41

[a] Source: Gilman (1964).
[b] Official gages only.

* Public interest is shown by the fact that on the island of Hawaii there are 7 official gages and 37 unofficial gages per 1000 km^2. Of the 450 active stations, 380 were not in the official network but were operated by sugar plantations (151 gages), cattle ranches, or other governmental agencies (Hawaii, 1970).

Bergeron (1960), from the results of "Project Pluvius," concluded that in no country does the official network of rain gages "give an acceptable picture of rainfall distribution at different wind directions, weather types, etc. Even less it may disclose the real structure of precipitation mechanisms and how the terrain affects them." Bergeron concluded that these networks "cannot even give a representative picture of the actual distribution of the average rainfall, owing to the loose network and the irrepresentative distribution of stations."

Variations Due to Terrain In the Pluvius network in Sweden, Anderssen noted significant differences in rainfall between cultivated plains and adjacent small uplands, 30–60 m higher, which were forested (Fig. IV-2). Some of this difference he ascribes to the roughness of the forest, some to the difference in elevation.

In terrain that is deeply dissected, storm rainfall is influenced by the steepness of slopes and their orientation with respect to the rain-bearing winds. The 0.5 1-hr intensity in winter storms in northern California, by data from 126 recording raingages (Linsley, 1958), was found to be related to six parameters: One was climatic, one a regional grouping, and the others were expressions of altitude, slope steepness, orientation, and barrier effects of the terrain. These six factors explain 0.77 of the total variance in rainfall intensity among the 126 stations.

Fig. IV-2. Distribution of frontal upglide rainfall over network Pluvius around Uppsala, Sweden, in the night of 14–15 October 1953. The solid lines indicate the rainfall (in mm); the points represent the reporting stations (Bergeron, 1961).

The fact that four parameters are necessary in order to represent the influence of land shape on storm rainfall probably arises from the fact that different terrain influences operate at different spatial scales. Steepness and orientation of a hillside—an area of 1 km^2 or less—affect only the streamlines of air near the earth, hence only the last few hundred meters of the trajectories of the raindrops or snowflakes. These factors represent terrain-bound wind systems that redistribute precipitation on a meso- (local) scale.

In a windy storm these trajectories might be as long as 10 or 20 km, and are influenced by the regional slope of the mountain range, which is indexed by altitude and major barriers, factors found in Linsley's study. Altitude effects on the vapor flux in the atmosphere, as described in Chapter II, produce an increase in storm activity with height, which augments the orographic effect on vertical motion in the storm. Barrier effects on the inflow of vapor into a storm are evident in the lee of many mountain ranges.

Effects of altitude on storm temperature, which were also noted in Chapter II, increase the probability that snowflakes will reach the ground. Their small falling speed makes them more subject to redistribution by terrain-bound winds, particularly the local factors of slope steepness and orientation.

Most studies on terrain influences deal with annual precipitation, as will be noted in Chapter V. In some regions, however, the yearly pattern is comprised principally of the numerous middle-sized storms rather than rare extreme events. An example comes from leeward Oahu, where mean annual rainfall varies from 3000 mm or more in the mountains to 500 or less at the beach. This rapid decrease in a distance of no more than 10 km is well known to Honolulu citizens. Those who need frequent rain for elaborate landscaping of their yards settle in mountain valleys; those who prefer drier conditions and lower humidity settle at lower altitudes.

In the major storm of March 1958, however, for which intensity data were presented in an earlier chapter, there was virtually no effect of orography. Vapor flux was from the south (hence reversing the usual windward–lee situation), and was steady with little increase of wind speed with height. Storm rainfalls in the city of Honolulu were as follows:

Weather Bureau office (near sea level)	450 mm
Stations below 60 m altitude (6 stations)	430 mm (s = 25 mm)
Stations higher than 60 m (8 stations)	405 mm (s = 25 mm)

(s is the standard deviation of station rainfalls in each group). Storms of this kind create a distinctive pattern of rainfall distribution with terrain, but are so rare as to have little effect on the actual annual mean, which displays a marked increase with altitude.

AREAL SYNTHESES

The best picture of the transfer of water from the atmosphere to the earth's surface in individual storms is provided by cartographic analysis of all available measurements. Such special collections and analysis of data are costly, and usually are made only for storms that produced large floods.

Such analyses unearth as many unofficial measurements as possible, to supplement the coarse official network. These additional measurements come, for example, from rain gages maintained by many farmers and small-town merchants for their own use or as hobbies. A field census of such gages in California found many with long records, sometimes being operated by the second generation of observers. As noted earlier, more than 80% of the active gages on the island of Hawaii are unofficial; these would be indispensible in the cartographic analysis of a major storm.

Immediately after a large storm "bucket surveys" are carried out to obtain more data for the storm map. Any open container holding water is measured to provide at least a rough approximation of total storm rainfall at places between gages. Farm buckets, washtubs, even soft-drink bottles, have yielded useful information in such surveys. Often they define localized bursts of rainfall, generating flood flows that were inexplicable from official rain measurements.

Such a survey was made of the unprecedented storm of 19–20 August 1969 in Virginia, which was "associated with the remnants of hurricane Camille" (Schwarz, 1970; Kinkead, 1971). The largest depth found was 680 mm, while a depth as great as 780 mm might have occurred. The probability of exceeding 680 mm in 24 hr in the Appalachian Mountains is 0.003 and of exceeding 780 mm is only 0.001 (Haggard *et al.*, 1973).

Storm Studies

In a thorough storm study, all measurements of total storm rainfall in tubs and buckets, those made at official or unofficial stations, and charts from continuously recording gages are assembled and checked for discrepancies and internal inconsistencies. From the resulting

Fig. IV-3. Isohyetal map storm of 20–23 September 1941 (Storm GM 5-19, Galveston District, U.S. Corps of Engineers, 1952); duration 78 hr.

body of total storm rainfall data, an isohyetal map is drawn (Fig. IV-3 for a storm in New Mexico in 1941, from U.S. Corps of Engineers, 1952), and time distributions of rain during the storm are graphed (Fig. IV-4). As well as showing the spatial pattern of storm centers, this map provides information on depth of rainfall over areas of

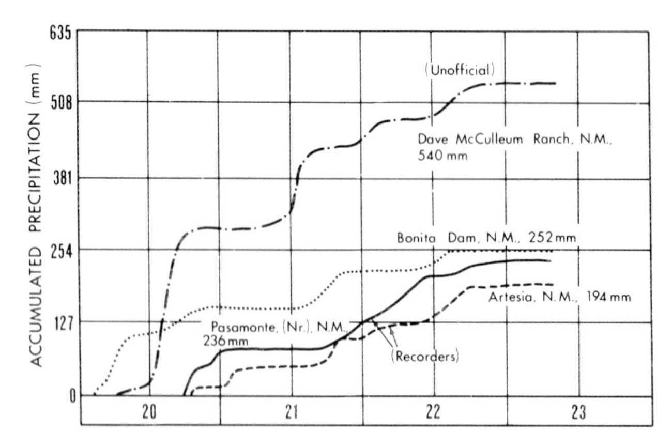

Fig. IV-4. Mass rainfall curves in storm of 20–23 September 1941 (U. S. Corps Engineers, 1952).

different sizes. When its isohyets are measured by planimeter, we obtain depth as a function of area.

The last row in Table III, labeled 78 hr, shows total storm rainfall in the New Mexico storm. For an area of 25 km^2, taken by convention equal to the greatest point depth recorded, the storm total was 538 mm. Over a larger area, 250 km^2, the storm total was 380 mm. Over the total area of the storm, which was 97,000 km^2 (Fig. IV-5), it was 140 mm.

By plotting the amounts of rainfall accumulated from the beginning of the storm by means of charts of the recording gages, the detailed time distribution of rainfall at these gaged places is found. Using these cumulated curves as guides, measurements of gages read only once a day can be expanded into an estimate of the intervening accumulation of storm rainfall at the nonrecording gages. This interpretation is aided by synoptic analysis of the movement of the fronts and zones of atmospheric convergence and instability in the storm. (See also Fig. III-2 in the preceding chapter.)

These graphs of accumulated rainfall against duration, which are called "mass curves" since they show mass of water, serve for scaling off information on rainfall during the heaviest 6 hr, heaviest 12 hr, 24 hr, etc., as also shown in Table III. Intensities can now be derived from these data. Over an area of 5000 km^2, the heaviest 6-hr intensity of rain is 11 mm hr^{-1}, over 12 hr it is 10 mm hr^{-1}, over 24 hr it is 6 mm hr^{-1}.

Depth–Area–Duration Analysis These two distributions, depth ver-

TABLE III

Depth–Area–Duration Data on Rainfall in the Storm of 20–23 September 1941 in New Mexico[a]

Duration (hr)	Area (km^2)					
	25	250	1250	5000	25,000	97,000
6	260	150	112	64	50	28
12	284	211	157	122	80	50
24	307	228	175	142	107	80
78	538	380	267	224	178	140

[a] Units: mm. Source: U. S. Corps Engineers. (1952, Pertinent data sheet, Storm Studies). The basic data for this storm study include 64 sheets of hourly rainfall, 45 of daily rainfall and miscellaneous records, and 76 of mass curves—185 in all.

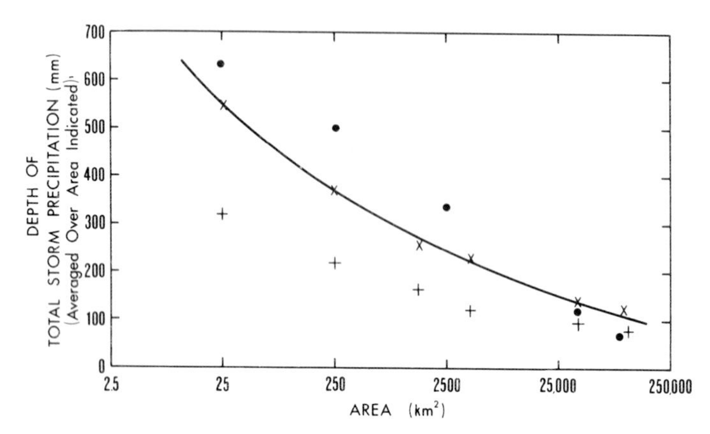

Fig. IV-5. Area–depth relations in the storm of 20–23 September 1941 in New Mexico over durations of 24 hr (+) and total storm (78 hr, ×), and in Camille, 19–21 August 1969 in Virginia (•) for 18 hr.

sus area and depth versus time, are then pulled together to establish a four-dimensional pattern of depth–area–duration of the storm rainfall. This D–A–D data, summarized into a small table, epitomizes the important time and space distributions of the rain delivered by this particular storm to the surface of the earth, condensing data from hundreds of record sheets.

This kind of summary of rainfall is more tangible and satisfying than either the analysis of the variation of point rainfall through time, as motion systems happen to pass near or over the gage, or descriptions of the geographic distribution of such a fiction as mean monthly rainfall. The D–A–D procedure summarizes data in the natural unit of the hydrologic storm. It combines areal and temporal distributions that portray the bursts of heavy rain and intervening lulls and the passage of innumerable precipitating cells over the several days' duration of a major storm as the cells move across the face of the earth.

The D–A–D technique is applicable to available data from storms that occurred years ago. Table III presents information selected from the depth–area–duration table of a storm in September 1941. It illustrates methods developed by the Corps of Engineers and Weather Bureau Hydrometeorological Section in storm-study analyses beginning about 1938 and now used in many countries (WMO, 1969) in standard and computer forms.

Down any column in Table III can be seen the slowing rate of accumulation of rainfall as increasing durations of time are considered. Across the rows of the table can be seen the decreasing depth of mean rainfall over increasing area. These variations of rainfall with time and area, separately discussed in earlier pages, are here seen as they occur in nature—interrelated. For example, the accumulation of rainfall at the storm centers increases only two times as the duration increases from the heaviest 6 hr to the total storm 78 hr. Over the entire expanse of the storm, on the other hand, the total storm amount (140 mm) is five times as great as that in the heaviest 6-hr rain over the entire area.

Use of D–A–D Data for Comparing Storms

A skeleton depth–duration table for a storm at Sydney was presented earlier. It was composed of information in the news broadcasts based on a large number of rainfall stations in the metropolitan area. The table showed that the storm was confined to a small area, and that rainfall intensity was sustained, without marked concentration in the heaviest hour. Even when information is incomplete, as in this case, arraying it in a depth–area–duration format brings out important hydrologic characteristics of the event.

Comparison of storms through their depth–area–duration parameters is facilitated if depths, instead of being expressed in millimeters, are expressed in terms of the 25-km^2, 24-hr precipitation as 1.00. For example, consider such data for the Sydney storm, the New Mexico storm, and Camille as shown in the accompanying tabulation. Camille's depth over any duration of time was twice or more that in the other two storms, and its heavy rain extended farther from the center, as shown by the fact that 0.83 of the 25-km^2, 24-hr depth fell over an area of 250 km^2. At an area of 5000 km^2, the proportions were about the same in Camille and the New Mexico storm, approximately 0.45 of the 25 km^2, 24-hr depth.

	Sydney, 1966 Area (km^2)		New Mexico, 1941 Area (km^2)		Camille, 1969 Area (km^2)	
Duration (hr)	25	250	25	250	25	250
6	0.42	—	0.82	0.49	0.72	0.58
24	1.00[a]	0.74	1.00[b]	0.71	1.00[c]	0.83

[a] 1.00 = 230 mm. [b] 1.00 = 307 mm. [c] 1.00 = 635 mm.

That useful fiction, the probable maximum storm discussed in the second chapter, can also be arrayed in the depth–area–duration format.

Such data for this hypothetical construct on Oahu are shown in the accompanying tabulation. The 1-hr intensity in Oahu is a greater fraction of the 24-hr total than it was in Sydney (0.12). The 6-hr depth of rainfall in Oahu is likewise a greater fraction of the 24-hr total than in Sydney, but less than in the New Mexico storm (which, however, was a longer, bigger event).

	Oahu, any time[a] Area (km^2)		
Duration (hr)	2	25	250
1	0.30	0.26	0.18
6	0.64	0.60	0.49
24	1.06	1.00	0.88

[a] Modified from Schwarz (1963); (1.00 = 1000 mm).

In this, as in the other analyses of storm rainfall, the valuable role of isohyetal maps is to be noted. Planimetering the areas within each isohyet provides spatial means. These means are more precise than an alternative method (use of Thiessen polygons), in which the value at a gage is taken to represent rainfall in the polygon of map area formed by lines enclosing the land nearer to this gage than to any other.

Cartographic analysis can be applied not only to measured amounts of rainfall but also to relative values, such as percentage of mean annual precipitation. Such isopercentual maps of individual storms are useful in mountains subject to similar vapor advection in different storms. Here the observed pattern of rainfall in a given storm is a mixture of a quasi-permanent pattern caused by elevation, slope, exposure, and barrier effects, and the individual pattern of convergence and vapor inflow of the particular atmospheric system. The use of relative or isopercentual values segregates the pattern in an individual storm and permits a closer analysis of its structure.

The new method of trend-surface analysis is, in effect, a computerized representation of the isohyetal map. This method can produce a linear, quadratic, or cubic surface, depending on the degree of faithfulness desired. In frontal rainfall on 8 days, for example, the accounted-for fraction of the sums of squares of readings at 17 gages in an area of 19 km^2 in England was:

Linear:	0.28
Quadratic:	0.46
Cubic:	0.73

Source: Mandeville and Rodda (1970).

Similar reproductions were obtained for annual sums of rainfall. This method appears to be a sound means of arriving at a storm total or a basin mean value of rainfall in a hydrologic storm.

Other Areal Analyses

Basin Means Without attempting to study individual storms, it is possible to analyze daily measurements alone, if a number are assembled from a large drainage basin. An example is the calculation of spatial averages of daily rainfall at a number of stations distributed over the upper basin of the Colorado River (area = 280,000 km^2) (Marlatt and Riehl, 1963). Days when this spatial mean was more than 2.5 mm were found to occur in groups of varying length, which were called "precipitation episodes" (Riehl and Elsberry, 1964), or could be called "hydrologic storms," as we have termed them here. About a quarter of all the days of the year occurred in these episodes.

On the average, 24 precipitation episodes occur per year in this large drainage basin. Those yielding the most water in general lasted the longest, up to more than a week for episodes yielding more than 32 mm.* Over the 30 years studied the amounts of water contributed each year by local or light rainfalls and snowfalls (those hydrologic storms for which the basin average was less than 2.5 mm) were found to vary randomly. These contributions bore little relation to years being classed as wet or dry, and in an examination of year-to-year variation they could be discounted as "noise." This noise level totaled about 155 mm yr^{-1}.

On the other hand, a quarter of the precipitation episodes (i.e., 6 yr^{-1}) brought half the storm precipitation of the whole year, as shown in the accompanying tabulation taken from Riehl and Elsberry (1964).

Fraction of total number of precipitation episodes	Fraction of total storm precipitation
0.25	0.50
0.40	0.67
0.50	0.75
0.75	0.90

The correspondence of 40% of the precipitation episodes (i.e., 10 yr^{-1}) with two-thirds of the storm precipitation was also found in the quite

* This relation suggests that the yield of water per storm day was relatively constant; it was found, in fact, only to vary between 3 and 4 mm day^{-1} in the various size classes of episodes.

different climate of Oahu (Riehl, 1949). The importance of a few storms in making up the yearly sum is also shown at single stations in Hawaii. The accompanying tabulation (Chang, 1963) shows the fraction of annual rainfall delivered on the heaviest 0.1 of the days, and so on. Almost all the year's rain comes on half of the rainy days, especially at the drier stations like Ewa.

	Station		
Rainy days	Ewa	Makiki	Manoa
Annual rainfall (mm)	520	960	4050
Heaviest 0.1	0.60	0.57	0.45
Heaviest 0.2	0.75	0.70	0.63
Heaviest 0.5	0.93	0.91	0.87

Regional Means Spatial averaging was applied in a comprehensive study of Hawaiian rainfall (Leopold *et al.*, 1951), one of the most searching investigations on regional rainfall ever carried out. Meteorologists and climatologists focused on a small area with an already comprehensive net of rainfall observations. They used areal averages of daily shower rainfall and of summations over seasons and years to identify circulation types that had gone unrecognized in the single-station records of scattered, apparently random occurrence of showers. Hydrologic analysis of water delivered to the cane and pineapple crop ecosystems was much improved for management purposes.

As mentioned, the spatial mean precipitation on a quarter of the days in the year in the Colorado basin had values exceeding 2.5 mm; in the wetter climate of West Germany the frequency of such days is much greater, 0.42. Other characteristics of the whole-national areal average rainfall in summer are shown in the accompanying tabulation (Flohn, 1966).

West Germany areal mean (mm)	Fraction of days in July and August
Nonzero days	0.90
Exceeding 1	0.60
Exceeding 2½	0.42
Exceeding 5	0.23
Exceeding 10	0.05

Areal means served to isolate the influence of the jetstream on rainfall in Australia. The purpose was to segregate "rainfall episodes

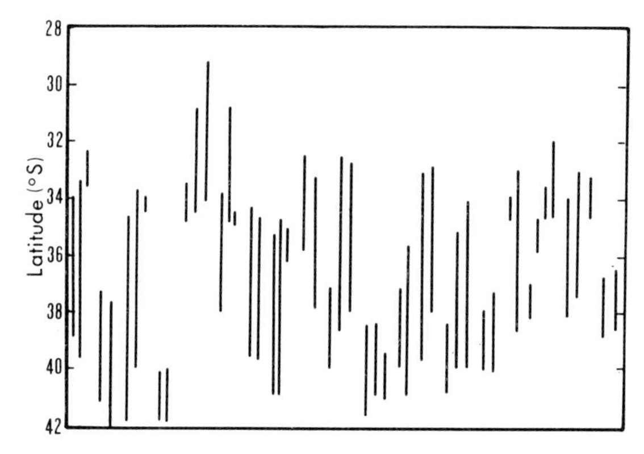

Fig. IV-6. Latitudes in Chile in which heavy rain fell during 19 storms between 1931 and 1941 (49 storm days), to illustrate the north–south extent of the precipitation shields of cyclonic storms (Husen, 1967).

in which falls occur over large areas" due to "widespread destabilization of the lower troposphere because upward motion is in progress" (Campbell, 1968).

In the Hunter Valley 12 rain episodes exceeding 15-mm areal average occurred yearly. Each episode averaged 32 mm. Total rain delivered each year in these episodes was well correlated with total annual rainfall from storms of all sizes.

The nation of Chile offers an interesting north–south section across the paths of storms moving in the circumpolar westerlies, hence daily rainfall data give a good idea of the latitudinal extent of hydrologic storms. The country is well exposed to the ocean, and sheltered on the east; rains in heavy storms coming off the Pacific between 1931 and 1941 are shown on a latitude graph in Fig. IV-6 (Husen, 1967). Many daily rains extended over 6–7° of latitude, i.e., 600–700 km.

THE EPISODIC OCCURRENCE OF HYDROLOGIC STORMS

The lapse of time between one storm and the next is important to water-converting ecosystems, depending on the size of the storage capacity each controls. If storage is small, as it is in sandy terrain, it is soon depleted and the system must go on short rations and a standby status until the next rain comes. The essential unpredictability of the time when this will happen makes the waiting period difficult for the farm manager, representing still one more area of uncertainty in his planning. Storms in the mobile atmosphere exhibit great variability in their formation and growth, and also in their movement. In a given region, their occurrence can be best described as "episodic."

Reports of daily catches of rainfall are usually of the appropriate time scale for agricultural operations, and we can use daily data to illustrate the lengths of storm periods and of the interstorm periods that separate them (Table IV). The table has been set in 18-day periods. (This duration is no more arbitrary, in view of what little we know about rainfall distribution, than a month would be.) About half of the 35 rain days were grouped in three clusters 5–6 days long; the rest come in smaller clusters, or singly. In the first 18-day period rainy days came in groups; in the third they were rather evenly spaced out. In the first and fourth periods rain fell on more than half the days. In the fifth period heavy rain fell on several days, and the mean amount per rainy day was twice as great as in any other period. There is little consistency in rainfall occurrence among the different periods, a result of sudden changes in vapor inflow and the passing reigns of various atmospheric lifting mechanisms. Modified forms of Markov analysis are frequently used to characterize these time distributions.

Rainfall does not occur evenly or uniformly at most places, nor does it form a random pattern (due to the tendency of rainy days to cluster). We saw earlier, however, that a degree of order is introduced if spatial averages are analyzed, and even more appears if analysis is applied to hydrologic storms in which the spatial means exceed such a threshold as 2.5 mm.

In this case relatively constant conditions begin to show up; for example, uplands customarily enhance convergence in the atmosphere, support upglide motion, and trigger convection. Topographic barriers that cut off the moist lower levels of the airstreams entering a region have a constant effect except on rare storms coming from an abnormal direction, like the Oahu storm of 1958 noted earlier.

The southern Appalachians, where there is no clear distinction between windward and lee sides, experience frequent rains from most atmospheric storms that come anywhere near them.* New England, the funnel for several major tracks of storms leaving North America, also experiences frequent rainfall. In the area of the Great Lakes winter storms are reenergized by direct injection of moisture and instability from the underlying water surface and by drawing in streams of Gulf

* Streams in the Coweeta Experimental Forest in western North Carolina experience about 100 rises per year—a fine place for hydrologic experimentation! Thus there are more than 100 hydrologic storms, compared with 24 over the basin of the upper Colorado River.

TABLE IV

Daily Rainfalls at Coshocton, Ohio, in Spring and Summer 1953[a]

1 May–18 May	0,	0,	0,	7,	2,	7,	37,	T,	0,	0,	0,	1,	7,	3,	8,	4,	9,	0		
19 May–5 Jun	0,	0,	8,	13,	0,	2,	0,	0,	0,	0,	0,	0,	0,	0,	0,	0,	0,	0		
6 Jun–23 Jun	21,	0,	0,	0,	T,	0,	1,	0,	0,	0,	3,	2,	0,	0,	0,	3,	0,	0		
24 Jun–11 Jul	0,	2,	11,	0,	7,	T,	8,	17,	7,	0,	0,	0,	T,	0,	12,	0,	T,	0		
12 Jul–29 Jul	0,	0,	0,	T,	0,	0,	30,	0,	11,	0,	24,	0,	0,	0,	0,	18,	0,	0		
Lengths of Rainy Spells[b] (days)																				
In chronological order:	5,	6,	2,	1,	1,	1,	1,	2,	1,	2,	5,	1,	1,	1,	1,	1,	1,	1,	1,	
In rank order:	6,	5,	5,	2,	2,	2,	1,	1,	1,	1,	1,	1,	1,	1,	1,	1,	1,	1,	1,	
Lengths of Rainless Spells[c] (days)																				
In chronological order:	3,	3,	3,	1,	12,	3,	1,	3,	3,	3,	1,	3,	1,	1,	4,	2,	1,	1,	4,	2
In rank order:	12,	4,	4,	3,	3,	3,	3,	3,	3,	3,	3,	2,	2,	1,	1,	1,	1,	1,	1,	1

[a] Units: mm. T = trace, less than ½ mm. Days with T have been counted as rainy (data from Harrold *et al.*, 1962).
[b] Totaling 19 spells taking in 35 days.
[c] Totaling 20 spells taking in 55 days.

air. This rejuvenation affects areal precipitation over the whole region (quite aside from the mesoscale "lake-effect" snow squalls). The eastern Mediterranean Sea tends to channel Atlantic storms far into the interior of the Old World land mass.

Nonetheless, these quasi-constant factors are not sufficient to produce rain at the earth's surface unless the conditions of vapor inflow and massive lifting occur, both being related to large-scale convergence in the atmosphere, plus the essential processes of cloud physics. Often these conditions are lacking; no rain falls. The fundamentally episodic nature of getting water from the atmosphere down to the earth's surface stands revealed. How long are these rainless intervals?

The Spacing between Hydrologic Storms

Analysis of water delivery to the earth's surface by a hydrologic storm helps answer three important questions about this water flux: How fast? How much? Over what area? The question of how often, or its correlative, how far apart, remains to be examined.

The Coshocton data on daily rainfall in Table IV show how the days on which no rain was measured were distributed with time. The rainless spells were mostly about 3 days long, which suggests that storms pass over the area about half a week apart. The 12-day spell that began on 25 May seems to be of a different breed, suggesting that the customary procession of storms was interrupted by some major shift in atmospheric flow.

Data of this kind can be compared with the frequency distribution of periods of rainless days that would arise from chance. At a large number of places in the United States, Blumenstock (1942) found that *short* rainless periods, that is from 1 to 5 days long, are more frequent than would be expected if rainy and rainless days occurred randomly. He associates this high frequency of short dry periods, which covers the numerous 3-day dry spells at Coshocton, with the usual spacing between extra-tropical cyclones. Longer runs of rainless days indicate either a customary lack of precipitation, as in California in summer, or else a "breakdown in the usual pattern of recurrent precipitation."

In summer, consecutive rainless days are periods of increasing moisture stress on plants. The degree of stress depends on the particular species and on the soil-moisture reserve, field by field, but is felt to some extent by all soil–vegetation systems. The length of these periods therefore is important in agriculture.

Curry (1962) shows how the concept of an "exhaustion period" between rains, defined for specific conditions of grass rooting and soil-moisture capacity, can be examined in a probabilistic sense. In this concept the amounts and spacing of rainstorms become the basis for decision-making by the managers of grassland farms in New Zealand. The manager's evaluation of the probability of rain and hence grass growth in the next week or two is his basis for deciding whether to try to save grass by selling or starving his animals, or to prepare for a flush of growth by purchasing more feeder stock (see Plates 5 and 6).

Weekly Summations of Rainfall Weekly or 10-day summation of rainfall has come into wide use for agricultural studies of the North American Midwest. This concept indicates the period between storms that becomes important for replenishing soil moisture. The spacing of rains *within* a week or 10-day period is of secondary significance.

An assessment of the moisture supply for corn from sandy soils of

Plate 5. The Waikato Valley late in a wet summer; an almost completely grassland agriculture. Growth responses of this forage crop to increases or decreases in soil moisture are marked, and the land manager's perception of them is applied directly to his problem of keeping livestock consumption in balance with photosynthetic production. The tree and hedge borders suggest the wind energy in the frontal systems that bring the rain.

Plate 6. Dairy farm in the Waikato Valley, showing cows coming into the herringbone milking shed. Cows in the background are returning to the pasture, lush after a wet summer. Mr. Scott's farm runs 200 cows on 100 hectares, and is operated by only two men since the production and processing of forage for seasonal storage is minimal compared with the typical Wisconsin dairy farm.

low moisture-holding capacity begins with the premise that in each of six critical 10-day periods corn transpires, if it can, 25 mm of water. This requirement of six sufficiently wet periods has never been met in central Wisconsin in 26 years. In one summer, 25 mm of rain fell in each of five of these 10-day periods; on the other hand, in eight summers it fell in only two of them. Yields of corn for silage in the wet summer were 30 tons hectare^{-1}, and in the eight drier summers only 20 tons (McNall *et al.*, 1952). Because the overhead costs of preparing land, planting, and cultivating are large, the farmer can lose much of his investment of time and money when many dry periods occur, in other words, when rains come too far apart.

Daily amounts of rainfall are not normally distributed statistically, because many days bring no rain at all. Rainfall in weekly periods is still not in the range of statistical normality. Figure IV-7 (Joos, 1964), which presents the frequency distribution of rainfall in the week of 5 to 11 July at Horton, Kansas, over 54 years, indicates that the median

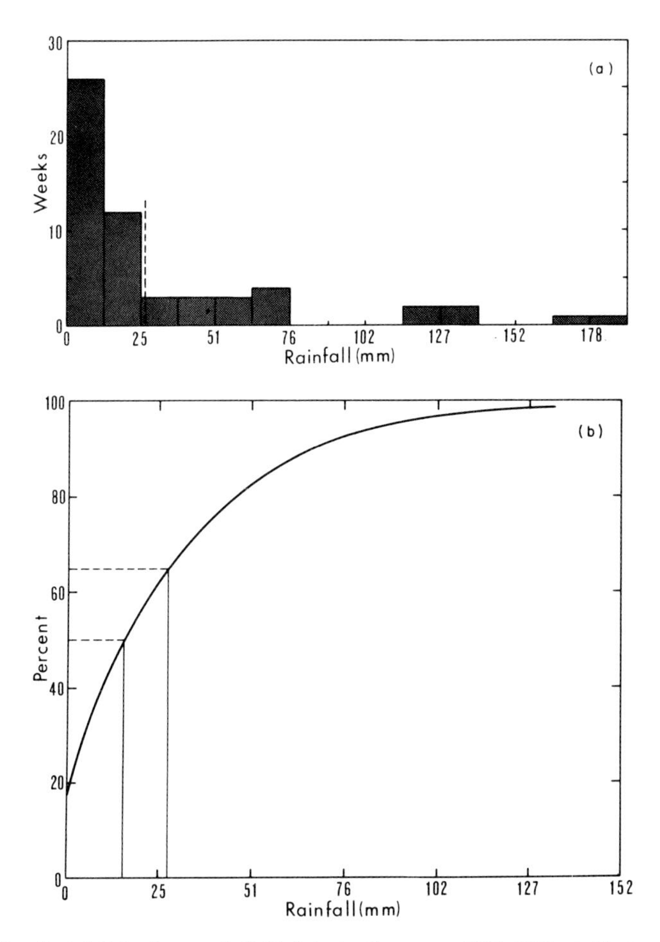

Fig. IV-7. Rainfall in the week 5–11 July in the years 1901–1954 at Horton, Kansas (40N, 96W). (a) Bar graph of occurrences; median = 16 mm, mean = 26 mm. (b) Smoothed curve of weekly rainfall probability (Joos, 1964).

(15 mm week^{-1}) is a more likely estimate of what a farm near Horton will receive in this critical week of any given summer than is the mean (26 mm week^{-1}). Clearly, this week is so dry in many summers as to bring corn under increasing moisture stress.

CLOSING

In this chapter we have considered hydrologic storms, that is, rainfall and snowfall areas on the earth's surface at areal scales up to about 10^5 km^2 and at time scales to about a week. The spatial analysis of gage reports by depth–area–duration methods or cartographic arrays provides our best present information, barring direct measurements of areal precipitation, and makes a marked improvement over the kinds of information to be gotten from single-station data as described in Chapter III. Hydrologic storms are real events in nature, with all the advantage in analysis that this implies.

Analysis on this spatial scale also elucidates some of the ways in which the terrain, i.e., the shapes of the underlying surface, affects storm rainfall, although it also reveals serious defects in the density at which we deploy rain gages in most regions of the earth. The time scale of hydrologic storms is excellent for the analysis of floods and suitable for that of the drying periods that set in after each storm passes. The longer drying periods that constitute droughts, the seasons of rain-often-enough and rain-not-often-enough that determine the crop year and the annual cycle of wildland ecosystems, and the yearly summed inputs of water to these systems will be our next concern.

REFERENCES

Bergeron, T. (1960). Problems and methods of rainfall investigation, *Am. Geophys. Un. Geophys. Monogr.* **5,** 5–25. Disc. 25–30.

Bergeron, T. (1961). Preliminary results of "Project Pluvius," *Int. Assoc. Hydrol. Sci. Publ.* **53,** 226–237.

Blumenstock, G. (1942). Drought in the United States analyzed by means of the theory of probability, U. S. Dept. Agric. Tech. Bull. 819.

Campbell, A. P. (1968). The climatology of the sub-tropical jet stream associated with rainfall in eastern Australia, *Aust. Meteor. Mag.* **16,** 100–113.

Carte, A. E. (1967). Areal hail frequency, *J. Appl. Meteorol.* **6,** 336–338.

Chang, J.-H. (1963). The role of climatology in the Hawaiian sugar-cane industry: an example of applied agricultural climatology in the Tropics. *Pac. Sci.* **17,** 379–397.

Court, A. (1961). Area–depth rainfall formulas, *J. Geophys. Res.* **66,** 1823–1831.

Curry, L. (1962). The climatic resources of intensive grassland farming: The Waikato, New Zealand, *Geog. Rev.* **52,** 174–194.

Dalrymple, T. (1964). Flood characteristics and flow determination. *In* "Handbook of Applied Hydrology" (V. T. Chow, ed.), Chapter 25-I. McGraw-Hill, New York.

Flohn, H. (1966). Bemerkungen zum Problem der Langfristprognose. *Arch. Meteorol. Geophys. Bioklimat. Suppl.* **1,** 134–139.

Gilman, C. S. (1964). Rainfall. *In* "Handbook of Applied Hydrology," (V. T. Chow, ed.), Chapter 9, pp. 1–68. McGraw-Hill, New York.

Haggard, W. H., Bilton, T. H., and Crutcher, H. L. (1973). Maximum rainfall from tropical cyclone systems which cross the Appalachians, *J. Appl. Meteorol.* **12,** 50–61.

Harrold, L. L., Brakensiek, D. L., McGuinness, J. L., Amerman, C. R., Dreibelbis, F. R. (1962). Influence of land use and treatment on the hydrology of small watersheds at Coshocton, Ohio, 1938–1957. U. S. Dept. Agric. Tech. Bull. 1256, 194 pp.

Hawaii, Department of Land and Natural Resources (1970). An Inventory of Basic Water Resources Data: Island of Hawaii, Div. Water Land Management Rep. R-34, 188 pp.
Hershfield, D. M. (1965). On the spacing of raingages, *Int. Assoc. Sci. Hydrol. Publ.* **67,** 72–79.
Huff, F. A. (1970). Spatial distribution of rainfall rates,*Water Resources Res.* **6,** 254–270.
Huff, F. A., and Shipp, W. L. (1969). Spatial correlations of storm, monthly and seasonal precipitation, *J. Appl. Meteorol.* **8,** 542–550.
Husen, C. van (1967). Klimagliederung in Chile auf der Basis von Häufigkeitsverteilungen der Niederschlagssummen, *Freiburger Geogr. Heft* **4.**
Joos, L. A. (1964). Variability of weekly rainfall. U. S. Weather Bur., Weekly Weather and Crop Bull., 29 June 1964.
Kinkead, E. (1971). Big rain, *The New Yorker,* 31 July 1971, pp. 66–74.
Leopold, L. B. *et al.* (1951). On the rainfall of Hawaii: a group of contributions, *Meteorol. Monogr.* **1,** No. 3, 55 pp.
Linsley, R. K. (1958). Correlation of rainfall intensity and topography in northern California, *Trans. Am. Geophys. Un.* **39,** 15–18.
Mandeville, A. N., and Rodda, J. C. (1970). A contribution to the objective assessment of areal rainfall amounts, *J. Hydrol. (New Zealand)* **9,** 281–291.
Marlatt, W., and Riehl, H. (1963). Precipitation regimes over the upper Colorado River, *J. Geophys. Res.* **68,** 6447–6458.
McNall, P. E., Anderson, H. O., Albert, A. R., and Abbott, R. W. (1952). Farming in the central sandy area of Wisconsin, Univ. Wisconsin Agric. Exp. Station Bull. 497, 24 pp.
Riehl, H. (1949). Some aspects of Hawaiian rainfall, *Bull. Am. Meteorol. Soc.* **30,** 176–187.
Riehl, H., and Elsberry, R. L. (1964). Precipitation episodes in the upper Colorado River basin, *Pure Appl. Geophys.* **57,** 213–220.
Roads, John O. (1973). A study of hailswaths by means of airborne infrared radiometry, *J. Appl. Meteorol.* **12,** 855–862.
Schwarz, F. K. (1963). Probable maximum precipitation in the Hawaiian Islands. U. S. Weather Bur. and Corps of Eng., Hydrometeorol. Rep. 39, 98 pp.
Schwarz, F. K. (1970). The unprecedented rains in Virginia associated with the remnants of hurricane Camille, *Mon. Weather Rev.* **98,** 851–859.
Smith, R. E. (1974). Point processes of seasonal thunderstorm rainfall. 3. Relation of point rainfall to storm areal properties, *Water Resources Res.* **10,** 424–426.
Thornthwaite, C. W. (1937). The life history of rainstorms, *Geog. Rev.* **27,** 92–111.
U. S. Corps of Engineers (1952). Storm studies, Pertinent Data Sheets (loose-leaf, continuation format), Washington.
Wilson, J. W., and Pollock, D. M. (1974). Rainfall measurements during Hurricane Agnes by three overlapping radars, *J. Appl. Meteorol.* **13,** 835–844.
WMO (1969). Manual for depth–area–duration analysis of storm precipitation. WMO No. 237, TP 129, 114 pp.

Chapter V

LARGE-SCALE ORGANIZATION OF RAINFALL

The reign of each type of atmospheric circulation at the surface of the earth is experienced as a period of more or less uniform occurrence of individual storms. In any one circulation period, storms pass at about equal intervals of time. For example, the Coshocton daily rainfall data presented in the previous chapter show a month in which interstorm intervals were about three days; then there occurred a sudden shift to a circulation pattern producing a long series of rainless days. These patterns represent groupings of rainfall events on a larger scale than that of the individual hydrologic storm, as large as some of these might be. This grouping can be extended in time to encompass seasonal and even annual rainfall; beyond the year, it can be extended to long-term averages embracing a number of years. The increase in time scale is often accompanied by an increase in spatial scale, which will be described in the latter part of this chapter.

ORGANIZATION OF STORMS IN TIME

In many regions a predictable change in atmospheric circulation comes with the change of seasons through the year. This gives the framework for an important means of organizing rainfall on the seasonal time scale.

Seasonality

Storm activity tends to be more vigorous in winter if energy is derived from contrasts in potential energy or airstream temperature, and in summer if latent heat is the major energy source. Vapor inflow into storms is generally greater in summer, when the vapor content of the atmosphere is high. There is a marked summer maximum in the

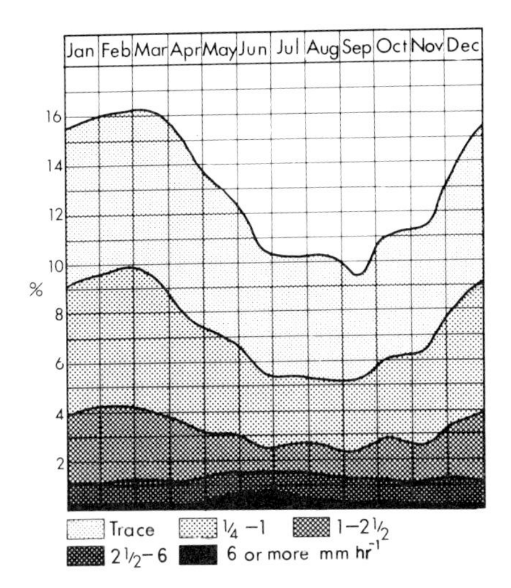

Fig. V-1. Frequency of hourly precipitation intensities in different months of the year at Washington, D.C. Curves show percentage frequency of intensities of 0.1, 0.25, 1.0, 2.5, and 6.3 mm hr^{-1} (Gilman, 1964, from U. S. Weather Bureau, 1949).

frequency of occurrence of record-breaking rainstorms (Schloemer, 1955).

Figure V-1 shows seasonal changes in the frequency of a trace or more of precipitation at Washington, D. C. ranging from 0.10 or less in late summer to 0.16 in early spring. A smaller seasonal range is evident in the frequency of intensities of $\frac{1}{4}$ mm hr^{-1} (so-called "measurable precipitation") and 1 mm hr^{-1}, although both are still most frequent in spring. When we consider intense rains (curves labeled $2\frac{1}{2}$ mm hr^{-1} and 6 mm hr^{-1} in the figure), however, the winter frequency has so shrunk that a definite summer maximum appears, which is a reflection of the more rapid vapor flux into storm cells in the humid atmosphere of summer.

In place of hourly intensities of rainfall, we can look at amounts totaled over a week as a useful agricultural unit. These were shown at the end of the preceding chapter for one particular week in early July each year at Horton, Kansas. Seasonal progression also is important. Figure V-2 shows that at Urbana, Illinois, in the middle of the corn country, the probability of receiving 25 mm or more of rain $week^{-1}$ is about 0.33 during April, May, and June. It drops slightly in early July; and declines more through the autumn. Even in midwinter, however, the probability of at least 25 mm $week^{-1}$ is appreciable, about 0.15. Seasonality is expressed as a doubling of summer rainfall probability over winter probability.

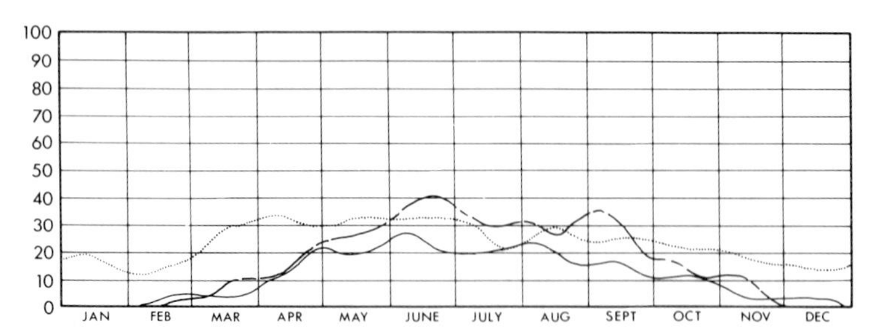

Fig. V-2. Probability in percents of receiving 25 mm or more precipitation per week at three places in the Middle West: Garden City, Kansas (—), Urbana, Illinois (···), and Spooner, Wisconsin (— – —). Note the higher expectation of this moderate amount of precipitation in Illinois, nearer the source of water vapor and the very small expectation of this amount in winter in Kansas and Wisconsin (North Central Region, 1960, p. 24).

The practice of summing amounts of rainfall in all storms by periods of a month is useful because a month is a convenient segment of the annual cycle. Monthly data are usually adequate to portray differences in seasons, and to show the coming and going of a rainy season if there is one.

Fourier analysis of monthly sums of rainfall identifies the importance of the first harmonic, and thus identifies a single cycle in rainfall through the year. Regimes with a strong first harmonic (Fitzpatrick, 1964) are shown in Table I. At these places most of the total variance in rainfall through the yearly cycle is associated with the alternation between one wet and one dry season. (Amounts of rainfall are given as millimeters per day to avoid the difficulty presented by months of varying length.)

At Perth winter is definitely the wet season, the variance of the first harmonic making up 0.90 of the total variance, and the maximum in the first harmonic occurring on 12 July, only three weeks after the winter solstice. At Townsville, diametrically opposite Perth on the Australian continent, the variance of the first harmonic is 0.81 of the total variance. Its peak at 12 February comes five months earlier than the peak at Perth.

Seasonless Rainfall In some places an annual regime is difficult to find. Rainy periods seem to occur almost at random as one calendar year follows another. In central Australia much of the rain comes in a few hydrologic storms; at Alice Springs, for instance, Slatyer (1962) estimates that rainfall necessary to start plant growth occurs on the average three times a year, and that two of these episodes are followed

TABLE I

Annual Regime of Rainfall at Places in Australia with Significant Seasonal Variation[a]

Location	J	F	M	A	M	J	J	A	S	O	N	D	Year
Perth	0.3	0.5	0.7	1.5	4.2	6.4	5.8	4.7	2.9	1.9	0.6	0.4	2.5
Adelaide	0.6	1.0	0.7	1.2	2.0	2.5	2.0	2.1	2.0	1.3	1.0	1.0	1.5
Townsville	9.0	9.9	4.7	2.2	0.8	1.3	0.6	0.5	0.4	1.0	2.1	4.0	3.0
Darwin	13.2	11.1	9.2	2.6	0.3	0.1	0	0	0.5	1.6	3.7	7.0	4.1

[a] Units: mm day^{-1}. Underline indicates high and low. Seasonal variation is for places at which the variance of the first harmonic is greater than 0.80 of the total variance of mean monthly rainfall in period 1911–1940 from Fig. 6 in Fitzpatrick (1964). Source: Climatic Averages/Australia, Australia Bureau of Meteorology (1965).

by sufficient rain to support continued growth.* These isolated periods of wet soil are the most important events in the climate of the region. Woody plants replace their leaves, and the seeds of herbaceous plants germinate. The follow-up rain, if it comes, sustains enough growth that woody plants can store up food and the herbaceous plants produce seeds ensuring survival in the ensuing dry period.

Animals have a more continuous requirement for food and water than plants. They suffer if the intervals between soil wettings are long, and in their hunger they damage the recuperative power of the vegetation.

From the standpoint of land management, these rainless intervals are of great importance. Yet each one has to be regarded as unpredictable in length. This unpredictability of duration means that a drought can get well established before it is recognized.

Long periods without rain are important where most unexpected. The dry summers of California and southern Australia are part of everyone's perception of the climate, and are compensated for. Dry periods in the Midwestern summer present quite a different problem. As noted earlier, even in western Germany 0.10 of all the summer days are completely rainless over the whole country, and a series of many such days would be severely felt.

Variations in Rainfall from Year to Year

Figure V-3 is a ranking of annual rainfall sums at several stations in the United States, from least to heaviest year in the period 1931–1960. It is plotted on probability paper on which a straight line indicates a statistically normal distribution (Hershfield, 1962). Some indication of skewness remains, although it is less than it would be for individual storms.

Variation in rainfall from year to year can be expressed by the standard deviation of annual sums, which runs about 200 mm. It varies from about 50 mm at some stations to as much as 400 mm at others, depending on the mean annual rainfall and the number of days on which rain fell. The relation to the mean annual total, i.e., the coefficient of variation, is approximately 0.2. The coefficient of variation is less than 0.2 when rain comes in more than 100 days yr^{-1}, and greater if the same amount comes in fewer days.

* Where separate soil-wetting events are so important, depth–area analysis of storms is of interest. How large an area gets enough water for its ecosystem to complete a cycle of growth? How far apart are these wetted areas? How does this compare with the distance that animals can migrate in search of food?

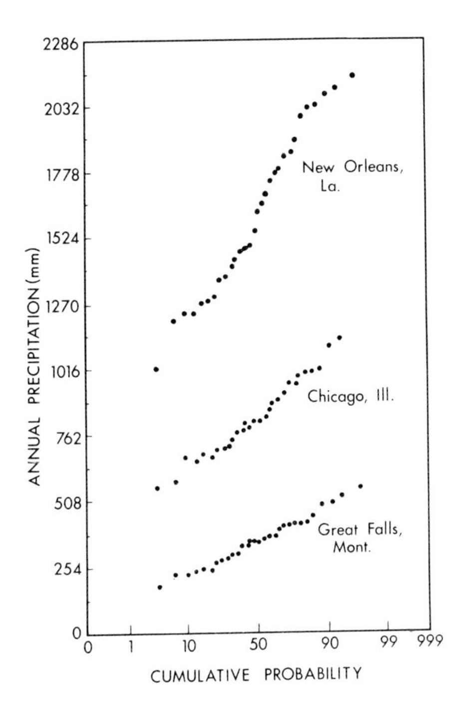

Fig. V-3. Cumulated frequency distribution of annual amounts of rainfall at three places in the United States (from Hershfield, 1962).

The variation of rainfall from year to year, as Wallén (1955) points out for Mexico, is often of greater practical importance than the more or less fictitious mean annual total. In Mexico, fortunately, reliability from year to year is greatest in the nucleus of the country, the eastern part of the Central Plateau where the relatively small amounts of rainfall are less variable than might be expected. At Puebla, for instance, the median yearly rainfall is 790 mm, from which the driest year and the wettest year of record depart by only 250 mm. The standard deviation is 130 mm.

The fact that a dense population in central Mexico depends on reliable rainfall makes even a rare failure into catastrophe. Figure V-4 shows the probability of a given amount of rainfall during an individual year. If, for example, the mean rainfall at a place is 790 mm and the farmer wants to be 90% certain, he will plan on receiving 580 mm.

Added to the uncertainty caused by year-to-year variation of rain at a place is the unreliable performance of rain gages as representative of the environing countryside. From analyses in "Project Pluvius" in

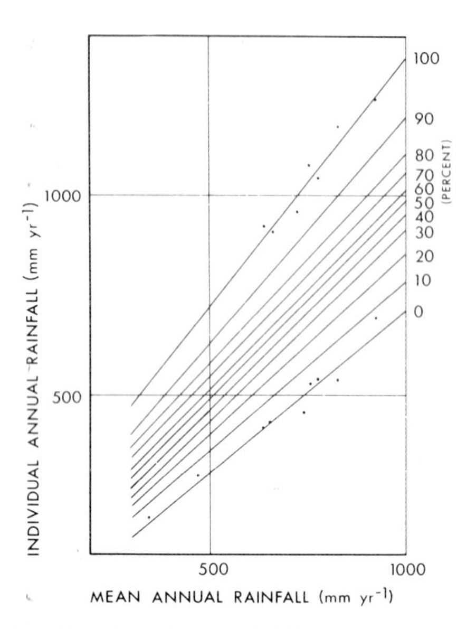

Fig. V-4. The probability of receiving rainfall up to the amount indicated on the ordinate (mm yr^{-1}) in any given year in relation to mean annual rainfall at that place (abscissa) in Central Mexico. At Puebla (mean = 790 mm yr^{-1}), rainfall in 10% of the years is less than 580 mm (Wallén, 1955).

Sweden over several years, Anderssen (1963) concluded that the recorded annual average at any one station could not be regarded as closer than 0.05 to the areal average over the surrounding 1000-km^2 area, until its record incorporated the data of 20 years of readings. To obtain an estimate within 0.05 of the *true* mean rainfall, a still longer period of readings is necessary.

Sustained Variation in Rainfall

Because it seems apparently easy to measure, rainfall in some countries has been recorded for two centuries or more. Even in new lands the rainfall record might be longer than the record of streamflow and will be examined in detail when investments are made in water-development projects. In some of these long records it is possible to discern long-term changes toward wetter or drier conditions that are of great concern to project planners. These changes probably are not regular or "cyclic," but nevertheless express real wet or dry periods that are important to the water economy of a region's ecosystems.

Many real long-term changes are subtle, however, and may be obscured by such disturbances as an altered exposure of a rain gage.

Over the years trees may grow up near, if not actually over, the gage, changing its exposure to rain-bearing winds. Often a gage is moved, without an attempt to obtain a parallel set of readings at old and new sites. Cities spreading around old gage sites have uncertain effects on the catch of rain.

The climatological group of the National Oceanic and Atmospheric Administration has identified stations in rural, slowly changing surroundings and given them a special protective status as "benchmarks." Such stations are not lightly to be moved, as most city and airport stations have been, for some transient operational reason, but are to be maintained in order to identify real changes over long periods of time in the highly variable delivery of rain (Landsberg, 1958).

As we look at rainfall over longer units of time, we are also looking at rain-bearing circulation patterns of increasing size. In examining the areal organization of rainfall, we begin with the simpler spatial groupings and then work toward the factors causing long-term large-area distributions.

SPATIAL GROUPING OF RAINFALL

We saw how rain and snow are spatially distributed inside individual hydrologic storms. These usually display an elliptical pattern, modified by terrain, with streaky patterns caused by passage of active rain cells in the general storm structure. Essential water-budget characteristics of the storm are summarized in depth–area–duration analyses that relate mass of water to area within each storm. We can now go on to examine the numbers of storms occurring over a given area, reinforcing each other's contributions, and then to the seasonal and annual accumulations of rainfall that many storms deliver to areas of the earth.

Spatial Grouping of Major Storms

Grouping basic data on the historic storms that have occurred over a specific part of the earth's surface provides one kind of spatial generalizing of this water flux. Such a composite has been made from the largest rainstorms in Illinois during the 10 years 1948–1957, in the form of depth–area–duration values that envelop the greatest depths recorded for each area and each duration (Table II). These data may be compared with the data of the single New Mexico storm shown in Chapter IV. At the 25,000-km^{-2} size, for instance, the short-period (6-

TABLE II

Maximum Depth of Rainfall for Different Areas and Durations in Storms in Illinois, 1948–1957[a]

Duration (hr)	Area (km²)				
	25	250	1250	5000	25,000
6	320	276	216	165	107
24	418	383	317	226	152

[a] Source: Huff *et al.* (1958).

hr) rainfall in Illinois was twice as intense as that in the New Mexico storm—107 mm compared with 50 mm.

To extend our generalizing further we look at the maximum values of rainfall for each area and duration in storms that have occurred over the eastern United States within the period of historical record shown in Table III, which was compiled by Gilman (1964) from several hundred individual storm studies carried out by the Corps of Engineers. A few extremely large storms contribute most of the entries in this table. For example, in the row for the 6-hr duration, the first item comes from analysis of a storm in Pennsylvania, the last item from a storm in Oklahoma, the next-to-last from two storms (in Oklahoma and Alabama), and the others in that row from several storms in Texas—truly a broad sampling.

More commonly, however, the southern states, particularly those along the coast of the western Gulf of Mexico, experience more big

TABLE III

Maximum Depth of Rainfall for Different Areas and Durations in Historic Storms in the Eastern United States[a]

Duration (hr)	Area (km²)						
	25	250	1250	5000	25,000	125,000	250,000
6	630	500	392	285	145	63	43
24	985	895	833	630	308	160	110
72	1148	1030	950	755	542	290	226

[a] Source: Gilman (1964) from Corps Engineers (1952–1960).

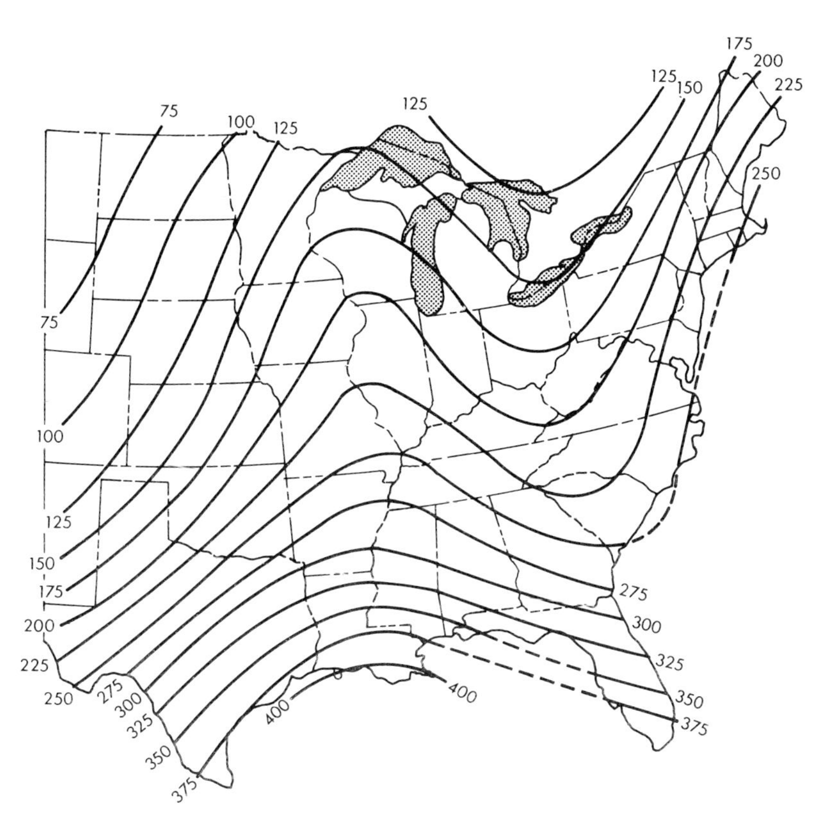

Fig. V-5. Isarithms of precipitation depths of equal probability of occurrence, in a storm of 72-hr duration and 50,000-km^2 area. All seasons are included. The depth in such a storm in southeastern Wisconsin is 190 mm (mean rate = 190/72 = 2.6 mm hr^{-1}), and at New Orleans 400 mm (mean rate = 5.5 mm hr^{-1}) (Schloemer, 1955).

storms than do the states farther north. The areal pattern of rainfall potential at a particular combination of area and duration can also be mapped from the assembled storm-study statistics. For example, Fig. V-5 (Schloemer, 1955) portrays the 72-hr rainfall depth over 50,000 km^2—great depths at the Gulf Coast, decreasing most rapidly toward the Appalachians and the Northern Plains, and showing a slower decrease up the Mississippi Valley. In southeastern Wisconsin the depth over 72 hr is about 190 mm, or half the depth at the Gulf Coast. The average intensity of such a storm in Wisconsin is slightly less than 3 mm hr^{-1}.

It would be interesting to extend such analyses of the fundamental parameters of big rains to the whole world, but in many countries the necessary but laborious depth–area–duration analyses of individual storms have yet to be made.*

Weekly and Monthly Rainfall The spatial distribution of rain during a particular week often represents the contributions for several storms crossing a region along tracks that are more or less parallel but not identical. An example is given from spatial averaging applied to the rainfall sums accumulated during the week ending 6 May 1973 in each of nine sections of the state of Wisconsin, from the Weekly Crop and Weather Report (Wisconsin Statistical Reporting Service, 1973). There were substantial differences among the nine sections, ranging from 38 mm in southeastern Wisconsin to 73 mm in west central Wisconsin. Even in the driest section, 38 mm was too much of a good thing; one farm reporter said "Some field work last week but now there are ponds in most of the fields." In the wetter sections comments like these were made: "Waiting for it to stop raining"; "stores doing good business in sale of sump pumps"; "you just can't get into the fields"; "farmers getting a little nervous, water very evident in most fields"; and so on. Spatial averaging takes out some of the farm-to-farm fluctuation in weekly rainfall totals, presenting a more representative idea of rainfall over large agricultural zones of the state.

The month as a conventional interval for summing up meteorological events also serves for rainfall. Figure V-6 shows the totals of precipitation during April 1973, just preceding the week discussed above. Large areas of the southwestern and intermountain United States were dry; most of the East received between 100 and 200 mm; and a tongue receiving more than 200 mm (about 7 mm day^{-1}) extends up the Mississippi embayment, with an outlier extending from southwest to northeast into Wisconsin. The pattern on this map is a composite of the maps of several large and many smaller hydrologic storms that crossed the United States during these 30 days.

A Representative Sample of Many-Storm Rainfall Although much of the areal nonrepresentativeness of a single rainfall station is reduced when time averaging is done, considerable spatial variability remains.

* It is interesting to calculate the volume of water delivered to the earth in such large hydrologic storms. Let us take the 24-hr duration. The storm at Sydney that was discussed earlier dumped about 0.2 km^3 of water on the city's business district (i.e., 200 $\times 10^6$ tons). In the same period the New Mexico storm delivered 8 km^3 to the ground. The enveloping curve in the U.S. (Table III), 110 mm over 250,000 km^2, gives a volume of 28 km^3. These tremendous volumes of matter, far greater than the volumes handled in man's economic activities, represent one cause of the high potential water has for destruction.

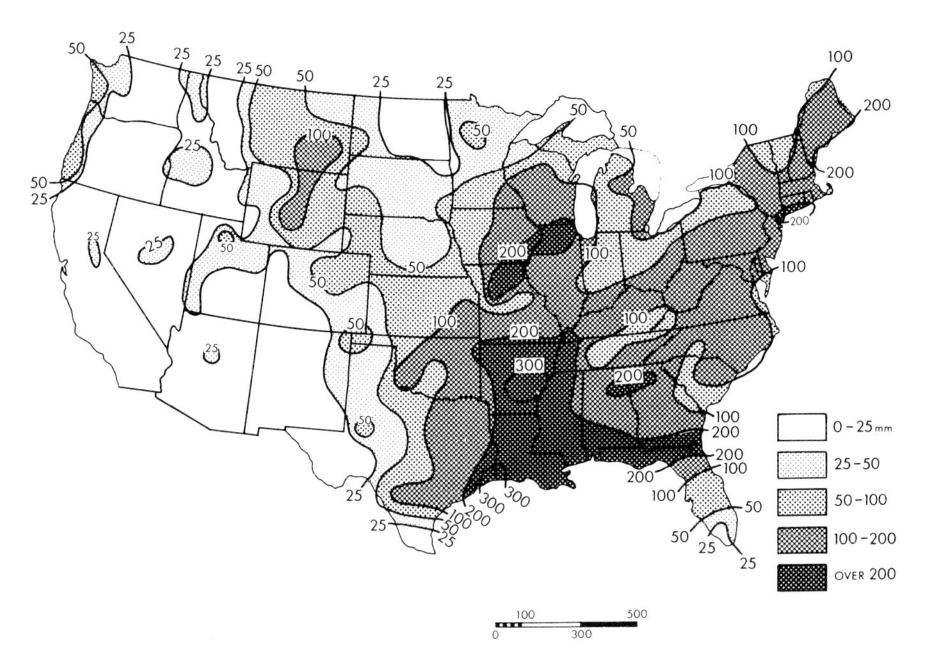

Fig. V-6. Total precipitation (mm) in April 1973 over the conterminous United States. Large areas of the west received less than 25 mm in this month, while the lower Mississippi Valley received from 200 to 300 mm. The track of a storm across Missouri and northern Illinois shows clearly (U. S. Weather Bureau, 1973).

In a dense gage network in very flat country in east central Illinois, 50 gages in an area of 1000 km^2 caught amounts in August 1958 that ranged from 30 to 138 mm each (the 50-gage mean was 64 mm) (Stout, 1960). Over a longer period, the growing season of 1958, gage catches ranged from as little as 550 mm over much of the network area to more than 700 mm in several areas. Over the year, a long period encompassing scores of rainstorms and snowstorms, the differences between highest and lowest gages in this and three other networks in Illinois remain large—of the order of 300 mm. This range is about a third of the network mean annual amounts, and is found between places that are only 10–15 km apart.*

Seasons Areal averaging of rainfall over the seasons of winter and spring 1965 was carried out for the reporting and forecast regions of New South Wales; coastal regions are shown in the accompanying

* Expressed in terms of gradients, these differences in annual rainfall are approximately 40 mm km^{-1}.

tabulation. These data show a marked shift from winter to spring, with the central regions in a hinge position. The northern districts became the driest instead of the wettest, and the southern districts likewise switched places. The difference between the eastern and western metropolitan districts in amount of rainfall is interesting, largely due to distance from the warm waters of the Eastern Australia current in the Tasman Sea.

Region[a]	Latitude (°S)	Winter rain		Spring rain	
		mm	Rank	mm	Rank
Upper North Coast	28	523	1	160	8
Lower North Coast	30	374	2	162	7
Manning	32	261	4	200	5
Hunter	33	164	6	165	6
Metropolitan East	34	327	3	312	1
Metropolitan West	34	184	5	233	4
Illawarra	35	158	7	281	2
South Coast	36	127	8	257	3

[a] Source: Australian Climatological Summary, Surface Data (Bureau of Meteorology, 1965).

Like moisture delivered to the earth's surface by successive storms, rainless periods also have a spatial expression. In most years, for instance, drought extends over large areas of Australia and successive drought years occur in some areas (such as western Victoria in 1959 and 1960) (Heathcote, 1969) (see Fig. V-7).

Spatial Grouping of Long-Term Rainfall Probabilities

Proceeding from rainfall in individual weeks, months, and seasons that are spatially averaged over areas of the appropriate size, we can examine spatial averages over decades, usually expressed in terms of probabilities. These are, of course, spatial extensions of the intensity and depth–frequency relations described in the preceding chapter.

Areal patterns of the probabilities of weekly or seasonal rainfall show the differences in opportunities open to land managers in different parts of the Midwest, for instance. Probability estimates give some guidance, under the conditions of extreme variability that characterize the rain produced by many rain-generating cells over a period of 30 years or more; maps presenting probability statements over a large area show how this guidance differs from place to place.

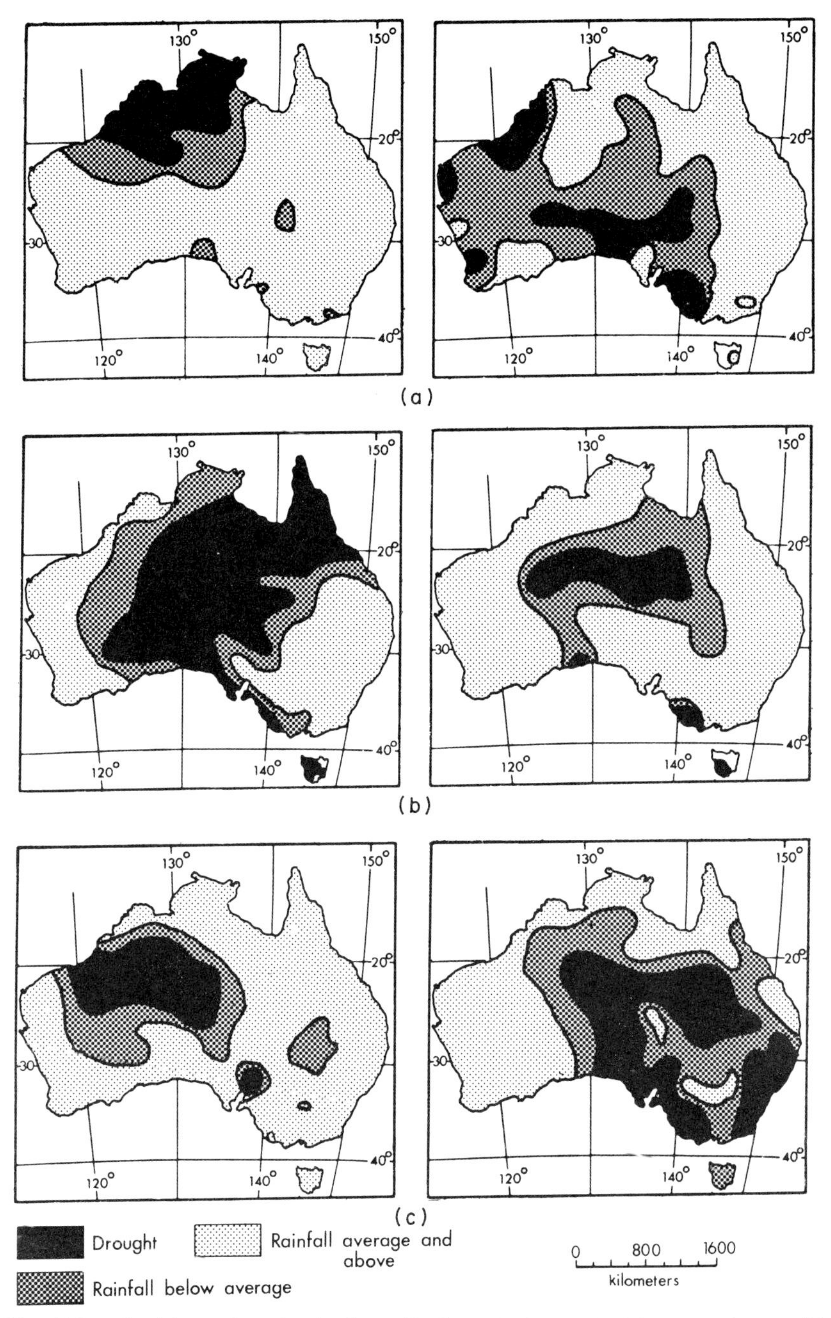

Fig. V-7. Significant droughts in Australia from 1958 to 1965: (a) 1958 and 1959; (b) 1961 and 1963; (c) 1964 and 1965 (Heathcote, 1969, from data of Gibbs and Maher, 1967). Reprinted, with permission, from the *Geographical Review*, Vol. 59, 1969, copyrighted by the American Geographical Society.

Weekly Rainfall We may take weekly rainfall as an example. The probability of receiving at least 25 mm of rain during the week of 26 July–1 August is about 0.2 in Horton, Kansas, and surrounding farmlands in Kansas and Missouri. The same probability is found over much of Illinois, but in Wisconsin it rises to more than 0.3 (North Central Region, 1960) (Fig. V-8). In the far corner of North Dakota it sinks to only 0.1.

The probability of a dry week is 0.15–0.20 in most of the area. The probability of a dry 3-week period is, however, near zero. This is fortunate because few soils can hold enough moisture to carry vigorous plants through so long a time. Table IV shows these frequencies for two areas at the fringes of the Corn Belt.

Areal averaging of weekly precipitation is carried out for the "Weekly Weather and Crop Bulletin" published by the U. S. Department of Agriculture and National Weather Service. These averages make it possible to keep track of growing conditions over areas of the order of 10,000–20,000 km^2.

Monthly and Seasonal Rainfall Maps of the probabilities or the average depths of precipitation totaled over a given calendar month or given season are familiar, showing the likelihood of a certain behavior of the water delivery function during its annual progression.

A less familiar mode of presentation is provided by taking the ratio of mean precipitation in one season to that in another, e.g., the ratio of winter to summer conditions. This ratio is mapped in Fig. V-9 (Phillips and McCulloch, 1972), which shows the typical dry winters of

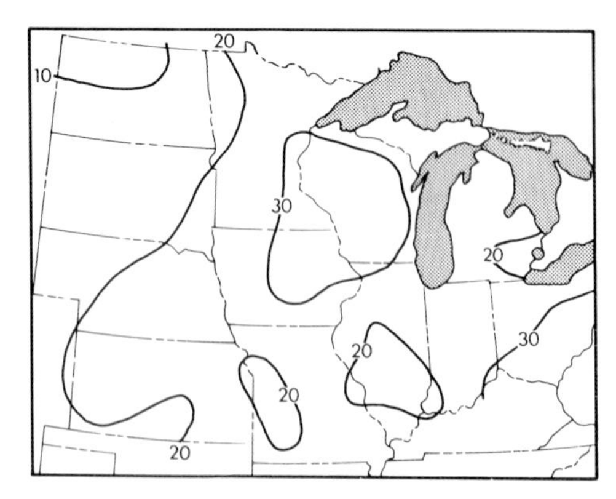

Fig. V-8. Percentage probability of receiving more than 25-mm of rain during the week from 26 July to 1 August. Wisconsin has more than a 30% chance of this amount (North Central Region, 1960, p. 25).

TABLE IV

Precipitation Probabilities in Two Regions of the Middle West[a]

Period	High summer	
	Wisconsin	Kansas
25 mm week^{-1}	0.3	0.2
50 mm in 3 weeks	0.5	0.4
Dry week	0.15	0.2
Dry 3 weeks	Near zero	Near zero

[a] Source: North Central Region (1960).

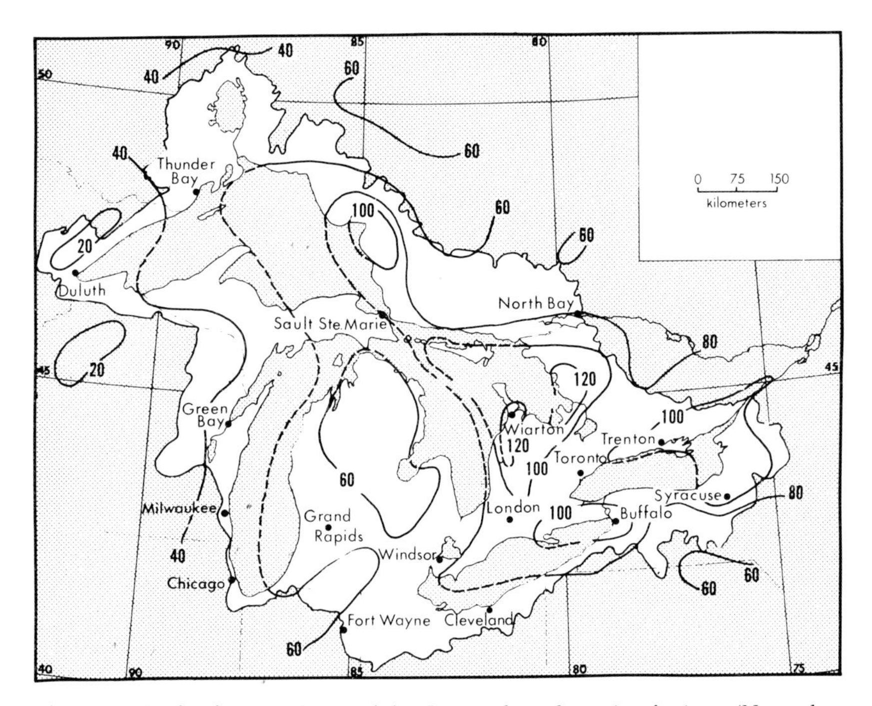

Fig. V-9. In the drainage basin of the Great Lakes, the ratio of winter (November–April) precipitation to that in summer (May–October) is 40% or less in the west. It increases toward the east, and reaches values of 100% or more (winter precipitation equals summer) to the lee of the lakes, indicating the destabilizing, moistening effects of the lakes on winter airstreams (Phillips and McCulloch, 1972, Chart 22).

the Middle West, receiving as little as 0.2–0.4 of the amount of water that comes in summer. To the lee of the Great Lakes, however, the proportion shifts as a result of winter lake-effect snowstorms that increase winter precipitation. Such snow belt areas as those to the lee of Lakes Erie and Ontario receive as much water in winter as they do in summer. More precipitation is also produced by large storms moving northeast as they pull in Gulf air and receive turbulent energy transfers of several hundred W m^{-2} from the Lakes.

Fourier analysis of seasonal regimes of rainfall, discussed under time organization of rain, can be extended to areal generalization if several such analyses are mapped. A map of Australia (Fitzpatrick, 1964) shows the size of regions of wet winters, for example; the south-facing coast of the continent, and especially parts of the coastline aligned across southwesterly airstreams are wet. The low latitudes, in contrast, show up as a large area dominated by rain in summer.

These regions can be objectively delineated. The gradients between them identify the role of mountains, coastline direction, and other factors governing the circulation types that bring rain. A zone of steep gradient in the map showing the date of maximum rainfall lies across southern Australia, and "coincides generally with the mean latitudinal position of centres of anticyclonic cells. Though in no sense a topographic barrier, this zone represents a major separation or 'climatic divide' . . ." (Fitzpatrick, 1964).

The average ebb and flow of the seasons of rain over the United States and southern Canada east of the Rocky Mountains has been calculated in connection with a large-area water budget. This cycle is shown by values in each month, expressed as millimeters of water per day (from Rasmusson, 1968):

D	J	F	M	A	M	J	J	A	S	O	N	D
1.4	1.3	1.6	1.8	1.9	2.5	2.9	2.9	2.3	2.4	1.9	1.6	1.4

The cycle has a minimum in winter, a principal maximum in midsummer, and a secondary maximum in the hurricane season. These features probably are real, although instrument malfunction and sampling error of the kinds mentioned earlier are undoubtedly present, according to Rasmusson, who considered the values to be 10% low, if not more.*

* Although poor snowfall-measuring techniques must produce deficient values in winter when snowfall is common, it is likely that a winter minimum would remain even after corrections were made.

SPATIAL PATTERN OF ANNUAL PRECIPITATION

The depth of the annual cycle in precipitation is greater than that of any other recurring variation that rainfall displays. Averages over such time periods as days, weeks, months, and seasons have operational value for different hydrologic purposes but do not represent natural units in the rainfall-generation processes, which are found only in the time of duration of the hydrologic storm and the cycle of the year. While the cycle of the year is a weaker entity than the hydrologic storm, it is useful as a higher level of generalizing the water deliveries from many storms. The annual cycle at a place sums up the contributions from a score to a hundred storms, as shown over the 10 km^2 of the Central Sierra Snow Laboratory in the winter of 1947 to 1948 (Hildebrand and Pagenhart, 1954) (Fig. V-10).

When annual totals at many places in an area are summed, we attain a further degree of generalization, which brings out differences between regions that are due to more or less permanent factors working on the processes of many storms (Fig. V-11) (U. S. Corps Engineers, 1956, Plate 4-5).

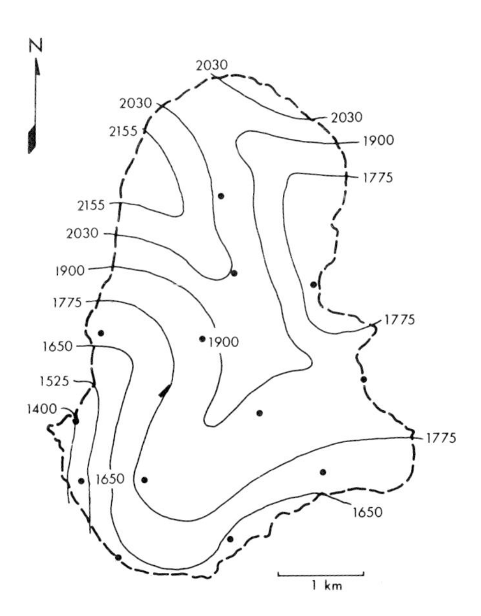

Fig. V-10. Precipitation in the basin of the Central Sierra Snow Laboratory in water year September 1947–August 1948. Mean over the basin was 1835 mm. The points indicate the station positions (from Hildebrand and Pagenhart, 1954).

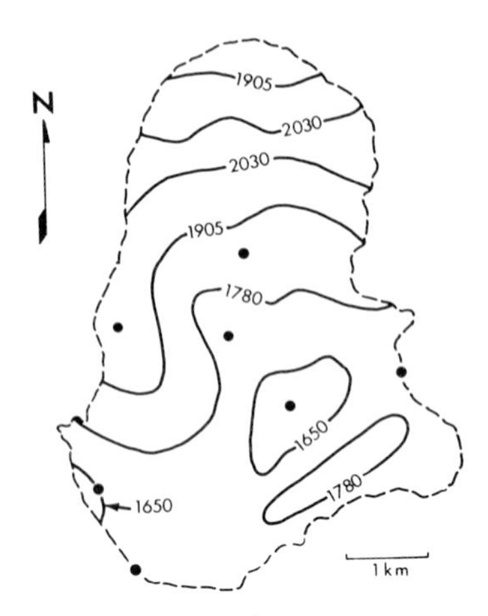

Fig. V-11. Annual precipitation (mm yr^{-1}) over the period 1946–1951 in the basin of the Central Sierra Snow Laboratory, California (area = 10 km^2). Circles indicate gage locations where rain and snow were measured and records adjusted for wind speed. The pattern in the northern part of the basin is calculated from measurements of mass of snow cover (from U. S. Corps Engineers, 1956, Plate 4-5).

Areal differentiation of annual rainfall is shown in a mean isohyetal map of the Hunter Valley, N.S.W. (Tweedie, 1963) (Fig. V-12). Planimetering this isohyetal map identifies the rainier parts of the Hunter Valley for a regional research and planning program, and shows how small a fraction of its area receives most of the basin's rainfall.

The comparison between the Hunter Valley and the whole continent (Table V) shows that about the same fraction of total area receives heavy rainfall, that is, more than 1100 mm yr^{-1}. This fraction of area is about 0.06. The modal classes differ, however. Most of the Hunter Valley receives 500–750 mm yr^{-1}, while most of the continent as a whole receives less than 300 mm. Figure V-13 makes this comparison by accumulated area curves.

Orographic Effects Persistent enhancement of vertical motion in the parts of storms that pass over uplands results in greater precipitation. The ratio of mountain to lowland precipitation may be stable if a large sample of storms is taken, but might vary in small samples.

In deeply dissected terrain the local differences in precipitation are large, even over short distances. Annual sums of rainfall at a large number of stations on the front slopes of the San Gabriel Mountains of southern California (San Dimas Experimental Forest) averaged over 10 years range from 660 to 1000 mm (Burns, 1953). The variation is

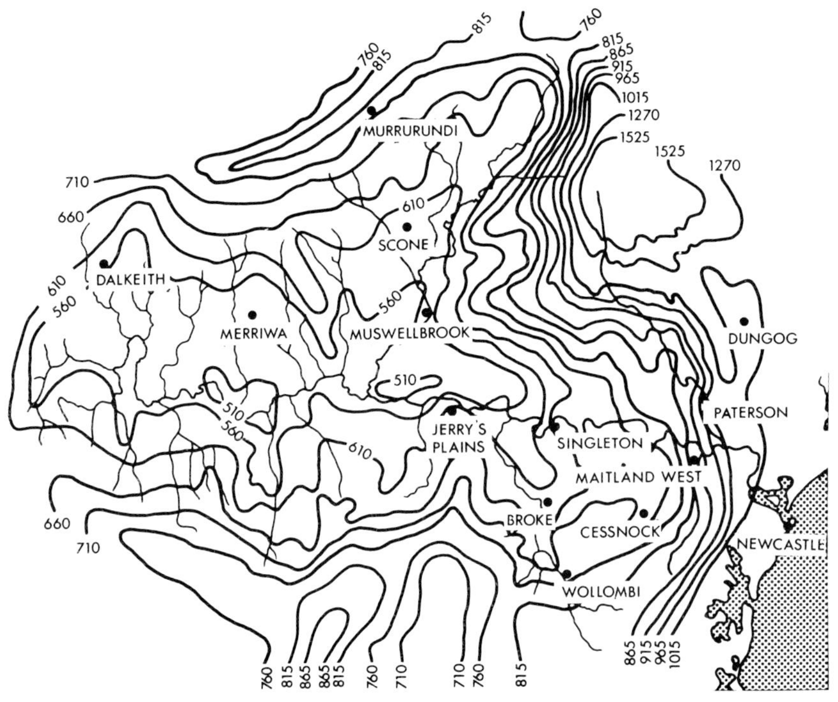

Fig. V-12. Annual mean rainfall (mm yr^{-1}) in the Hunter Valley on the Pacific coast of Australia (drainage area 21,000 km^2) (Tweedie, 1963).

TABLE V

Fraction of Area of the Hunter Valley and of Australia Receiving Certain Amounts of Mean Annual Rainfall[a]

Class (mm)	Fraction of area: Hunter Valley	Fraction of area: Australia
<300	—	0.51
300–500	0.01	0.20
500–625	0.31	0.08
626–750	0.29	0.07
751–875	0.17	0.03
876–1000	0.10	0.03
1001–1125	0.08	0.03
1126–1250	0.02	0.02
>1250	0.02	0.03

[a] Sources: McMahon (1964); Atlas of Australian Resources.

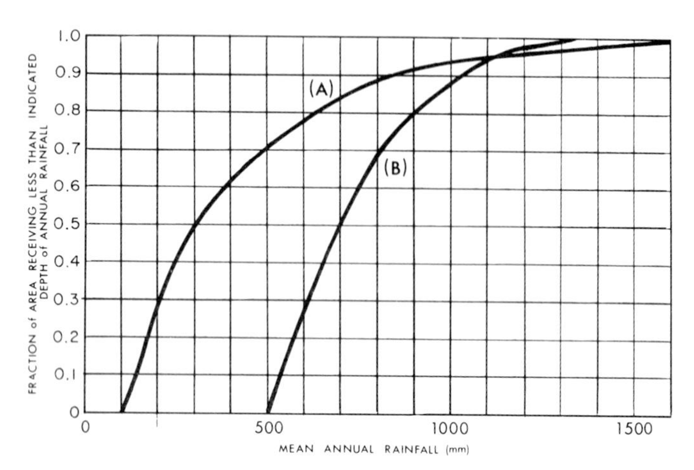

Fig. V-13. Cumulated area curves at increasing depths of mean annual rainfall, for the Hunter Valley (curve B) and for all of Australia (curve A). A depth less than 600 mm yr^{-1}, for example, is found over 0.78 of the country but only over 0.27 of the area of the Hunter Valley.

associated with several topographic parameters, all small scale or mesoscale within this area of only 70 km^2 (actually smaller than some of the Midwestern networks discussed earlier).

One parameter, of course, is altitude; another is the direction faced by the slope on which the gage is located. Three parameters are expressions of terrain inclination: the steepness of the slope facet, the rise to the highest point within 8 km to the northeast (which is the downwind direction in most rainstorms there), and a term that indexes the effects of upwind barriers to airflow. The combined effect of these five parameters is indicated by a multiple-correlation coefficient $R = 0.84$. The standard error of estimate in annual rainfall is reduced to 30 mm. This is 0.04 of the overall mean of 750 mm.

In a basin of 10-km^2 area in the Sierra Nevada, snowfall accumulates to depths of about 1000 mm water equivalent. Measurements of snow cover at 25 stations show that the standard deviation around this overall average is about 250 mm. The effects of the important terrain parameters here can be expressed in terms of the range in snow cover associated with the standard deviation in each parameter as shown in the accompanying tabulation (Miller, 1955). Although the parameters are not independent of one another, the tabulation shows that each produces an appreciable fraction of the total variance in precipitation from place to place. The parameters associated with wind shelter might be more important here than in the southern California case.

Snowfall is more subject to wind effects in the last 100 m of its descent than rain is. See Figs. V-10 and V-11 for spatial patterns.

Parameter	Standard deviation	Associated variance in precipitation (mm)
Elevation	110 m	100
Slope steepness	0.12	125
Slope aspect	80°	100
Land-surface curvature	One class of a five-class ordering	200

The major ridge of the Koolau Range on the island of Oahu has a strong effect on the many trade-wind rain cells crossing the mountains. Leeward of the ridge, annual rainfall increases for about 1 km to a depth of 4800 mm; farther leeward it decreases logarithmically (Mink, 1962).

Figure V-14 shows the pattern of long-term annual rainfall on the island of Hawaii, a land mass built up by five volcanoes, of which four are topographically prominent (shown by × on the map). The areas of

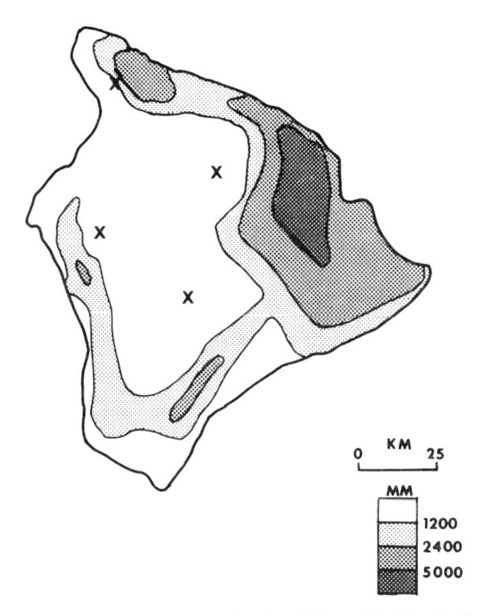

Fig. V-14. Mean annual rainfall on the Island of Hawaii. The influence of altitude on rainfall is evident in some places, but not beyond a certain altitude. The highest mountains (×) receive relatively small depths of rain, less than 1200 mm.

heavy rainfall here do not coincide with the mountains, but in general lie along the windward slopes, facing the incoming marine air in the easterlies. Rainfall is less at elevations above the marine air. The difference in rainfall amounts within short distances is as great here as anywhere in the world.

Differences in inflow of moisture also account for differences in rainfall. The upper basin of the Colorado River, for example, has about the same regime of large precipitation episodes in different seasons of the year as the eastern United States (Riehl and Elsberry, 1964), but the amounts per episode are less, mainly for the reason that airstreams entering the basin are truncated, losing their moist lower layers. Most airstreams entering the basin have surmounted effective barriers of 2-km elevation, which cuts off approximately half their vapor content.

Unsampled Areas A discussion of the areal distribution of rainfall must include the admission that our data are poor in many areas. The heterogeneous nature of the trails of rain and snow that sweep across the earth suggest that sampling rainfall under the most favorable conditions for observation on uniform terrain presents problems. Local differences result from slight elevation, forest cover, or an urban surface. The descending trails of raindrops and especially of snowflakes are difficult to measure in windier places. We have seen that measuring snowfall in mountains and on the tundra is difficult; measuring it on the polar ice plateau is not even attempted.

Problems are introduced by altitude and exposure in mountains. Valley stations do not represent rain or snow on adjacent slopes, ridges, and plateaus. This difficulty, however, can be overcome to a degree by the kind of research that has been discussed earlier on terrain influences. If records are made at stations encompassing a good range of slope steepness and aspect, elevation, upwind exposure, and so forth, the influence of these topographic factors on precipitation can be evaluated. These relations can then be applied to estimate precipitation at ungaged places where slope steepness and other variables can be determined from maps.

This method was applied to southwestern Colorado (Spreen, 1947). The resulting isohyetal map (Gilman, 1964, p. 32) is more detailed than one based on unadjusted records. Also, when planimetered, the map indicates that an areal average based only on valley stations badly underestimates actual precipitation. This underestimate is about 0.23, or 9 km^3 of water over the whole area.

Measurement of precipitation into any water body is made difficult by the wind effect and the lack of a stable platform. Thus precipitation input to the North American Great Lakes is known only in gross terms, leaving a gap in the water budget.* Without an accurate budget it is difficult to adjust the conflicting demands of navigation, power production, riparian ownership, and urban and industrial uses. This is especially the case in years of very low lake levels (as in the early 1960s) or of high lake levels (as in the early 1970s). Precipitation to the oceans is still harder to determine. The global water budget is therefore somewhat in doubt.

We have to include the forested regions of the earth, even in nonmountainous regions, among those with poorly known precipitation. A gage is seldom placed at the active surface of the forest, although this is where raindrops and snowflakes are impacted on foliage. A gage at the bottom of a hole in the forest is certain not to represent anything in its neighborhood.

Vast areas of high-latitude lands remain *terra incognita* with respect to precipitation; the problems of measuring snowfall in wind-swept terrain, as described in an earlier chapter, have not yet been solved. By applying regional water and energy budgets in combination, it is possible to estimate approximately what annual precipitation should be (Hare, 1971; Hare and Hay, 1971). The recorded values fall far short.

Our lack of accurate information about the delivery of water to various parts of the earth's surface—mountains, lakes and oceans, windy areas, cities, and forests—reflects the fact that insufficient interest has been felt in the water budget in these sites. Because water input to croplands has been recognized as important for a long time, the observing networks attempt to sample rainfall in agricultural regions carefully. Areal averaging of water inputs to regions, nations, and continents is therefore most reliable where agricultural lands are extensive.

Continental Means Advancing to the continental scale we sum, partly from observation, partly from estimate, the rainfall delivered in innumerable individual hydrologic storms (Table VI). These overall averages transcend great expanses of space and long periods of time. In them the initial variability of rainfall has been smoothed to a range that for major divisions of the earth runs between 150 and 1630 mm yr^{-1}, one order of magnitude.

* "Whether more, or less, precipitation falls directly on the lake as compared to the land basin is a matter of controversy among researchers studying over-lake precipitation" (Phillips and McCullough, 1972, p. 20). As we have seen, our data on land precipitation itself is not very reliable.

TABLE VI

Areal Means of Precipitation over Continents and Oceans[a]

Area	Mean (mm yr^{-1})
Arctic Ocean	240
Eurasia	610
North America	660
Atlantic Ocean	890
Pacific Ocean	1330
South America	1630
Africa	690
Indian Ocean	1170
Australia	470
Antarctica (est)	150

[a] Source: Budyko (1971).

About 1000 mm yr^{-1} of rain and snow reach the surface over the earth as a whole. This is equivalent to 19 mm $week^{-1}$, 2.7 mm day^{-1}, or 0.1 mm hr^{-1}. These rates form the baseline below which we see long periods without precipitation pass by. Above this line rise the bursts of rain and snow left in the brief lives of the traveling motion systems of the atmosphere.

AREAL PATTERN OF LONG-TERM CHANGES IN RAINFALL

The atmospheric circulation patterns that produce the areal distribution of rainfall in different seasons do not necessarily remain fixed over long periods. In southeastern Australia, Kraus (1954) found that the 300-mb wind speed was related to the precipitation mechanisms, which may account for a shift in the ratio between winter and summer rainfall.

Summer rainfall has risen since about 1895. There has been a small decline in winter, and great declines in fall and spring rainfall, with radical effects on agriculture. High wind speed at 300 mb favors upslope and frontal precipitation in winter on the mountains and western piedmont. It reduces convective rainfall in the amorphous nonfrontal depressions that drift along the east coast to the lee of the mountains, probably by shearing off the upper parts of the convection columns in these depressions. This reduces summer and fall rains.

The low mountains of southeastern Australia thus operate in conjunction with shifts in the general circulation of the atmosphere to change rainfall. Surface features can delineate regions in which changes in vapor fluxes and movement of storms occur, as registered in rainfall at the underlying surface.

A long-term fluctuation also appears in the more featureless terrain of southwestern Australia, where variation in winter rainfall, which makes up most of the annual total, is associated with the long-studied Southern Oscillation of the planetary atmosphere (Wright, 1974). The association of rainfall patterns with circulation patterns here is amenable to investigation because rain is delivered by systems in a relatively few circulation types to a simple regional terrain; moreover, the westerlies of the Southern Hemisphere are more regular than those of the Northern.

On the basis of these different responses, shown in trends of annual rainfall over many decades, Gregory (1956) identified precipitation regions in the British Isles. These fall into four general classes, which differ chiefly in their exposure to westerly or northwesterly winds. Since long-term changes in total volume of rainfall tend in opposite directions in different regions, he recommends that for a reliable supply a large city should import its water from catchments in more than one rainfall region.

Agricultural regions do not have this option of drawing on sources with compensating responses to a change in atmospheric circulation, but have to take what the atmosphere brings. For this reason changes such as those found in Mexico (Wallén, 1955) are very important. In the early 19th century the trend near Mexico City seemed to be downward. From the latter part of the century until the mid-1920s, it was upward. More recently another turndown came. The range of these changes in the yearly income is about 150 mm, of the same order of size as the standard deviation among years.

Slow changes like these are usually associated with changes in the general circulation that are of large scale both in area and in time. The decrease in rainfall from the 1930s to the 1950s "seems to be connected with a weakening of the Trades associated with the southward and easterly shift of the subtropical high pressure over the Atlantic (Wallén, 1955)."

Features of long-term trends in rainfall are seen more clearly by charting 5- or 10-yr running means, which smooth out year-to-year fluctuations. Such a graph is presented in Fig. V-15 for the basins draining into Lakes Superior and Michigan. Differences appear between the two curves, as might be expected from the geographic

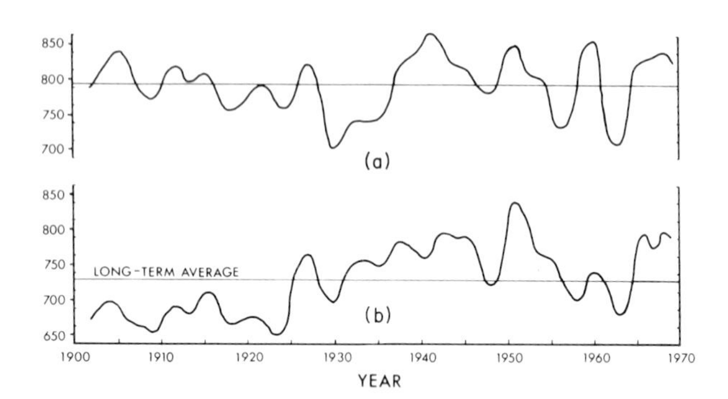

Fig. V-15. Five-year weighted means of rainfall from 1900 to 1970 into the drainage basins of (a) Lake Michigan and (b) Lake Superior. Variations of rainfall with time are complex and do not always occur the same way even over adjacent pieces of the earth's surface (Phillips and McCulloch, 1972).

distances between the two basins. After about 1960, however, there appears a degree of correspondence among the two; a period of dry years brought about low levels in the lakes, which reached critical stages about 1964. Generally higher rainfall after the mid-1960s produced a relentless rise in lake level over the ensuing decade, so that by 1974 the level of wave attack on lake bluffs and on structures built imprudently near the water was high enough to result in heavy damage in every large storm.

The essential unpredictability of the rainfall over large areas, as well as small ones, means that we have little certainty about whether the trends of high rainfall and rising lake level will continue, hold steady, or reverse in any given year. While much of the concern about possible changes in the earth's climate have centered on its energy aspects, these are bound to have concommitants in long-term, large-area rainfall. Regions now marginal in their water inputs might slip into semiarid status; other regions nearly swamped by excess water might become unproductive as far as particular crops are concerned. Because these long-term changes are felt over great areas of the earth, they will surely create serious disturbances in biological production systems that are geared to operate with certain probabilities of rain and that would give way to other ecosystems if these probabilities shifted. The rate of investigation of these changes must be accelerated.

Ecosystems have as a feature of their environment the changeability of rainfall from week to week and year to year, and are adapted to it. A long-term fluctuation in climate, such as a change in year-to-year variability, exerts pressure on ecosystem functioning and ultimately on species composition and survival. Although long-term fluctuations are so masked by year-to-year variations that we can scarcely discern them, considering the shortness and imprecision of the records we keep, it is clear that attempts to modify weather represent long-term alterations by man in one direction. They shift the rainfall expectancies in ways that affect ecosystems and of which we are unaware. For example, proposals have been made to attempt to increase the snowfall in the Sierra Nevada in winters that are forecast to be dry. If we could trust early forecasts of total winter snowfall, if we knew whether seeding would increase storm intensity or duration, and if seeding of storms actually succeeded, we would still not know the consequences of such meddling with snow climate, which would reduce winter-to-winter variability and eliminate the occasional winter of light snow.

ASSOCIATED MASS FLUXES

The spatial pattern of the fluxes of other forms of matter that are associated with rainfall diverges from the pattern of the water flux itself. We can take chlorides as an example. The ocean is the source for most of the chlorides in rain and snow (Junge, 1963, p. 321), which decline inland from the coasts. At places remote from the ocean, chloride deposition is small; at the South Pole the annual deposit has been measured as 2 mg m^{-2} (Wilson and House, 1965). Junge's map (1963, p. 319) of chloride in rain indicates 50–100 mg m^{-2} over central North America, several times greater in coastal lands. Douglas (1972) notes that exposed coasts in Australia receive about an order greater concentrations of salt than do inland sites.

A detailed study of the chloride budget of small drainage basins in the White Mountains of New Hampshire (Juang and Johnson, 1967, part of a comprehensive program of mass budgets in this area), found an input of 270 mg m^{-2}. This input of chloride in rain can be compared to an output in the streams of 410 mg m^{-2}, additional chloride being received by so-called "dry fallout" (between storms) and by impaction directly on ecosystems.

The amounts of chloride in rainfall are often compared with basin yields in order to identify other possible sources, such as impaction. Douglas (1968) found that rainfall brought 0.6–0.7 of the input to basins in Queensland in eastern Australia, and even more—probably exceeding the basin outputs—to areas on the Southern Tablelands of New South Wales near Canberra.

In basins very near the ocean, Baldwin (1971) found that only 0.2 of the "cyclic" or atmospheric chloride came during rain. Much more came in dry fallout and in fog drip, an impaction of cloud droplets on exposed ecosystems in mountainous terrain, which will extract salt along with water from the atmosphere near the sea.

Mention was made earlier of the acid rains of parts of Canada and the United States, and of Scandinavia, as representing sulfate transport in the atmosphere. Inasmuch as the sulfate sources—industrial areas, smelters, or power plants—are fixed in location, they will generate continuous inputs to all ecosystems in a general downwind half-circle. This mass-flux pattern represents a tradeoff; we get smaller ecosystem production but more conversion of fossil energy into electricity by the cheapest method. It is possible to have both trees and power plants, but only by an act of public will that forces clean operation of industries. Otherwise we will have initiated a permanent alteration in the precipitation climate over a large area. This process is also seen in a doubling of the input of nitrogen in rainfall on the croplands of the United States between 1930 and 1969 (Viets and Hageman, 1971, p. 15).

TIME AND SPACE ORGANIZATION OF THE WATER DELIVERED TO ECOSYSTEMS

Much of our study of water has to do with apparent maldistributions in time and space. Maldistributions in time refer to different scales: to short times, to the durations of storms, to weeks or 10-day periods, to seasons and to long-term expectancies, and to possible changes in climate.

Maldistributions in space involve, first, difficulties in obtaining measurements that truly represent an ecosystem or larger area of the earth's surface. Problems of instrumentation and sampling what often is a highly heterogeneous areal pattern are serious at this spatial scale.

At a larger scale the distribution of precipitation within the area of a storm is important, as is the pattern produced by terrain. At still larger scales are the areal distributions over river-basin regions that may be the sites of water-development works, and over whole nations and continents.

Even in wet parts of the world, precipitation occurs only a small fraction of the time, perhaps 1 hr in 20. In a rain hour there may arrive as little as a fraction of a millimeter or as much as 100 mm. A range of variation over three orders of magnitude presents difficulties in description, to say nothing of analysis.

Difficulties in presentation and the lack of an overriding intellectual organization of the observational data are due in part to the fact that observations have not been made of the descending trail of water itself as it moves across the surface of the earth. We have only the sample a precipitation gage happens to catch from each moving shaft of raindrops or snowflakes that happens to pass above it. Perhaps we may be able to develop a solid theory for the distribution of rainfall in time, when we shall have information of the kind that might be collected by radar probes of precipitating cells.

At present, with no such command of the time distributions, we have to rely on empirical accounts of the frequencies of different intensities, the daily, storm, and weekly totals, and those more abstract indexes (monthly and yearly sums), and the even more abstract long-term means. The sampling method complicates the actual phenomenon, and the artificial grouping of the yields of many individual storms into long time periods may bring a specious simplification. Caution in interpretation of mean rainfall data is enjoined upon us. Data grouped by individual hydrologic storms, on the other hand, represent a style of abstracting that is consistent with real objects in nature, and has corresponding value for climatologic analysis.

From the standpoint of the earth's surface, the water output and input are as different as black and white. The output invisibly diffuses upward from large expanses of surface at rates that regularly wax and wane. The input, collected from broadly diffused streams of atmospheric vapor into small cells of vertical motion, is precipitated toward the surface in great shafts of water that cross the landscape. The injection of this water into the water budget at the surface is rare, erratic, unpredictable, and wholly indispensable.

REFERENCES

Anderssen, T. (1963). On the accuracy of rain measurements and statistical results from rain studies with dense networks (Project Pluvius), *Ark. Geof.* **4,** 13, 307–332.

Australia, Bureau of Meteorology (1965). Australian Climatological Summary, Surface Data, Melbourne.

Australia, Commonwealth Dept. Natural Resources, Atlas of Australian Resources. loose-leaf sheets, various dates.

Baldwin, A. D. (1971). Contribution of atmospheric chloride in water from selected coastal streams of central California, *Water Resources Res.* **7,** 1007–1012.

Budyko, M. I. (1971). "Klimat i Zhizn'." Gidromet. Izdat., Leningrad, 472 pp. (Translated (1974) as "Climate and Life," D. H. Miller, ed., Academic Press, New York, 508 pp.)

Burns, J. I. (1953). Small-scale topographic effects on precipitation distribution in the San Dimas Experimental Forest, *Trans. Am. Geophys. Un.* **34,** 761–767.

Douglas, I. (1968). The effects of precipitation chemistry and catchment area lithology on the quality of river water in selected catchments in eastern Australia, *Earth Sci. J.* **2** (2), 126–144.

Douglas, I. (1972). The geographical interpretation of river water quality data, *Progr. Geog.* **4,** 1–181.

Fitzpatrick, E. A. (1964). Seasonal distribution of rainfall in Australia analysed by Fourier methods, *Arch. Meteorol. Geophys. Bioklimat. B* **13,** 270–286.

Gibbs, W. J., and Maher, J. V. (1967). Rainfall deciles as drought indicators. Australia, *Commonwealth Bur. Meteorol. Bull.* **48,** Melbourne.

Gilman, C. S. (1964). Rainfall. *In* "Handbook of Applied Hydrology" (V. T. Chow, ed.). Chapter 9, pp. 1–68. McGraw-Hill, New York.

Gregory, S. (1956). Regional variations in the trend of annual rainfall over the British Isles, *Geog. J.* **122,** 346–353.

Hare, F. K. (1971). Snow-cover problems near the Arctic tree-line of North America, *Kevo Subarc. Res. Sta. Rep.* **8,** 31–40.

Hare, F. K., and Hay, J. E. (1971). Anomalies in the large-scale annual water balance over northern North America, *Can. Geog.* **15,** 79–94.

Heathcote, R. L. (1969). Drought in Australia: a problem of perception, *Geog. Rev.* **59,** 175–194.

Hershfield, D. M. (1962). A note on the variability of annual precipitation, *J. Appl. Meteorol.* **1,** 575–578.

Hildebrand, C. E., and Pagenhart, T. H. (1954). Determination of annual precipitation Central Sierra Snow Laboratory. U. S. Corps. Eng., Snow Invest., Res. Note 21, 18 pp. + 4 pl.

Huff, F. A., Semonin, R. G., Changnon, S. A. Jr., and Jones, D. M. A. (1958). Hydrometeorological analysis of severe rainstorms in Illinois 1956–1957 with summary of previous storms, State Water Surv. Rep. Invest. 35, Urbana, Illinois, 79 pp.

Juang, F. H. T., and Johnson, N. M. (1967). Cycling of chloride through a forested watershed in New England, *J. Geophys. Res.* **72,** 5641–5647.

Junge, C. E. (1963). "Air Chemistry and Radioactivity." Academic Press, New York, 382 pp.

Kraus, E. B. (1954). Secular changes in the rainfall regime of SE. Australia, *Quart. J. Roy. Meteorol. Soc.* **80,** 591–601.

Landsberg, H. E. (1958). Trends in climatology. *Science* **128,** 749–758.

McMahon, T. (1964). Hydrologic Features of the Hunter Valley, N.S.W. Hunter Valley Res. Foundation Monogr. 20, Newcastle, 158 pp.

Miller, D. H. (1955). "Snow Cover and Climate in the Sierra Nevada, California." Univ. Calif. Publ. Geog. **11,** Berkeley, California, 218 pp.

Mink, J. F. (1962). Rainfall and runoff in the leeward Koolau Mountains, Oahu, Hawaii, *Pac. Sci.* **16,** 147–159.

North Central Region, Agricultural Experiment Stations (1960). Precipitation probabilities in the North Central States. Univ. Missouri Agr. Exp. Sta. Bull. 753, North Central Regional Publ. 115. 72 pp., maps.

Phillips, D. W., and McCulloch, J. A. W. (1972). The Climate of the Great Lakes Basin. Atmos. Environ., Climat. Stud., No. 20, Toronto, 42 pp., 57 pl.

Rasmusson, E. M. (1968). Atmospheric water vapor transport and the water balance of North America. II. Large-scale balance investigations. *Mon. Weather Rev.* **96,** 720–734.
Riehl, H., and Elsberry, R. L. (1964). Precipitation episodes in the upper Colorado River Basin, *Pure Appl. Geophys.* **57,** 213–220.
Schloemer, R. W. (1955). An empirical index of seasonal variation of intense precipitation over large areas, *Mon. Weather Rev.* **83,** 302–313.
Slatyer, R. O. (1962). Climate of the Alice Springs area. *In* Lands of the Alice Springs Area, Northern Territory, 1956–57. Commonwealth Sci. Ind. Res. Org., Land Res. Ser. No. 6, pp. 109–128.
Spreen, W. C. (1947). A determination of the effect of topography upon precipitation. *Trans. Am. Geophys. Un.* **28,** 285–290.
Stout, G. E. (1960). Natural variability of storm, seasonal, and annual precipitation, *Proc. Am. Soc. Civil Eng. J. Irr. Drain. Div.* **86** (IR1), 127–128.
Tweedie, A. D. (1963). Climate of the Hunter Valley, Commonwealth Sci. Ind. Res. Org., Melbourne. Land Res. Ser. No. 8, pp. 62–80.
U. S. Corps of Engineers, Snow Investigations (1956). Snow Hydrology. Corps of Engineers, Portland, Oregon, 437 pp.
U. S. Weather Bureau (1949). The Climatic Handbook for Washington, D. C. Tech. Paper 8.
U. S. Weather Bureau (1973). Climatological Data, National Summary.
Viets, F. G., Jr., and Hageman, R. H. (1971). Factors affecting the accumulation of nitrate in soil, water, and plants. U. S. Dept. Agriculture, Agric. Handbook 413. 63 pp.
Wallén, C. C. (1955). Some characteristics of precipitation in Mexico. *Geog. Ann.* **37,** 51–85.
Wilson, A. T., and House, D. A. (1965). Chemical composition of South Polar snow. *J. Geophys. Res.* **70,** 5515–5518.
Wisconsin Statistical Reporting Service (1973). Weekly Crop and Weather Report.
Wright, P. B. (1974). Temporal variations in seasonal rainfalls in southwestern Australia. *Mon. Weather Rev.* **102,** 233–243.

Chapter VI

RECEPTION OF WATER BY ECOSYSTEMS

The water delivered to the earth's surface by the sporadic rain cells that wander across it passes through a sequence of hydrologic events in and near the surface. The events set in action by a rainstorm may be looked on as an alternation of fluxes and storages, as the water moves from one level of the ecosystem to another level. At each level, storage (or very slow movement) affords an opportunity for redistribution of water into competing outflows. This series of partitions of water into different outflows represents a sequence of branching processes, each of which depends on external or environmental conditions. The theme of variability that runs through the preceding chapters on delivery of water from the atmosphere to ecosystems therefore continues, though with different emphases, through this and succeeding chapters on water dynamics *within* ecosystems.

ECOSYSTEM HYDRODYNAMICS

Figure VI-1 shows in simplified form the major storages of water in a liquid or solid state at and near the earth's surface in regions bearing terrestrial vegetation. All the storages depicted in the figure are mixtures of water with other forms of matter. Water is stored or moved in environments of plant foliage, ground vegetation, rock particles and organic matter that comprise soil bodies, or rocks that harbor ground water. In depicting the dynamics of water moving from one zone or storage to another within the ecosystems, the name "ecosystem hydrodynamics" is applicable, both to the diagram and also to the description of the sequences of processes at all levels from the outer active surface to the bottom of the root zone.

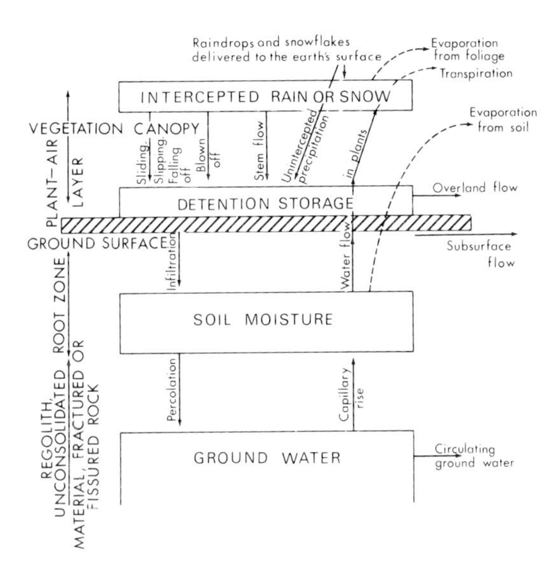

Fig. VI-1. Diagram of levels in an ecosystem. Boxes represent sites where water is stored for longer or shorter periods; arrows represent fluxes of water substance in liquid or solid state (solid lines) or vapor state (dashed lines).

Partitioning of Water

The four principal storage zones (and others that might be added in more elaborate schemes) are stacked up more or less in vertical profile. The topmost zones receive rain and snow and pass it down to the lower zones.

While gravitational force is ever present, not all the forces that move water from one storage to another act in a downward direction. Vapor might diffuse upward, and water in all states moves horizontally under the appropriate forces.

The important feature of each storage is that it receives a certain input (or inputs) and provides opportunity for branching outputs. For instance, rainwater caught on plant leaves can drip off vertically, be blown some distance away, run down the bark, or evaporate. The partitioning or branching process simply says that several outflows can move out of a given storage zone. The exact manner of the partition among them depends on the forces causing the movement.

The opportunities for partition might be brief, however. Some outflows operate only intermittently because the applications of energy that cause them to move are intermittent. Some of them are brief because the storages themselves might have short life spans. The storage of intercepted water occurs only a small percentage of the time, for example.

The principal characteristic of the delivery of water to the earth's surface is its variability. This characteristic is transferred to the water fluxes that occur following a rainstorm. Variability is sometimes increased by the fact that water moves only on the application of energy, itself a variable input to an ecosystem.

Some storages tend to reduce the initial variability, just as a farm pond accumulates bursts of inflow from each rainstorm and delivers a smooth flow of water to the stock-watering trough. It might, therefore, be thought that the several storage zones depicted in Fig. VI-1 would produce a steady output of water. This is not necessarily so, however, because the branching processes from each storage zone tend to increase variability. Most rainstorms that deliver water to a corn field, for example, do so at rates that allow all the water to soak into the soil, but in a few storms rain falls so hard that water runs off the surface. This sporadic outflow of water is much more variable even than the bursts of rain that descend on the field. The already high variability of the incoming water flux is increased by the branching process at the detention-film storage on the corn-field surface.

The Environments of Water

These storages and fluxes of water occur in a set of environments. Water intercepted and held in the outer canopy of the vegetation occupies a particular, rather peculiar, environment that is familiar by sight to all of us but for which the student of interception will find almost no useful quantitative description, that is, the foliage surface area. Snow in forest canopy, for example, melts, evaporates, or blows away in response to exchanges of energy between it and various parts of this complicated environment in which it occurs. The same statement applies to water at the soil surface, in the soil body, and in the rocks. Its energy state and movements are determined by its relation to these environments.

Unfortunately, their characteristics are not well known. In listing research needed in the area of water resources, a report to the Committee on Natural Resources of the National Academy of Sciences (Wolman, 1962) calls for better "quantitative characterization of land-use and land-cover parameters. . . ." These environments are landscapes we have grown up in, but we have never measured them, except in the dimensions of some narrow economic purpose. The broader perspective of the International Biological Program and its associated work in ecology may in the future result in better data on the ecosystems through which water moves at the surface of the earth.

Conversely, each environment is affected by the presence of water in it. Tree form is sometimes affected by the recurring loading of intercepted snow, for instance. The canopy must meet not only the usual biological requirements of ventilation to carbon dioxide and exposure to light, but also the intermittent imposition of a heavy burden. In its long genetic history the canopy has been modified and has changed the way it receives and holds water. The result is seen in the shape and flexibility of trees of the Snowy Mountains in Australia, for instance. On the other hand, many of the beautiful spreading trees of Honolulu that give such wide areas of welcome shade would not stand up an hour in a good snowstorm. Thus the existing system of water and its several environments is the result of long mutual interaction over biological time.

The ecosystem balance equation for water is

$$r = E + f_w + G,$$

where r is the rainfall, E the evapotranspiration, f_w the runoff at or near the surface, and G the deep percolation of water to underlying layers of rock. We have considered factor r in the preceding chapters; now we shift our view to the aftermath of an individual delivery of rain or snow to a terrestrial ecosystem.

In the next chapters we will consider how r creates reservoirs of stored water in each of several environments within an ecosystem—on the foliage, on the ground, and in the soil. Flux E from the outer surfaces of leaves and ground-level storage is discussed in these chapters, but that from the soil and the insides of leaves is so large a subject that it warrants full treatment in four later chapters. Flux G into local rock formations is considered next, and then the horizontal outflows of water from local rocks and from surface and near-surface runoff f_w, which represents the off-site yields of water away from the ecosystems to which water was initially delivered.

DELIVERY OF RAIN AND SNOW TO VEGETATION

Water in the Zone of Influence of the Earth's Surface

Precipitation, descending from an atmospheric updraft where large raindrops and snowflakes have grown, is moving with the storm wind, which grows more irregular and turbulent in the boundary layer near the surface of the earth. The effect of the surface on the wind field is felt to a considerable height, especially if the surface is rough and the wind strong. The descending particles of precipitation come increasingly under the dominance of the surface.

The nearly horizontal trajectory of the smaller particles exposes them to surface influences for an appreciable period while they still are airborne, a time during which local redistributions occur. The irregularities of the earth's surface produce accelerations in the wind field over convex land forms and slackening over concave forms. These accelerations redistribute the falling particles away from convex forms and into concave ones.

A ridge commonly receives only half as much rain or snow as the immediate leeward slope, particularly if a horizontal eddy forms behind it. The familiar cornices formed in snowstorms, which in favorable locations may reach a depth of 5 to 10 m, illustrate an effect that occurs also, although less visibly, with raindrops (see Plate 7).

Experimental cutting of openings in dense coniferous forest in Idaho to increase the trapping of snowfall obtained the greatest effects when

Plate 7. Snow cornice formed by horizontal axis eddy at lee of crestline of Boreal Ridge, Central Sierra Snow Laboratory. The snow surveyors indicate the depth of snow accumulated in this aeolian form, as do the snow rollers that have fallen from it (from CSSL, U. S. Forest Service photo).

the opening was cut in forest on the upper part of a windward slope (Haupt, 1973). Aerodynamic forces that operate here are different from those in an opening in flat terrain, and produced a large cornice to the lee of the ridge. Its mass of 600 kg m^{-2}, which represents in part eroded and redistributed snow particles and in part the effect of the ridge on falling snow, is an increase of four times over the pretreatment deposition of snow.

Wind in the Delivery Mechanism The common connection between precipitation and wind is seen in the overlapping meanings of the word "storm." In storms part of the latent heat generated as precipitation occurs is converted to kinetic energy, some of which moves the air all through the boundary layer, so that it is usual for precipitation particles to reach the surface in violently moving airstreams.

Usually the atmosphere is in motion all the way down to its lower boundary. Raindrops and snowflakes make their approach to the earth in a turbulent, swiftly moving medium. Rather than descending gently in vertical paths, they are flung at the irregularities of the surface. As they swirl over the surface they are removed by gravitational settling, but even more by impaction on foliage and ground surfaces. The final state of their delivery is an aerodynamic process, often a violent one.

The paths of water particles delivered to any point on the earth's surface are affected by the irregularities and roughnesses of that surface—the numbers, shapes, flexibility, and attitude of the leaves of vegetation—and by small variations in height, slope, steepness, and slope orientation of the land. The result is a fine-textured pattern of distribution of water delivery in which local topographic and vegetational differences are very important. Where topography and vegetation have been modified by man, the pattern of water delivery is modified also.

Delivery of Different Particles

Delivery as an aerodynamic process is also determined by the sizes of the water particles: fog droplets are small and are carried easily by a strong wind. Snowflakes respond somewhat more to gravitational forces; and raindrops respond most of all.

Fog Droplets Impaction of water particles is illustrated where fog or cloud droplets are present in air blowing across mountain ridges. What is called "fog drip" is in part the mechanical screening of these small droplets from moving fog-filled air; in larger part, as Japanese experimenters have shown (Hori, 1953), it is the impaction of droplets caught in the turbulent motion of air moving over a rough forest canopy. As vertical motions in this turbulent flow throw the water particles downward, many are impacted on tree leaves and removed from the air.

This effect is so rapid that a layer of fog driven across a forest belt, continually losing droplets from its lower layers, is cleared of them in traversing 1–2 km (Hori, 1953). The impacted water droplets accumulate on the leaves, from which they drip to the ground below.

Mountain ridges projecting into the cloud decks of middle-latitude storms receive heavy deposits of "horizontal rainfall," as supercooled droplets form rime on all exposed surfaces in limited areas. Such water inputs have been studied in Germany, where the loads of rime have serious effects on high-altitude forests. In Bavaria the area experiencing heavy rime deposition during storms (i.e., the fraction that was constantly within the storm clouds) amounted to 0.02 of the total area of the mountain (Baumgartner, 1958). Rime in trees of mountains in New Mexico has been determined to be equivalent to an areal depth of 9 mm of water in a single storm (Gary, 1972).

Ridges projecting into persistent west coast stratus clouds are favorable locations for belts of trees that increase roughness and accelerate the impaction of droplets. The hill behind Berkeley, California is crowned by eucalyptus forest planted since the 1920s, which increases the water delivered to the surface by as much as 200–300 kg m^{-2} during a summer in which little rain occurs (Parsons, 1960).

The island of Lanai in Hawaii, not high enough to produce much shower activity, projects into nonprecipitating trade-wind clouds that sweep droplets over the central highland of the island. Trees planted at about 800-m altitude catch these droplets and in trade-wind weather drip water to the ground at a rate of approximately 4 mm day^{-1} (Ekern, 1964). The input, 750 mm yr^{-1}, is a valued addition to the groundwater reservoir of the island.

Wind-tunnel studies of simulated leaves in Hawaii, done to make possible quantification of the pertinent factors (Merriam, 1973), verified an equation for the buildup of storage S relative to leaf storage capacity S_0:

$$S/S_0 = 1 - \exp(-F/S_0)$$

in which F is the fog input term, accumulated up to the time in a storm for which S is desired. (F is a function of wind speed w, the liquid-water content of the fog u, and the product E of two efficiency factors: one dependent on drop sizes and speeds, and one on area, sizes, and spatial pattern of the leaves.) Figure VI-2 shows that the

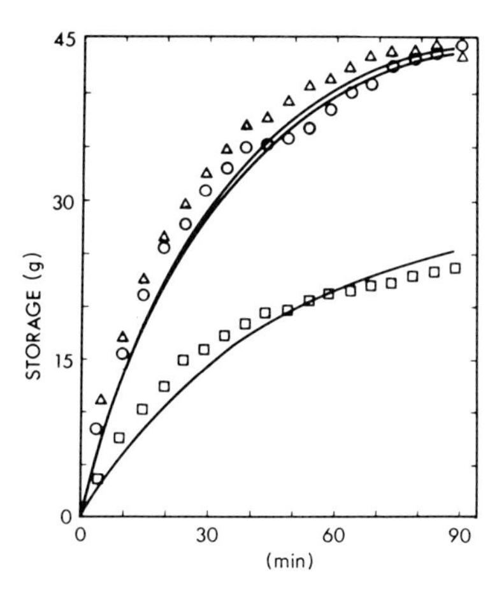

Fig. VI-2. Increase of stored water on clusters of plastic pine needles from fog droplets impinging on them at a constant speed. Curves represent calculated buildup of storage; points represent measured mass of water; ○ and △ represent nine clusters of needles in two experiments, and □ represent four clusters of needles. Merriam, R. A., *Water Resources Research,* Vol. 9, pp. 1591–1598, 1973, copyrighted by American Geophysical Union.

measured buildup follows the predicted buildup closely. Dripping off the leaves accelerates as storage S approaches the storage capacity S_0, and reaches a steady rate $P = wuE$. At $w = 0.4$ g m^{-3} for the liquid-water content of the fog, speed $u = 2$ m sec^{-1}, and combined efficiency E as 0.04, the steady rate of fog drip from the leaves is

$$(0.4 \text{ g m}^{-3})(2 \text{ m sec}^{-1})(0.04) = 0.032 \text{ g m}^{-2} \text{ sec}^{-1},$$

which is equivalent to 115 g m^{-2} hr^{-1} or 0.115 kg m^{-2} hr^{-1}, or 0.12 mm hr^{-1}.

Higher parts of the Canary Islands and of Margarita, off the coast of Venezuela, also serve as water-intake areas. "Leaf drip supplies water to numerous springs which, in turn, combine to form small mountain streams" (Alexander, 1958).

These instances of horizontal precipitation by the removal of already condensed cloud droplets on ridges that project into low cloud layers are restricted in size. *In toto* they represent a minor means of extraction of water from the atmosphere, but make an interesting addition to the more usual way in which water is transferred from atmosphere to surface. They illustrate one of the least understood links in the delivery process—the final step that depends on the aerodynamic properties of the earth's surface.

Delivery of Snowflakes Snowflakes, like cloud droplets, are affected by turbulence in the air moving over an underlying surface. Think of the angle at which a snowflake approaches the surface; its falling speed is less than 1 m sec^{-1}, which means that even in a wind moving nonturbulently at 5 m sec^{-1} it will move 5 m across the surface for each meter it descends. The tangent of the angle of approach is 0.2, and the angle is 11°. Now consider the fact that air moving at this speed over a rough surface usually is highly turbulent with many eddies carrying air both downward and upward.

The motion of the snowflake combines a mean slow low-angle descent with the turbulent motions of the air in which it is embedded. Resultant motions can be seen if you watch a snowstorm in progress outside your window. They are even plainer if you can watch snowflakes impinging on the irregular top of a forest canopy. Climb high enough in a tree during a snowstorm to look out over the canopy, and you will see that the delivery of snow to the forest is more a matter of impaction of flakes on foliage than of gravitational settling.

Now, climb down from your tree and make your way to the nearest precipitation gage, which is attempting to sample particles in a highly turbulent wind stream by measuring their gravitational settling. You will wonder how good a sample it is obtaining. Look also at whether

Plate 8. Impacted snow on vines on a wall facing the wind direction in a wet spring snowstorm (Milwaukee, March, 1972).

or not the wind-shielding device affixed to the gage, if there is one, seems to be improving the aerodynamic situation. The author has done this many times, and cautions the reader against putting implicit faith in the last millimeter of a measurement of snowfall!

The role of impact in delivery is easily seen in a wet snowstorm (Plate 8). Even leafless trees carry deposits of snow on the windward side of branches and trunk.

Delivery of Raindrops Two examples of raindrop impaction can be given. One has to do with forest, the second with structures.

Compared with the canopies of lower vegetation, a rough forest canopy is conducive to greater impaction of rain, yet little attention has been given to measuring precipitation delivered to this aerodynamically complex layer. Our information is mostly indirect. For example, Rauner (1972, p. 160) concluded, from research on forest energy and water budgets, that forest bodies in the western USSR receive, vaporize, and percolate to groundwater much more water than croplands do. In the latitude of 47N, for instance, the differentials are:

80 mm more evapotranspiration,
20 mm more off-site yield, and
100 mm more precipitation.

The excess precipitation, equivalent to about 0.2, comes mostly as rain, only to a minor extent during snowstorms.

Rain impact also is visible in architectural protection for walls of buildings. Eaves, which are a common feature of houses, afford no protection when winds are strong and rain is driven nearly horizontally; the walls must be waterproof. This property is not generally found in stone, or in the older type of brick construction, which had to be sheathed in stucco (Freeland, 1968, p. 76) or weatherboards (Freeland, 1968, p. 23). The Australian invention of cavity walls about 1885 gave a "non-structural brick skin encasing the building." While this skin becomes saturated with water in rainstorms, it does not transmit it into the house, the water simply running down the inner face of this outer wall (Freeland, 1968, p. 188).

Supercooled raindrops are a special form of water particle occasionally delivered to ecosystems, which spread into a film that freezes tight to a tree branch, car windshield, or sidewalk. This form can build up to crushing weights on trees and shrubs, and even thin deposits on flat surfaces are highly dangerous for their slipperiness.

INTERCEPTION OF WATER BY VEGETATION

When rain or snow is delivered to vegetation-covered portions of the earth's surface, how is it received? How much of it comes to rest on foliage? How long does it remain there?

Unfortunately, quantitative information is scanty. Any analysis of water delivery in the environment of a plant canopy requires data on three factors: (a) the environment, i.e., the pertinent characteristics of foliage and branch structure, (b) impaction conditions in the storms that bring the rain or snow, and (c) measurements of the buildup of rain or snow. These three kinds of data are rare.

For point (a), the environment that determines the form of storage, the strength and stability of its physical support, its thermal surroundings, and the conditions under which input and outflows of water take place, are in general not known.* For point (b), data on the rate of snowfall, storm temperature, and wind speed are necessary. For point (c), measurements of the mass of stored water are sporadic, and only rarely are associated with readings of storm conditions.

A few years ago we examined 110 studies of the interception of snowfall, representing virtually the entire recorded literature at that time, and including studies that had run many years with substantial expenditure of research funds. Of all these, only one reported concurrent measurements of the environment of the canopy, the input of snow, and the amounts intercepted. Only this one study provided data that might be utilized to make a predictive statement about the interception process. No other study of the 110 reported data suitable for inclusion in a general statement. Having recognized the sparseness of data on interception storage of rainwater or snow, let us proceed with a qualitative discussion.

Low Vegetation

Interception of precipitation by low vegetation such as shrubs or grass, or by litter on the ground, results in a redistribution of the water flux on a small scale. Amounts of stored water might be large, and in the sheltered location water is not rapidly removed by evaporation, most of it eventually going as liquid to the soil.

* Volumes of data describe the trunks of forest trees, the length of logs that can be cut from them, and the number of board feet on an acre of land. In contrast, little attention has been given to the biologically active zone of the forest—the numbers, sizes, flexibility, and strength of the different orders of branches, the areas and attitudes of the leaves. While leaf areas of some plants are beginning to be recorded, more information on the environment they provide for water is needed.

Deformation under load can result from rain. If the plants do not straighten up afterward, as when grain lodges, the harvested yield is reduced considerably (see Plate 9).

Interception has no definite upper limit when snow falls into low vegetation. The intercepted snow merges almost insensibly into the snow mantle on the ground, although the mantle often contains large voids, as Wilken (1967) showed for manzanita in the Sierra Nevada. This merging is favored by the behavior of the plants themselves, which are flexible and tough, easily deforming under load but not breaking, and continuing to support much snow. Voids in the mass of snow and vegetation may total 0.4 of its volume (Wilken, 1967).

Tall Vegetation

Trees extend their catching surfaces upward in the wind, presenting many leaves and branches for the impaction of the arriving precipitation particles. The location of the impact zone depends on the wind field, but in turbulent motion most foliage surfaces, even those deep within the crown, or on the lee side, are hit by snowflakes or raindrops. How tightly these particles adhere depends on surface

Plate 9. Lodging of oats in a field in southern Minnesota. The difficulty in harvesting the balance of the crop can be visualized.

smoothness, hydrophobic or hydrophilic character of the leaf surfaces, and the stickiness of the particle, as indicated by the contact angle of raindrops on a leaf or adhesion of snowflakes, which, in turn, depends on storm temperature and snowflake type.

Additional particles have a base layer to build on after the first layer has adhered. This step comes later for snow than for rain; thus in small storms rain is largely intercepted while snow is not. On the other hand, snowflakes begin to bridge across openings, increasing the catchment area, a process that cannot happen with rain. The sequence of events in rain is best described by Horton (1919, p. 604) from watching the process in trees near his hydrology laboratory:

> When rain begins, drops striking leaves are mostly retained, spreading over the leaf surfaces in a thin layer or collecting in drops or blotches at points, edges, or on ridges or in depressions of the leaf surface. Only a meager spattered fall reaches the ground, until the leaf surfaces have retained a certain volume of water, dependent on the position of the leaf surface, whether horizontal or inclined, on the form of the leaf, and on the surface tension relations between the water and the leaf surface, on the wind velocity, the intensity of the rainfall, and the size of the falling drops. . . .

Foliar interception of many substances other than water is common. Dry fallout of such materials as chloride and sulfur is a major link in their cycling in nature, and is enhanced by impaction on rough vegetation. Radioactive particles like ^{90}Sr are impacted on foliage during rain (Menzel, 1963), as is also radioactive iodine, which, accumulating on pasture grass, enters the grass–cow–milk–child food chain rapidly. Collecting devices in a krummholz ecosystem on a New Hampshire mountain that caught impacted as well as gravitationally delivered rain and aerosols registered 4.5 times as much water and 5–6 times as much magnesium, sodium, potassium, and lead (Schlesinger and Reiners, 1974). The impaction process enriches the cation deposits, especially calcium (8 times).

STORAGE OF RAIN AND SNOW ON FOLIAGE DURING STORMS

Rain on Foliage

The adherence and storage of rainwater depends on the nature of leaf surface, the angle at which leaves hang, and their total area. To indicate the order of magnitude, the wetting of forest foliage seems to involve 1–3 kg of water over 1 m^2 of the area of the stand of trees, and on litter about the same amount (Lull, 1964).

In a thick canopy (Plate 10), most raindrops are intercepted and join the temporary storage on leaves; relatively few fall through the canopy without contact. The size of storage to be filled is shown by the delay after rain starts before dripping from leaf tips begins. This may be 10 min or longer. Outflow eventually reaches a steady state in which the incoming rain entering foliage storage is equaled by water moving by drip and stemflow, and to a small degree by evaporation. Weighing a branch held in steady artificial rain showed that the buildup of storage to a steady rate took less than 10 min (Grah and Wilson, 1944).

The amounts of storage probably are not greatly different on tree foliage than on low vegetation (Reynolds, 1967). This might be expected, considering that leaf areas of both have between three and six times the ground-surface area.

Duration of storage is more often measured than its volume, because free water on leaves forms an excellent culture for airborne spores and other pathogens. The period that leaves of potato plants stay wet is closely related to the spread of potato blight at times of high atmospheric humidity and wet foliage. Duration of liquid water

Plate 10. Dense canopy of mesic deciduous forest near Baraboo, Wisconsin, at the peak of the growing season in July 1965.

on grass has been recorded as four times the duration of rain, even in summer (Steinhauser, 1964).

Water on leaves is sometimes visible as an enhanced radar echo, although no operational use of this fact has been made yet. Such a film after a heavy rain in Texas seems to have been thick enough to show as a dark streak in a satellite photograph (Hope, 1966). Although first interpreted as an area of new plant growth following an earlier rain, the streak was considered after further study to represent interception storage itself.

Storage of Intercepted Snow

In its relation to vegetation, snowfall represents a more complex phenomenon than rain. The impaction process is more important than with raindrops and because of the low density of snowflakes depends more on the aerodynamic characteristics of vegetation. Impaction is greatest on obstacles of small diameter, if these provide enough surface friction for adhesion.

Visual observations of the interception of snowflakes indicate that favored locations for the base layer to form are within clumps of needles; from these sites a little body of snow builds outward, between and around needles that serve as structural elements. On such broadleaf species as eucalypts and manzanita the accumulation of snow shows a similar initial preference for leaf bases and joints of twigs before it extends over the leaf surface itself.

If snowflakes are dendritic and finely branched, or if they are wet, initial cohesion is large. They rapidly build into shapes that demonstrate wind packing. Further additions bridge the gaps between snow masses on needles or twigs until they form large slabs. In extreme cases a coat of snow sheaths the tree from top to bottom.

Wind and air temperature, as the principal atmospheric influences on snowfall impaction and adhesion, interact in a complex fashion; both have nonlinear effects on interception storage (Miller, 1964) (Fig. VI-3). The plastering effect increases with wind speed if the temperature of the snowflakes is near 0°C. At lower temperatures, high wind prevents heavy deposition. Light wind does so also, since some air movement is needed to transport snow into the interior of tree crowns to fill the available storage capacity.

Tree Parameters Most trees native to regions of heavy snowfall are flexible in one or another structural member (see Plate 11), such as the

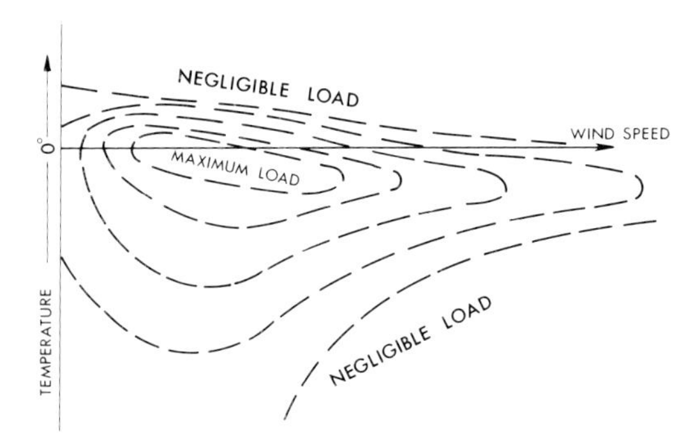

Fig. VI-3. Relative weight of snow load in tree crown as a function of snowstorm temperature and wind speed (unscaled) (from Miller, 1964).

needles of pines or the branches of cryptomeria and fir. Branches of cryptomeria that initially stand at an angle of 60° from the trunk are depressed to 110° by snow loading. This behavior greatly increases the mechanical ability of the tree to support loads of sticky snow that amount to 10 or even 20 kg m^{-2} of ground area (Watanabe and Ozeki, 1964).

Plate 11. Snow loads down a flexible branch of a conifer in Milwaukee on 22 February 1974. The unloaded branch would be 10 ft above the ground.

Magnitudes A small sheltered pine in the Sierra Nevada yielded 6 kg m^{-2} of crown projectional area when it was beaten with a pole (Miller, 1964), and not all the snow it was holding was dislodged. Better measurements were made in a Japanese study in which cut trees were placed on a scale that was read at intervals during and after snowstorms. Their exposed situation probably did not, in this case, work against representative measurements because at the low altitude of the experimental site most snowfalls are wet and clinging. In a number of storms, median snow loads on a 32-yr-old cryptomeria, which had a total value for length of branches of 86 m, were 10.5 kg m^{-2} of crown projectional area (which was decreased by about a third by the tendency of the branches to bend down) (Watanabe and Ozeki, 1964). A different variety of cryptomeria, which has stiffer branches, carried a snow load about 3 kg m^{-2} greater.

This method was later applied in the United States in research in the northern Rocky Mountains, from which Fig. VI-4 is taken (Satterlund

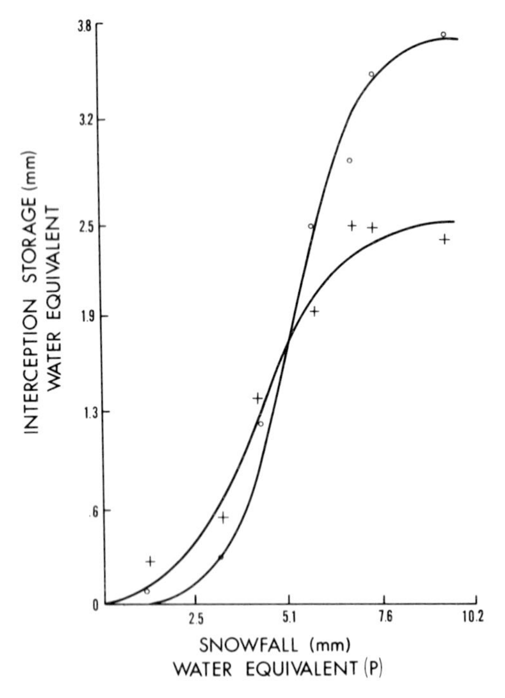

Fig. VI-4. Interception storage during progress of snowstorm on 12 January 1967. Calculated curve has been fitted to field measurements for Douglas fir (○) and white pine (+) (from Satterlund and Haupt, 1967).

and Haupt, 1967). It shows storage building to 2.5 kg m^{-2} on white pine and 3.8 kg m^{-2} on Douglas fir, with the most rapid increase about midway along the sigmoid curve.

THE OUTFLOWS FROM INTERCEPTION STORAGE OF RAIN AND SNOW

Liquid Water

The drops of water on leaves grow into films as more raindrops join them. When these are thick enough to overcome internal friction and surface tension, bulk flow begins. Whether this flow collects at the leaf tips and drips off, or runs as rivulets down the branches and the tree trunk depends on the geometric structure of the leaves and branches and the friction they exert on fluid flow.

Some of the liquid outflow enters drainage channels on the outer integument of the plant—the leaves and bark. In some trees like beech, these surfaces are smooth and speed water downward with little skin friction. As a result stemflow begins at as little as 0.3 mm of rain, and continues heavily, reaching the ground at the base of the trunk often more than at any other place under the canopy and as much as 2½ times the amount of rain (Lull, 1964).

Some of the liquid outflow drips off the leaves in amounts that depend on their shape and smoothness as well as the attitude at which they hang. It reaches the ground in large drops that, if falling from a height greater than 6 or 8 m, hit the soil as hard as if they had fallen from the clouds.

Drip water joins raindrops that have come directly through the foliage without touching it, forming a mixture that is hard to sample because of its large microscale variation. For this reason Brechtel set out 1206 gages in a small area of pine forest and obtained the means and dispersions at sites classed by their interception of clear-day solar radiation shown in the accompanying tabulation (Brechtel, 1965, modified from Tables 2 and 4). Variation, as indexed by standard deviation, is greatest under the canopy that is most closed.

	Dense			Open stand
Transmission of solar radiation	0.5	0.6	0.7	0.8
Water flux (mm day^{-1})	4.0	4.3	4.8	5.2
Standard deviation (mm day^{-1})	0.5	0.5	0.4	0.3

Problems in sampling throughfall and drip, as ecosystem water fluxes having large spatial variability, occur also in sampling the associated cation fluxes. In forest systems in British Columbia Kimmins (1974) found that a very large number of gages is needed to obtain a 95% confidence interval as small as 0.05 of the mean flux: 272 gages on a 30 × 30-m plot for the water flux and 2991 gages for the potassium flux.

In a study of *Acacia aneura,* Slatyer (1965) reports that measurable catch in gages below the tree canopy did not start until 0.8-mm depth of rainfall had accumulated, and that no measurable stemflow occurred until 2.5 mm had accumulated. Both outflows (drip and stemflow) then continued at a high rate during the rest of the storm, interrupted at times by gusts of wind that increased the rate of dripping.

The rate of outflow by dripping and wind action is suggested from a laboratory study (Grah and Wilson, 1944) in which a pine branch under a spray was weighed (Fig. VI-5). Of the accumulation of 1 mm of water, a third dripped off in 10 min after the spray ended. Another third could then be shaken off, in a simulation of wind action. The last

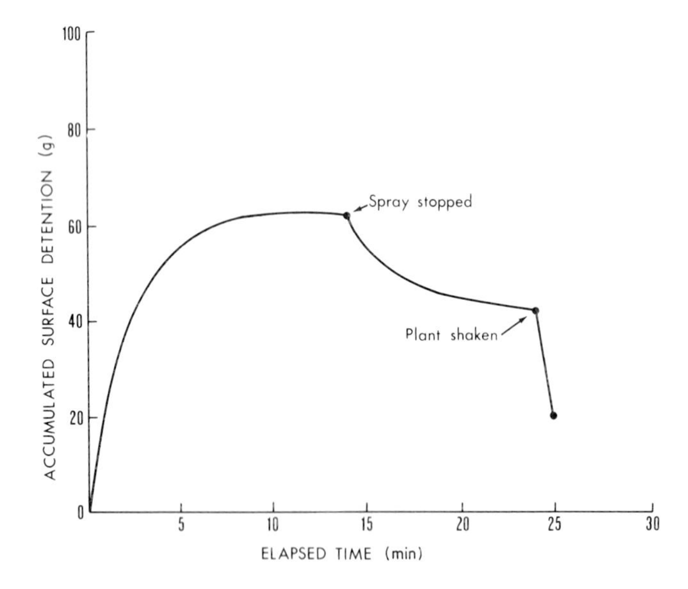

Fig. VI-5. The rise and fall of storage of intercepted simulated rain on a branch of *Pinus radiata*. Approximately a third of the intercepted water was lost by dripping and evaporating after the rain stopped; another third was lost under conditions simulating wind action on the branch; and a third was retained and presumably would later evaporate (from Grah and Wilson, 1944).

third of the water appeared to be held tenaciously by surface tension and presumably could be removed only by evaporation.

Magnitudes of Drip Plus Stemflow The amount of water reaching the ground during and after a rainstorm is what remains after canopy storage is filled. Storage is a rather stable amount for a given species, varying with windiness but not with the total depth of rain in a storm. The residue of water reaching the ground is therefore small or even zero in a small storm, and large in a big one. It cannot be expressed as a constant fraction of storm rainfall.

The total volume of water in all the episodes of outflows from the canopy over a year depends on how many small storms and how many large ones occur. It has no one percentage value, but must be estimated by determining the amount of storage in each rain, then subtracting the sum of these extracted amounts from the total water input.

Some measurements of total delivery of water to the ground under a plant canopy are shown in Table I from various sources. The fraction delivered as stemflow, as contrasted with drip or straight-through fall of raindrops, is quite variable with species. Different ecosystems bring about different spatial redistributions of this downward flux of water.

Mass Recycling by Drip and Stemflow

The leaf surfaces from which intercepted water drips or flows off are not necessarily pristine pure. They carry various metabolites exuded from the plant, and, just as they collect airborne droplets of water, also are impacted by particulates in the atmosphere. The "dry fallout" of

TABLE I

Water Delivery under Canopy

Vegetation	Total input (mm season^{-1})	Fraction as stemflow
Canary Island pines[a]	463	0.02
Eucalyptus regnans[b]	141	0.04
Bracken[c]	158	0.19
Alfalfa[c]	174	0.10

[a] Source: Kittredge *et al.* (1941).
[b] Source: Brookes and Turner (1964).
[c] Source: Haynes (1937), cited by Lull (1964).

salt and sulfates, for example, which might be as great as their delivery in rainwater, remains on leaves until a rain washes it off. Moreover, some very soluble compounds are leached out of leaves by rainwater, and this represents a very fast recycling of nutrients within an ecosystem (Ovington, 1968).

Water reaching the ground under a hardwood forest carried 2.5 g m^{-2} more potassium than rainwater in the open, 1.4 g m^{-2} more calcium, and 0.7 g m^{-2} more magnesium (Ovington, 1968), to take one example. Such large inputs of nutrients have definite effects on the soil component of an ecosystem, as Zinke (1962) has shown. These tend to be greatest at the drip line and near the stem. Stemflow is greatest from trees such as beech, and shows the greatest impact, including organic carbon, exchangeable potassium, and a lower pH, on the soil properties around such trees (Gersper and Holowaychuk, 1971). The outflows of water from tree crowns thus recycle nutrients and produce variability in soil characteristics of an ecosystem.

Whether intercepted water originated as raindrops or impacted fog droplets does not seem to affect foliar leaching of nutrients; the ratios of cation concentration in the drip water to the concentration in rainwater are about the same (Azevedo and Morgan, 1974). For example, the concentration of potassium in dripping fog water in coastal California is three times its concentration in rainwater, similar to enrichment ratios found in rain situations.

The flows of water thereby help tie the ecosystem together. In fact, the nutrient cycles are so closely geared to water fluxes that there is considerable methodological advantage in investigating them together (Bormann and Likens, 1969). This was done in the comprehensive Hubbard Brook basins in New Hampshire. "The small watershed approach, utilizing measured parameters of hydrological and chemical input, output, and net change [of storage], is a powerful tool for the study of biogeochemical relationships of individual ecosystems" (Bormann and Likens, 1969). As we progress through the sequence of hydrologic processes in ecosystems, we will also take note of the fluxes of other forms of matter associated with water.

Snow

In a gusty storm, masses of snow accumulated in the branches over a period of several minutes may be suddenly attacked by an eddy reaching down from the turbulent air above the trees, and blown to pieces. Snow accumulations are particularly subject to being blown away before ice bonding increases the cohesion among flakes enough to hold a mass together. The resulting downwind transport of intercepted snow enriches deposition in adjacent forest openings and may bias measurements made in them. "Possibly the greater snowpack in the open is partly a result of much of the intercepted snow eventually blowing out of the trees into the open" (Wilson, 1954).

An evaluation of aerial observations of snow cover in the basin of the Kings River (Court, 1963) suggests that a large excess of snowfall is to be found in the places where it is customary to make snow surveys. This bias is still present when such sites are compared with measurements of snow on the forest floor. When snow surveys were expected to provide only "index" or relative data, this bias could be overlooked. As water budget methods and models come into use, however, absolute data on water deliveries to the ecosystems of a drainage basin are imperative and an overestimate of 200 mm, as in the Kings River basin (Court, 1963), is not acceptable. The transport of intercepted snow out of forest stands into openings represents a monitoring problem for the land manager as well as a sharp contrast in water dynamics between forest and grass ecosystems in the same region.

In the Japanese experiments, wind effect on intercepted snow became evident at speeds about 1 m sec^{-1}. At a speed of 3 m sec^{-1} half the intercepted snow in exposed trees was blown out of them (Japan, 1952).

Snow falling in warm storms has more initial cohesion and is less subject to wind removal; however, it is vulnerable to even brief rises of air temperature above freezing or pulses of solar radiation that come during breaks in the storm. These pulses of thermal energy and the resulting melting of the base layer of a snow mass may release the potential energy of the mass and let it slide off a sloping support.

Removal of snow masses by sliding means that the melting of only 0.2 of the original snow releases the often precarious hold on sloping branch supports. "By using potential energy stored when snow lies on sloping branches or imposes elastic loading on them, this mode of mass transport requires only a small amount of thermal energy" (Miller, 1966, p. 6). Analysis of data from a field experiment (Japan, 1952) indicates that this movement of snow from crowns to ground runs from 1 to 2 kg m^{-2} hr^{-1}, not an inconsiderable increment to the snow cover, which is often pitted by these falling bodies of dense snow and becomes less uniform in structure and density.

If the snow mass is securely balanced in place on a branch, it simply melts away when heat is applied. The meltwater drips down or flows along the surface of the branch like rainwater.

Rowe and Hendrix (1951) report stemflow from pines equivalent to 1.5 kg m^{-2} of stand area, deriving almost entirely from melting of canopy snow. Meltwater from intercepted snow does not go as far as snow blown out of tree crowns, and represents a different kind of spatial redistribution.

One set of measurements of several modes of outflow of water from snow intercepted by Douglas fir and white pine in the northern Rocky Mountains, averaged over two relatively warm, wet winters, is as follows:

Washed off by rain	17 mm per winter
Released in large masses (sliding)	9
Minor masses and drip	6
Evaporation	5
	37

(Satterlund and Haupt, 1970) (11 storms per winter). It is interesting that even in warm weather so small a fraction of the intercepted snow evaporated.

In a region with more radiant energy, Arizona, the movement of intercepted snow from pine canopy was observed by time-lapse photography, which allowed ranking the different fluxes (Tennyson *et al.*, 1974). Most of the intercepted snow was moved by sliding, blowing, or melting; only a little evaporated, either from the snow masses themselves or the meltwater films on the branches.

Residual Storage of Snow after Storms

In forested regions where heat advection and radiation surpluses provide too little energy to remove snow from the tree crowns and where increased cohesion after a storm eventually renders the snow almost invulnerable to wind force, its arboreal existence continues for a long time. It remains in the canopies through most of the periods between storms and therefore through much of the winter, and becomes an ecologically important component of the forest ecosystem (Pruitt, 1958). Such snow-bearing boreal forests are reflected in an Eskimo word for snow in trees (*qali*), a different root than that for snow on the ground. Arboreal snow also is common in some middle-latitude coastal ranges and on poleward slopes in some interior ranges. It is common in the conifer plantations of the Allegheny Plateau (Muller, 1966).

Snow is quickly cleared from the trees after storms by wind action or melting in regions where the energy level is high. Clearing often occurs during breaks in storms, and in any case shortly thereafter. As John Muir said of the Sierra, "the forest soon becomes green again."

EVAPORATION AS A MODE OF OUTFLOW FROM INTERCEPTION STORAGE

One major outflow of water from its transient storage place on leaves takes place in the form of vapor, and is favored in high-energy environments. Vapor may move downward to condense on the ground or snow cover beneath the trees, or it may diffuse upward into the atmosphere. The first attempts to measure rain interception, in the 1860s, derived from an interest in its role in the meteorological problem of the sources of atmospheric humidity, and in certain conditions this evaporation is a major flux in the water budget of the forested region. In other conditions it is small and has been greatly overestimated, as has generally been the case with snow.

Evaporation of Intercepted Snow

Snow in trees is visible and highly photogenic, as its depiction on millions of Christmas cards demonstrates. It is natural to think that forests of mountain drainage basins are vaporizing snow at a tremendous rate, and robbing thirsty downstream irrigators and cities. From pictures to folklore is only a step. We find that the idea that trees waste snow is firmly embedded in hydrologic doctrine (Miller, 1961; Sartz, 1969). The question of how to reduce this supposed large outflow of vapor appears easy to answer: Cut down the trees! If a land manager also has other reasons to log off the forest, he can now adduce a water-saving rationale: Cut the trees and increase the off-site yield of liquid water.

This rationale, however, is based on slender evidence. Measurements of actual interception are scarce and difficult to generalize; furthermore, meteorological analysis suggests that most of the snow held transiently in canopy storage never evaporates, because the necessary conditions of energy supply and dry air do not usually coincide (Miller, 1967).

Energy supplied by warm air often is accompanied by high vapor pressure; canopy snow in mountains invaded by marine airstreams is likely to melt, slide out of tree crowns, or drip to the ground. These modes of outflow removed almost all intercepted snow in an experimental basin in the Cascades of Oregon. Evaporation from canopy snow into the moist air was less likely to occur than condensation of vapor from the air onto the snow.

Satterlund and Haupt (1970) in two warm, humid winters in northern Idaho (0.7-km altitude, 48N latitude), with mean snowfall of 120 mm per winter, found that only 0.15 of the detained snow evaporated. About one-third of the evaporation occurred from the snow in place, and two-thirds at times of drip and release of small masses, i.e., probably from films of meltwater, rather than from the snow itself.

Forests of interior mountain ranges lie in drier but colder air. Canopy snow evaporates in sunny weather although the low temperature is inhibiting. Absorption of solar energy by those branches that are bare tends to warm some microclimatic niches in the canopy. If their warmth reaches a level permitting melting, much crown snow moves as meltwater (although locally the warmed films of meltwater on branches can reach a high vapor pressure and evaporate). Hoover and Leaf (1967) feel that in general evaporation is minor in the Colorado Rockies. This is shown also by the long duration of snow in the trees.

Boreal forests receive much less solar radiation than the maritime and interior highlands just mentioned. Since they often carry canopy snow over from one storm to the next, there seems to be no more tendency for this snow to evaporate than the snow on the ground.

The combination of warmth and dryness required to evaporate snow is not found in marine air, which is the most common warm airstream in winter. It is most likely to occur in air descending from a high level in the atmosphere, either as free foehn caused by dynamic processes, or as leeward foehn caused by the drawing downward of air after it passes the crest of such ranges as the Rocky Mountains. In these situations snow evaporates, since an input of energy coincides with a gradient of vapor pressure away from the snow into the dry air.

Snow does not evaporate as easily as liquid water. For a given temperature difference the vapor-pressure gradient is smaller at low temperatures than at high; furthermore, the vapor pressure of snow cannot exceed 6.1 mb. While local intensification of incoming radiation may provide heat to evaporate meltwater or snow, the most usual heat supply is sensible-heat flux from the air. Unless low vapor pressure obtains in this air, however, the sensible-heat flux* will be

* Sensible heat is the heat content of an object or medium (like the air) that we sense directly as warmth and that is measured as a product of specific heat and temperature. A sensible-heat flux therefore proceeds from a higher temperature to a lower one.

accompanied by a latent-heat flux* *to* the snow, which will then melt rather than evaporate.

In general, the large amount of energy required to move snow by evaporation rather than by melting, sliding off, or blowing out of the crowns militates against much evaporation (Miller, 1967). In addition, the requisite environment of dry air—vapor pressure less than 6 mb—is rare, especially where energy in the environment might be sufficient.

Evaporation from Wet Leaves

Liquid water on foliage differs from snow in that there is no upper limit on its temperature and, hence, vapor pressure. It can evaporate into air of moderate humidity and in particular into maritime airstreams, which do not accept vapor from snow.

Another difference between intercepted snow and rain lies in the greater absorption of solar radiation by water films as compared with snow. Wet surfaces are dark and absorptive—snow is not. Both potential sources of energy, the atmosphere and the sun, work more effectively on intercepted rain than on snow. In addition, rainwater is found on foliage in the warm season of the year rather than the cold.

Evaporation is rapid after a rainshower when the air is unstable and winds are high, and after sunrise as sunshine is absorbed by the leaves. Figure VI-6 (Rauner, 1958) shows the heat flows to and from the foliage of mixed forest on dry days and on a day following rain. The radiation surplus (i.e., the sum of all the flows of radiant energy) of the wet foliage is larger than that of dry foliage and most of the heat goes into evaporation, which does not meet as much diffusion resistance as transpired water does. The fraction of interception storage that returns to the atmosphere as vapor is large after a rainshower on a warm day; it is zero when insufficient heat reaches the vegetation.

The yearly sum of evaporation of intercepted precipitation depends on the heat supply and the frequency with which films of intercepted water are formed. If rains are frequent, filling the interception storage many times, total evaporation can be large. The duration of water films on leaves is probably shorter for trees than for grass because trees are more open to turbulent heat exchange.

* Latent (or hidden) heat is manifested as the greater energy content of vapor compared with liquid water. The flux of latent heat therefore proceeds from higher vapor pressure to lower.

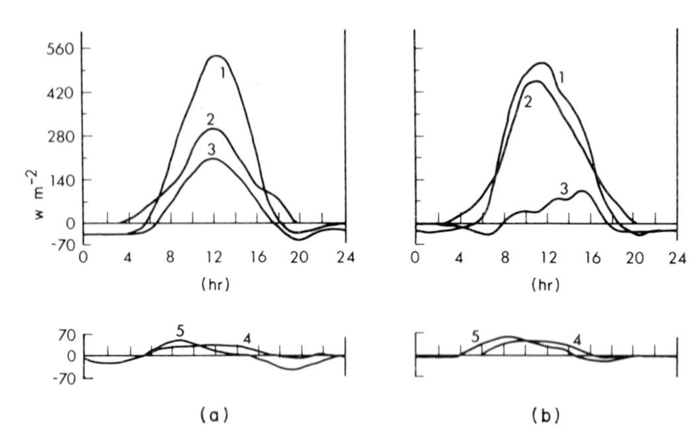

Fig. VI-6. Daily march of heat fluxes from crowns of birch–aspen forest in late summer 1957 near Moscow. (a) Days in dry clear weather, several days after a rain; (b) a clear day just after 15-mm cyclonic rainfall. Curve 1: surplus or deficit of whole-spectrum radiation. Curve 2: latent-heat flux upward from crowns. Curve 3: sensible-heat flux upward from crowns. Curve 4: heat exchange with the soil. Curve 5: heat exchange with biomass (Rauner, 1958).

Wet leaf surfaces, like wet soil particles, are exposed directly to the atmospheric water-vapor sink without any additional resistance to the diffusion of vapor intervening. The only resistance met as vapor diffuses from wet leaves is therefore atmospheric, a function of wind speed and turbulence, i.e., the aerodynamic properties of the surfaces. Determinations of atmospheric resistance from pine needles are, for example (sec cm^{-1}), shown in the accompanying tabulation (Source: Rutter, 1967).

Resistance	Mean	Range of values
For intercepted water on needle surface (atmospheric resistance only r_a)	1.2	0.6–3.4
For transpired water from needle interiors (atmospheric + stomatal resistance $r_a + r_l$)	9.1	4–22

Like a psychrometer's wet bulb, wet leaf surfaces support an outward movement of latent heat that draws in an inward flow of sensible heat. In a study of water movement from pines on rainy days Rutter (1966) found the highest incidence of evaporation of intercepted water to occur in autumn, perhaps because in this season while radiation sources are weak, the air is still warm. Evaporation rates on rainy days averaged as large as 2 mm day^{-1}. These were about half of the mid-summer maximum rates for transpiration and about the same as transpiration in spring and autumn.

In these circumstances it is useful to transfer models from evapotranspiration and energy-budget research that have been successful in reconstructing micrometeorological profiles in vegetation canopy, and to employ detailed specifications of leaf area, angles, and absorptivity for short-wave and long-wave radiation, with aerodynamic parameters that determine turbulent heat exchanges in the canopy. Murphy and Knoerr (1975) made such a model and found that rain-wetted leaves (upper side only) evaporate intercepted water 1.5–2 times as fast as unwetted leaves transpire it. The difference is larger in moderate than in high-energy weather. Under typical rainy-day conditions, evaporation from water films on leaves is calculated as 0.08 mm hr^{-1}, equivalent to 2 mm day^{-1}. While transpiration is in general independent of wind speed, the evaporation of intercepted water is not, increasing somewhat as wind speed increases.

Szeicz (1970) determined values of the ratio of evaporation from wetted and unwetted surfaces from studies of vapor flux from wetted and unwetted leaves. For pines this ratio was about 3.5; for grass 1.1–1.2; for root crops and tall clover 1.4. The greater aerodynamic exposure of tree foliage is important in this situation, reducing the resistance to diffusion of vapor molecules from leaf surfaces into the air. We have, however, seen the defects of the means of measuring and analyzing the arrival of rain and snow at the complicated zone of the vegetation mantle, and determining how much remains there. These problems carry over into attempts to evaluate how much of the rainwater and snow in canopy storage is vaporized. Our techniques for measuring snow delivery to a forest canopy are inadequate, and "evidence concerning the interception of snow is confused" (Ward, 1967, p. 70).

Measurements of water that moves in solid, liquid, or gaseous state from the canopy are virtually nonexistent except in a few sites. Even here, "the main source of error tends to arise from inadequate sampling" (Ward, 1967, p. 66) or the heterogeneous pattern of these outflows. It is hoped that as research on ecosystems leads us to a better understanding of the dynamics of water within their complex structures, we will learn more of the outflows of water from their foliage.

WATER INTERCEPTED BY LITTER

Organic material shed by the aerial parts of an ecosystem covers the mineral soil and forms still another intercepting layer in the downward hegira of water. Its storage capacity is small, some water being held on its surfaces and some absorbed within the litter itself, especially by its finer components.

Absorption is potentially relatively large but slow, taking up water at an areal equivalent rate of less than 1 mm day^{-1}. Therefore rainfalls at rates higher than this have little further effect on moisture uptake (Fosberg, 1972). The duration of rain is the important dimension, rather than intensity or total depth.

Much of the rain caught on forest litter does not penetrate to the mineral soil but is held tightly until it evaporates when the litter dries out after the rain. The amount of evaporation is 1–2 kg m^{-2} (Helvey, 1964) from the surface film. Absorbed water evaporates as it moves outward to the surface.

Rates of evaporation depend on the coarseness of the litter. Fine material has a large surface area and a high surface-to-volume ratio, so that water in surface films is soon evaporated. Coarser material dries more slowly. In evaluating the hazard of wildfires, coarseness of the fuel material on the forest floor has to be considered; weighing of fuel sticks of various sizes gives an idea of the rate of drying of the actual litter.

The yearly evaporation from litter can be appreciable in climates of high energy and frequent rains. At the Coweeta Experimental Forest in North Carolina estimates run about 75 kg m^{-2} annually (Helvey, 1964), equivalent to about 0.03 of the total rainfall. In most places the amount would be considerably smaller. This zone, however, performs a useful short-term service in storing short bursts of heavy rainfall. It feeds the water out more steadily, at a rate at which the mineral soil is more able to take it in and prolongs the evaporation opportunity.

AREAL REDISTRIBUTION OF WATER BY VEGETATION ABOVE THE SOIL

Whether intercepted rain or snow is moved out of tree crowns as vapor, as stemflow, as drops falling off leaf tips, or as clumps of snow from a branch in a gust of wind, spatial redistribution of the initial areal pattern of water supply is inevitable. The water reaches the surfaces of lower vegetation, litter, or soil in a pattern that is different than that in which it was delivered to the upper surface of the vegetation mantle.

The process of stemflow illustrates this redistribution because it concentrates water into a small area near the tree trunk. *Acacia aneura* in central Australia not only produces large amounts of stemflow—as much as a third of the total rainfall in storms greater than 10 mm—but also directs it into mounds of permeable soil around the trunks where infiltration capacity is twice that of other parts of the forest floor (Slatyer, 1965).

In acacia groves windy storms sometimes produce amounts of water reaching the ground by drip and stemflow that are in excess of the amounts reaching the ground in the open. Slatyer ascribes this excess to the vertical shape of the trees, which abstract water descending at a slant. Strips of land between the acacia groves, being in the rain shadow of the trees upwind, are deprived of rain. Areal redistribution of the incoming water flux thereby favors the groves and starves the intergrove belts.

Excess catch by *Eucalyptus niphophila* (snow gum) also has been reported in the Australian Alps (Costin and Wimbush, 1961). In this wet region the additional water directed into the root zone probably is little needed for tree growth but contains nutrients, particularly potassium, leached from the tree leaves, which are beneficial for lower vegetation. The drip ring at the circumference of the canopy is associated with a proliferation of roots under some trees.

This water flux from foliage, or from organic litter lying on the ground, is thus altered in composition from the rain or snow initially delivered. In its brief contact with foliage it picks up organic chemical compounds and translocates them to the soil. Where these have nutrient value, the dripping of intercepted water facilitates cycling of metal ions from the deep layers from which tree roots have extracted them to the shallow soil horizons. Some substances influence the formation of the soil. Others seem to make it repellent to infiltrating water, as is the case under chaparral cover in southern California (Krammes and Debano, 1965), and thus illustrate an interesting feedback of one level of ecosystem hydrodynamics to another level.

Water falling from foliage may reach the soil in larger drops than the original raindrops and hit with greater impact. Snow that has remained in tree crowns even briefly usually has undergone some transformation and agglomeration into larger bodies. When these fall they represent the transport to the ground of a different kind of material, wetter and denser, than that initially delivered directly from the clouds.

The hydrologic importance of interception is that it creates a pattern of delivery of water to the soil that differs from the initial pattern of delivery from the motion systems of the atmosphere—different in state or condition, different in chemical content, different in spatial pattern.

It is usually smaller in magnitude, but uncertainties about the true distribution of the movements of liquid and solid water from the intercepting mantle of vegetation prevent us from knowing how much smaller.

Bearing in mind that much of the earth's surface is covered with tall vegetation in continuous areas, groves, or belts that interfere with the near-horizontal movement of particles approaching the earth's surface, we know that changes in characteristics of delivered water are likely, and can be certain that its spatial pattern is changed. The concentration of rain water into stemflow is the most obvious example of this modification of an initial flux into a variegated pattern that favors certain locations and slights others. Considering the irregular initial pattern in which water is delivered to vegetation systems, and the systematic and random irregularities imposed on it by vegetation foliage, it is clear that inflow of water to the soil surface of an ecosystem has a highly intricate pattern.

REFERENCES

Alexander, C. S. (1958). The geography of Margarita and adjacent islands, Venezuela, *Univ. Calif. Publ. Geog.* **12,** 85–192.

Azevedo, J., and Morgan, D. L. (1974). Fog precipitation in coastal California forests, *Ecology* **55,** 1135–1141.

Baumgartner, A. (1958). Zur Höhenabhängigkeit von Regen- und Nebelneiderschlag am Gr. Falkenstein (Bayer. Wald), *Int. Soc. Sci. Hydrol. Publ.* **43,** 529–534 (Toronto Vol. 1).

Bormann, F. H., and Likens, G. E. (1969). The watershed–ecosystem concept and studies of nutrient cycles. *In* "The Ecosystem Concept in Natural Resource Management" (G. M. van Dyne, ed.), pp. 49–76. Academic Press, New York.

Brechtel, H. M. (1965). Methodische Beiträge zur Erfassung der Wechselwirkung zwischen Wald und Wasser, *Forstarchiv* **35,** 229–241.

Brookes, J. D., and Turner, J. S. (1964). Hydrology and Australian forest catchments. *In* "Water Resources Use and Management," pp. 390–398. Melbourne Univ. Press, Melbourne.

Costin, A. B., and Wimbush, D. J. (1961). Studies in catchment hydrology in the Australian Alps. IV. Interception by trees of rain, cloud, and fog, Div. Plant Ind., Tech. Paper 16, CSIRO, Australia, 16 pp.

Court, A. (1963). Snow cover relations in the Kings River basin, California, *J. Geophys. Res.* **68,** 4751–4761.

Ekern, P. C. (1964). Direct interception of cloud water on Lanaihale, Hawaii, *Proc. Soil Sci. Soc. Am.* **28,** 419–421.

Fosberg, M. A. (1972). Theory of precipitation effects on dead cylindrical fuels, *Forest Sci.* **18,** 98–108.

Freeland, J. M. (1968). "Architecture in Australia, a History." Cheshire, Melbourne, 328 pp.

Gary, H. L. (1972). Rime contributes to water balance in high-elevation forests, *J. Forestry* **70,** 93–97.

Gersper, P. L., and Holowaychuk, N. (1971). Some effects of stem flow from forest canopy trees on chemical properties of soils, *Ecology* **52,** 691–702.

Grah, R. F., and Wilson, C. C. (1944). Some components of rainfall interception, *J. Forestry* **42,** 890–898.

Haupt, H. F. (1973). Relation of wind exposure and forest cutting to changes in snow accumulation, *Int. Assoc. Hydrol. Sci.,* **Publ. 107,** 1399–1405.

Helvey, J. D. (1964). Rainfall Interception by Hardwood Forest Litter in the Southern Appalachians, U. S. Forest Serv. Res. Paper SE-8, 9 pp.

Hoover, M. D., and Leaf, C. F. (1967). Process and significance of interception in Colorado subalpine forest, *In* "Forest Hydrology" (W. E. Sopper and H. W. Lull, eds.), pp. 213–224. Pergamon, Oxford.

Hope, J. R. (1966). Path of heavy rainfall photographed from space, *Bull. Am. Meteorol. Soc.* **47,** 371–373.

Hori, T. (ed.) (1953). "Studies on Fogs in Relation to Fog-Preventing Forest." Tanne, Sapporo. (*Summarized in* Coastal fogs and clouds, D. H. Miller, *Geog. Rev.* **47,** 591–594, 1957.)

Horton, R. E. (1919). Rainfall interception, *Mon. Weather Rev.* **47,** 603–623.

Japan. Government Forest Experiment Station (1952). Study of the fallen snow on the forest trees (snow crown) (First Report), *Govt. For. Exp. Sta. (Meguro) Bull.* **54,** 115–164.

Kimmins, J. P. (1974). Some statistical aspects of sampling throughfall precipitation in nutrient cycling studies in British Columbian coastal forests, *Ecology* **54,** 1008–1019.

Kittredge, J., Loughead, H. J., and Mazurak, A. (1941). Interception and stemflow in a pine plantation, *J. Forestry* **39,** 505–522.

Krammes, J. S., and Debano, L. F. (1965). Soil wettability: a neglected factor in watershed management, *Water Resources Res.* **1,** 283–286.

Lull, H. W. (1964). Ecological and silvicultural aspects. *In* "Handbook of Applied Hydrology" (V. T. Chow, ed.), Chapter 6. McGraw-Hill, New York, 30 pp.

Menzel, R. G. (1963). Strontium-90 accumulation on plant foliage during rainfall, *Science* **142,** 576–577.

Merriam, R. A. (1973). Fog drip from artificial leaves in a fog wind tunnel, *Water Resources Res.* **9,** 1591–1598.

Miller, D. H. (1961). Folklore about snowfall interception (abstr.), *J. Geophys. Res.* **66,** 2547.

Miller, D. H. (1964). Interception Processes During Snowstorms, U. S. Forest Serv. Res. Paper PSW-18, Berkeley, 24 pp.

Miller, D. H. (1966). Transport of Intercepted Snow from Trees During Snow Storms. U. S. Forest Serv. Res. Paper PSW-33, Berkeley, 30 pp.

Miller, D. H. (1967). Sources of energy for the thermodynamically-caused transport of intercepted snow from forest crowns. *In* "Forest Hydrology" (W. E. Sopper and H. W. Lull, eds.), pp. 201–211. Pergamon, Oxford.

Muller, R. A. (1966). The effects of reforestation on water yield: a case study using energy and water balance models for the Allegheny Plateau, New York, *Publ. Climat.* **19** (No. 3), 251–304.

Murphy, C. E., Jr., and Knoerr, K. R. (1975). The evaporation of intercepted rainfall from a forest stand: an analysis by simulation, *Water Resources Res.* **11,** 273–280.

Ovington, J. D. (1968). Some factors affecting nutrient distribution within ecosystems. *In* "Functioning of Terrestrial Ecosystems at the Primary Productivity Level" (F. E. Eckhardt, ed.), pp. 95–105. UNESCO, Paris, 516 pp., pp. 95–105.

Parsons, J. J. (1960). "Fog drip" from coastal stratus, with special reference to California, *Weather* **15,** 58–62.

Pruitt, W. O., Jr. (1958). Qali, a taiga snow formation of ecological importance, *Ecology* **39,** 169–172.

Rauner, IU. L. (1958). Nekotorye rezul'taty teplobalansovykh nabliudenii v listvennom lesu, *Izv. Akad. Nauk SSSR Ser. Geog.* No. 5, 79–86.

Rauner, IU. L. (1972). "Teplovoi Balans Rastitel'nogo Pokrova." Gidrometeoizdat, Leningrad, 210 pp.

Reynolds, E. R. C. (1967). The hydrological cycle as affected by vegetation differences, *J. Inst. Water Eng.* **21,** 322–330.

Rowe, P. B., and Hendrix, T. M. (1951). Interception of rain and snow by second-growth ponderosa pine, *Trans. Am. Geophys. Un.* **32,** 903–908.

Rutter, A. J. (1966). Studies on the water relations of *Pinus sylvestris* in plantation conditions. IV. Direct observations on the rates of transpiration, evaporation of intercepted water, and evaporation from the soil surface, *J. Appl. Ecol.* **3,** 393–405.

Rutter, A. J. (1967). An analysis of evaporation from a stand of Scots pine, *In* "Forest Hydrology," (W. E. Sopper and H. W. Lull, eds.), pp. 403–417. Pergamon, Oxford.

Sartz, R. S. (1969). Folklore and bromides in watershed management, *J. Forestry* **67,** 366–371.

Satterlund, D. R., and Haupt, H. F. (1967). Snow catch by conifer crowns, *Water Resources Res.* **3,** 1035–1039.

Satterlund, D. R. and Haupt, H. F. (1970). The disposition of snow caught by conifer crowns, *Water Resources Res.* **6,** 649–652.

Schlesinger, W. H., and Reiners, W. A. (1974). Deposition of water and cations on artificial foliar collectors in fir krummholz of New England mountains, *Ecology* **55,** 278–286.

Slatyer, R. O. (1965). Measurement of precipitation interception by an arid zone plant community (*Acacia aneura* F. Muell), *Arid Zone Res.* **25,** 181–192.

Steinhauser, F. (1964). Eine statistische Untersuchung an die Andauer der Benetzung durch Tau, Nebel und Regen, *Agric. Meteorol.* **1,** 184–200.

Szeicz, G. (1970). Comments on the value of some wetting experiments in studies of crop evaporation, *Water Resources Res.* **6,** 347–348.

Tennyson, L. C., Ffolliott, P. F., and Thorud, D. B. (1974). Use of time-lapse photography to assess potential interception in Arizona ponderosa pine, *Water Resources Bull.* **10,** 1246–1254.

Ward, R. C. (1967). "Principles of Hydrology." McGraw-Hill, New York, 403 pp.

Watanabe, S., and Ôzeki, Y. (1964). Study of fallen snow on forest trees. II. Experiment on the snow crown of the Japanese cedar, *Govt. Forest Exp. Sta. Bull.* **169,** 121–139, Japan.

Wilken, G. C. (1967). Snow accumulation in a manzanita brush field in the Sierra Nevada, *Water Resources Res.* **3,** 409–422, photos.

Wilson, W. T. (1954). Analysis of winter precipitation observations in the cooperative snow investigations, *Mon. Weather Rev.* **82,** 183–199.

Wolman, A. (1962). Water Resources. A Report to the Committee on Natural Resources of the National Academy of Sciences—National Research Council, U. S. Nat. Acad. Sci. Publ. 1000-B, 35 pp.

Zinke, P. J. (1962). The pattern of influence of individual forest trees on soil properties, *Ecology* **43,** 130–133.

Chapter VII

WATER DETAINED ON THE SOIL SURFACE

Water in the form of snow or rain arrives at the surface of the earth directly from the clouds of a storm or indirectly after a sojourn in a foliage canopy. When it arrives as snow on cold soil or on earlier snow cover it forms a new mantle. When it arrives in liquid state faster than the surface can take it in, it piles up as a film called "detention storage," a new element in an ecosystem.

These new mantles are water in storage. They have dimensions of duration, areal extent, mass per unit area, depth relative to vegetation height, and also exhibit such characteristic properties as density, temperature, reflectivity, and vertical structure. Let us begin by considering the longer-lived kind of storage, the snow cover.

SNOW COVER

Snow forms a mantle on the earth's surface that survives the snowstorm by a few hours, a few days, or a few months, or for millennia as it does in the icecaps. We will consider here chiefly the snow cover that forms in winter and disappears in spring, which covers some 61 million km^2, mostly in the Northern Hemisphere* (Kotliakov, 1968, p. 86).

Most precipitation particles in middle-latitude storms begin life as snowflakes, but not all arrive at the ground as snowflakes. The necessary condition for snowfall is a below-freezing temperature in the lower atmosphere. This occurs in cold continental air beneath the

* This area is greater than the sum of the areas of snow cover on the ice plateaus (16 million km^2) and on sea ice (38 million km^2) (Kotliakov, 1968, p. 86). Its average mass, 140 kg m^{-2}, is slightly less than the yearly accumulation of snow on the ice caps and sea ice.

precipitation-generating airstream of a frontal storm, or at high altitudes (Chapter II).

After the snowstorm ends the survival of the snow deposited depends on how much energy the new snow receives. Energy comes to it from solar radiation (of which a small fraction is absorbed), from the soil, as atmospheric (long-wave) radiation from the poststorm airstream, and by convection from it. The snow surface, by reason of the upward limit on its temperature, cannot radiate away more than 315 W m^{-2}. If the energy inputs listed above add up to more than 315 W m^{-2}, the snow will start to melt. If thin, it soon vanishes, if thick, it may hold out until cooler weather or a new snowstorm comes. In the latter case, the formation of a winter snow cover has begun.

Environments of Snow Cover

Considering the survival of a layer of new snow, we see that in addition to considering terrain factors that affect deposition and subsequent drifting we must also consider factors that affect exchanges of thermal energy between snow mantle and environment.

Some of these environmental factors are atmospheric: transparency of the air, allowing short-wave radiation to reach the snow mantle; clouds that emit long-wave radiation; and turbulence that determines downward fluxes of sensible and latent heat from air to snow. These energy flows to the snow are also affected by terrain: the sheltering in valleys, the exposure to wind and sun, and the trees towering above the snow mantle that both reduce the incoming short-wave (or solar) radiation and augment downward long-wave radiation to the snow.

On a large scale, environmental conditions favoring survival of a snow cover are found in regions of a generally cold atmosphere, or weak solar radiation, or both. There are several classes of such regional environments.

Maritime Mountain Ranges In many mountains near cooler parts of the ocean, snow is deposited in amounts so great that in spite of warmth in interstorm periods a large fraction of each deposit survives to form a base for the next layer. The snow mantle of winter represents a series of large accretions separated by the effects of melting periods during which translocation processes produce changes in crystal size and shape tending toward large rounded grains. Many of these changes are accelerated by percolating meltwater, some of which reaches the ground and becomes an outflow from this particular water storage.

This type of storage requires large snowstorms. Typical climatic conditions also include the likelihood that some storms may bring rain instead of snow. The warmth of the intervening periods when maritime airstreams remain dominant is also characteristic. For these reasons the snow mantle in maritime mountains is closely associated with the variation of temperature with height in the atmosphere.

The temperature conditions in storms affect the probability of receiving snow rather than rain, as we saw earlier. The warmth of the air in interstorm periods affects poststorm melting. Windward slopes of coastal mountains display a rapid change from zero snow cover at their bases to great depths that last all winter at the higher elevations where the air during and between storms is cold.

Altitude of the lower edge of the snow cover fluctuates during the winter. Between storms it recedes upward; cold storms bring it back to lower altitudes. The snow line lies close to sea level in the Alaskan Panhandle, and at about 1.2-km altitude in the latitudes of central California.

Such mantles, often called "snowpacks" because of the high density of the stored water, are typical of the new mountain ranges around the Pacific Ocean—the ranges of North America, the Andes, the Australian and New Zealand Alps, and the Japanese Alps. Vapor from the North Atlantic Ocean nourishes the snow covers of the European Alps, Scotland, and Scandinavia.

Field measurements of the water equivalent (or mass) of the snow cover at 2.2-km altitude in central California (Table I and Fig. VII-1) illustrate the way the amount of water in cold storage, nearly zero until the winter solstice, steadily grows until early April, about 100 days after the solstice. After April the storms are small, and the mantle melts away faster than new snow is added to it. In late winter and spring it is several meters thick, very dense, and affords good skiing for several months. It forms a new landscape that differs in almost every visual feature from the landscapes where no snow cover comes into existence, and this contrast is a further attraction to visitors from outside the snow zone (see Plates 12 and 13). A model of the accumulation and ablation processes (Riley *et al.*, 1973) gives satisfactory reconstitutions for both the California and Montana basins of Table I, and also for another snow laboratory basin, in Oregon.

In maritime mountains, with their heavy snowfall and steep rise above adjacent lowlands, the snow cover possesses enormous potential energy. Consider its mass and altitude—a ton m^{-2} poised 2 km above sea level. The energy equivalent, renewed every year, is 20 MJ

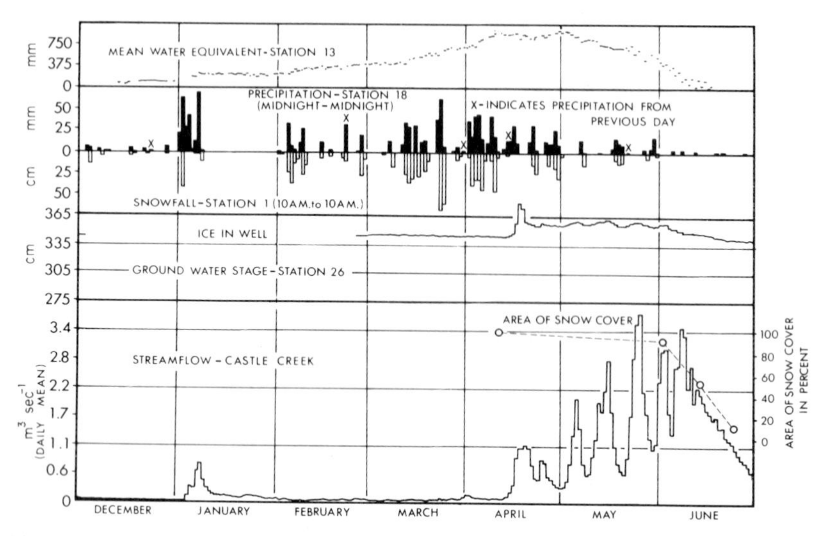

Fig. VII-1. Courses of basic hydrometeorological elements during the period of snow-mantle accumulation and melting in 1947–1948, at the Central Sierra Snow Laboratory, California (U. S. Corps Engineers, 1952, p. 13).

Plate 12. Shelter hut in Lower Meadow, Central Sierra Snow Laboratory, California, in July 1957. The projection is called a "Santa Claus chimney" (U. S. Forest Service photo).

TABLE I

Duration, Mass, and Density of the Snow Cover of Experimental Basins in a Maritime Mountain Range (the Sierra Nevada) and an Interior Range (the Northern Rockies)[a]

	Depth (cm)	Mean duration of snow cover (days)	Last date
Castle Creek basin, California (Central Sierra Snow Laboratory) Latitude 39N, altitude 2.2 km			
	>25	166	10 May
	>100	120	30 April

Date:	21 Nov	21 Dec	21 Jan	21 Feb	21 Mar	21 Apr	21 May	21 Jun
Mass (kg m^{-2})	60	195	500	645	760	460	70	0
Density (kg m^{-3})	200	260	330	340	370	400	460	—

	Depth (cm)	Mean duration of snow cover (days)	Last date
Bear Creek basin, Montana (Upper Columbia Snow Laboratory) Latitude 48N, altitude 1.7 km			
	>25	177	16 May
	>100	129	28 April

Date:	21 Nov	21 Dec	21 Jan	21 Feb	21 Mar	21 Apr	21 May	21 Jun
Mass (kg m^{-2})	65	230	340	405	485	380	65	0
Density (kg m^{-3})	160	230	260	280	310	380	430	—

[a] U. S. Corps Engineers (1956).

(or 6 kwh) from each m^2 of a mountain ecosystem. The economic equivalent is usually far greater than the value of any other product of these ecosystems.

In the Sierra Nevada this potential energy was put to use a century ago to power one of the most advanced technologies of the time, hydraulic gold mining. When excessive sediment outputs stopped this enterprise, the hydropower was converted to hydroelectric power that supported one of the first electrified economic regions of the country, centered on urban activities, and irrigation from pumped ground water.

In the Southern Hemisphere, most of the 2 million km^2 of transient snow cover lies in such maritime mountains as the New Zealand Alps, interior Tasmania, and the Snowy Mountains of Australia. Its potential energy gives it far greater economic importance than its area indicates, and in addition all three highlands provide attractive recreation.

Plate 13. Same shelter hut on 9 April 1958, when the snow cover is near its deepest (U. S. Forest Service photo by Kenneth Knoerr).

Interior Mountains Airstreams reaching mountains in the interior of the northern continents are depleted in moisture and bring less snow than they did to the maritime ranges. As a result the altitudes of snow cover are higher in these mountains than in those nearer the oceans at the same latitude: for instance, 2.5 km in the Colorado Rockies versus 1.2–1.5 km in the Sierra Nevada.

Snow is deposited under much colder conditions, which has significant effects on sizes and characteristics of its crystals. Densities, as shown in Table I, are smaller than in maritime snow covers,* and high winds produce more drifting. These make it feasible to attempt to control final placement of the snow being added to the mantle, for example, by erecting snow fences or taking other measures that enhance topographic "snow trapping efficiency and capacity" (Martinelli, 1967).

Interstorm periods, although sunny, are cold; the snow suffers little melting and remains low in density. The winter's accumulation builds to appreciable totals. The Rockies and the high mountains of central

* At altitudes of 2.5–3.0 km, snow-cover density in the Tian-Shan and Pamir Mountains averages 280 kg m^{-3} as against 340 in the eastern Alps (Kotliakov, 1968, p. 132).

Asia enter spring with considerable water storage, which has great value to the surrounding dry lowlands.

Interior Lowlands The snow mantle in the interior of the Northern Hemisphere continents, as in the mountains, often is not established much earlier than the winter solstice since the ground must first be cooled to the freezing point. This cover often has a shorter life than mountain snow. It is more likely to disappear and to be reinstated in correspondence with the alternating reigns of tropical and polar airstreams, especially nearer to the Atlantic.

Individual snowfalls usually are light; any depth greater than 10 cm, which scarcely would be noticed in the mountains, has a crippling effect on lowland activities. The difference in snowstorms is the difference between a system of vertical air motion anchored in one area and a system that moves several hundred kilometers a day, spreading its snow thinly over a wide expanse. Heavy snowstorms can occur, however, leaving up to 20- to 30-cm depth. Such a deep deposit disrupts communications and may last for months. Figure VII-2 (U. S. Weather Bureau, 1964) shows the 0.1 frequency (once in 10 years) of water equivalent in the upper Midwest and Plains in early March. Note how the mass increases to the east. Also note that compared to mountain snow cover the values are small, even in this snowy year.

The interior lowlands of North America are interrupted by one feature that tends to steer and strengthen passing snowstorms and create new ones and so to create areas where snow is deeper and lasts longer than would otherwise be the case. This feature is the tremendous heat storage formed by the Great Lakes, which produces a special intensification of snowfall downwind from them. The duration of six major snow belts to their lee (Lake Superior supports two belts) (Muller, 1966) is shown in Fig. VII-3 (Phillips and McCulloch, 1972). Snow in these belts lasts up to 150 days, whereas in adjacent locations without lake effect it may last as little as 100 days. The lake effect also appears outside the true snow belts; snow lasts 100 days at Muskegon, to the lee of lower Lake Michigan, about 50 days at Milwaukee directly across the lake. Farther east, frequent storms that draw in Atlantic vapor bring heavy snow. Melting periods occur near the coast, and altitude effects are marked.

The final disappearance of the snow mantle of continental interiors, with 100–200 kg m^{-2} mass, is a less well-defined event than that of a deep mountain snowpack with 1000 kg m^{-2}. Melting rates are high in strong southerly airflow in spring and the stored water may be released in a short time. In some years melting is sporadic and

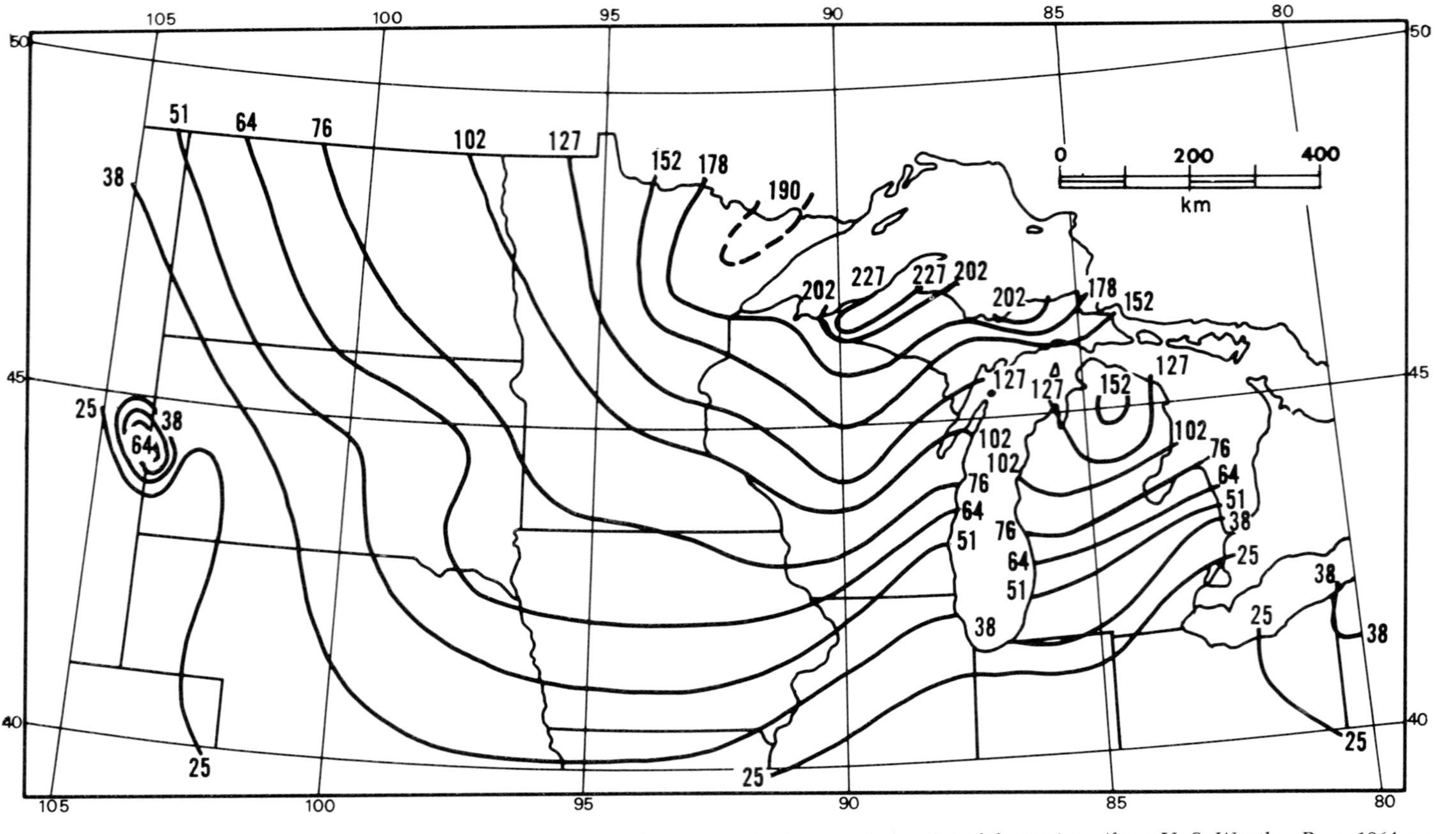

Fig. VII-2. Maximum mass of water stored in snow cover (mm) in early March, exceeded in 0.1 of the springs (from U. S. Weather Bur., 1964, p. 17).

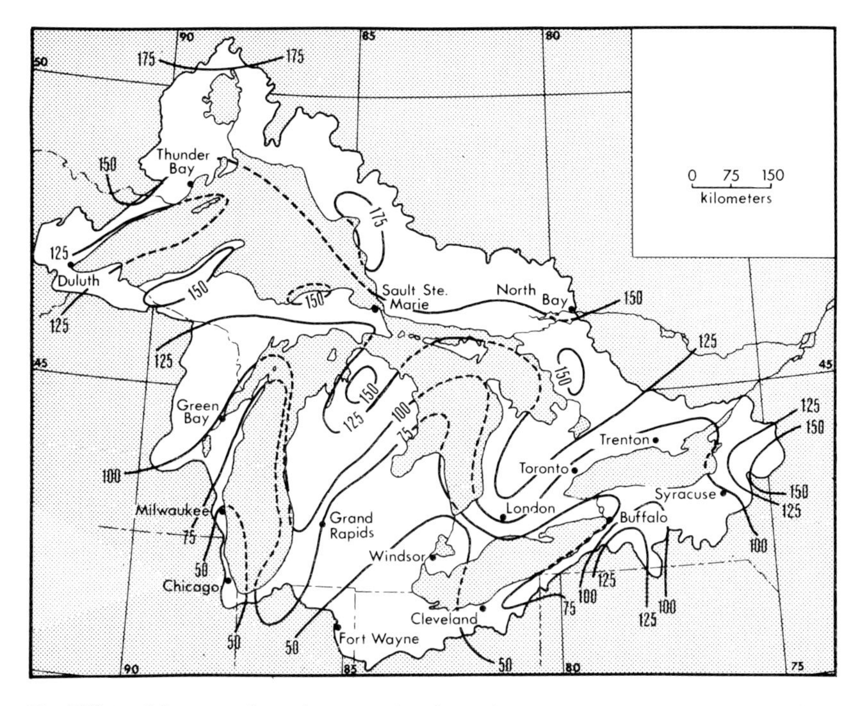

Fig. VII-3. Mean number of consecutive days of snow cover 2.5 cm or more in depth in the basin of the Great Lakes, showing major snowbelts to the lee of the lakes (from Phillips and McCulloch, 1972, Chart 32).

interrupted by periods of cold cloudy weather, and extends over several weeks.

On the drier portions of the interior plains the extensive snow mantle is, in spite of its thinness, an important contributor to the water needs of summer crops. In such areas various practices are used to retain this water storage on crop land. For example, standing wheat stubble has been observed to trap enough snow to improve spring soil moisture by 50 mm, an important amount in Dakota circumstances. Similar methods of holding the snow mantle in place have long been standard practice in drier parts of European Russia and western Siberia.

In range country of eastern Montana, snow fences trap drifting snow on small (8 × 8 m) catchment basins covered by butyl rubber, the runoff from which is stored for stock water in the summer. Catchment without snow fences produced an average of about 25 kg m^{-2} (Saulmon, 1973), whereas catchments with fences produced an average of 195 kg m^{-2} (12.5 tons of meltwater from each catchment).

Heavy, extensive snowfalls early in the season establish a large snow-covered territory that steers later storms in such a way that cold air is more than normally present over the snow cover. No melting then occurs until late in spring. Such a winter was 1842–1843 in the upper Midwest (Rosendal, 1970). Snow cover was established about 10 November and in most areas remained until February, when it was deepened by heavy snowfalls that spread "a deep and extensive snow cover . . . over much of the eastern two-thirds of the country." The severe cold of February and March (10°C below the 1931–1960 normal) enabled the snow cover to remain in Wisconsin until mid-April. Its duration was about 155 days, which is nearly as much as that of the mountain snow covers. In a farming region, it occasioned widespread suffering, and many cattle starved to death. It is far from impossible for such winters to come again, with even more severe effects in our transport-dependent economy.

Other Regional Environments of Snow Cover On Arctic lowlands snow is thin because the snowstorms are poor providers. The meager deposits do not, however, evaporate or melt during the long dark winter, but are incessantly shifted about the landscape by the winds. The earth's surface becomes one of streamlined wind-sculptured shapes. Wind action reduces the crystals to small grains that pack into a cover of moderately high density and good bearing strength. From a water-storage standpoint this snow cover is significant in a regional water budget in which all the water fluxes are small.

Snows deposited on the high ice caps, which have been built-up 2–3 km above the level of the polar seas, are likewise thin and windblown, but they do not melt in summer and next year are covered under the snows of later storms. Transformed into firn and then ice, and sinking into the continent itself, the deposits of each annual cycle retain such individuality that in drill cores the snows of 10,000 years ago can be distinguished.

Low-latitude mountains that are high enough to be well within below-freezing air accumulate snow in the wet season; this cover melts or evaporates in the dry season. The response pattern follows the precipitation regime since there is little annual change in energy supply. The environment is dry, cold air; weak incoming long-wave radiation and intense short-wave radiation from a sun nearly at the zenith produce sharp points of snow facing the sun. Evaporation of snow is more important here than it is in any other environment of snow cover.

Ecosystem Environments of Snow Cover As we saw in Chapter VI, snow accumulates in shrub ecosystems differently than in ecosystems of grass or annual species, because aerodynamic deposition processes are affected by the greater roughness. These aerodynamic characteristics, little studied, also affect the turbulent energy fluxes that bring about metamorphosis and melting of the snow cover. Characteristics of these ecosystems also affect radiation components; the intimate mixture of branches, foliage, and snow increases absorptivity and accelerates melting.

Snow cover under forest canopy lies in an environment that is characterized by dominant diffuse-source radiation and weak turbulent transfers. Deposition is less an aerodynamic and more a gravitational process; organic debris is mixed into the snow cover, and the snow cover is pitted by snow clumps and drip water falling from the canopy overhead. In a south-slope cedar-hemlock forest in Idaho, Haupt (1972, p. 9), using a special lysimeter in the snow, measured 104 mm of dripwater percolate during winter. Aerodynamic and radiative processes in metamorphosis and ablation are also modified in the forest environment.

Snowfalls in urban ecosystems were discussed in Chapter II; these systems also form a special environment for snow cover. Obvious characteristics are deposition of contaminants, patchiness resulting from snow removal activities, and the effects of urban aerodynamics on accumulation and melting. These effects operate even in small settlements. For example, Barrow Village on the Arctic Coast of Alaska experiences "meltout" 2 weeks earlier than adjacent tundra ecosystems, due to lower albedo of the urban snow cover, its patchiness, and the aerodynamic roughness of the village. These factors increase both the radiative components of the snow energy budget and its turbulent fluxes in warm-air advection. Computer simulation of the energy budget of the urban and tundra ecosystems (Outcalt *et al.*, 1975) indicates that the radiative effects are more important in ablation than the aerodynamic.

Characteristic Changes in Snow Cover

Snow has many properties that are rare in nature: a high albedo or reflectivity of solar radiation; small thermal conductivity; and a small net loss of energy by exchange of long-wave radiation because the temperature of its surface cannot exceed 0°C. The upper limit on its surface temperature also affects its exchanges of energy with the atmosphere, which carry heat or vapor downward more commonly than is true over most types of surface. Furthermore, some of these properties can change radically.

Metamorphosis Ecosystems change with age; rock formations metamorphose; water bodies age and die; but few natural bodies age as fast as a snow cover. Albedo and bulk density are qualities that epitomize the history of accumulation and weathering of the deposits of snow made during preceding weeks or months. Snow crystals undergo rapid metamorphosis into a large-grained snow of high density and low albedo where winters are relatively warm. In the Sierra Nevada of California, the density* of spring snow reaches 500 kg m^{-3} and the albedo 0.45 or lower. In the coastal mountains of British Columbia snow-cover density increases to 420 kg m^{-3} (standard deviations of 50 kg m^{-3}) by mid-April, as compared with 330 kg m^{-3} in the Great Lakes area, 260 kg m^{-3} in the taiga, and 330 kg m^{-3} in the wind-packed snow of the tundra (McKay and Findlay, 1971).

One equation for density of lowland snow cover relates it to factors that express winter melting or energy availability, its own mass (gravitational settling), wind packing and wind-powered translocation of mass within the snow cover, and effects of new-snow density and of rain:

$$\rho = 10 \sum T + 0.1 \sum h + 24n - \Delta\rho_0 + 100rh^{-1} + 210,$$

in which ρ is density, kg m^{-3}; ΣT accumulated sum of positive air temperature; h depth of snow cover, cm; n number of days when wind speed exceeded 6 m sec^{-1}; ρ_0 density of new snowfalls on the cover; and r rain amounts in kg m^{-2} (Kotliakov, 1968, p. 109). The fluffiest snow lies in the quiet cold forests of Canada and Siberia, where no rain or warm air intrudes and wind-packing is minimal. In eastern Siberia density in open sites is 180–190 kg m^{-3} and in sheltered forest ecosystems only 160–170 (Kotliakov, 1968, p. 111). The equation shows that mechanical and thermodynamic forces are more important than gravity in increasing snow density; they are the factors that bring about metamorphosis.

Figure VII-4 depicts how a winter snow cover is built up from successive layers of individual storms. The heavy lines show how each layer of stratum shrinks as time passes, and gets denser (stippling) under conditions of temperature (light lines) that are generally high.

* High density means relatively high thermal conductivity. Because the entire mantle is warm, however, the gradient of temperature through it is small and little heat moves by conduction.

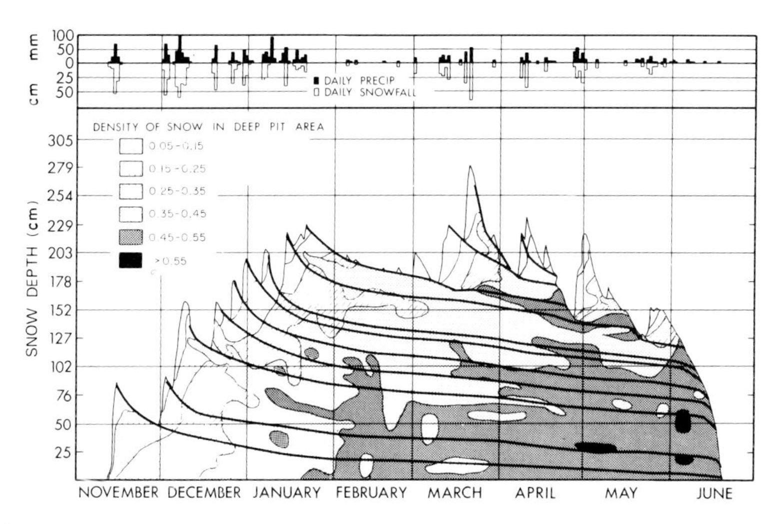

Fig. VII-4. Structure of the snow cover at Lower Meadow, Central Sierra Snow Laboratory, California, in the winter of 1952–1953. Density is shown by patterning, and the horizons are shown (solid curves) between strata deposited in each snowstorm (data from U. S. Corps Engineers, 1956). (Density is in g cm^{-3}.)

Figure VII-5 relates the albedo of the snow surface to its age, the aging process going faster in the high-energy environment of spring. Changes in albedo indicate changes in grain size, shape, and wetness of the snow. Radiation absorption and scattering theory show how albedo is determined by the area of ice–air interface per unit volume of snow. This parameter depends on grain size and density (Bergen, 1975); for snow of density 450 kg m^{-3} and grain size 3 mm, albedo is about 0.68; lower values might result from trapping of solar radiation within the water film on the ice grains. Low albedo characterizes shallow snow, in which some of the incoming solar radiation reaches the underlying soil and is absorbed. The limiting depth depends on scattering in the snow mass and may be as great as 20 cm (Whiteley *et al.*, 1973). The typically lower albedo tends to accelerate the metamorphosis of lowland snow cover.

In contrast, metamorphosis of the cold snow goes slowly in the Colorado Rockies, taking place less by liquid translocation than by vapor movement and condensation in different layers of the snow. Condensation might form new crystals that become unstable layers within the mantle and act as separation planes along which avalanches can break off. One classification of snow metamorphosis distinguishes "equitemperature" from "temperature-gradient" conditions (Sommerfeld and LaChapelle, 1970). The latter typify interior locations where

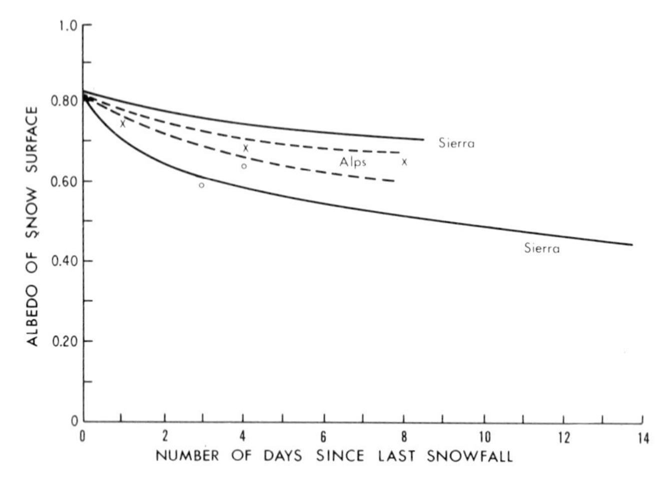

Fig. VII-5. Decline in albedo with increasing age of the snow surface in the Sierra Nevada (solid line) and the Alps (dashed line and ○). The crosses are for data from central Russia (Miller, 1955, p. 129). Upper curve for each location is winter; lower is spring.

ground and air bounding the snow below and above form a great contrast in temperature.

In maritime mountains the snow cover passes much of its life at a temperature at or close to 0C. This condition also characterizes continental snow covers in spring. In such an environment, ice grains, vapor, and liquid-water films coexist during daytime hours on most days. On small crystals with large total surface area, liquid water is found in a proportion up to about 0.05 of the total mass. On large, coarse grains of old snow, this fraction is 0.02–0.03, as in sand. Larger fractions of liquid-state water are found during rains and in periods of rapid surface melting, and represent water in transit from the surface to the lower layers of the snow cover and on into the soil, a draining process that takes several hours.

Wind storms often shake down needles, twigs, and bark fragments onto forest snow, covering as much as a third of its surface (Miller, 1955, p. 130). As the debris is leached by percolating meltwater, the chemical composition of the snow cover changes. In one of the Hubbard Brook studies, Hornbeck and Likens (1974) found that the snow cover was less acid than snowfall by several tenths of a pH unit and contained substantially more potassium (up to 20 mg m^{-2}).

Snow Cover as Environment

The presence of a snow cover changes the outer surface of the earth in winter not only in physical ways, like albedo, but also in such conditions as trafficability for animals, man, and vehicles. As the snow cover deepens during the winter, it affects the over-winter survival of plants, food searching by large animals, the habitat of small ones, and the movement of man.

The proposal of a scheme for augmenting snowfall in the Sierra Nevada to overcome design defects of a state water plan gave rise to studies of possible environmental impacts of such augmentation (if it worked). Studies showed that a great number of things would be affected—survival of young red fir, migration and winter range of deer herds, costs of snow removal on roads, accident rates, and so on.

Snow cover, porous and full of air, equable in temperature and humidity, offers a vital shelter from winter rigors for both animals and plants, the value of which is plain in winter when it is absent or delayed. The time cross section in Fig. VII-4 represents the environment of fir and pine seedlings through several months of the year in the Sierra; when they grow taller, they encounter a much more rigorous climate, with extremes of temperature, radiation impact, and wind abrasion. Shallow snow is permeated with isotropic light and many perennial plants begin spring growth while still buried (Iashina, 1960). Deeper snow patches produce a patchy ecological effect. In the Russian steppe, deeper snow in meadows fosters an ecosystem with northerly species, and in gullies it promotes tree species.

Figure VII-6, a diagram of an Alpine ridge on which snow is thinner on top than on the lower slopes, indicates the ecosystems to be found in each location, depending on wind exposure and soil temperature associated with snow of varying depth. The diagram also indicates the kinds of damage that occur in each ecosystem from snow pressure or lack of shelter, and the planting measures and times for pine and larch if reforestation is attempted in this high zone (Aulitzky, 1963). This composite of information on the relations of ecosystems of low vegetation and forest with snow as one form of water in the environment is the product of a long interdisciplinary investigation of the severe zone of the upper tree line, in which biologists as well as meteorologists took part.

The soil is often thawed although cold under snow, and yields some water to roots during the winter. Its slow warming in spring, however, which cannot start until the snow cover is gone, means that ecosystem uptake of water and nutrients also begins late, shortening the period for growth and nutrient cycling.

Snow cover in lowlands assures us that the growing season will begin with a full reservoir of soil moisture—a role extremely important where summer rains are spotty or unreliable. Manipulations of snow-cover depth and density are carried out for many purposes, including

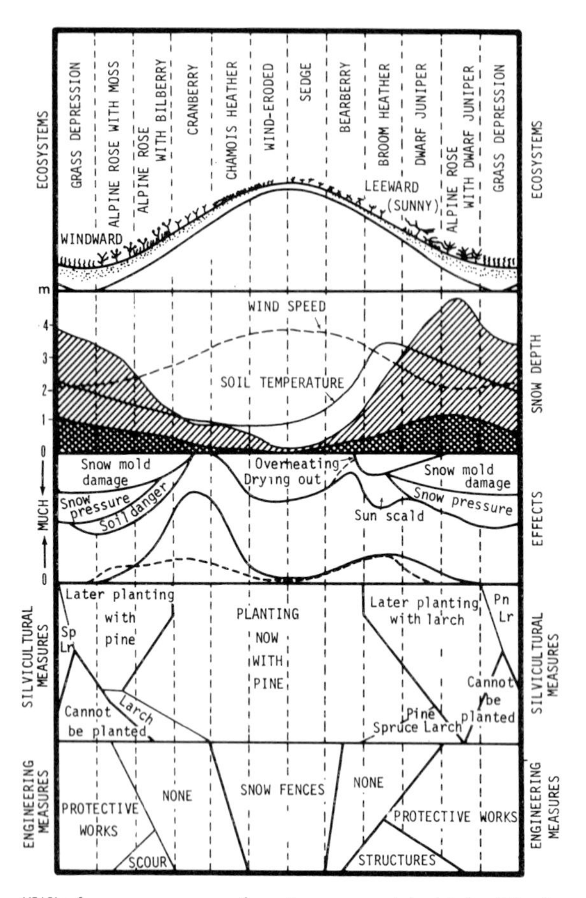

Fig. VII-6. "Wind–snow ecogram for site-appropriate high-altitude reforestation in deforested areas of the subalpine zone of the Inner Alps" (Aulitzky, 1965). A sequence of ecosystems across a ridge is shown in the top diagram; in the next diagram are shown wind speed, soil temperature, and maximum and minimum depths of snow cover. In the third diagram are shown the effects of these site factors on tree reproduction, both damaging and favoring the named species. In the bottom two diagrams are shown silvicultural and structural measures to make possible forest regrowth on these harsh sites, each measure being site-specific to a particular ecosystem.

that of managing soil temperature in the environment of the roots of winter grains, but perhaps most often for the purpose of optimizing soil moisture.

Spatial Changes in Snow Cover The snow cover embodies large place-to-place differences that sum up the micrometeorological pattern of deposition in many storms, that is, the aerodynamics of topography and forest. To a lesser degree, the snow cover is affected by differences in persistent patterns of heat supply between storms, as on south slopes.

Local differences in snow cover over small drainage basins demonstrate the influences of elevation, orientation, and steepness of slopes, and their concavity or convexity. An intensive network of snow surveys and aerial photographs provide measurements that enable us to assess quantitatively the aerodynamic and thermodynamic influences of these different aspects of terrain. For example, the variation in mass is about 2000 mm km^{-1} of change in altitude in the basin of Castle Creek. The density of the snow cover is more uniform; its standard deviation over this small basin was 26 kg m^{-3}, which is only $\frac{1}{20}$ of the mean (Miller, 1950).

Other differences are associated with forest or open exposure. Where winter melting from advection is common, as in Wisconsin or western Russia, the forest protects the snow from a large input of heat. Where high winds occur during storms or soon after them, the shelter of the forest produces additional accumulation, particularly at the lee edge of large open areas. Accumulation of snow cover in ponderosa pines in Arizona is related to such forest attributes as basal area, sum of diameters, number of trees, and volume per hectare (Ffolliott and Thorud, 1972). These characteristics index the interception of falling snow, the interception of direct-beam solar radiation, and the flow of long-wave radiation from trees to snow.

The size of openings is critical to snow deposition in them and in bordering forest, since it determines how easily gusts from the air stream can reach down to the ground. The aerodynamics of openings of different sizes has not been worked out, however. Empirical studies suggest that openings in an extensive forest do not induce more overall delivery of snow but serve to redistribute it. For example, a small opening in a lodgepole pine forest in Wyoming caught about 75 kg m^{-2} more snow at the expense of the forest downwind from it (Gary, 1974). The magnitude of this redistribution of mass was approximately 40,000 tons.

Formozov (1961) points out that snow depth and structure (ice lenses and crusts) in given ecosystems have an important bearing on the ecology of small animals. The effect on large animals, in contrast, is behavioral, because they can choose among many ecosystems that differ in snow depth, density, and structure.

Area of Snow Cover Snow cover has an areal dimension as well as depth. Like depth it increases by jumps through the winter as each storm extends snow over adjacent bare lands. The area shrinks back between winter storms as thin cover melts at the edges, and in spring it decreases in a regular way that expresses correspondence with the accumulated amounts of meltwater derived from the snow cover.

Figure VII-7 shows this relation for Castle Creek basin in spring 1947 (Miller, 1953; U.S. Corps Engineers, 1956), with a characteristic break in slope that is found in all melting seasons. This break separates the early and mid-season period, appearing in the upper left of the diagram, when runoff is generated more by vertical loss of the deep snow cover than from areal shrinkage. After about 0.6 of the meltwater has been generated, the bare areas of the basin (then 0.4 of the total basin area) are large enough to speed the attack on the snow remaining, which is soon reduced to patches. The snow is thinner than at the beginning and shrinks away rapidly in this late season situation, providing less meltwater from each hectare of land becoming free of snow. The zero intercept in the diagram represents water yet to come out of the soil and rocks after no more snow remains in the basin.

These regular relations between runoff generated from melting and

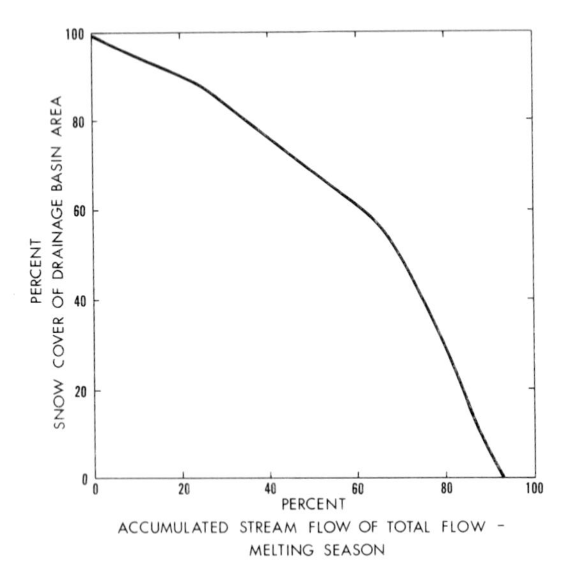

Fig. VII-7. Depletion in area of the snow mantle in the basin of the Central Sierra Snow Laboratory during spring 1947, as related to accumulation of meltwater runoff from the basin in Castle Creek (Miller, 1953).

the area of snow cover in a drainage basin permit hydrologists to determine how much meltwater remains on the slopes of a basin at any date during spring, by measuring the remaining area of snow cover (Leaf and Haeffner, 1971). Such measurements can be made by air photographs, from satellites (Rango, 1975), by aerial reconnaissance in a light plane, and occasionally by sketch mapping from a high ridge.

Water Storage and Its Measurement Most westerners are familiar with the picture of the snow surveyor skiing miles through mountain country to reach the course where he will take samples from the snow cover with his tube. They probably have heard of such improvements as snowmobile or helicopter transportation to, or radio-transmitting gages in, remote sites. These efforts, carried out at 2000 snow courses in the western U.S. and Canada at intervals during the later part of the snow-accumulation season, provide relative data on snow-cover mass in comparison with that in earlier years. Some of these are now equipped with pressure pillows that give readings that can be telemetered to the forecast office (Davis, 1973).

From these comparisons each spring it is possible to estimate meltwater runoff in the following summer. Figure VII-8 shows a

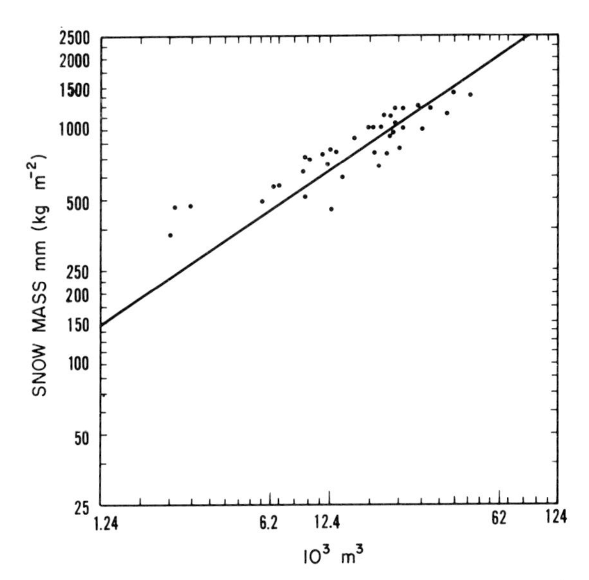

Fig. VII-8. Relation between mass of snow cover (mm) at Blue Lakes (altitude 2.4 km) on 1 April in 40 years and subsequent runoff in Carson River near Fort Churchill, Nevada (10^3 m^3) (from Richards, 1959).

typical relation between water equivalent at a mountain survey course (Blue Lakes, altitude 2.4 km) and subsequent streamflow in the Carson River from 1 April to the end of runoff (Richards, 1959).

These observations, and the use of them, represent an index concept rather than a genuine water budget. No one knows the true water storage in a mountain basin, or the rates of input and outgo of water. Church (1933), the persuasive professor and mountaineer who developed the snow-survey procedure,* considered employing a water budget as long ago as 1910. He concluded, however, that the necessary measurements of all the water fluxes in the budget were beyond his reach, doubtless a correct conclusion for the time. The situation is improving now, however. Integrated basin values of snow-cover mass can be calculated by taking account of snow accumulation and density in different ecosystems in the mosaic landscape encompassed in the area of a drainage basin. In the basin of the Central Sierra Snow Laboratory we stratified the terrain and, finding for instance that a given topographic facet constituted 0.09 of the total area, assigned it an appropriate number of snow courses and air-photo grid points. Stratification reduces sample variance, which "increases confidence about the sample mean, reduces the number of snow courses required, [and] increases the probability that snowcover [sic] is similarly distributed over each areal unit comprising each class" of ecosystem (Steppuhn and Dyck, 1974, p. 319). For instance, in the basin of Bad Lake, Saskatchewan, six kinds of ecosystems were mapped and measured. Density of snow cover varied from 140 kg m^{-3} in upland stubble systems to 270 in lowland pasture.

Aerial and satellite imagery of snow-covered areas (Rango, 1975) is made with high resolution and measurements stratified by ecosystem and topographic facets can be aggregated into drainage-basin units. Increased telemetering makes possible the sampling of remote places and also can provide a close time breakdown if desired. Upper-air data and cloud satellite measurements can make possible energy-budget calculations during the melting season. With these aids we will be able to work out the water budgets of these high, empty, but very productive sources of water.

Radiation Melting of Snow Cover

When heat is taken in by snow at rates exceeding about 315 W m^{-2}, the excess produces changes in the physical state of snow by melting

* It would be a fascinating study to see how much our knowledge of the mountains—and their preservation from despoilers—owes to the spare-time activities of academic people who studied their surroundings as Church did the central Sierra Nevada near Reno, where he was a professor of classics in the University of Nevada.

it. The high reflectivity of winter snow cover, however, limits the absorption of solar energy and indicates the intrinsic survival powers of snow cover. In spring, in contrast, when reflectivity declines as snow grains get older, and when the days lengthen and the sun climbs higher in the sky, the income of solar heat can bring the total supply above the 315 W m^{-2} threshold for several hours a day on several days per week. A typical spring day displays a marked cycling between melting and nonmelting conditions.

Daily Regime of Heat Flows On 16–17 April 1946, values at 2-hr intervals of the flows of radiant energy in the Sierra were as shown in the top part of Table II. Total radiation absorbed from the air during the night was approximately 255 W m^{-2}. It rose rapidly after sunrise as solar radiation added to the input. It leveled off around 695 W m^{-2} in the middle of the day, then declined during the afternoon and evening. The radiant-energy loading exceeded the critical value of 315 W m^{-2} for many hours.

This daily regime followed the course of the sun. The surface temperature of the snow mantle, however, did not follow the pattern of morning rise and afternoon decline that might be expected in plant or soil-surface temperature. Rather, it rose only a few degrees, then held steady at 273 K for about 15 hr. Upward long-wave radiation, the energy flux that is a function solely of surface temperature, perforce behaved in the same way, holding at 315 W m^{-2} during most of the day. Other outward flows of energy, which are functions of temperature or humidity gradients in the air, rather than of surface temperature alone, behaved differently than upward long-wave radiation. The turbulent heat fluxes never changed to an upward direction as they normally do if the temperature of a surface is free to rise. Rather, through the day and much of the night, they continued to move heat downward from the air to the snow. These downward fluxes (an average of 45 W m^{-2} over the 24-hr period) were not large, but did make a contribution to the heat surplus that supported the melting process.

The peculiar regime of surface temperature also affected the movement of heat into the deeper layers of the snow. Intake of heat into the snow mantle lasted only until all layers were warmed to 273 K, then it ended. Heat cannot move by conduction without a gradient of temperature.

Melting The principal energy conversion taking place at the snow surface during the day being examined has about the same magnitude as the algebraic sum of the energy fluxes we have been discussing. The conversion associated with the phase change from snow to

TABLE II

Energy Budget of Open Snow Fields of the Crest Region of the Sierra Nevada 39N lat., 2.2-km elev., 16–17 April 1946[a]

Energy flux	2-hr intervals												Average
	06	08	10	12	14	16	18	20	22	24	02	04	
Temperature-independent energy fluxes													
Incoming short-wave radiation	+10	+500	+880	+1040	+920	+425	+135	0	0	0	0	0	+340
Reflected short-wave radiation[b]	−10	−290	−505	−575	−510	−235	−70	0	0	0	0	0	−200
Absorbed short-wave radiation	0	+210	+375	+465	+410	+190	+55	0	0	0	0	0	+140
Incoming long-wave radiation[b]	+240	+240	+260	+265	+270	+280	+280	+260	+250	+250	+250	+260	+265
Total loading by radiant energy	+240	+450	+635	+730	+680	+470	+335	+260	+250	+250	+250	+260	+405
Surface temperature (°C)	−3	0	0	0	0	0	0	0	−1	−1	−2	−2	
Surface vapor pressure (mb)	5	6	6	6	6	6	6	6	6	6	5	5	
Temperature-dependent energy fluxes													
Emission of long-wave radiation[b]	−300	−315	−315	−315	−315	−315	−315	−315	−315	−315	−300	−300	−310
Sensible-heat flux	0	0	+25	+90	+100	+115	+55	+25	+15	+15	0	0	+37
Latent-heat flux	0	0	0	+25	+25	+25	+15	0	0	0	0	0	+7
Substrate-heat flux	+60	−80	0	0	0	0	0	0	0	+15	+15	+15	+1
Freezing of meltwater into a crust	0	0	0	0	0	0	0	+30	+50	+35	+35	+25	+15
Melting	0	−55	−345	−530	−490	−295	−90	0	0	0	0	0	−155
Total													−405

[a] Units: W m^{-2}; source: modified from Miller (1955, Tables 25, 73, and 75). Basic data from Cooperative Snow Investigations, 1947, U. S. Corps of Engineers and Weather Bureau.

[b] Notes: Emissivity of the snow very close to 1.00. Albedo = 0.56.

meltwater, the latent heat of fusion, is 335 J g^{-1} of water substance, or 335 kJ m^{-2} for a 1-mm layer of water over any given area.

This process requires only that the meltwater be removed, which is easily done through gravity drainage. The energy fluxes that in other situations remove excess energy from a surface under strong radiation loading are small here, or even reversed, because the surface temperature fails to respond to the daytime increase in radiant-energy input to the surface. Most of the large energy input is therefore available to change the physical state of water.

In the middle of the day the rate of this conversion of energy exceeded 400 W m^{-2}. At night, when radiation loading declined to a level that failed to equal the energy emitted as long-wave radiation from the snow cover (315 W m^{-2}), the day's melting ended. The daily regime of this process is therefore a 13-hr period of quiescence and an 11-hr burst of activity, which peaked soon after noon.

Meltwater This regime of melting, modulated as the water moves from the snow surface to the stream channels, becomes the daily flood wave in Castle Creek, and in other streams that drain the snow-covered basins of the Sierra. In mass terms, the amount of meltwater was 36 kg m^{-2} of snow surface, or 36-mm depth of water substance over the area of the snow field. A rainstorm that delivered this much water in 11 hr would qualify as being at least moderate in intensity.

Figure VII-9 shows as bar graphs the distribution of 33 mm of meltwater generated in this region on a two-day period in April in a different year. The wave of liquid water reaching the ground beneath the snow mantle (and drained off to a measuring tank) is shown by the smooth curve that rises steeply just after noon, peaks at 1500, and recedes through the evening and night hours. Little or no snow evaporates because vapor pressure in the air exceeds that at the cold surface of the snow (6.1 mb). The lag between the two curves indicates the 2–3 hr required for meltwater to travel from the snow surface to the ground.

Where there is little transport of energy, radiative melting is interrupted every time darkness falls. This interrupted, prolonged course of spring melting is welcome to people living near rivers that drain extensive, snow-covered lowland and hill country, as in Wisconsin, New England, or central Russia. Daily melting pulses give daily flood waves of meltwater, separated by periods of receding flow that stream channels can handle over a 24-hr period. The situation is quite different from that in a period of southerly winds bringing an uninterrupted energy supply.

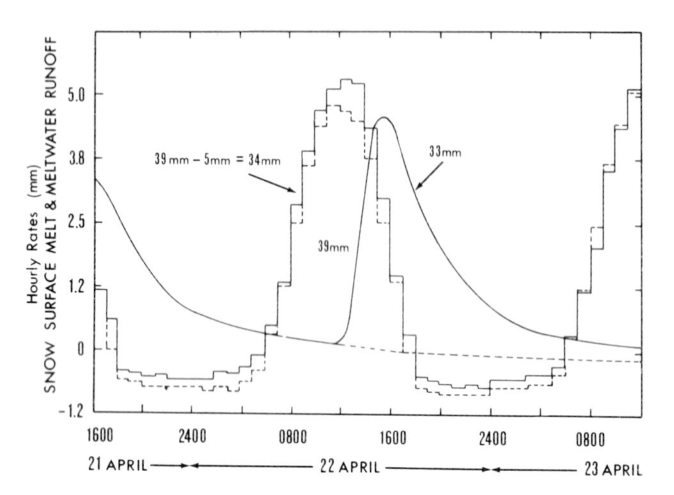

Fig. VII-9. Diurnal march of energy input to snow surface of lysimeter at Lower Meadow, Central Snow Laboratory, 21–23 April 1954 (bar graphs), and of runoff of meltwater into lysimeter tank (smooth curve) (from U. S. Corps Engineers, 1956, Pl. 3-8). The weather was clear with noon rates of solar radiation 1030 W m^{-2}, and nocturnal net loss of energy by exchange of longwave radiation 70 W m^{-2}. Daytime energy supply was equivalent to the generation of 39 mm of meltwater; nocturnal heat deficit equivalent to 5 mm; computed 24-hr melting was 34 mm. Measured volume of meltwater was 33 mm. Parts of bars above broken lines show melting due to turbulent energy fluxes.

The seasonal onset of radiation-melting weather is more regular than the seasonal variation in advective melting, which can in fact occur even in winter. Figure VII-10 portrays the mean regime of factors affecting melting from November, when the Sierra snow cover begins to build up, to June, when it is virtually gone. Each energy-supply mode is expressed as equivalent daily rates of melting (mm or kg m^{-2}). Melting from heat in the ground and in occasional rains is always small, and through winter and spring declines to zero. Melting due to turbulent sensible- and latent-heat flux from air to snow is negligible until late April; in May, the peak of the melting season, it accounts for about 4 mm day^{-1} on the average. Little of this, however, is genuine advection in the lowland-weather sense of the word, but represents a short-path travel of heat from the sun-absorbing foliage of mountain forest ecosystems. The largest spring increase is seen in melting due to the net sum of all the fluxes of radiation, which becomes a positive force about the middle of March and then rapidly rises to a rate of 25 mm day^{-1} at the end of May.

This smooth curve is interrupted from time to time by spring storms. The actual course of meltwater yield in one spring in the Sierra is shown in Fig. VII-11. In these graphs both the observed yield and that calculated by the heat-budget method developed in the Snow Investigations is presented (Rantz, 1964).

Advective Melting

This term is applied to situations of strong warm airflow that melt snow cover more or less continuously. When warm rainstorms sweep over snow-covered drainage basins, rain intensities are augmented by melting supported by a supply of energy that comes not from solar radiation but from the warm humid airstream. Atmospheric energy is being imported, or advected, into the snow-covered region in tremendous quantities, and transferred to the snow cover in several ways: (a) by long-wave radiation emitted by water vapor in the warm air and

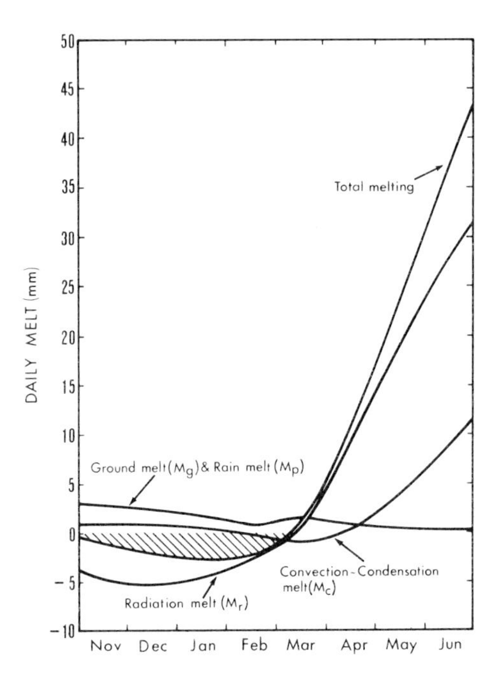

Fig. VII-10. Monthly means of energy fluxes at the snow surface expressed in terms of melting rates (mm day^{-1}) at the Central Sierra Snow Laboratory. Total melting is equal to radiation component M_r, turbulent heat transfer by convection and condensation (sensible and latent heat fluxes M_c), and heat flow from the ground M_g and rain M_p. Shaded area represents deficit of heat from winter, which must be compensated by springtime warming before streamflow can be generated from the entire basin (U. S. Corps Engineers, 1956, Pl. 5-7).

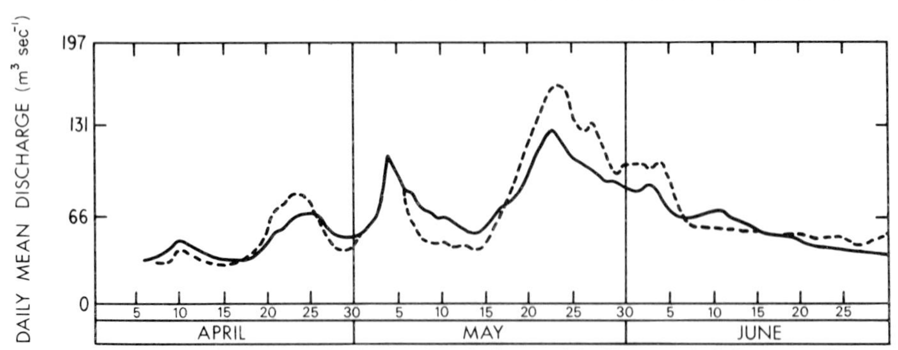

Fig. VII-11. Calculated (dashed curve) and observed (solid curve) daily streamflow in North Yuba River, California during the snow-melting season of 1956 (from Rantz, 1964).

the bases of the precipitating clouds, (b) by the warmth of the rain itself, and (c) by downward turbulent heat fluxes in the lower layers of the airstream, supported by high wind speeds over rough terrain. Mechanical turbulence is strong enough to overcome the thermal stability of warm air overlying snow at 0°C, as found by Sverdrup (1936) in an analysis of turbulent heat fluxes over a smooth snow cover, and even stronger where tree stands roughen the earth's surface.

It was thought that warm rain was a major factor in melting snow, but calculations show that its heat content, 40–60 J g^{-1} of rainwater, can account for only an insignificant part of the observed rate of melting. This quantity of heat would melt only 0.2 g of water from the snow per gram of rainwater.

Another plausible idea is that the balmy warmth of the air, which an observer can feel for himself, is the primary cause of the high rates of snow melting that occur. The question can be put this way: How much of the undoubtedly large turbulent heat flux from airstream to snow-covered drainage basin moves as sensible heat, how much as latent? It turns out that condensation of vapor on the snow melts more of it than direct transfer of sensible heat does.

The fact that neither of these high rates of heat transfer warms the surface means that the gradients of temperature and humidity down which heat flows from air to snow do not diminish as the storm continues, but remain large. The total transfer is a function of wind speed, which produces mechanical turbulence over the rough terrain, and of duration of the inflow of warm, humid air.

In a major rain-on-snow storm, this total might be 500 W m^{-2}, 100–150 of which comes as sensible heat (U.S. Corps Engineers, 1956, pp.

227, 362), the larger fraction as latent heat. At times of strongest wind speed the rates are greater since mechanical turbulence depends on wind speed.

An analysis of warm advection (Treidl, 1970) measured the loss of energy in an airstream during a 9-hr traverse across snow in lower Michigan in January, 1969. The airstream delivered the following heat fluxes to the underlying snow cover:

Long-wave radiation	325 W m^{-2}
Latent-heat flux	60
Sensible-heat flux	80
Total	465 W m^{-2}

Absorption of a weak flux of solar radiation coming through the clouds added about 15 W m^{-2} to the energy budget of the snow, making a total of 480. This amount was available to support long-wave radiation by the snow (315 W m^{-2}) and leave a surplus (165 W m^{-2}) for melting. This figure agreed with the observed melting of 15.2 kg m^{-2}.

For comparison, the design rate of maximum daily melting (at frequency = 0.01) in the Moscow region is 56 kg m^{-2} (Kuz'min, 1960), which is equivalent to a mean energy-conversion rate of 220 W m^{-2}. The design rate in the maximum hour is 7.0 kg m^{-2}, an energy-conversion rate of 650 W m^{-2}.

Condensation of vapor on snow supplied appreciable fractions of the energy used in melting snow and ice on the Kårsa Glacier in northern Sweden (Wallén, 1948/49), as shown in Table III. In favorable

TABLE III

Melting at Surface of the Kårsa Glacier Due to Condensation of Vapor[a]

Month	Flow of energy (W m^{-2})	As fraction of total
May	−1.2 (evap.)	−0.01
June	+3.2	+0.04
July	+16.0	+0.15
August	+13.0	+0.16
September (1–21)	+1.6	+0.05
Average	+6.5[b]	+0.10

[a] Calculated mean for 1930–1939 period. Source: Wallén (1948/49, p. 177).

[b] Equivalent to condensation of 0.23 mm water day^{-1}.

weather,* during late July and early August 1946, the latent-heat flux averaged 21 W m^{-2}. The amount of water condensed onto the snow was 0.7 mm day^{-1}. The mass of snow melted by the heat of condensation was seven times as great.

The roles of radiative and turbulent energy fluxes to a melting snow cover, which at the meadow in the high-radiation climate of the Sierra (Table II) were 95 and 60 W m^{-2}, respectively, are changed in forest ecosystems. As noted earlier, the direct-beam radiation is depleted (Miller, 1959), and long-wave radiation from the branches is large. The turbulent energy fluxes are reduced.

In the radiation climate of the Sierra, however, substantial solar heating of pine canopy supports large fluxes of sensible heat to the air and snow (Miller, 1956), which supplement solar radiation that penetrates the canopy and long-wave radiation emitted by it. As a result, melting of snow in these open ecosystems proceeds at about 0.7 of the rate of melting snow in meadows (which also gain by heat transported from the sunlit canopy of adjacent pine stands) (Miller, 1955, pp. 117 and 165). In advective climates, on the other hand, damping of turbulence in forest systems reduced melting rates. In lowland forest of density 0.4 in the western Soviet Union, melting of snow in bare-crowned deciduous stands is 0.7 of that in the open, and in needleleaf stands only 0.5 (Kuz'min, 1960). In stands of density 0.75 the snow melts at rates 0.6 and 0.3, respectively, of the rate in sites open to advection. Significant shelter is provided by forest ecosystems.

Of the various methods proposed for calculating or predicting the rate of generation of meltwater, the most successful is the one based on the elementary concept of the heat of fusion of water—335 J g^{-1}. When all avenues of heat supply and loss of snow cover are balanced, the amount available for melting is determined. This method was developed by Scandinavian glaciologists in the 1920s and applied to mountain snowfields in the United States in a program beginning in 1945 (U.S. Corps Engineers, 1956). The heat flows can be calculated with sufficient precision from data on other atmospheric conditions and these calculations handled by computer (Leaf and Brink, 1973). As incorporated in a simulation model for streamflow synthesis and reservoir regulation in the basin of the Columbia River (Rockwood, 1972), they are bases of a model that has been developed over 15 years to provide projections for 67 subbasins, 47 channel reaches, 28 reservoirs, and 68 downstream control points on a real-time basis.

* Mean vapor pressure at 1.7-m height was 7.6 mb, mean air temperature +5.7°C, mean wind speed 3.3 m sec^{-1}.

This whole development illustrates the application of fundamental laws of physics to the geophysics, engineering, and water-resource management of snow-mantled landscape mosaics and regions.

Evaporation from Snow Cover

The principles governing evaporation in terms of the energy available for it are well illustrated by snow cover. The snow surface has little available energy in many sites and incurs little evaporation. The wind may howl across Arctic snow for months, but at the end of winter the mantle still contains most of the snow that winter storms deposited, demonstrating the incompetence of wind without an energy supply to make water evaporate.

Three sources might supply energy to a snow mantle under conditions permitting evaporation, and all three are usually small. (1) The snow substrate is a poor conductor and yields little heat to its upper surface. (2) Airstreams above the snow that are dry enough to accept vapor from it usually are too cold to supply much energy to it as sensible heat. (3) Radiation at the snow surface is usually small. When one of these three sources happens to grow to a significant size, it does so in circumstances that for other reasons are often unfavorable to evaporation of the snow. Let us look at them one by one.

Substrate Heat Snow is a substance of very low density and thermal conductivity. Its layers in winter are porous, storing little heat themselves, and conducting little from the ground beneath. They yield little heat to the surface unless the surface temperature drops very low indeed. At such times the surface humidity becomes vanishingly small, 1 mb or less, so that the driving force or gradient along which vapor might move away from the surface is almost nonexistent.

Spring snow is denser and more conductive, but the temperature gradient within it is usually lacking. The underlying ground is cold by this season and can supply little heat. Little or no heat is supplied to the surface by the substrate-heat flux because no vertical gradient exists.

Sensible Heat Competition between the latent heats of fusion and vaporization springs to life if energy is supplied to the snow surface in substantial amounts from radiation or the atmosphere. Evaporation is then in competition with melting. The partition of available energy between them depends on the vapor pressure in the overlying atmosphere.

When atmospheric vapor pressure is very low, evaporation may claim a sizable fraction of the total heat available, but pressure must be less than 6 mb, which is the vapor pressure at a surface of melting snow, if there is to be net upward movement of water molecules from the snow surface.

At times when the air is warm enough to support a flow of sensible heat to the snow, it is likely also to be humid and will therefore supply vapor to the snow. Additional melting results from this condensation, and evaporation is negative, as we saw for intercepted snow in Chapter VI.

The major source of warmth in the snowy maritime mountains of middle latitudes is the ocean. Oceanic airstreams are both warm and humid relative to the 0°C and 6-mb limits of a snow surface. The snow melts instead of evaporating.

Atmospheric warmth and dryness are not usually found together in winter and spring except in situations of mountain-lee foehn or free foehn. Otherwise, evaporation based on the sensible-heat flux from air to snow is a rare event.

Radiation These limitations on sensible-heat flux as a source of energy for evaporation of snow, and the absence of a large substrate-heat flux, make us look to radiation as a possible source of energy. The radiation surplus at a snow surface is, however, seldom large because of the high albedo.

Upward removal of vapor from an irradiated snow surface may be inhibited by the high humidity of the air and weak development of turbulence in a stable atmosphere above a smooth surface. The local atmospheric humidity increases on days of strong radiation in situations where warm, moist surfaces, such as tree leaves, can evaporate water. In the Sierra, midday vapor pressure in spring is far above 6 mb, causing condensation to prevail over evaporation from the snow (Miller, 1950, 1955).

Evaporation from Snow Cover in Different Environments Measurements of evaporation from snow cover in the Sierra Nevada illustrate conditions in maritime mountain ranges. West and Knoerr (1959) measured evaporation from February to June as 22 mm in a small opening in the forest, 8 mm in a forest of 0.7 density. Evaporation still proceeded slowly in this relatively warm environment. Much of the total occurred in one period of easterly winds having the requisite combination of warmth and dryness mentioned earlier. Toward the end of the season, condensation on the snow outweighed evaporation.

In an expansion of this study (West, 1962), 3-yr means for the entire snow-cover period were calculated as 33 mm in the opening and 16 mm in the forest site. Considering conditions of terrain and forest over the entire basin of Castle Creek (2.1-km altitude, 39N latitude), West estimated total evaporation of 50 mm—a minor fraction of the total precipitation (almost 1900 mm yr^{-1}).

In an interior mountain range, exemplified by the Fraser Laboratory in Colorado (2.7-km altitude, 40N latitude), daytime evaporation averaged about 1 mm on selected, usually sunny, days in early spring. This figure is partly offset by condensation of 0.2 mm at night (Bergen and Swanson, 1964). Differences among days followed differences in inputs of solar energy.

In late spring and summer, condensation begins to outweigh evaporation in most places. Invasions of humid airstreams from the Gulf of Mexico (Martinelli, 1960) produce heavy condensation in the Rockies. Condensation is dominant after mid-spring in the Swiss Alps (Quervain, 1951).

Snow cover in the Arctic would be expected to evaporate little in the winter darkness when no solar energy is available at all. Evaporation in spring is small also. Theoretical and laboratory wind tunnel studies (Diunin, 1961) show significant rates of evaporation from snow particles in blizzards but little from snow on the ground. Snow on the ice caps is considered to have virtually zero net evaporation, although it is exposed to the atmosphere and sky during the entire yearly cycle.

The apparent vulnerability of a snow cover to attack by evaporation is an illusion. In spring the available heat produces little evaporation because the air has grown humid. In winter, when the air is dry enough to accept vapor at the low pressure obtaining at the snow surface, the supply of heat to the snow is too small to vaporize it.

Low evaporation from snow, except in rare incursions of warm, dry airstreams, has been confirmed by all careful measurements. The reasons are the infrequency of a strong gradient of vapor pressure from snow to air, and the low-energy nature of the environment of snow in most climates. When one condition is met the other is usually absent. The extensive snow covers of the Northern Hemisphere continents in winter and spring generally lose only small amounts of mass by evaporation, and their exposure has less effect on regional water budgets than might be expected from their area and long life. In this respect, as in others, snow cover presents a classic contrast with bodies of water in liquid state.

Freezing of Water in the Snow Cover

We have been looking at a process in which solar and other forms of energy are transformed into the heat of fusion of water. In the reverse process water goes from the liquid state to the solid state. An example, noted in the table presenting Sierra data for 16–17 April 1946, was the refreezing of meltwater clinging to the grains of the top layers of the snow cover. The total energy released during this night was 540 kJ m^{-2}.

The amount of liquid water in films on snow grains and in the pores between them makes up 0.05–0.10 of the mass of the top layers of a melting snow mantle. Some of it drains out, but as night falls the demand for energy catches some of the water before it escapes. It freezes and releases the usual 335 J g^{-1}.

In a 10-cm layer of snow with density 400 kg m^{-3} and liquid-water content 0.10, there is 4 kg m^{-2} of liquid water. If half of this freezes, it releases 670 kJ m^{-2} as latent heat of fusion. Under the low-energy conditions of night this is an appreciable source of heat and keeps the night temperature from dropping very low.

Freezing forms a crust that reaches 10 cm thickness by morning and has substantial bearing strength to provide easy walking or skiing. Under a stand of trees, in contrast, the snow receives enough heat by long-wave radiation from the foliage that it might not freeze at all. Such soft snow is not expected by an inexperienced skier, who spills when he suddenly runs into it.

The Saturated Layer

Meltwater, rainwater, and dripwater percolating from the surface of the snow cover to the soil surface accumulates there if the ground is frozen or poorly drained, saturating the lowest layers of snow, a medium with porosity 0.5 or larger. Horton (1936), though ascribing more capillary action to water movement in this saturated layer than is now believed to be the case, found that building up this storage caused as much as a day lag in generation of off-site flow.

This storage zone is often of long duration, being protected from evaporation and freezing, and prolongs the input of water to the underlying soil. Lateral flow off the site also occurs in this layer, especially in deep snow lying on steep slopes (Colbeck, 1974). More information is needed about density, ice content, and liquid water content as functions of depth in the snow cover, so that porosity and permeability can be calculated.

In shallow snow the saturated layer may occupy the entire depth. We then see a layer of slush of liquid-water content almost great enough to flow except for being restrained by what remains of the original structure of the snow cover. In cities the mass of this slush layer, which constitutes a unique form of storage of water at the earth's surface, may reach 40–50 kg m^{-2}. It is a large but unstable reservoir of water and represents a transition to the purely liquid detention film.

LIQUID WATER ON THE GROUND

Water is detained on the surface of the earth in liquid form as well as in the form of snow. Its mobility as a liquid is so much greater than it is as snow that the water films have a much briefer life span and are much thinner than snow, but are initiated by a more violent process.

Rain Impact on Soil

The direct impact of heavy, fast-falling raindrops on exposed soil is serious in dry regions and at certain times of year on cultivated land. In dry regions, plants are so far apart that they afford only minor protection to the soil, and this might be further reduced by heavy grazing or fire.

Bare land is exposed during several months of the year where such row crops as corn and cotton are grown. During the time from the disappearance of the snow mantle in spring and the time when the crop plants cover the whole ground surface, often several months, the land is vulnerable. This is a time when the seasonally increasing water-vapor content of the air can supply increasingly large rainstorms. Wisconsin crop reports in 1974 trace some of the problems of corn in July back to wet weather in May; a hard crust on the rain-packed soils also caused emergence difficulties, according to a crop reporter.

In the southeastern United States, where no stable snow cover protects the soil in winter, the period of exposure is long, and comes at a time when long cyclonic storms deliver great volumes of rain over a period of several days, completely saturating the soil.

At raindrop impact, the kinetic energy of the drops knocks soil aggregates to pieces. The release of kinetic energy depends on the raindrop's size (hence also their falling speed) and intensity (number of drops per unit of time). If expressed on a basis of unit area (1 m^2) and unit depth of rain (1 mm or 1 kg m^{-2}), the conversion of energy (in joules) for increasing rates of rainfall is as shown in the accompanying tabulation, modified from Wischmeier and Smith (1958). Thus 5 mm of rain delivered in an hour (at a rate of 5 mm hr^{-1}) releases 100 J of mechanical energy on 1-m^2 surface in 1 hr.

Rainfall rate (mm hr^{-1})	Energy conversion on area of 1 m^2 (J m^{-2} per mm rain)
1	13
2½	16
5	20
10	24
25	27
50	29

The impact of drops of intercepted rainwater falling from pine canopy 8 m above the soil released as much energy as did raindrops falling from the sky, because the dripwater was delivered at as high intensity as the rain and in larger drops (3 mm in diameter instead of 1–2 mm) (Chapman, 1948). As a result, energy release that varied little with rainfall intensity amounted to 25 J m^{-2} per mm of delivered water.

The sealing effect of raindrop impact on bare soil is evident both in a dense layer about 0.1 mm thick, which "contained no visible pores under high magnification" (McIntyre, 1958), and in a layer of reduced permeability immediately beneath. This is a layer about 1½ mm thick in which the larger pores are plugged by in-washing of fine material, much of which was broken up by the impact of drops on soil particles. In this experiment, permeability of the underlying soil was about 35 mm hr^{-1}, that of the washed-in region about 3 mm hr^{-1}, and that of the soil crust formed by impaction of raindrops about 0.3 mm hr^{-1}. This sealing effect in a bare soil surface in summer can outweigh the effect of vegetation and soil organisms in keeping the soil open. An example will be noted when we look at the process of infiltration in Chapter VIII.

The Environment and Growth of Detention Storage

The Surface Film When liquid water is delivered to the surface faster than it can enter the soil, it ponds into a film or layer called "detention storage." This is smaller in depth than a snow mantle and often exists only during the most intense part of a rainstorm. In its short life it is, however, an important regulator of subsequent fluxes of water, since it provides an opportunity for partitioning to take place.

Another storage, called "depression storage," is sometimes distinguished in theory from detention storage. It is that portion of the liquid-water layer that is trapped in microdepressions and cannot flow away. It remains trapped until it eventually infiltrates into the soil, or evaporates. Its approximate size is 1–5 mm on flat pasture and 50 mm or more on plowed land (Musgrave and Holtan, 1964). Depression storage of 40 mm in contour furrows at 1-m vertical intervals on pasture land near Wagga Wagga, N.S.W. held an additional 460 mm of water yr^{-1} in the forage–plant ecosystem (Adamson, 1974). Annual runoff events were reduced from 26 to 6, and sediment outflow from 3630 to 39 g m^{-2}; dry-matter production doubled to 580 g m^{-2}, and the stock-carrying capacity likewise doubled. On a "relatively smooth undulating" surface, a determination of depression storage put it at 1.5 mm (Langford *et al.*, 1973). Similar depths have been found on typical urban surfaces.

Mass of the Film The depth of the film of water on the soil depends on the slope of the land surface and the obstructions to movement of water in a film or rills, as well as on the rate of rainfall and infiltration into the soil. On smooth bare land the detention layer may be less than 1 mm thick, while a layer of water lost amid the tangle of a dense grass sod that impedes it at every turn may accumulate to a thickness of several millimeters. Estimates of this depth are shown in Table IV.

Field measurements of this storage are few. It is generally lost to view in the grass or litter covering the soil surface, although it certainly can be felt as something that soaks our shoes when we walk in the rain. It might be defined operationally in terms of the subsequent water flux, as, for example: "Detention is the time-varying volume of runoff water which must be built up to induce flow over lots, in gutters, and in conduits" (Jens and McPherson, 1964), being evaluated in terms of its effect in delaying surface runoff. A thin

TABLE IV

Detention of Water on the Land Surface in Terms of Resulting Runoff[a]

Detention storage (mm)		Rate of surface runoff (mm hr^{-1})	Ratio of storage to flow (hr)	
Grass[b]	Corn[c]		Grass	Corn
0.7	0.1	0.5	1.4	0.2
1.2	0.2	1.0	1.2	0.2
1.7	0.3	1.5	1.1	0.2
2.0	0.4	2.0	1.0	0.2
2.4	0.4	2.5	1.0	0.16

[a] Data from American Society of Civil Engineers (1949).
[b] Grass: Bluegrass of good density, not recently grazed.
[c] Corn: Rows of corn in direction of the slope.

detention film is experienced as a faster runoff of rain water, a change that takes place when formerly rural land is covered with roofs and street pavements.

Water is continuously infiltrating from this film into the soil body throughout the storm, and rain must fall at a rate higher than the rate of infiltration if the film of detained water is to accumulate to any depth. For this reason knowledge of the intensities of rainfall over short periods is necessary for understanding how this storage of water builds up. This is one of the reasons we emphasized rainfall intensity in Chapter III.

The detention film is one of the storages in the sequence of alternating fluxes and storages that mark the progress of water through ecosystems on the earth's surface. Like the other storages, it fluctuates in response to changes in its inflow and outflow. On a briefer time span it resembles snow-mantle storage, which grows with each snowstorm and wanes with subsequent melting. The mass of water in the detention film increases from a burst of heavy rain and decreases from outflows of water filtering into the soil or moving away from the site. Unlike the snow mantle, it does not persist long after rainfall ends.

Changes During a Rainstorm Table V shows the cumulated totals of rainwater on an experimental plot at Edwardsville, Illinois, in 1942, summed at irregular intervals of time from the beginning of the storm until a hundred minutes had elapsed. The heaviest bursts of rain were delivered between minutes 3 and 6 (when the mean rate was 100 mm hr^{-1}), and between minutes 14 and 29 (when it was 68 mm hr^{-1}).

The response of the earth's surface to these bursts of input was not to accelerate the rate of infiltration (a process we shall discuss in detail later) but to build up water in the detention film. This storage took form in the period between minutes 3 and 6, growing to a depth of 3 mm. The second burst of rain built it up further, until at minute 17 it reached a critical depth of 5 mm, at which lateral movement of water downslope began. The cumulated totals of this off-site flow of excess water are shown in the table after the plus signs in the last column. At the end of the 100 min, 11 mm had moved down the slope and 19 mm had moved into the soil body.

The detention storage responds immediately to a change in input of rainwater. It supports one steadily changing outflow, infiltration, which proceeds at a rate of 27 mm hr^{-1} early in the storm and decreases to about 5 mm hr^{-1} at minute 100. When the film fills the depressions on the surface and gets deep enough to overcome surface

TABLE V

Extimated Cumulated Sums of Water Fluxes at Experimental Plot in Illinois, 21 June 1942[a]

Time (min)	Accumulated precipitation (mm)	Accumulated infiltration (mm)	Detention film (mm)
1	0	0	0
3	1	1	0
6	6	3	3
11	7	5	2
14	9	6	3
17	12	7	5
20	15	8	5 + 2
24	19	9	5 + 4
27	25	10	5 + 10
29	27	11	5 + 11
32	27	12	4 + 11
39	28	14	3 + 11
100	30	19	0 + 11

[a] Source: Holtan *et al.* (1966, data selected from Table 1).

friction, water starts to flow downslope fast enough that the storage grows no larger.

Generation of Runoff The detention storage of water on the surface of the earth has an important role in the partition of rainwater or meltwater between two destinations: the fraction that enters the soil body, and the fraction moving away off-site. This partition between water that stays on the site, remaining in the ecosystem to which the atmosphere delivered it, and water that moves as off-site flow into other ecosystems is fundamental in the spatial patterning of systems on the earth's surface. This off-site component depends on the depth of detention storage.

Table IV indicates how fast water stored in the detention film will run off. The deeper the film, the less frictional drag is exerted by the roughness of the ground and vegetation, and hence the greater the runoff. If there is much friction between water and ground the film has to be fairly thick to start moving, as will be shown quantitatively in a later chapter. A method for developing a relation between runoff rate and depth of detention storage is shown by Musgrave and Holtan (1964, pp. 12–15), with runoff increasing as the $\frac{5}{3}$ power of the depth. A given depth of detention film supplied less outflow during rain, when raindrop impact might have increased the hydraulic resistance to flow, than it did after the rain, when flow was undisturbed (Langford *et al.*, 1973). In heavy rain the film was 3–4 mm deep.

In developing a nonlinear model relating runoff with rainfall Chiu and Huang (1970) derive four expressions for the response of the drainage basin. In one example, the storm-runoff event of 1 January 1948 in Chartiers Creek basin, Pennsylvania, they show that the equivalent of a mean storage over the basin was 1.5 mm. In a larger flood a year later the equivalent storage was about 10 mm. Some of the difference might result from more water being held in creek channels and from a change in roughness of the land surface, but it is likely that much of it represents the thicker detention film that logically follows heavy rainfall and produces rapid runoff.

Modifying Detention Storage

In efforts to alter this important branching process at the land surface, land managers sometimes change the frictional relation between depth of detention storage and runoff. This can be done by roughening the surface by furrows or pits (which also increase the depression storage), by decreasing local slope of the surface, and by adding such obstacles to flow as are presented by dense low vegetation, litter, or crop residue.

One trial at making depressions in the surface succeeded in increasing the storage capacity by 18 mm (Bruce *et al.*, 1968), but the roughness wore down with time; a month later the treated surface held only 5 mm more than the smoother, untreated plot. Nevertheless, the greater storage, holding a deep film long enough to permit most of it to infiltrate into the soil, had an appreciable effect on soil moisture and the yield of corn.

Shaping the land slopes into benches also holds detention storage longer and allows more meltwater or rainwater to infiltrate. This practice increased alfalfa yields from 4 tons ha^{-1} on unbenched land of the northern Great Plains to 9 tons on benched land (Willis and Haas, 1969).

Several kinds of modification of the surface of rangelands of the northern Great Plains hold water on the surface and prolong the opportunity for infiltration, to improve the efficiency with which precipitation is converted by the range plants into biomass. These modifications, which include pitting, scalping, and contour furrowing, are of greatest value on soils having relatively slow rates of infiltration (Wright and Siddoway, 1972), suggesting the importance of the duration of detention storage of water.

A contrasting situation is found where no water infiltrates the soil, as in a wet spot near a creek or at the foot of a steep slope. At such a place detention storage accommodates all the rain, and rapidly builds up to the critical depth at which surface runoff begins. With a steady rate of rainfall it will support an equal rate of outflow as surface runoff.

Detention storage occurs at the interface between a melting snow cover and the soil body, although in this location it is little studied. The lower surface of the snow, which often is several millimeters above the soil, is in part eroded away by water in this interfacial zone. The saturated layer discussed earlier provides a form of detention storage.

OUTFLOWS FROM DETENTION STORAGE

Detention of snow on the surface of the land represents long storage, from which a small fraction evaporates but most melts. The meltwater percolates through the snow mantle and enters a second storage, directly on the surface.

Raindrops arriving at bare soil enter the same kind of storage, directly on the ground. Detention storage gives rise to outflows of water, which can be divided into what goes *down* and what goes *off*.

Movement downward into the soil body is an "on-site" process. Because it is largely pulled into the soil by forces exerted by the attraction between water and soil molecules, it is best discussed as a soil process. This will be done in the next chapter.

The residual water builds up a film deep enough that it begins to move laterally downhill and off the site. This process, too, will be discussed in a later chapter. Both movements of water, downward or laterally, are outflows from water detained on the surface of the land, a storage that provides the opportunity for an important partitioning of water at the earth's surface.

REFERENCES

Adamson, C. M. (1974). Effects of soil conservation treatment on runoff and sediment loss from a catchment in southwestern New South Wales, Australia, *Int. Assoc. Hydrol. Sci. Publ.* **113,** 3–14.

American Society of Civil Engineers (1949). "Hydrology Handbook." Am Soc. Civil Eng. Manual of Eng. Practice No. 28, New York, 184 pp.

Aulitzky, H. (1963). Bioklima und Hochlagenaufforstung in der subalpinen Stufe der Innenalpen, *Schweiz. Z. Forstwesen* **N.12,** 1–25.

Aulitzky, H. (1965). Waldbau auf bioklimatischer Grundlage in der subalpinen Stufe der Innenalpen. *Cbl. Ges. Forstwesen* **82,** 217–245.

Bergen, J. D. (1975). A possible relation of albedo to the density and grain size of natural snow cover. *Water Resources Res.* **11,** 745–746.

Bergen, J. D., and Swanson, R. H. (1964). Evaporation from a winter snow cover in the Rocky Mountains forest zone. *Proc. Western Snow Conf.* 52–58.

Bruce, R. B., Myrhe, D. L., and Sanford, J. O. (1968). Water capture in soil surface microdepressions for crop use, *Int. Congr. Soil Sci., 9th, Trans.* **1,** 325–330.

Chapman, G. (1948). Size of raindrops and their striking force at the soil surface in a red pine plantation, *Trans. Am. Geophys. Un.* **29,** 664–670.

Chiu, C.-L., and Huang, J. T. (1970). Nonlinear time varying model of rainfall–runoff relation, *Water Resources Res.* **6,** 1277–1296.

Church, J. E. (1933). Snow surveying: its principles and possibilities, *Geog. Rev.* **23,** 529–563.

Colbeck, S. C. (1974). On predicting water runoff from a snow cover. *In* "Interdisciplinary Symposium on Advanced Concepts and Techniques in the Study of Snow and Ice Resources" [Proc.], pp. 55–66. Nat. Acad. Sci., Washington.

Davis, R. T. (1973). Operational snow sensors, *Proc. Eastern Snow Conf., 30th,* pp. 57–70.

Diunin, A. K. (1961). "Isparenie Snega." Izdat. Akad. Nauk Sibir. Otdel, Novosibirsk, 119 pp.

Ffolliott, P. F., and Thorud, D. B. (1972). Use of forest attributes in snowpack inventory—prediction relationships for Arizona ponderosa pine, *J. Soil Water Conserv.* **27,** 109–111.

Formozov, A. N. (1961). O znachenii struktury sneznogo pokfora v ekologii i geografii mlekopitaiushchikh i ptits. *In* "Rol' Snezhnogo Pokrova v Prirodnykh Protsessakh; k 60-letiiu so Dnia Rozhdeniia G. D. Rikhtera," (M. I. Ivernova, ed.), pp. 166–209, Izdat. Akad. Nauk SSSR, Moscow.

Gary, H. L. (1974). Snow accumulation and snowmelt as influenced by a small clearing in a lodgepole pine forest, *Water Resources Res.* **10,** 348–353.

Haupt, H. F. (1972). The Release of Water from Forest Snowpacks during Winter, U. S. Forest Serv. Res. Paper INT-114. 17 pp.

Holtan, H. N., England, C. B., and Shanholtz, V. D. (1966). Concepts in hydrologic soil grouping, *Trans. Am. Soc. Agric. Eng.* **10,** 407–410.

Hornbeck, J. W., and Likens, G. E. (1974). The ecosystem concept for determining the importance of chemical composition of snow. *In* "Interdisciplinary Symposium on Advanced Concepts and Techniques in the Study of Snow and Ice Resources" [Proc.], pp. 139–151, Nat. Acad. Sci., Washington.

Horton, R. E. (1936). Phenomena of the contact zone between the ground surface and a layer of melting snow, *Trans. Internat. Comm. Snow and Glaciers;* reprinted *Internat. Assoc. Hydrol., Bull.* **23** (1938), 545–561.

Iashina, A. V. (1960). Rol' snega v formirovanii rastitel'nogo pokrova. *In* "Geografiia Snezhnogo Pokrova" (G. D. Rikhter, ed.), Izdat. Akad. Nauk SSSR, Moscow, pp. 90–105.

Jens, S. W., and McPherson, M. B. (1964). Hydrology of urban areas. *In* "Handbook of Applied Hydrology" (V. T. Chow, ed.), Chapter 20. McGraw-Hill, New York, 45 pp.

Kotliakov, V. M. (1968). "Snezhnyi Pokrov Zemli i Ledniki," Gidrometeoizdat., Leningrad, 478 pp.

Kuz'min, P. P. (1960). Itogi issledovanii teplovogo i vodnogo balansov snegotaianiia i zadachi dal'neishikh rabot v etoi oblasti. *In* "Geografiia Snezhnogo Pokrova," (G. D. Rikhter, ed.), Izdat. Akad. Nauk SSSR, Moscow, pp. 138–150.

Langford, K. J., Mayer, R. J., and Turner, A. K. (1973). Studies of infiltration and overland flow for natural surfaces. *In* "Results of Research on Representative and Experimental Basins," Vol. 1, pp. 645–652, Internat. Assoc. Hydrol. Sci.–UNESCO, Paris.

Leaf, C. F., and Brink, G. E. (1973). Computer Simulation of Snowmelt within a Colorado Subalpine Watershed. U. S. Forest Serv. Res. Paper RM-99, 22 pp.

Leaf, C. F., and Haeffner, A. D. (1971). A model for updating streamflow forecasts based on areal snow cover and a precipitation index, *Proc. Western Snow Conf. 39th*, pp. 9–16.

Martinelli, M., Jr. (1960). Moisture exchange between the atmosphere and alpine snow surfaces under summer conditions (preliminary results), *J. Meteorol.* **17,** 227–231.

Martinelli, M. (1967). Possibilities of snowpack management in alpine areas. *In* "Forest Hydrology" (W. E. Sopper and H. W. Lull, eds.), pp. 225–231. Pergamon, Oxford.

McIntyre, D. S. (1958). Permeability measurements of soil crusts formed by raindrop impact, *Soil Sci.* **85,** 185–189.

McKay, G. A., and Findlay, B. F. (1971). Variation of snow resources with climate and vegetation in Canada, *Proc. Western Snow Conf., 39th,* pp. 17–26.

Miller, D. H. (1950). Insolation and snow melt in the Sierra Nevada, *Bull. Am. Meteorol. Soc.* **31,** 295–299.

Miller, D. H. (1953). Snow-cover depletion and runoff, U. S. Corps Eng., Snow Invest., Res. Note 16, 64 pp. [Much of this material was reprinted in Chapter 7 of Snow Hydrology. U. S. Corps of Engineers, Portland, Oregon, 1956.]

Miller, D. H. (1955). "Snow Cover and Climate in the Sierra Nevada, California." Univ. Calif. Press Publ. Geog. **11,** Berkeley, California, 218 pp.

Miller, D. H. (1956). The influence of open pine forest on daytime temperature in the Sierra Nevada, *Geog. Rev.* **46,** 209–218.

Miller, D. H. (1959). Transmission of solar radiation through pine forest canopy, as it affects the melting of snow, *Mitteil. Schweiz. Anst. forstl. Versuchswesen* **35** (Festschrift Hans Burger), 57–79.

Muller, R. A. (1966). Snowbelts of the Great Lakes, *Weatherwise* **19,** 248–255, map.

Musgrave, G. W., and Holtan, H. N. (1964). Infiltration. *In* "Handbook of Applied Hydrology" (V. T. Chow, ed.), Chapter 12. McGraw-Hill, New York, 30 pp.

Outcalt, S. I., Goodwin, C., Weller, G., and Brown, J. (1975). Computer simulation of the snowmelt and soil thermal regime at Barrow, Alaska, *Water Resources Res.* **11,** 709–715.

Phillips, D. W., and McCulloch, J. A. W. (1972). The Climate of the Great Lakes Basin. Atmos. Environ. Climat. Stud. No. 20, Toronto, 1972, 42 pp, 57 pl.

Quervain, M. de (1951). Zur Verdunstung der Schneedecke, *Arch. Meteorol. Geophys. Bioklimatol. B* **3,** 47–64.

Rango, A. (ed.) (1975). "Operational Applications of Satellite Snowcover Observations," Nat. Aeron. Space Agency, Washington, 430 pp.

Rantz, S. E. (1964). Snowmelt Hydrology of a Sierra Nevada Stream, U. S. Geolog. Surv. Water-Supply Paper 1779-R, 36 pp.

Richards, H. B. (1959). Blue Lake snow vs. Carson River, *Proc. Western Snow Conf. 27th,* pp. 21–26.

Riley, J. P., Israelsen, E. K., and Eggleston, K. O. (1973). Some approaches to snowmelt prediction, *Internat. Assoc. Hydrol. Sci., Publ.* **107,** 956–971.

Rockwood, D. M. (1972). New techniques in forecasting runoff from snow, *Int. Assoc. Hydrol. Sci. Publ.* **107,** 1058–1061.

Rosendal, H. E. (1970). The unusual general circulation pattern of early 1843, *Mon. Weather Rev.* **98,** 266–270.

Saulmon, R. W. (1973). Snowdrift management can increase water-harvesting yields, *J. Soil Water Conserv.* **28,** 118–126.

Sommerfeld, R. A., and LaChapelle, E. (1970). The classification of snow metamorphism, *J. Glaciol.* **9,** No. 55, 3–17.

Steppuhn, H., and Dyck, G. E. (1974). Estimating true basin snowcover. *In* "Interdisciplinary Symposium on Advanced Concepts and Techniques in the Study of Snow and Ice Resources," [Proc.], pp. 314–328. Nat. Acad. Sci., Washington.

Sverdrup, H. U. (1936). The eddy conductivity of the air over a smooth snow field—Results of the Norwegian–Swedish Spitzbergen Expedition in 1934, *Geofysiske Publikasjoner* (Oslo) **11,** No. 7, 69 pp.

Treidl, R. A. (1970). A case study of warm air advection over a melting snow surface, *Bdy.-Layer Meteorol.* **1,** 155–168.

U. S. Corps of Engineers (1956). "Snow Hydrology," Portland, Oregon. (Reprinted by U. S. Govt. Printing Office, 1958).

U. S. Corps of Engineers and Weather Bureau. Cooperative Snow Investigations (1947). Hydrometeorological Log of the Central Sierra Snow Laboratory 1945–46. Corps of Engineers Tech. Rep. 5. San Francisco, California, 62 pp.

U. S. Corps of Engineers and Weather Bureau, Cooperative Snow Investigations (1952). Hydrometeorological Log of the Central Sierra Snow Laboratory 1947–1948 Water Year. Corps of Engineers, San Francisco, California, 220 pp. + app. (The figure is from p. 13.)

U. S. Weather Bureau (1964). Frequency of Maximum Water Equivalent of March Snow Cover in North Central United States. U. S. Weather Bureau Tech. Paper 50, Washington, D.C., 24 pp.

Wallén, C. C. (1948/49). Glacial–meteorological investigations on the Kårsa Glacier in Swedish Lappland 1942–1948, *Geog. Ann.* **30,** 451–672.

West, A. J. (1962). Snow evaporation from a forested watershed in the central Sierra Nevada, *J. Forestry* **60,** 481–484.

West, A. J., and Knoerr, K. R. (1959). Water losses in the Sierra Nevada, *J. Am. Water Works Assoc.* **51,** 481–488.

Whiteley, H. R., Dickinson, W. T., and Core, T. (1973). The usefulness of standard hydrometeorological data for snowmelt calculations, *Proc. Eastern Snow Conf., 30th,* pp. 1–14.

Willis, W. O., and Haas, H. J. (1969). Water conservation overwinter in the Northern Plains, *J. Soil Water Conserv.* **24,** 184–186.

Wischmeier, W. H., and Smith, D. D. (1958). Rainfall energy and its relationship to soil loss, *Trans. Am. Geog. Un.* **39,** 285–291.

Wright, J. R., and Siddoway, F. H. (1972). Improving precipitation-use efficiency on rangeland by surface modification, *J. Soil Water Conserv.* **27,** 170–174.

Chapter VIII

INFILTRATION OF WATER INTO THE SOIL OF AN ECOSYSTEM

Rainwater or meltwater stored in the detention film on the surface supplies two major outflows. One of these is horizontal runoff over the surface or in the porous upper few centimeters of the soil body, and is the residue after the partition of water at the surface. The other, and prior, claimant is the downward pulling of water into the soil body. If this pull is slow, the detention film deepens and supports off-site runoff; if it is rapid, water enters the soil as fast as it arrives as rain or meltwater, forestalling both detention storage and off-site runoff.

Water in the soil can be considered in two modes: (1) as it is pulled into small pores, where it remains until extracted by plant roots or, occasionally, evaporation; or (2) as it enters big pores and keeps going downward, draining rather promptly out of the soil itself into lower layers of rock material. This is the topic of a later chapter. Since soil pores come in all sizes, there is also a gray area including pores from which gravitational drainage is slow but not negligible.

Water in the small pores of the soil body can be analyzed rather simply by input–output methods. The amount of moisture in soil storage at any given date is the result of earlier intake of water from the surface, so-called "infiltration," filling the small pores; subsequent withdrawal by plant roots reduces the amount of water in storage. This steady output will be the topic of Chapters XI and XII; here we will consider the episodic inputs of water into the below-ground part of an ecosystem and the system's capacity to take water in.

This capacity, an important variable in the coupling of an ecosystem with its hydroclimatic environment, was introduced by the great Robert Horton in 1933 as the "maximum rate at which a given soil can absorb rainfall when the soil is in a specified condition." The new concept filled a demand for improved approaches to the problem of estimating flood flows, because the old assumption that off-site flow always took the same share of storm rainfall, whatever the kind of storm, had clearly proven inadequate. The new infiltration theory, which defined off-site flow as a residual left after the soil had taken in as much as it could from the water delivered to it, provided a sound basis for analyzing runoff. It met great success in methods for flood forecasting and estimating runoff in the design of dams and other structures for river control, which were then being planned in large numbers in response to an increased Federal responsibility for reducing flood damage. This period in hydrology was correctly called "the era of infiltration" in Cook's (1946) comprehensive stock-taking paper. The phrase suggests how hydrologists turned their attention to the land and to the soil and other characteristics of ecosystems.

THE SOIL AS ENVIRONMENT OF WATER

In studying the role of the soil in the water budget, we see it as a layer of weathered rock materials intermingled with the living and dead products of life processes in and above the soil, and enclosing interstices and pores of many sizes. These pores hold air enriched in CO_2 and water as vapor and liquid, sometimes as ice also. Soil air, plant roots, pores, and the soil-particle surfaces constitute the environment of soil water.

A pertinent characteristic of the soil body as environment of water depends on how finely the rock particles are divided. Particles present a large area of active surface, which may total many square meters within a gram of soil, where complex molecular and chemical phenomena take place, and on which molecules of liquid water adhere. In fine-textured soils this surface area is multiplied even more, with still greater opportunity for physical and chemical activity and for the holding of water.

Another important characteristic is the presence of roots in the soil. They not only accelerate weathering processes, but also retrieve water from pores that are small enough to retain water against gravitational forces. The upward movement of water in root systems contrasts with gravitational forces, which, in most other environments we have examined, do the work of moving water from place to place. These and other actions of roots distinguish soil from simple weathered material as an environment for water.

"The soil cover," in L'vovich's (1973, p. 38) words, "is a kind of arena in which many processes . . . interact: soil moisture, formed in the process of infiltration, goes into evaporation and transpiration and [recharges] groundwater." This environment is critical to the functioning of all these processes.

In connection with the movement of water, we think of the environment as composed of big pores, large enough to be dominated by gravitational forces, and small pores and surfaces in which water is controlled more by attractive than gravitational forces. Marshall (1959) separates big from little pores at a radius of 15 μm.

The relation of soil environment to water is shown in a quotation from the scientist in charge of a large network of measuring stations (1500 or more) in the Soviet Union. She says that

> soil is a wetting body. For this reason capillary moisture in the soil has a concave surface and is invariably under supplementary negative pressure [or suction]. Its magnitude is governed by the surface tension of the water and the radii of the curves, which depend on the size and shape of the interstices, i.e., in the final analysis on the dispersion and structure of the soil (Razumova, 1965, p. 492).

Another characteristic of a soil body as an environment for water is its irregular stratification. Typical soils are composed of layers that have quite different sizes and numbers of pores, and particles with differing amounts of active surface area. They represent contrasting environments of water, each layer accepting and holding it differently and harboring a different population of roots. Deeper layers are penetrated by fewer roots and water stored in them is withdrawn less rapidly than that from upper layers, a factor which is important in casting a water budget for the whole soil body (Mather, 1974, p. 135).

To the extent that many soil layers were formed by past movement of soil water, the water–soil relationship is a mutual one. This is true in a chemical as well as in a physical sense. Infiltrating water carries the "soluble products of decaying organic matter . . . into the soil, where they tend to coat the particles, thus improving the structure" (Pillsbury, 1968, p. 45). The aerial parts of the ecosystem thereby influence water dynamics in the below-ground parts.

INFILTRATION OF WATER INTO THE SOIL

When liquid water from rain or melting snow arrives at the soil surface it is pulled into the pores of the soil in a process somewhat misleadingly called "infiltration." Like the interception of rain and snow by leaf canopies above the surface, infiltration of water through the surface and into the soil body is to a large degree dependent on the environment of the water, that is, of the texture and structure of the soil body.

The Infiltration Process

Water is not drawn into a soil body principally by gravity, as the term "infiltration" might suggest, but by forces that reflect the soil environment and the kinds of energy available in it. This environment, the interstices between soil particles, i.e., the large and small pores, is comprised of the surfaces of the soil particles, to which water molecules are attracted. The sizes of pores are all-important because they determine the relations of the surface areas to the mass of water.

Capillary Potential These parameters of the soil body as environment affect the force that acts most strongly to move or hold water, the capillary potential ψ. This is an expression of the surface tension in the soil–air–water* system. The gravitational force becomes important only when the mass of water is large relative to the area of the rock–water interface, i.e., in big pores, or cracks in the soil body. Other forces in the soil–water system are of minor importance to infiltration.

The flow of water downward from the soil surface is expressed by the same gradient relation that characterizes other flows in nature:

$$\text{flow} = K\ \partial\psi/\partial x,$$

in which x is the distance. The proportionality coefficient K is called "capillary conductivity" (American Society Agricultural Engineers, 1967) and is again a function of soil moisture and soil structure and porosity.

* The curious phenomenon of water repellency in some soils apparently results from the presence of substances that affect the surface tension of water and hence the contact angle at the soil–air–water junctions in pores. Some of these substances are organic, being products of certain soil fungi (Bond, 1964). Others are leachates from litter of chaparral, which, for example, produce a hydrophobic reaction in as much as 60% of the soils of the San Dimas experimental basin in southern California. Often the impermeable layer is a few centimeters deep in the soil (DeBano and Rice, 1973), with the result that the "hydrologically active" layer is so shallow that it can store but little water. "Even a modest-size storm can saturate this thin layer and cause overland flow and severe erosion" (DeBano and Rice, 1973).

Suppose that water is made available at the surface of a fairly dry soil body in which, below the shallow-wetted top layer, the capillary potential ψ is large, and capillary conductivity K small because the soil is dry. Between the wetted layer and the main soil body, however, there is a large difference in capillary potential, hence a steep gradient $\partial\psi/\partial x$. This gradient, combined with a large capillary conductivity K in the moist top layer, pulls water rapidly into the soil body. Such initial rates of infiltration are large in most soils, being of the order of 10–100 mm hr^{-1}.

This high initial rate of infiltration decreases with time during a rain. Formerly this decrease was thought to result from sealing of entry pores by the washing-in of fine particles or the shrinkage of their walls. While such effects do occur, particularly in bare soils exposed to rain impact and in certain clays, the decrease in infiltration is experienced also in nonshrinking soils covered by vegetation that protects them from raindrop impact forces. In experiments that excluded any possible effects of impact and in-washing, Bodman and Colman (1944) showed that the wetting process in the soil body involves a decrease in the gradient of capillary potential $\partial\psi/\partial x$. This decrease occurs as the zone through which water was being transmitted downward grows thicker, i.e., as x in the gradient term increases. This zone is shown as having approximately 35% moisture content in Fig. VIII-1. At 173 min after the surface entry of water it is about 8 cm thick; at 615 min it is 20 cm thick.

This moisture-related decrease in infiltration rate occurs in addition to the effects Horton (1939) identified—colloidal swelling in clay soils, raindrop impaction, and in-washing—and also to the mechanical effect when bare soil settles in rains of high intensity. For example, Pillsbury (1968, p. 46) recommends that sprinkler irrigation not exceed a rate of 5 mm hr^{-1}. A "low application rate can be most important in preserving high water-entry rates into the soil and, possibly, in preventing crusting that would be detrimental to seeding emergence." This limiting rate is less than many of the rain intensities discussed in Chapter III; even at Seattle it is no more than the intensity over a 4-hr duration at 0.5 frequency of occurrence. Periods in cropping regimes when the soil is bare are, in most climates, periods of system vulnerability to loss of soil and of infiltration capacity.

Decrease in infiltration rate as a storm continues, whether from mechanical or soil-moisture reasons, means that the accumulated sum of water infiltrated from the beginning of the storm is a nonlinear function of time. It can be approximated as $S(t)^{1/2}$ where t is time and S is a function of soil-moisture shortage (Dunin and Costin, 1973). Even

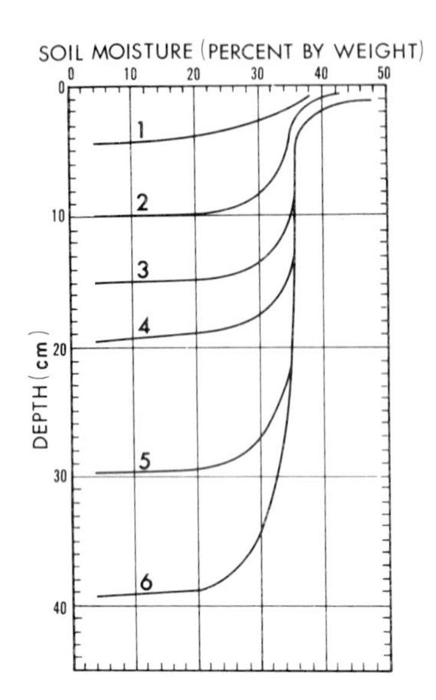

Fig. VIII-1. Distributions of moisture in Yolo silt loam during an infiltration episode at the following times: 1—after 12 min; 2—after 76 min; 3—after 173 min; 4—after 280 min; 5—after 615 min; 6—after 1020 min. The moisture content of air–dry soil is 4%; pore-space saturation occurs at 42%. The descent of the wetting front is indicated by the horizontal parts of the moisture curves, e.g., 4 cm at 12 min, 10 cm at 76 min, 19 cm at 280 min, and 39 cm at 1020 min (from Bodman and Colman, 1944). Reproduced from Soil Science Society of America, Vol. 8, p. 118, 1943, by permission of the Soil Society of America, Inc.

when rain begins at moderate rather than high intensities and is totally infiltrated so that no detention layer forms, the infiltration rate in most ecosystems declines to a lower rate that persists into later hours of the storm. Such a sequence calls for a two-phase model (Mein and Larson, 1973).

Zones in the Soil Body This transmission zone occupies most of the wetted depth of a soil body that is taking in water. It exhibits relatively uniform conditions of moisture with time, or with depth. Separated by a transition layer, it underlies a saturated zone in the top 1–2 cm of the soil body, which is in contact with the film of detention storage on top of the soil, described in Chapter VII.

Below the transmission zone and beyond another transition zone lies the wetting front. This is the advancing edge of the space in which the small pores are filled with water (see Plate 14). As this wetting front advances deeper into the soil body, the transmission zone through which its supply of water from above must come gets ever longer. As a result, the gradient of capillary potential $\partial\psi/\partial x$ in the transmission zone decreases. The rate of movement of water slows down and the entry of water into the soil body decreases until it reaches a constant rate that is much smaller than the initial rate of infiltration. In short rains, therefore, a cornfield infiltrates 10–20 mm hr^{-1}, while in an artificial 5-hr rain the rate averaged only 7 mm hr^{-1} (Musgrave and Holtan, 1964).

This gradient relation in the zone of transmission in a soil body explains why a moist soil takes in water more slowly than a dry one. For example, whole-storm rates of infiltration on Ralston Creek basin in Iowa have a median in spring, 12 mm hr^{-1}, that is only half of the median rate in summer (Johnson and Howe, 1956). Unless the soil has a chance to dry out, as is more likely in summer, the rates of infiltration will be small. The infiltration process is inherently intermittent rather than continuous.

The gradient relation also explains why the infiltration rate is only slightly affected by the depth to which water is ponded on the soil surface. Variation in depth of ponded water "from 1 to 5 cm. produces only a 4 per cent change" in the total gravitational + capillary potential (Philip, 1954).

Plate 14. The wetting front in granitic soil of the New England Plateau in Australia, following a winter rain of approximately 30 mm. The transmission zone above is 30–40 cm thick.

Layers in the soil body that differ in structure or texture naturally have different values of capillary potential and capillary conductivity, even at the same moisture content. Water tends to move faster or slower through them than in adjacent layers. This stratification of the environment of water in a soil body, which is a normal feature in most soils, complicates the movement of water. The rate of infiltration at the surface may change suddenly when one of these deeper layers begins to act as a damper. If the restricting layer is at shallow depth, infiltration may drop to small values even early in a storm.

Determining the Rate of Infiltration Infiltration is sometimes estimated by measuring how long it takes for a given quantity of water ponded or sprinkled on an enclosed soil surface to disappear. This method gives relative values for different soil types, but for various reasons, including boundary effects, it overestimates the actual rates of infiltration that occur at surfaces over a large area. For the same reason, laboratory measurements of this process in different soil columns are difficult to transfer to the field.

Infiltration is more rapid into a sandy soil than a clay, but beyond this statement, soil texture data do not serve as good predictors of infiltration capacity (Krimgold and Beenhouwer, 1954). Both capillary potential and capillary conductivity at all levels in the soil profile are important, but for many soils these parameters are not known, even in relation to other soil properties.

Infiltration that occurred during a storm may be deduced after the fact by analyzing runoff during a period in which the time distribution of rainfall is accurately known. The amount of water infiltrated into the soil to any given time is determined by adding the amount of rain accumulated to that time and then subtracting runoff accumulated to that time and also the depth of the detention–storage layer on the surface. This procedure yields the general time distribution of infiltration, with its typical decrease with time. Data on the total volume of infiltration during a storm give a general idea of the effects of different soil–vegetation systems in a drainage basin on this process. This reversed analysis of water dynamics during a storm, however, is not precise because we lack data about water detained at the ground surface. Rates of outflow from this storage into the soil thus cannot be determined accurately.

Uncertainties "Infiltration capacity is thus a hydrological parameter whose importance equals that of precipitation, evapotranspiration, or runoff" (Ward, 1967, p. 181), about which present data are scanty. Field measurements are not easy to interpret in terms of drainage-basin conditions, although they do show the kind of changes in infiltration capacity that are produced by cultivation and other radical alterations of the soil–vegetation ecosystem. The problems of determining infiltration are well summed in this quotation:

> There have been thousands and thousands of empirical measurements of infiltration on many, many different kinds of soils. We know that rate of infiltration may change markedly during the course of a given storm; that it may change greatly over the course of a growing season; that it is affected by vegetative cover, surface mulches, surface conditions, texture, nature of base saturation, and antecedent moisture; and that cracks and wormholes may have a dominant effect. . . . It still is not possible, however, to predict infiltration rate from a knowledge of the physical properties of the soil even though notable progress is being made. . . . Difficulty also has arisen in making valid projections from measurements made on one square foot or 20 square feet of soil surface to infiltration performance on a 40-acre field or a 10-square-mile watershed (*Soil Science Society of America, 1966*).

INFLUENCES OF VEGETATION ON INFILTRATION

The Forest Floor

The capacity of the soil to take in water varies with the activity of vegetation, both in the root zone and above the soil surface, which directly affects water movement and indirectly conditions the soil body to pull in water. The statement has been made that the greatest hydrologic benefit of a forest is that it produces a forest floor. Such a statement suggests the importance of vegetation in developing soil with high organic content, good aggregation, and a large population of microscopic and larger organisms that keep its structure open and its infiltration capacity high.

Table I shows high rates of infiltration capacity at a forest floor, even after it has lost some surface layers of litter and humus by mechanical removal or burning. Such large values of infiltration capacity suggest why many forested slopes in experimental basins, which have been closely observed for many years, have never been seen to produce surface runoff. Never have rates of rainfall exceeded the high rates of infiltration capacity into their upper soil layers.

Values of infiltration capacity change during ecological plant succession, as is shown in Table I at four stages along a sequence common in

TABLE I

Infiltration Capacity[a]

Ecosystem	Capacity (mm hr^{-1})
Undisturbed forest floor	60
Forest floor without litter and humus layers	49
Forest floor burned annually	40
Pasture, unimproved	24
Succession vegetation	
Old pasture	43
Pine forest, 30 yr old	75
Pine forest, 60 yr old	63
Oak–hickory forest	76

[a] Source: Lull (1964, pp. 6-14, 6-15).

the southeastern United States. Abandoned pasture is colonized by pines, which, in turn, give way to shade-tolerant hardwoods. During the 100 years of this particular succession, the infiltration capacity gradually increased.

Managed Ecosystems

Amounts of infiltrated water during the first hour of rain on several crop systems planted on a group of similar soils are shown in Fig. VIII-2. The declining rate of intake of water is visible in the curvature of the lines.

Values of infiltration rate during 5 hr of rain on shallow and deep soils occupied by corn and bluegrass in the Middle West of the United States are given in Table II. The contrast between these values and those of uncultivated land as shown in Table I are quite marked and typify a major problem that land managers have to contend with. Farmers and foresters try to minimize methods of cultivation that result in tight soil layers or expose soil to rain impact, and foster plant growth that has beneficial effects. A healthy, dense grass cover takes in twice as much water as a poor one.

Stubble mulching, practiced in parts of the Midwest receiving high rain intensities, has a good effect on the surface by preventing damage and the washing of fine particles into pores. The beneficial effect of this surface treatment is demonstrated by rates of infiltration (mm

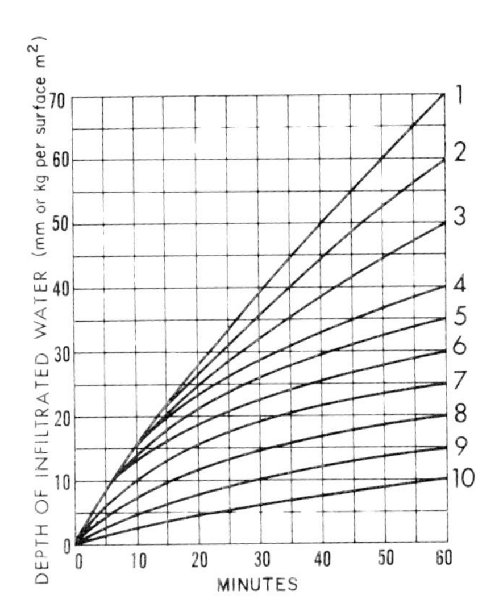

Fig. VIII-2. Curves of accumulated water infiltrated into Piedmont soils under different types of vegetation: 1—old permanent pasture; 2—permanent pasture 4–8 yr old; 3—lightly grazed permanent pasture 3–4 yr old; 4—moderately grazed permanent pasture; 5—hay; 6—heavily grazed permanent pasture; 7—strip-cropped land; 8—grain; 9—clean-tilled land; 10—crusted bare soil (from Holtan and Kirkpatrick, 1950).

hr^{-1}) in the first hour of rain shown in the accompanying tabulation (McCalla and Army, 1961). The protected field maintained a higher rate of infiltration than the unmulched field. The changes in rates of infiltration into the unmulched cornfield indicate the value of the corn plants themselves in promoting good soil structure in summer, as

TABLE II

Rates of Infiltration into Silt–Loam Soil under Pasture and Corn, under Simulated Rainfall at a Rate of 45 mm hr^{-1} [a]

	Mean rate over 5 hr (mm hr^{-1})		Mean rate in last hour (mm hr^{-1})	
Soil series	Bluegrass pasture	Corn	Bluegrass pasture	Corn
Muscatine (deep, high in organic matter)	27	7	15	3
Berwick (intermediate)	18	6	8	3
Viola (shallow, low in organic matter)	4	7	4	2

[a] Source: Musgrave and Holtan (1964).

compared with the bare field of April and the sparsely covered field of October.

Surface	Infiltration (mm hr^{-1})		
	April	June	October
Unmulched cornfield	10	22	6
Mulched with wheat–straw residue	41	42	39

The vegetative mulch that remains on top of and within the upper soil layers in zero-tillage land management produces much of its beneficial effects via its improvement of infiltration capacity of the soil–vegetation system. Not only does this material protect the surface from raindrop impact and reduce the slaking and sealing of the surface, but it also provides structural stability (Baeumer and Bakermans, 1973, p. 90). Infiltration meets less resistance and the moisture status of the soil is improved.

Total Infiltrated Water

The effects of different kinds of land use—a term including soil-cultivation practices as well as the kind of plant cover—are shown in the disposition of the total amount of water delivered at the soil surface during the 3–4 weeks of snow melting in the forest steppe zone of Russia (Plate 15). Natural systems include oak forest and virgin prairie grass, the prairie having been long ago set aside from agricultural use because it was pasture for Cossack cavalry.

The numbers in Table III show total amounts during the spring, not rates, to suggest the general ranking of different soil–vegetation systems over an important part of the year. The supplies of water differ with vegetation type. Differential catch of snowfall and trapping of snow drift during the winter results in different storages of water in the snow mantle at the beginning of spring.

The disposition of the available water, however, differs more among ecosystems than does the supply. Half the supply is infiltrated into mowed or grazed steppe, but only a quarter into the wheat field, which at this season is bare and low in infiltration capacity. As a result, less water goes to building up the store to be drawn on by the summer crop.

Fall plowing enhances infiltration, which is one reason for the adoption, since the 1930s, of this now common practice, which makes

Plate 15. Virgin steppe vegetation in the Alekhin Nature Reserve of south central European Russia in August 1969.

TABLE III

Water Fluxes at Natural and Cultivated Land Surfaces at the Alekhin Central Chernozem Reservation, Kursk Oblast, USSR. Mean of Spring Conditions, 1962–1965[a]

Land type	Supply[b] (mm)	Increase of soil moisture (mm)	Fraction of supply that is infiltrated
Oak forest	153	153	1.00
Virgin prairie, unmowed	132	111	0.84
Virgin prairie, mowed	110	57	0.52
Virgin prairie, pastured	109	58	0.53
Bare, plowed in fall	94	69	0.73
Winter wheat	110	30	0.27
Stubble	120	42	0.35

[a] Source: Grin (1967).

[b] Supply is the liquid equivalent of snow cover and ice crusts at the beginning of the melting period, plus precipitation received during that period. This figure, the sum of winter and early spring precipitation, is the principal accretion to soil moisture during the whole annual cycle. It is largest in the ecosystems that hold snow the best.

"the topsoil more friable to a considerable depth, and therefore more permeable during snowmelt. Soil moisture resources have thus increased and surface runoff from the fields has been reduced. At flood time less snowmelt [water] flows into streams. . ." (L'vovich, 1973, p. 170). Deep fall plowing "has increased annual soil moisture by approximately 15 cubic kilometres since World War II" (p. 173), a sizable fraction of the additional water needed in the Soviet Union's zones of insufficient or variable water supply.

Over the whole year, L'vovich (1973, p. 91) determines total amounts of water infiltrated into natural ecosystems of the major vegetation zones of the Soviet Union as shown in the accompanying tabulation. Forest ecosystems take in largest amounts of water; those in drier climates and in the far north take in less. The second column shows the fraction of the total water entering the soil of these ecosystems that goes into the small pores, to be held there until evaporated or transpired. The rest of the water moving into the soil remains in the big pores and eventually percolates deeper, as will be discussed in a later chapter.

Zone	Yearly volume (mm)	Small-pore uptake
Tundra zones	300	0.65
Taiga or boreal forest	350	0.80
In permafrost regions	250	0.90
Mixed coniferous and deciduous forest	380	0.88
Forest steppe zone	380	0.96
Steppe	290	0.99
Semidesert	200	0.99

TIME DIFFERENCES IN INFILTRATION

In Different Storms

Just as the characteristics of a soil–vegetation ecosystem vary through the year, so also does infiltration. This can be shown from mean rates of infiltration during individual rainstorms determined by comparison of rainfall and runoff intensities. A sample from 29 yr of storm records in an Iowa drainage basin gives a frequency distribution presented in part in Table IV. Individual items in this frequency distribution are storm events, each one integrating the infiltration of water over the entire area (8 km^2), which was covered about equally

TABLE IV

Cumulated Frequency Distribution of Storm Mean Infiltration Rates in the Basin of Ralston Creek, Iowa[a]

Percentage of storms	Infiltration rate exceeded in given percentage of storms (mm hr^{-1})	
	All storms	Spring
0	45	30
20	30	20
50	20	12
85	10	6
100	3	2

[a] Figures are the average hourly rate above which the volume of rain in each storm is equal to the volume of runoff. Source: Johnson and Howe (1956).

by annual and perennial vegetation. Values range from as little as 2 mm hr^{-1} in some storms to as much as 45 mm hr^{-1} in others. This range of values shows the effects of variations in soil dryness at the time when a storm began, of porosity, and of other changeable soil conditions.

By Season

Table IV also shows a frequency distribution for storms during spring, when infiltration rates were about 15 mm hr^{-1} smaller than in summer and fall. Such a seasonal regime is typical of many regions. In the spring the soil is saturated. The capillary potential gradient is small and infiltration is slow. During the growing season the activity of vegetation and soil fauna opens up the soil structure and also dries it out; water is rapidly infiltrated. The infiltration rates cited earlier for an unmulched cornfield show a similar seasonal variation.

That this seasonal regime is not the only possible one, however, is shown in a study of sparsely vegetated shale-derived soil in western Colorado (Schumm and Lusby, 1963). This soil is poorly protected from the impact of rain or the freeze–thaw cycle by vegetation or plant litter, and does not harbor the large population of soil fauna that helps open the soil of humid regions in summer.

Summer rains are heavy and batter the bare surface. Fine particles knocked loose by rain impact fall into the pores and the impact crust reduces the infiltration capacity. In an average storm of spring, 6.5 out of 6.6 mm of rain infiltrates; but in an average autumn storm 6.7 out of 8.7 mm of rain infiltrates, showing that a substantial loss of infiltration capacity occurred during the summer.

Long-Term Changes

Infiltration capacity varies with changes in soil–vegetation systems brought about through fire or excessive grazing. The deterioration of the soil is accompanied by a reduced intake of water. These shifts also may be brought about by gradual changes in land use and agricultural technology.

On the basis of field experiments carried on by the Institute of Geography of the USSR Academy of Sciences in central Russia, the accompanying estimates of yearly sums of infiltrated water have been made, assuming no secular change in precipitation (480 mm yr^{-1}), but taking account of long-term changes in ecosystems (source: Grin, 1965, p. 136). A slow recovery from the low infiltration associated with plow agriculture is evident in these figures, but complete restoration of presettlement conditions does not seem to be anticipated.

Period	Infiltrated water (mm yr^{-1})
Before the 10th century	460
End of the 19th century	410
1925–1950	415
Early 1960s	425
Near future (1980s)	440
More distant future	450

INFILTRATED WATER IN ECOSYSTEMS

For perspective, it is interesting to look at some large area means for infiltrated water calculated by L'vovich (1972). The world average is 630 mm yr^{-1} (or 1.8 mm day^{-1}). This amount is approximately 0.76 of the rate at which rain and snow are delivered to ecosystems at the surface of the earth.

Spatial means for two continents, from L'vovich, are:

Australia	400 mm (1.1 mm day^{-1})
North America	467 mm (1.3 mm day^{-1}).

The greater availability of water, rather than any greater capacity of the soil to infiltrate water, accounts for the higher figures in North

America. The implication is that North American ecosystems in general can work with greater amounts of on-site water than Australian ecosystems. North American ecosystems infiltrate more water and store more water in their below-ground sectors than do those in Australia.

A new technology, the land treatment of urban waste water, is focusing renewed attention on the infiltration process because the type of system chosen by a city depends on the site characteristics that determine infiltration and percolation of water through the soil body in which renovation takes place. In general, if sites have low rates of infiltration, an overland flow system is recommended; on ordinary or permeable soils that can infiltrate 20–100 mm week^{-1} beyond natural rainfall, low-rate application systems are appropriate (Bouwer and Chaney, 1974). These rates of application are about equal to depletion of soil moisture by evapotranspiration. On permeable sites, high-rate application systems may be chosen that supply from 500 to 2000 or more mm week^{-1}. This gives an overall average (including drying periods in the rotation) of 3–15 mm hr^{-1}. Typical features of the storage and budgeting of infiltrated water in ecosystem soil will be considered in the next chapter.

REFERENCES

American Society of Agricultural Engineers (1967). Modern infiltration theory in hydrologic analysis: a symposium; *Trans. Am. Soc. Agric. Eng.* **10,** 378–404.

Baeumer, K., and Bakermans, W. A. P. (1973). Zero-tillage, *Adv. Agron.* **25,** 77–123.

Bodman, G. B., and Colman, E. A. (1944). Moisture and energy conditions during downward entry of water into soils, *Proc. Soil Sci. Soc. Am.* **8,** 116–122.

Bond, R. D. (1964). The influence of the microflora on the physical properties of soils. II. Field studies on water repellent sands, *Austral. J. Soil Res.* **2,** 123–131.

Bouwer, H., and Chaney, R. L. (1974). Land treatment of wastewater, *Adv. Agron.* **26,** 133–176.

Cook, H. L. (1946). The infiltration approach to the calculation of surface runoff, *Trans. Am. Geophys. Un.* **27,** 726–747.

DeBano, L. F., and Rice, R. M. (1973). Water-repellent soils: their implications in forestry, *J. Forestry* **71,** 220–223.

Dunin, F. X., and Costin, A. B. (1973). Analytical procedures for evaluating the infiltration and evapotranspiration terms of the water balance equation. *In* "Results of Research on Representative and Experimental Basins, Vol. 1," pp. 39–55. Internat. Assoc. Hydrol. Sci.–Unesco, Paris.

Grin, A. M. (1965). "Dinamika Vodnogo Balansa Tsentral'no-Chernozemnogo Raiona." Nauka, Moscow, 147 pp.

Grin, A. M. (1967). Issledovanie vodnogo balansa estestvennykh i sel'skokhoziaistvennykh ugodii lesostepi. *In* "Geofizika Landshafta" (D. L. Armand, ed.), pp. 67–73. Nauka, Moscow.

Holtan, H. N., and Kirkpatrick, M. H. Jr. (1950). Rainfall, infiltration, and hydraulics of flow in run-off computation, *Trans. Am. Geophys. Un.* **31,** 771–779.

Horton, R. E. (1933). The rôle of infiltration in the hydrologic cycle, *Trans. Am. Geophys. Un.* **14,** 446–460.

Horton, R. E. (1939). Analysis of runoff-plat experiments with varying infiltration-capacity, *Trans. Am. Geophys. Un.* **20,** 693–711.

Johnson, H. P., and Howe, H. W. (1956). Infiltration frequency on Ralston Creek watershed, *Trans. Am. Geophys. Un.* **37,** 593–594.

Krimgold, D. B., and Beenhouwer, O. (1954). Estimating infiltration, *Lab. Climat. Publ. Climat.* **7,** 421–432.

Lull, H. W. (1964). Ecological and silvicultural aspects. *In* "Handbook of Applied Hydrology" (V. T. Chow, ed.), Chapter 6. McGraw-Hill, New York, 30 pp.

L'vovich, M. I. (1972). "Vodnye Resursy Budushchego." Izdat. Prosveshchenie, Moskva.

L'vovich, M. I. (1973). "The World's Water: Today and Tomorrow." Mir, Moscow, 213 pp.

Marshall, T. J. (1959). Relations between water and soil. Commonwealth Agric. Bureaux, Tech. Commun. 50, Farnham Royal, England.

Mather, J. R. (1974). "Climatology: Fundamentals and Applications." McGraw-Hill, New York, 412 pp.

McCalla, T. M., and Army, T. J. (1961). Stubble mulch farming, *Adv. Agron.* **13,** 125–196.

Mein, R. G., and Larson, C. L. (1973). Modeling infiltration during a steady rain, *Water Resources Res.* **9,** 384–394.

Musgrave, G. W., and Holtan, H. N. (1964). Infiltration. *In* "Handbook of Applied Hydrology" (V. T. Chow, ed.), Chapter 12. McGraw-Hill, New York, 30 pp.

Philip, J. R. (1954). An infiltration equation with physical significance, *Soil Sci.* **77,** 153–157.

Pillsbury, A. F. (1968). "Sprinkler Irrigation," FAO Agric. Devel. Paper 88, Rome. 179 pp.

Razumova, L. A. (1965). Basic principles governing the organization of soil moisture observations, *Int. Assoc. Sci. Hydrol. Publ.* **68,** 491–501.

Schumm, S. A., and Lusby, G. C. (1963). Seasonal variation of infiltration capacity and runoff in western Colorado, *J. Geophys. Res.* **68,** 3655–3666.

Soil Science Society of America, Water Resources Committee, C. H. Wadleigh, chairman (1966). Soil characteristics in the hydrologic continuum, *Proc. Soil Sci. Soc. Am.* **30,** 418–421.

Ward, R. C. (1967). "Principles of Hydrology." McGraw-Hill, New York, 403 pp.

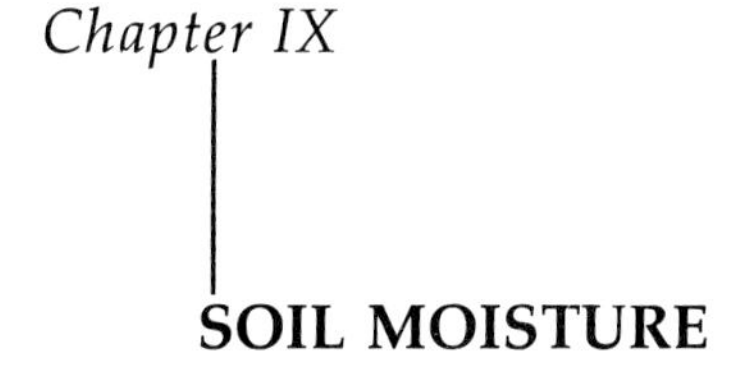

SOIL MOISTURE

We discussed the entry of infiltrating water into a soil body in terms of five layers (Bodman and Colman, 1944)—the thin saturated layer at the surface, the deep zone of transmission, the zones of transition above and below it, and the wetting front that advances downward into the depths of the soil with the logistical support of the transmission zone.

The depth to which the wetting front advances and the length of the supporting transmission zone affect the ultimate amount of water infiltrated by a soil and vegetation system during a rain storm.

Many farmers judge the size of a rain by the depth to which it has wet the soil, because this is the store of moisture on which plants will live during the succeeding rainless days. The amount of stored water depends on the depth to which infiltrating water penetrates, and on the characteristics that make one soil body good for storing agriculturally useful water, another soil body so poor in this function that it cannot be farmed at all. These characteristics relate mostly to the small pores in a soil and the capacity of their surfaces to hold water, i.e., the amount of capillary water that is held in the soil against gravity.

SOIL-MOISTURE BOOKKEEPING

Available Moisture

Upper and Lower Limits Richards (1960) says that

> when the thickness of adsorbed water films is reduced to 6 or 8 monomolecular layers of water, the soil water is so tightly bound that crop growth ceases. All agriculture is conducted in a soil–water film thickness range from this value up to two or three times this thickness.

> The total specific surface of Chino clay is about 220 m^2/gm.* If we take the 15 bar retentivity as a rough indication of the lower limit of water content that will permit plant growth, we see from the retention curve that this corresponds to 0.22 cm^3 of water per cm^3 of bulk volume of soil. At this rate, 45 grams of Chino soil would contain 8 cm^3 of water . . . on a surface of 1 hectare or 10^8 cm^2. Eight cubic centimeters of water spread evenly over this surface would give a film thickness of about 8×10^{-8} cm. It is clear that at the 15 bar retentivity value, the adsorbed water film is sufficiently thin to be strongly attracted by forces originating in the soil matrix.

It is therefore virtually unavailable to plant roots.

A suction of 15 atm (15 bars) is commonly taken as the drying limit, beyond which plants cannot extract water from the soil, and 0.1–0.3 atm for the point at which gravitational forces moving water in the soil are equal to capillary forces. At a lower suction, water fills the large pores and is held relatively weakly by the soil. Since there is little room for air in the soil in this condition, the roots of many plants suffer from lack of oxygen and cease to take in water.

The moisture content of the soil at these two values of suction, determined, respectively, by the capabilities of plant roots and of gravitational forces, has been determined experimentally in many soils, of which two are shown in Table I.

The suction pressure required to pull water out of the soil at field capacity is 10^4 N m^{-2} or, as expressed in the table, 0.1 atm. The figure varies in different soils and with different plants. As plant roots extract water from the largest of the micropores and have to resort to successively smaller micropores, they must exert more force. When the suction has reached 15 atm (15×10^5 N m^{-2}) most plants are taking in water very slowly. Some have wilted beyond the point at which a return to higher humidity can revive them; others, more resistant to wilting, continue slowly to extract water even at higher tensions.

Amounts of Water Available to Plants The range between the two cited values of moisture content represents the water customarily regarded as being available to plants, i.e., the amount removable by plant roots. This difference usually is of the order of 0.1–0.2 of the dry weight of the soil (Table I). It is greater in soils of medium texture like

* In a soil layer 1 m deep and of bulk density 1.5 g cm^{-3}, the total surface area thus would be 300×10^6 m^2 under a 1-m^2 area of land. The water-holding surface is very extensive.

TABLE I

Moisture Content of Two Australian Soils at 0.1 and 15 atm[a]

Soil type	Bulk density	Moisture content ($g\ g^{-1}$ soil) At 0.1 atm	At 15 atm	Diff.	Difference in mm of water in layer
Tatchera sandy loam	1.45	0.176	0.084	0.092	10
Grenville clay	1.24	0.324	0.146	0.178	17

[a] The layer is 0–8 cm. Source: Leeper (1964, p. 97).

silt loams than in either coarse-textured soils like sands or fine-textured ones like clay. Sand holds less water at low suctions because it has relatively small surface areas. Clay holds a great deal of water on its extensive surface areas at low suctions, but also at high suctions hangs onto much of it so tightly as to make it unavailable to plant roots.

The available range of moisture in a soil body is described by Marshall (1959, p. 59) as "the total volume of pores which are of the right size to be full at field capacity and empty at permanent wilting point." This is the storage drawn on primarily by evapotranspiration and to a smaller extent by downward drainage, water fluxes that will be discussed in later chapters. It is later replenished by infiltration of water through the atmosphere–soil interface, and so represents a revolving fund that receives intermittent inputs and supports two more or less continuous outputs. Furthermore, in a typical systems feedback, the level of this storage affects the initial rate of infiltration in the next storm to come. This association with flood potential accounts for much of the interest in soil moisture.

The soil-moisture level at which withdrawal by plants begins to decrease is not a definite value. The threshold value of moisture tends to be high on days of potentially heavy withdrawal—dry, hot, and sunny—and relatively low on days of smaller potential demand. Much of a long controversy about the effect of soil-moisture level on the rate of evapotranspiration seems to have risen from experiments made under different conditions that produced different plant demands for water.

Water is more easily available when suction is low than when it is high, and Marshall (1959) shows that growth and yield of many crop plants tend to decline at even moderate suction long before any threshold is reached. When suction is small, not only is the yield greater but plant growth also is more efficient. Plants produce more biomass yield per unit of water transpired.

Controlling Soil-Moisture Inflows and Outflows Efforts to keep soil water within the available moisture range form a major objective of agriculture, which is attained mainly by controlling inflows and outflows. Where moisture is scarce, efforts are made to maximize the amount infiltrated into the soil. Cultivation that maintains good soil aggregation and open structure and protects the surface from rain impact is particularly important. Promoting vigorous growth by the vegetation component of the soil–vegetation system also helps increase infiltration capacity.

It also is possible to increase the capacity available for moisture storage. This is done by developing a deeper soil layer, i.e., breaking up hardpans that are too close to the surface or modifying soil texture, for instance, by adding organic matter to sandy soil. Encouraging plants to root more deeply by fertilization or spacing of waterings forms part of every gardener's expertise. As noted in Chapter VIII, organic compounds in the infiltration water help aggregate clay and other fine particles. While these aggregates are large enough to resist erosion they contain large areas of internal surfaces that hold water for plant use.

To reduce the outflow of water is one reason for killing weeds. It is useful also to select crop plants that will produce the most dry matter per unit of water transpired, that is, to get maximum biological production for each millimeter of water leaving the soil via this particular outflow.

Outflow of excess water in the large pores of the soil is encouraged by many land-management practices, some of great age, in order to ensure the presence of oxygen needed by plant roots and soil fauna. Perhaps farmers devote as much effort to keeping the film of water on soil particles from getting too thick and blocking oxygen flow in the soil as they devote to irrigation. Both drainage and irrigation require technical skill and understanding of a soil–vegetation ecosystem in order to attain the optimum mixture of rock particles, organic matter, air, and moisture.

Irrigation is directed to maintaining this level of soil moisture. Irrigation handbooks go into great detail about the size of each application of water and the intervals between irrigations, which depend on such complex factors as labor cost, water cost and availability at a specific time, soil conditions, and mode of application of water.

Irrigation frequently has to be combined with drainage. Many expensive irrigation projects have failed because they paid too much attention to water inputs and not enough to water outflows from the soil. Irrigation and drainage are counteracting controls; too much water fills the big pores and keeps out oxygen, too little puts plants under excessive stress.

Soil moisture is as important a climatic element as soil or air temperature and plays a central role in the work of such climatologists as Thornthwaite. While the moisture content of the plants themselves, which is difficult to determine but affects the functioning of leaf cells, is important, it tends (except at midday stress periods) to reflect the moisture in the root zone. Moreover, the soil as an environment for bacteria, microfauna, and larger animals is, depending on its moisture, hostile or hospitable to them.

Soil-Moisture Bookkeeping and Measurement

Thornthwaite showed that the bookkeeping of income and withdrawal of water in a soil with a certain storage capacity, i.e., a known range of available moisture, can be used to show how the irregular input of rainfall is converted into actual output as evapotranspiration. Such an accounting is necessary in the realistic case of a crop growing on soil that is not always fully moist. Even when the method is simplified by assuming an arbitrary rate of withdrawal, it usually gives a good reproduction of soil-moisture storage and moisture availability to crops and other ecosystems.

An illustration of this bookkeeping method is given in Fig. IX-1 for part of 1960 at Waipio, Hawaii (Chang, 1963). The available moisture in this sugar cane field is 65 mm, a moderate amount of storage. In this high-energy climate the withdrawal rates are consistently high, 4–5 mm day^{-1} (see Plate 16).

Rains of 10 mm or more recharge the bank of soil moisture substantially. Rains of 5 mm day^{-1}, e.g., around 20 August, just about hold the line against current withdrawals. Applications of 65 mm of irrigation water were made on 1 August, 25 August, 14 September, 14 October (coming after several days when soil moisture was depleted), and 7 November (after a still longer period of dry soil). Following 10 November, rains came often enough that the level of soil moisture remained moderately high without irrigation.

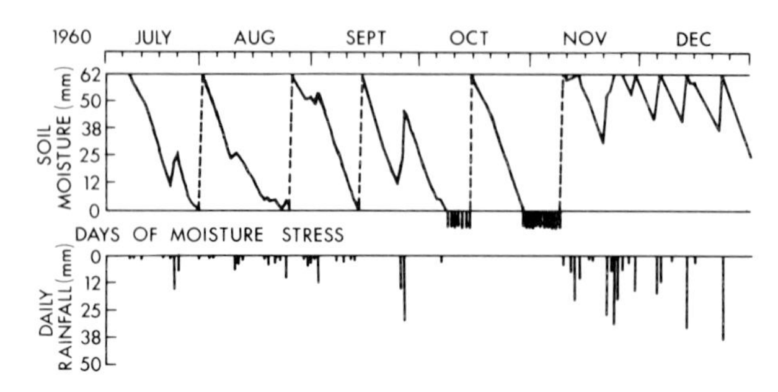

Fig. IX-1. Running budget of soil moisture (mm) at Waipio, Oahu. Daily rainfall is shown on the lower graph and replenishment of soil moisture by irrigation is shown by vertical dashed lines, e.g., on 1 August, on the upper graph. Days when available soil moisture was zero are indicated as stress days, as in October and November. After rains set in following 10 November, soil moisture on most days remained above 30 mm (from Chang, 1963).

In many bookkeeping calculations of soil moisture the vegetation may change more during a period of several months than did the cane in the foregoing situation. Expanded roots draw water from a greater volume of soil; an expanded canopy intercepts more radiation and covers more of the soil surface. These changes can be included in the bookkeeping procedure if desired.

In this example it was assumed that depletion of soil moisture did not decrease when the moisture tension was large, otherwise the descending segment of each curve would have tended to flatten out in its lower end. The calculated curves of soil moisture agree in general with the readings of gypsum blocks in the soil.

Procedures for keeping books on the stored moisture in the soil were extensively developed and applied to a variety of problems by Thornthwaite (1945, 1948) and his co-workers at the Laboratory of Climatology in the intensive farming region of southern New Jersey (see issues of Publications in Climatology from this Laboratory, and Mather, 1974, especially pp. 99, 108, 133 *seq*.). Their adaptability is demonstrated by their success in reconstructing the water conditions in a coffee and an irrigated tea plantation in East Africa. Pereira (1973, p. 129) states that "the development of computer techniques has made the soil-moisture budget a practical tool both for control of irrigation and for study of watersheds." It is particularly applicable to situations in which a proposed change in land use (e.g., converting bamboo to pine plantations that would provide more employment) might change the off-site yield of water, and also its economic value.

The generalizing power of the soil-moisture budget derives from the fact that of all the environments in an ecosystem where water can be stored, the reservoir of the soil is the most important. It is usually one of the largest storages, as compared for instance with the small transient storages of water in the leaves or on top of the ground. Although often smaller in volume than the storage in the groundwater body, its contents are more accessible to biological utilization; its fluctuations both reflect and control the moisture status of the plants rooted in it. The bookkeeping of soil-moisture storage provides a key to the variations of the moisture factor in plant tissue and in the functioning of ecosystems. For instance, one method for determining actual evapotranspiration employs iterative month-to-month bookkeeping of soil moisture through the annual cycle (Budyko, 1974, p. 98).

The input term in many calculated budgets has been taken as gross precipitation, even though investigators might be aware of the existence of surface and near-surface runoff that would reduce the amount of water actually entering the soil body. In a month (or a day, if this

Plate 16. Sugar cane at Kunia experimental station, Oahu. Irrigation flume is in foreground; behind the men are four lysimeters measuring evaporation. On a mast at left of center is an all-wave radiometer to measure surplus of radiant energy (April, 1970).

time unit is used) when rain is so heavy as to fill the soil-storage capacity, the excess is counted as surplus and it matters little, for bookkeeping purposes, whether this surplus percolates to groundwater or leaves the scene via the nearest stream channel.

Surface runoff, however, in heavy but short rains that do not fill up the soil storage represents a discrepancy in the bookkeeping, which can be corrected by calculating infiltration and applying it as an input. When infiltrated water was used as the input into the soil-moisture budget from which actual evapotranspiration was determined (Albrecht, 1971), an improved relation was found between the actual-to-potential evapotranspiration ratio and the yields of corn, soybeans, and wheat in the Missouri Valley. Other refinements and improvements in the bookkeeping procedure also have been made.

Moisture Sensors The wide application of bookkeeping methods for determining soil moisture suggests that perhaps direct measurement has not been considered easy or adequate. This deduction is correct. A serious problem lies in the complicated structure of many soils; texture, organic matter, stone content are all highly variable in spatial distribution and yet all-important in the capacity of the soil body to store water.

The basic measurement of moisture in the soil is gravimetric. The soil body is cored, the core is removed and weighed, then dried and weighed again, all under standardized procedures. This is precise although laborious. The sampled location is, however, destroyed for later measurements at the same place, which might be wanted in order to establish the variation with time. The alternative is to take enough samples at each visit to represent the mean water content of the site in general. The gravimetric method, however, is "the only one which can be generally recommended despite its many limitations," according to a WMO Technical Note (1968).

Other methods for measuring soil moisture utilize blocks of plaster-of-Paris (in dry sites), fiberglass (in wet), or other materials that are in equilibrium with the moisture of the soil in which they are buried. These can be read weekly to discern time variations.

Soil-moisture blocks are usually installed in a vertical stack that extends to a depth of 1 m or more, to determine variations with depth and to make it possible to calculate storage through the whole soil depth. The necessity to dig a pit to facilitate installation creates some disturbance in the soil-moisture distribution. Sometimes the blocks (or tensiometers for high levels of moisture) part company with the surrounding soil, ceasing to follow the changes in its moisture content. Sometimes chemical reactions eat them away; even consumption by gophers is not unknown. Often such changes are not evidenced by a sudden shift in the readings, which might continue for some time before their unreal nature is recognized.

Neutron Probes In the mid-1950s it seemed as though technology had provided the final solution to the problem of measuring soil moisture. An emitter and a receiver of neutrons given off by one or another radioactive substance are lowered together in an access tube in the soil. They are held at each desired depth long enough to count the emitted neutrons that are moderated by protons of hydrogen atoms (chiefly in water molecules) in the surrounding soil and come back to the probe's receiver. This count can be used to estimate the water in the soil, *if* the tricky problem of calibration has been solved. In half an hour or less, samples at several depths in the soil can be taken; in a few hours profiles can be obtained from several access tubes and combined into a mean value for the date and site.

This method became highly popular almost overnight. Its rapid, uncritical acceptance perhaps represented a hope that for every problem there exists somewhere a technological quick fix. The radioactivity (requiring special safety precautions), the nuclear terminology, and flashing lights on the counter made the probe seem the very model of a modern technical device. Unfortunately, not all this promise has been fulfilled. The method is still very much in use, but technical difficulties of calibration and reliability remain serious. Operating times and initial investment remain large. The WMO Note on soil moisture rates the probe as the fourth choice of four methods discussed, but "promising."

Difficulties in Sampling Along with the problems with moisture sensors is the fact that the withdrawal of moisture by plant roots is heterogenous in pattern. Moisture in plant rows usually is extracted before that in the furrows between rows. Most plants draw heavily on the moisture in a shallow layer of soil before they extend their roots deeper, if indeed they have the genetic capability to go deeper. Vertical zonation of soil characteristics that affect moisture-holding capacity already has been mentioned.

Lateral movement of moisture toward dense root populations tends to be slow and is limited as to distance. Upward capillary movement of water is likewise too limited to bring about a uniform redistribution of moisture within a soil body.

Accordingly, sampling the vertical profile of soil moisture below a single point on the surface shows the depth-related differences only at that point. It is usually necessary to replicate sensors at several points. One investigator comments that "soil moisture content in the field is so variable that amounts of moisture which are important to the crop are nevertheless small compared with the standard error of sampling. In Saskatchewan half an inch of stored moisture will increase the yield of wheat by about 2 bushels per acre and yet 30 or more samples taken to root depth are needed to establish this difference in storage between two treatments" (Staple, 1964). In other words, a 12-mm difference in soil moisture between two practices of land management cannot be identified with fewer than 30 measurements.

Another investigator says that "a major limitation to estimates of water-use by sequential gravimetric samples lies in the error from field variability" (Ekern, 1967). Even though the coefficient of variation is small, "these variations do introduce error into the determination of the average moisture content for a specific field . . . .When a Student's t value of 3 and a coefficient of variability of 5% are assumed for a small sample at 19:1 odds, with the intent to distinguish a 1% difference in soil-moisture content near the 15-bar suction, a minimum of 20 samples is required. This sample number is all too often not used in the assessment of field moisture content." Fortunately, the gross regime of soil moisture with season of the year is generally more easily measured, although for most budget computations 8–10 gravimetric samples per site are recommended (WMO, 1974, p. 62).

Networks of soil-moisture stations have been set up in many countries, usually as parts of their hydrometeorological observation programs. In the United States the major soil-moisture network is operated as a part of the mountain snow survey, with about 200 stations (Barton, 1974).

SOIL-MOISTURE DISTRIBUTIONS IN SPACE

Local Variation in Soil Moisture

Differences among Ecosystems We have seen that small-scale spatial differences cause trouble in measuring even in an apparently uniform field. They also occur at a mesoscale between different ecosystems. Depletion of soil moisture by woody vegetation, which usually is deep rooted, reaches a large total over the year because the trees or shrubs experience fewer days of limited moisture supply than does herbaceous vegetation. Green trees amid an expanse of baked, sun-dried grass are a familiar sight, representing continued availability of soil moisture at depths greater than grass roots can reach; ecosystem moisture storages are quite different.

As a result, forest evapotranspiration may be 80 mm more than that of field crops where the length of the growing season is the same. Where forest has a longer growing season than crops, the excess may be as large as 250–300 mm, according to a comprehensive study by Rauner (1965). Emptying and refilling of the larger reservoir under forest results in a larger turnover of water in the ecosystem. Fig. IX-2 shows soil moisture to a depth of 3 m under a small belt of oaks. The trees have depleted moisture to 1.5-m depth (Rode, 1969).

Many land-management practices create soil-moisture differences among crop fields, although the relations do not always turn out as expected. For example, the practice of stubble mulching in the northern Great Plains seems to produce little advantage in soil moisture as compared with fallow land (McCalla and Army, 1961, p. 145). Only in summers of heavy rain or where snow is trapped can a difference be ascribed to stubble mulch in this dry region.*

Figure IX-3 shows soil-moisture content through the year 1961 at several depths in lysimeters in Coshocton, Ohio (Dreibelbis, 1963). The soil was occupied by meadow up to 24 May, when corn was

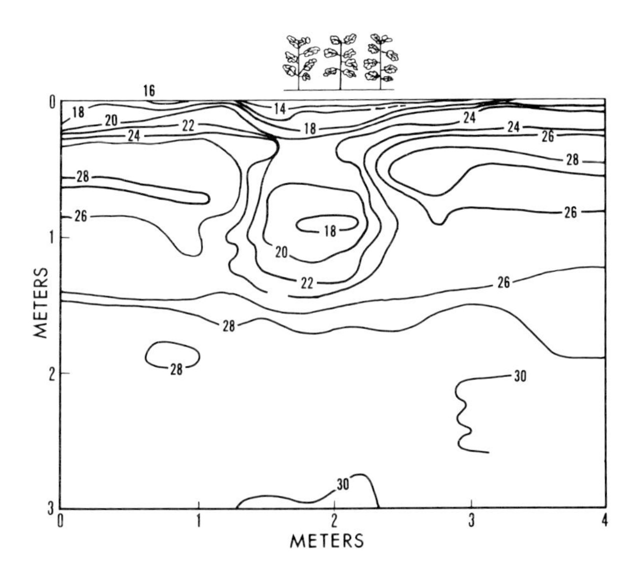

Fig. IX-2. Soil moisture (percent) sampled across a line of oak trees in a fallow field. Note depletion of moisture by the trees, especially in the top $1\frac{1}{2}$ m of soil (from Rode, 1969).

*Stubble mulch in corn-growing regions, on the other hand, which usually get heavy rains in summer, tends to increase soil moisture by several percent.

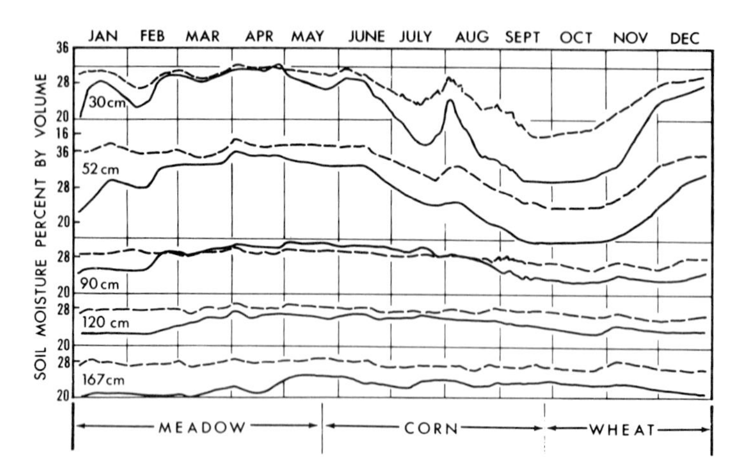

Fig. IX-3. Soil moisture at five depths in Keene silt loam at Coshocton, Ohio in 1961. Corn replaced meadow in May; after the harvest in September wheat was sowed. Dashed lines show moisture (percent by volume) under crop management practices prevailing in the 1930s and solid lines show moisture under "improved" practices intended to hold soil (Dreibelbis, 1963). Adapted from Soil Science Society of America Proceedings, Vol. 27, p. 457, 1963, by permission of the Soil Science Society of America, Inc.

planted. After the corn was harvested, wheat was started on 26 September. The two lines for each level show soil moisture under agricultural practices prevailing in the region and improved practices aimed at higher productivity and soil stability. The improved practices resulted in greater plant activity and growth, and deeper penetration of plant roots into the soil. Water was extracted by corn roots under improved practice cultivation at a depth as great as 70 cm.

Where moisture storage is strongly drawn down as a result of changed agricultural technology, "the absorption and storage opportunity for water is correspondingly greater. This tends to influence infiltration, surface runoff, and percolation" (Dreibelbis, 1963). In other words, a change in one water flux, transpiration, causes a change in water storage in the soil. This change in storage then causes changes in other water fluxes.

Differences Due to Topography Soil moisture varies not only from one soil–vegetation ecosystem to another, but from ridge to valley as slope affects soil depth and the infiltration and drainage of water. In Table II measurements of soil moisture in the top 60 cm of soil under hardwood forest in the rugged country of southwestern Wisconsin show less stored water on a slope facing the sun than on one turning from it. More soil moisture in the lower slopes than the upper suggests downslope movement of water in the soil. Downslope movement of soil particles also has occurred, forming deeper, finer-textured soil. Soil moisture under two pine plantations on the lower slopes also was less than under the native hardwoods, reflecting the greater biological activity and biomass of the pines (Fig. IX-4).

In Coweeta Hydrologic Laboratory in the southern Appalachians, a spatial analysis of the species in a hardwood forest revealed an ordination in terms of three terrain variables: distances from the stream, from the ridgeline, and altitude. All three are associated with a gradient of soil moisture (Day and Monk, 1974), although the relation is likely to be complex.

Spatial differentiation in soil moisture and other factors affecting the infiltration capacities of ecosystems when a storm comes is evident when it is seen that some ecosystems generate off-site flow while

TABLE II

Soil-Moisture Storage in the Upper 60 cm of Forest Soil on Valley Bottom and Slopes in Southwestern Wisconsin[a]

Location	Density of vegetation[b]	28 June	18 July	15 Aug	25 Sept
Top	4	223	220	189	151
South slope					
Upper[c]	1	106	96	83	64
Middle[c]	3	97	108	96	76
Lower[c]	3	119	117	95	70
Lower[d]	12	88	97	80	86
Valley bottom[e]	11	180	179	168	157
North slope					
Lower[f]	6	204	236	190	182
Lower[g]	13	132	127	112	73
Middle[f]	8	188	196	180	169
Upper[f]	5	181	173	153	140

[a] Precipitation during July, August, and September amounted to 270 mm. Source: Stoeckeler and Curtis (1960). Units: mm.

[b] Expressed as timber volume, in 1000 board ft acre^{-1}.

[c] Scrubby oak stand.

[d] Scotch pine plantation.

[e] Bottomland ash, elm, and basswood.

[f] Basswood, sugar maple, elm, and oaks.

[g] White pine plantation.

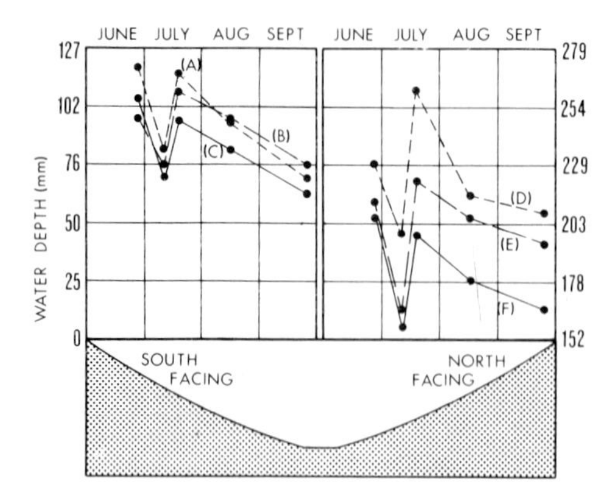

Fig. IX-4. Cross section of moisture measured in the upper 60 cm of soil across a valley in southwestern Wisconsin. (A) Lower slope, south-facing; (B) middle slope, south; (C) upper slope, south; (D) lower slope, north-facing; (E) middle slope, north; (F) upper slope, north (from data of Stoeckeler and Curtis, 1960). Note that the scales of soil-moisture are different for north and south slopes.

others do not. Ecosystems that occupy only 0.1 of a drainage basin (Dickinson and Whiteley, 1973) may produce most or all of the flood runoff, a situation recognized by Horton in 1937 and recently the object of revived interest.

TIME VARIATIONS OF SOIL MOISTURE

Days of Soil-Moisture Shortage

Measured or calculated values of soil moisture as they change from day to day identify days of sufficient moisture for plant growth as well as days when moisture is short. On such days transpiration is reduced and plant growth hampered.

Stress-Day Frequencies Drought or moisture-stress days determined by calculating the level of soil moisture can be treated like any other climatic measure for which long records are at hand. For example, frequencies can be determined for each part of the growing season in a particular region, to define the risk facing the manager.

These calculations can be made with different values of soil-storage capacity, a factor that is likely to vary from field to field even within one farm, depending on soil type and depth. Deep-rooted crops on deep soils of high moisture-storing capacity encounter fewer days with unfavorable moisture conditions than shallow-rooted crops on sand, given the same spacings between rains.

Such frequency data show drought hazard as it changes during the growing season. They are useful in deciding whether or not it is economic to install pumps and sprinklers for supplementary irrigation, and when such facilities are likely to be needed.

In the example cited in Table III, the vicinity of Tuskegee University in Alabama, drought hazard is greatest in May and September. On deeper soils, it is small in mid-summer. In a field in which the soil has a storage capacity of 50 mm of water, an irrigation system designed to provide adequate supplementary water in nine out of ten summers would have to supply 85 mm in May and 117 in June at the early summer peak of demand, and 85 in September at the late peak.* In any particular summer, bookkeeping methods such as those systematized by Palmer (1968) (Mather, 1974, pp. 169–172) keep track of the current moisture conditions so that the insidious, quiet setting-in of a drought can be detected.

In some regions, rainstorms large enough to charge the soil with moisture are isolated events, spaced so far apart that plant growth is only intermittent (see Plate 17). Such a region is central Australia,

Plate 17. Dry pasture country south of Broken Hill, Australia. The gallery forest along the creek bed only emphasizes the dryness of the upland soil (September 1966).

*Note that 85 mm of water is equivalent to 85 kg m^{-2} of land. This is equivalent to 850 tons (or m^3) of irrigation water per hectare, a number useful in deciding how large a storage pond is necessary if surface water is applied.

TABLE III

Minimum Number of Drought Days To Be Expected in Each 10-Day Period during the Growing Season in Macon County, Alabama, with Probability = 0.5 and Two Soil-Moisture Storage Capacities[a]

Storage capacity (mm)	10-Day periods starting:																					
	11 A	21 A	1 M	11 M	21 M	31 M	10 J	20 J	30 J	10 J	20 J	30 J	9 A	19 A	29 A	8 S	18 S	28 S	8 O	18 O	28 O	17 N
25	4	3	5	7	7	7	6	5	4	4	3	4	3	6	6	6	7	6	7	8	4	2
50	0	0	0	5	6	6	6	5	3	2	0	0	0	4	4	2	5	2	5	6	1	0

[a] Source: Ward *et al.* (1959).

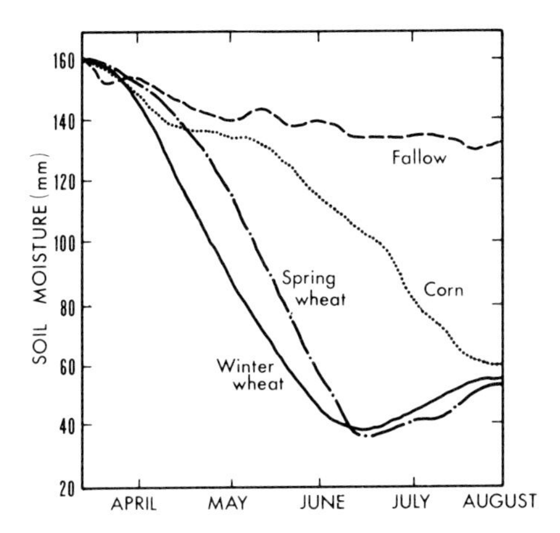

Fig. IX-5. Mean value of available ("productive") soil moisture (in mm) in the top meter, averaged over many years and many stations in the zone of deep spring wetting and deep groundwater table (from Kulik and Sinel'shchikov, 1966).

where the periods of dry soil are long and of unpredictable duration, plants go into standby status and wait until the next rain comes (Fitzpatrick *et al.*, 1967). Periods when soil moisture is favorable for plant growth are few. At Alice Springs the probabilities show that for median conditions about 100 days of growth occur in a calendar year,* usually grouped in several distinct periods of adequate soil moisture.

Seasonality The Coshocton graphs presented earlier in connection with the discussion of land-management practices also show seasonal changes. Highest moisture content comes in April, followed by a decline to low levels by fall. The annual cycle is most marked in the shallow levels of the soil, but can be seen as deep as 2 m.

Fig. IX-5 (Kulik and Sinel'shchikov, 1966) shows a Russian example of mean moisture in the top meter of the soil under winter and spring wheat, corn, and fallow during the growing season. Summer rain is helpful to all crops, but by itself cannot maintain a high level of soil moisture. The decrease in soil moisture comes early under wheat, with recharge after harvest; it comes later under corn, which is planted later.

In many climates the early part of the growing season is characterized by adequate soil moisture, resulting from spring rains, rains of

* This total is about the same growing season as obtains between spring and fall freezes in northern Wisconsin.

the previous winter, or melting of winter snow cover or frozen ground. Conditions are favorable for germination and early growth at the time the income of radiant energy is increasing and the temperatures of the soil and air are rising.

The critical question is whether, after the initial store of soil moisture—say 80–100 mm—has been used up, more water will be forthcoming. In unirrigated hill lands of California, growth begins in February in moist soil and is sustained by rains that continue at a declining rate through March and April. After this the grass has only the diminishing store of soil moisture to carry it to maturity.

In Wisconsin Corn planted when the soil has warmed sufficiently can count on adequate soil moisture from spring rains and residual melt water to carry it through May and June in Wisconsin. In early summer the countryside presents an equatorial luxuriance of vegetation, but as summer wears on the soil moisture often is used up if replenishing rains do not come. By August in many years the countryside begins to look dry, sometimes so much so that dairy cows cannot be pastured and have to be fed stored fodder.

An example of a summer when rains, although not absent, were small and far apart, occurred in 1973 (Fig. IX-6). The area of the state experiencing a shortage of soil moisture was small at the beginning of July, but by the end of the month encompassed most of the state. Spotty rains in August reduced it to about half the state, but it was not eliminated until the end of October.

On the other hand, a few summers end up very wet, as in 1972, when the soil remained very wet all fall. The usual filling of silos with chopped corn for winter dairy feeding went very slowly because farmers had to add extra power to pull the wagons of corn out of the

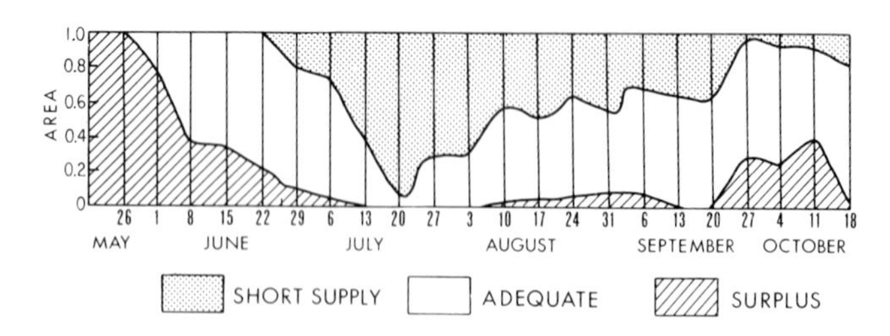

Fig. IX-6. Soil moisture in Wisconsin through the growing season of 1973, expressed in terms of the fraction of state area reported as being deficient, adequate, or surplus in soil moisture (from data in weekly crop and weather reports, Wisconsin Statistical Reporting Service, 1973).

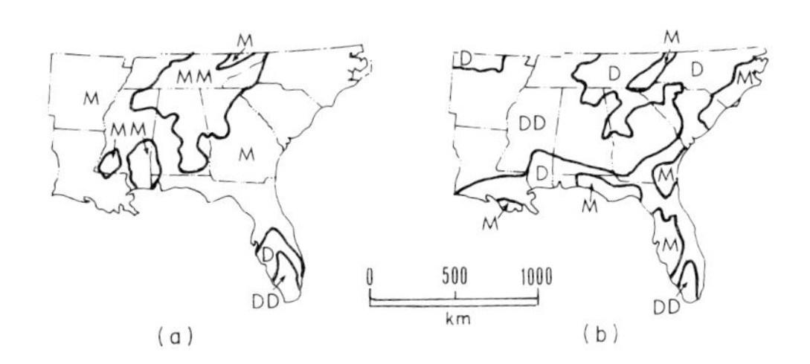

Fig. IX-7. Mean soil-moisture conditions in the southeastern United States as they affect trafficability for off-road vehicles in (a) March and (b) September. DD, very dry; D, dry; M, moist; MM, very moist, translatable into trafficability if soil texture is known. Note that at the end of summer very dry conditions prevail over large areas, moist conditions have shrunk, and very moist have vanished (Thornthwaite and Mather, 1955).

muddy fields. By 10 October only a third of the silo filling had been done, in contrast to the same date in 1971 when nearly all the job was done. Local comments were: "The corn has frozen but it cannot be ensilaged because of the wet soil condition." "Some farmers filling silo, 2 tractors on choppers." "Hasn't rained for a couple of days, silo filling going full blast." Many crops of vegetables such as sweet corn had to be abandoned in the fields (Wisconsin Statistical Reporting Service, 1972).

Tractionability Seasons Moisture in small and big pores of the soil affects its bearing strength, especially as expressed by its strength in shear under the wheels or tracks of vehicles. Texture also plays a role: sands are stronger if wet, up to a certain point, while clays tend to be stronger when dry. Calculating the water content, both capillary and gravitational, of a soil body can be done by bookkeeping methods (Mather, 1974, p. 368) and the results interpreted in terms of tractionability.

For example, over the southeastern United States in March most of the land surface is classed "M" (Fig. IX-7) (Thornthwaite and Mather, 1955), meaning that the "tractionability of plastic soils deteriorates rapidly. The soil becomes softer and offers considerable difficulty for heavy, wheeled vehicles. Track-laying vehicles have no trouble except after heavy showers." Much of the rest of the land is classed "MM": "very poor tractionability on plastic soils. Adhesion of plastic soils to foreign objects [tires, for instance] becomes greater than the cohesion between soil particles. Traffic of practically every kind is extremely difficult on bare soil but men and light animal-drawn vehicles can travel over grassed areas without too much difficulty until the surface is cut up."

The map for September, at the end of the soil-moisture depletion season, shows a radical difference from March. Most of the Southeast is now in the dry (D) and very dry (DD) classes. Off-road travel is much easier, up to the time when fall rains wet up the soil again, an occurrence that can be stated in probability terms for any particular place. Much of the history of the campaigns of the Civil War is related to this seasonality in soil trafficability. General Mud has always been a major military factor.

Soil Drought and Biological Productivity

The changes in soil moisture as the seasons pass have a clear effect on the vigor and biological productivity of ecosystems dependent on these stores of water. To associate crop yields with rainfall as the sole moisture factor, as has often been attempted in the past, is to limit the validity and accuracy of any explanation in physical or statistical terms. It is plain that plants obtain most of the water they need from the soil pores and very little from the rain that bathes their leaves. Thus the state of soil moisture needs to be brought into any yield investigation. This is done either in the form of the calculated shortfall it causes in moisture use, i.e., the difference between actual and potential evapotranspiration (to be discussed in Chapter XII), or simply in terms of soil moisture itself. The fact that reduced soil moisture is as important in a mild climate as a hot, thirsty one is illustrated in the following studies made in New Zealand.

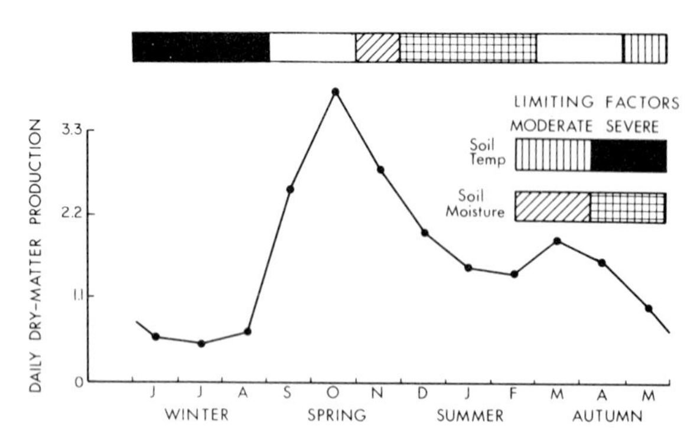

Fig. IX-8. Daily production of forage (dry weight) on unirrigated pasture at Winchmore, New Zealand through the growing season from June to May. Soil temperature is a limiting factor from May to August (Southern Hemisphere!) and soil moisture is limiting from November through February) (Rickard, 1968).

TABLE IV

Soil-Climate Limitations to Pasture Production and Pasture Yields, Winchmore Irrigation Research Station, New Zealand[a]

Limitation	J	J	A	S	O	N	D	J	F	M	A	M	J
Solar radiation (W m^{-2})	70	70	115	165	235	300	315	300	235	170	130	80	70
Soil temperature (°C)[b]	4	4	5	8	12	15	18	19	18	16	12	7	4
Limitation		Severe										Mod. Severe	
Unirrigated pasture													
Soil moisture (%)[c]	28	30	30	27	19	16	15	13	14	19	21	27	38
Limitation							Mod. Severe						
Production (g m^{-2} day^{-1})	0.6	0.5	0.6	2.3	3.5	2.5	1.8	1.4	1.3	1.7	1.5	0.9	0.6
Mean number of drought days[d]	—	—	—	0	0	3	5	10	10	6	4	1	—
Irrigated pasture													
Soil moisture (%)	34	34	34	30	26	27	27	26	26	28	29	32	34
Production (g $m^{-2}day^{-1}$)	0.6	0.5	0.7	2.5	4.0	4.2	4.4	4.7	4.4	3.3	2.2	1.1	0.6

[a] Length of record 10 years, except radiation 7, and drought days 40. Pasture sward is ryegrass and subterranean clover. Source: Rickard (1968, Table 1).
[b] At −5 cm.
[c] In layer 0 to −10 cm.
[d] Rickard and Fitzgerald (1969).

New Zealand Unirrigated Pastures On the Canterbury Plains of the South Island of New Zealand, unirrigated loess soils used for pasture (Table IV) have two seasons of biological limitation (Rickard, 1968): (a) from May through August, low soil temperatures limit plant growth—the Southern Hemisphere winter. (b) From November through February, in spite of the fact that monthly rainfall averages 70–80 mm, low values of soil moisture limit growth.

Severe limitations (Fig. IX-8) occupy six months of the year, and moderate limitations two more months. Periods of satisfactory soil temperature and moisture are brief, and the later of these two satisfactory periods, in March and April, finds pasture plants in poor condition after the three preceding hot, dry months.

The peak of pasture production comes in spring (October) and there is a slight resurgence in fall (March). This annual regime of growth is

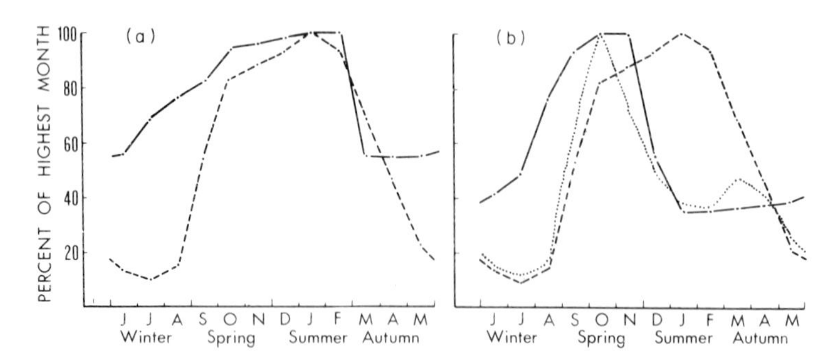

Fig. IX-9. (a) Forage production of irrigated pasture (dashed line) and requirement of beef cattle herd (—·—). (b) Forage production of unirrigated (dotted line) and irrigated pasture (dashed line) in New Zealand, in percentage of highest monthly rate, and forage required by ewe flock (—·—) (from Rickard, 1968).

in phase with the feed requirements of a lamb-producing economy, if lambing is scheduled early. The young animals reach salable size by the time pasture production begins to drop under the midsummer depletion of soil moisture (Fig. IX-9).

Drought days for unirrigated pasture were identified by casting soil-moisture budgets over a 40-yr period that included 16 yr of gravimetric soil-moisture measurements as a check. Their mean seasonal distribution is shown in Table IV. The incidence of drought days varies from none in a few seasons to 88 days in one year (Rickard and Fitzgerald, 1969).

Drought days tend to occur in sequences of consecutive days, some of them ruinously long. Half of the drought days of the 40 yr studied occurred in sequences longer than 14 days. (Remember, this is not the length of time between replenishing rains, but the period counting from the date when moisture stress set in.)

Extreme-value theory applied to the longest sequences in each season gave the following results:

Frequency:	0.5	0.2	0.1	0.05
Length (days):	18	30	38	45

In any season there is a 50–50 chance of experiencing at least 18 consecutive drought days.

An equation relating pasture production to the number of drought days in a season *(N)* (Rickard and Fitzgerald, 1969) is

$$\text{Production g m}^{-2} = 670 - 4N.$$

According to this equation, the 18-day period of drought days that can occur with 0.5 chance in any year reduces the seasonal production of pasturage, in dry-matter units, by (4) (18) = 72 g m^{-2}, or about 10% of the total. This reduction is added to the effects of shorter sequences of drought days.

New Zealand Irrigated Pasture Alteration of the water supply of these pasture soils by irrigation illustrates further associations of water and crop yield (Plate 18). Irrigation changes the regime of soil moisture, and permits vegetation growth to follow the cycle of energy supply more closely, as shown in the lower part of Table IV. Peak production can now be maintained for a long period at a rate exceeding the 3.5 g m^{-2} daily rate that on the unirrigated land was reached only at the spring peak (October).

Under irrigation high production extends from the vernal to the autumnal equinox. Production responds well to solar radiation, if water is available. The amounts of added water are not large, 8 irrigations of 30 mm each, and the total of 240 mm can be compared with the 375 mm of average rainfall during the 6 month period from September to March. The yield of biomass from the irrigated pasture is about double that from the unirrigated, 1150 g m^{-2} compared with 600 g m^{-2}.

Plate 18. Irrigated pasture and wheat stubble at Ashburton experimental farm, New Zealand, with beef cattle (late summer 1966).

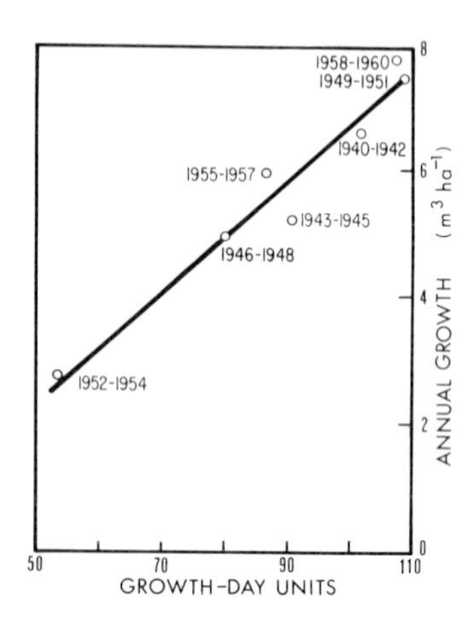

Fig. IX-10. Annual growth of pine wood in Arkansas related to soil moisture and energy supply expressed as growth-day units (from Bassett, 1964), plotted as 3-yr means. Reproduced from Soil Science Society of America Proceedings, Vol. 28, p. 438, 1964, by permission of the Soil Science Society of America, Inc.

The regime of biological production under irrigation, with its sustained yield all through the summer, is less suited to lamb production than was the dryland regime (Fig. IX-9). "The problem that faces the irrigating farmer is that of adjusting his farming system to this change in environment" (Rickard, 1968). One possibility is to switch to beef cattle production (Fig. IX-9).

In addition to the changes in total production and cropping regime that come with irrigation, the greater reliability eases many problems of farm planning. Unirrigated pasture has produced, in different Januarys of the ten recorded, from as little as 0.4 g m^{-2} day^{-1} to as much as 4.0. The production from irrigated pasture varied only from 3.6 to 5.8 g m^{-2} day^{-1}.

This steadiness is valuable to the farm manager, who can plan purchases and sales of livestock more accurately in his never-ending effort to keep the production and consumption of grass in balance. The consequences of irrigation are not startling when expressed as an increase in evapotranspiration, which probably is less than 1 mm day^{-1}. However, the increased yield of grass, a modified annual regime, and the reliability of grass fodder for a steady turnoff of animal products are important aspects of the regional water and biomass economy.

Soil climate, i.e., the changing conditions in the unseen yet vital part of an ecosystem below the ground surface, is important for the production of grass, which is the basis of New Zealand's grassland economy that exports meat and wool, butter and cheese. The major American crop ecosystem, i.e., the one receiving the greatest inputs of energy and fertilizer and growing in the best soil in the world, is corn. It is equally vulnerable to conditions of soil climate. A combination of a wet, late spring (soil temperature) and a summer month of spotty rains (soil moisture) together with the ever-present possibility of a cold fall season with an early frost causes worry among farmers, stock feeders, meat eaters, grain exporters, and oil importers. America's diversified economy also has its links with the soil components of ecosystems.

Forest A growth-day index that has been developed for forest studies (Bassett, 1964) represents a combination of energy supply and soil moisture similar to the number of days without moisture stress in the New Zealand case. Over the 155-day typical growing period, 22 May–13 October, for pines in southeastern Arkansas, growth-day units were calculated for comparison (in three-summer means) with mean volume of growth in the experimental forest (Table V). These data are shown in Fig. IX-10. The regression equation is

$$Y = 0.09Z - 2.2,$$

in which Y is calculated annual growth ($m^3\ ha^{-1}$) and Z the number of growth days per season (max = 155). The lowest period shown in Fig. IX-10, 1952–54, included two severe summers, which decimated many forests of the south central United States. It is

TABLE V

Average Growth in Volume of Pine Trees and Growth-Day Units at Crossett, Arkansas, 1940–1960[a]

Period	Volume growth ($m^3\ ha^{-1}$)	Growth-day units
1940–1942	6.6	102
1943–1945	5.2	92
1946–1948	4.9	81
1949–1951	7.4	108
1952–1954	2.7	54
1955–1957	6.0	87
1958–1960	7.9	109

[a] Source: Bassett (1964).

interesting to note that two three-year periods near the end of the study show actual growth greater than calculated, suggesting improvement of timber-management practices. Such a time effect is found in many yield studies.

Employing data by Bassett and others, Manogaran (1973) calculated the gains in pulpwood production that would ensue if Southern pines were irrigated. The high operating costs of irrigating them at the present price of pulpwood, however, make the idea infeasible, although the cost–benefit comparison differs a great deal in different regions from Texas to Virginia.

Large-Scale Variations in Soil Moisture

Areas of different degrees of soil-moisture stress are delineated in farming practices. In England and Wales the pattern of stress in late August is associated with the cropping type (Walker, 1965). Where moisture deficit is frequent, grassland tends to be replaced by alfalfa that roots more deeply, by cereals that mature early, and by orchards that also root deeply.

Where vegetation is dormant in winter, the reservoir of soil moisture is usually recharged in this season or in the subsequent melting period in spring. A very practical indication of the annual cycle of this critical factor in biological productivity shows the basic coherence of the entire eastern and central part of North America, an area of 6.4 million km^2.

Eastern North America Over this large area winter snowstorms are frequent but not always large, and soil moisture might or might not increase very much in this season. Spring melting and heavier precipitation result, however, in large infiltration into the soil if it is not frozen. Spring ends with soil moisture at high levels. This is the store of water that is drawn on during the summer to support evapotranspiration in amounts larger than precipitation. Such evapotranspiration supplies the flow of vapor off the continent with volumes higher than its flow into the continent, as described in an earlier chapter. By the end of summer soil moisture is near the low point of its annual cycle over most of this large area.

In connection with a study of atmospheric water-vapor flux, Rasmusson (1968) calculated the changes in stored water over the yearly cycle (Table VI) for a large sector of North America. The values do not represent changes in soil moisture alone, since the volumes of water stored in the snow cover, the ground water, and lakes and rivers are also changing. These latter storages are, however, minor, and the

TABLE VI

Annual Regime of Moisture Storage over the Plains and Eastern Portions of the United States and Southern Canada[a]

Month	Change of storage during month: Total (mm)	Change of storage during month: Rate (mm day^{-1})	Running 3-month means[b] (mm)	Cumulated change from 1 September to end of current month (mm)[c]
Sept	+13	+0.44	+1	13
Oct	+5	+0.15	+8	18
Nov	+7	+0.23	+7	25
Dec	+9	+0.31	+6	33
Jan	+2	+0.05	+9	35
Feb	+18	+0.62	+8	52
March	+5	+0.15	+4	57
April	−10	−0.35	−4	47
May	−6	−0.20	−9	40
June	−10	−0.32	−11	30
July	−17	−0.54	−13	14
Aug	−14	−0.44	−6	0
Sept	+13	+0.44	+1	13

[a] Area: 6.4×10^6 km^2. Period: 1958–1963. Sign convention: + indicates water going into storage. Source: Rounded from Table 4, Rasmusson (1968).
[b] Running means centered at month noted.
[c] Buildup occurs from Nov–March; decline from March–Aug. Compare rainfall over same area, on p. 106.

major seasonal variation is in the widespread increase in soil moisture in fall and winter and the decrease from April–August.

The variation is not symmetrical. Drawdown occupies only 0.4 of the total cycle, and buildup of storage 0.6. The buildup is irregular, even when averaged over a 5-yr period. A lull in January separates the fall and late winter peaks of input into storage. The peak in the fall probably is soil-moisture accretion, but that in February might reflect accumulation of snow mantle over much of the studied area.

The withdrawal segment of the annual cycle, in contrast, displays great regularity after the fast start in April. The maximum output from the soil-moisture reservoir takes place in July. Primarily it represents evapotranspiration occurring at a rate so high as to exceed by $\frac{1}{2}$ mm day^{-1} the substantial rainfall of that month (3 mm day^{-1}, greatest of any month of the year).

The total range of the water-storage cycle, 57 mm, comes between the end of winter and Labor Day, and indicates the size of the areally averaged turnover in soil and snow storage of water. Of this amount, soil-moisture storage makes up 40–45 mm, a cyclic variation upon which are superimposed the episodic inputs during storms and the drawdowns between storms.

While damaging periods of low soil moisture are frequent in eastern North America, the rigidities of our thinking about water-resource development are such that we have given little heed. One economist comments that "in regard to irrigation we are doing far too much in the dry areas, and probably not enough in the moist ones" (Boulding, 1964). Thornthwaite's name for these periods, "hidden drought," is highly appropriate.

A sign of changing attitudes is evident in the establishment in the National Oceanographic and Atmospheric Administration of Palmer's method of keeping a running soil-moisture inventory. This is calculated over most of the country and maps are drawn every week (Palmer, 1965, 1968) (Fig. IX-11). Originally used to guide the proclamation of drought–disaster areas qualifying for Federal subsidies,

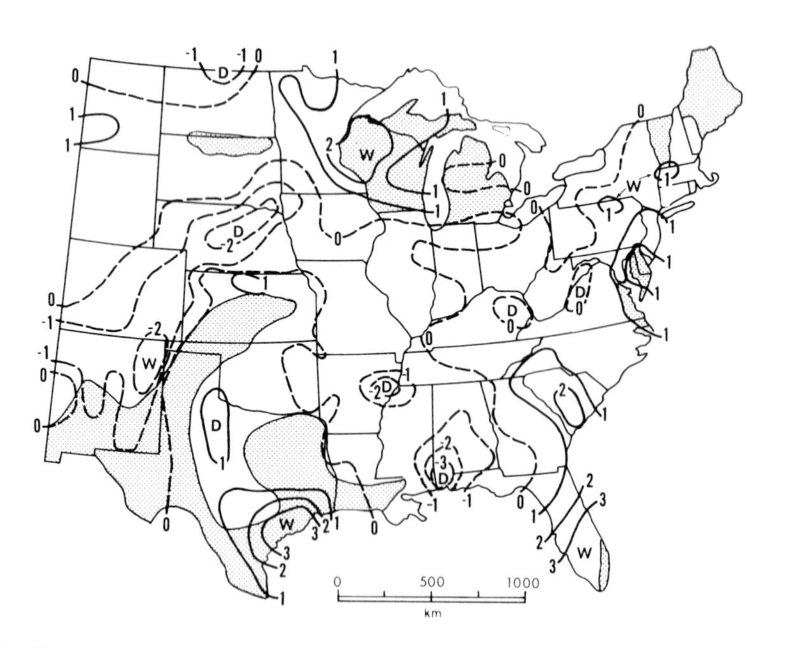

Fig. IX-11. Crop–moisture index derived from Palmer drought index. Isolines of the index in dry areas are shown by dashed lines, in wet areas by solid lines. Unshaded areas are drier than they were on the previous Friday. (Data of U.S. Environmental Sciences Service Administration, Environmental Data Service, *In* Palmer, 1968.) 21 June 1968.

these maps, which are transmitted over the regular facsimile network of the National Weather Service's forecasting program, are widely used by those concerned with crop prospects. In 1974, data on soil moisture gave some lead time ahead of the harvest estimates made by agronomists and crop reporters in the fields, when the short but severe drought of early summer so reduced the corn crop from which we had hoped to realize foreign exchange, oil imports, and availability of grain-fed meat.

Global Values Changes in the gross amount of water held at and near the earth's surface represent coming and going of snow cover, changes in the level of the water table, and changes in soil moisture, the latter changes being the largest. Averaged over the zone from 40 to 50N latitude, the total storage is greatest in March, when it amounts to 473×10^{10} tons, and least in September, when it amounts to 223×10^{10} tons (van Hylckama, 1956; Mather, 1974, p. 153). The effect of the growing season in drawing down the large end-of-winter storage is obvious.

Even if averaged over the whole globe, including lands of the Southern Hemisphere with a reversed change from March to September (18–38 $\times$ 10^{10} tons), the effect of the growing season of the ecosystems in the Northern Hemisphere is still present:

High:	March	2940×10^{10}	tons
Low:	September	2180×10^{10}	tons
	Difference	760×10^{10}	tons

This last figure is compensated by a counter change in the volume of water in the ocean basins.

FREEZING AND MELTING OF SOIL WATER

Soil moisture sometimes ceases to be "moist" when its physical state changes from liquid to ice. This aspect of seasonality, like the seasonal changes in moisture availability, also affects the value of water in ecosystems. Furthermore, frozen soil becomes an obstacle to drainage and consequently to oxygen entry into the soil body.

Diurnal Cycles

The diurnal freeze–thaw cycle occurs only during the parts of the year when soil temperature passes through the 0° level. In the middle

Plate 19. Churning exposed soil by nocturnal growth of crystals of needle ice in the Snowy Mountains, Australia at an altitude of 1 km. The snow grass (*Poa caespitosa*) does not cover the soil, and more palatable plants between its tussocks are vulnerable to overgrazing.

latitudes this amounts to a few score days in fall and in spring,* but in some mountain areas several scores of days. Frequent occurrence of the diurnal freeze–thaw cycle causes alternation of the mechanical stresses that are developed when water freezes in pores with the daytime fluidity of the thawed layer (see Plate 19). This cycle has significance for soil-forming processes and also for movement of the fragmented, saturated semifluid material.

Annual Cycles

The soil water at latitudes higher than about 40N in continental interiors usually remains frozen for a considerable period in winter. If

* Numbers of freeze-thaw cycles at the soil surface are related with the number of days in which the temperature of the air at 1.5 m crosses the freezing point. Some values of the latter quantity are: 10 days yr^{-1} at New Orleans, 20 at Baton Rouge, 60 at Memphis, 90 at St. Louis, and 90 at Milwaukee (Hershfield, 1974). Westward, the values rise from 100 days at Omaha to 140 at Denver, as the concentration of water vapor in the atmosphere decreases and emits less downward longwave radiation at night.

moisture content is low when freezing occurs, the soil structure remains fairly porous and does not become impermeable. On the other hand, if the large pores are full when freezing occurs, the resulting matrix of ice and rock may have the consistency of concrete and be virtually impermeable to the free drainage of water from surface melting when it begins the following spring.

In this situation most of the precipitation that builds on the ground surface during the winter is converted into runoff into the streams, little of it being added to the soil-water reserve (Baker, 1972). Fig. IX-12 shows the relation of winter snowfall to spring runoff in 7 yr from a basin in southern Minnesota.

In Permafrost Where permafrost, i.e., deep perennially frozen layers of soil exists, as shown for North America in Fig. IX-13 (Stearns, 1966), the depth of summer thawing does not penetrate beyond the frozen ground. Only a thin skim of soil, relatively speaking, experiences the summer thawing and winter refreezing—usually less than 1 or 2 m.* Fig. IX-14 (Stearns, 1966) shows the downward advance of the zero isotherm and the thawed soil from May to September. Soil

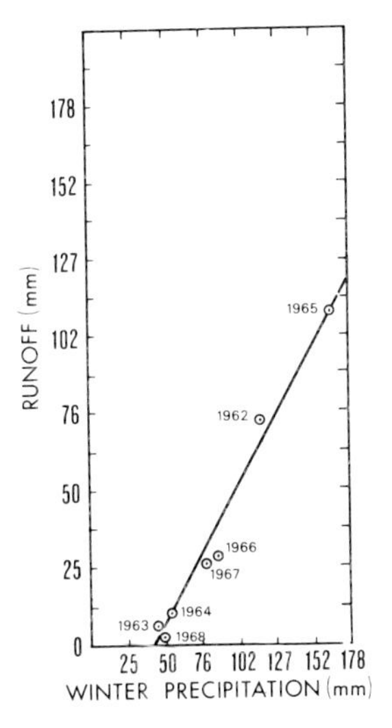

Fig. IX-12. Winter precipitation and spring runoff in the Cottonwood River at New Ulm, Minnesota, 1962–1968 ($r = 0.99$) (from Baker, 1972).

* This depth is less than the depth of snow and ice melted off a perennial snowfield. The difference reflects the relative slowness of heat transport from the irradiated active surface to the thawing front deep in the soil.

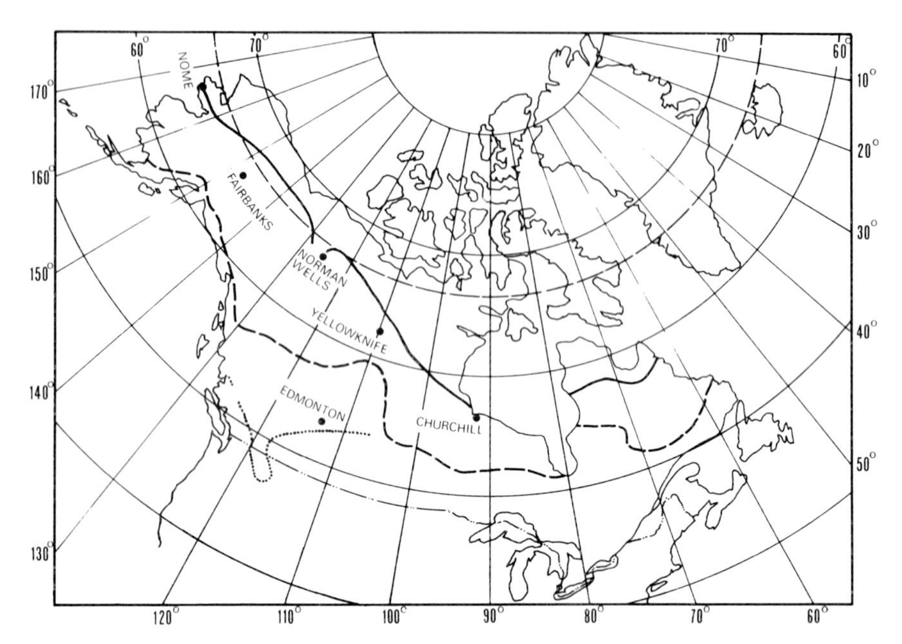

Fig. IX-13. Distribution of permafrost in North America: southern boundary of continuous (solid line), boundary of discontinuous (dashed line), and uncertain boundary of discontinuous permafrost in the Rockies (dotted) (from Stearns, 1966, p. 7).

conductivity affects depth of thawing, a peat soil thawing to only 33 cm after absorption of 30 kJ m^{-2} net surplus of whole-spectrum radiation, while a silty loam thawed about 60 cm (Moskvin, 1974). Free water at the surface has a low albedo, and promotes an increase in thawing beneath it of about one quarter. Density of vegetation cover has a similar effect in increasing the depth of the melted layer; bare soil thawed to 130 cm and more.

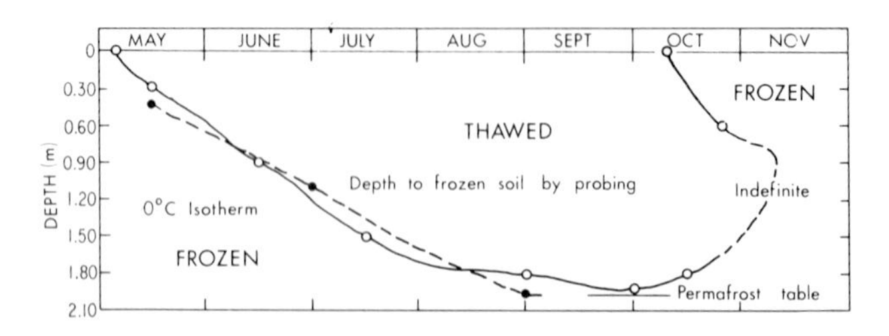

Fig. IX-14. Penetration of summer thawing of frozen soil at Fairbanks, Alaska in 1948. Depth of 0°C isotherm (solid line) was determined by thermocouples and depth of ice-cemented soil (dashed line) by rod probings (Stearns, 1966, p. 24).

It is often noted that where the vegetation cover of the tundra soil has been destroyed, as it easily is by vehicles, more heat is conducted into the soil. Reduction of the bearing strength of a deep layer of soil produces instability in slopes and land forms. The melted lines along a vehicle track fill with water; lowlands undergo a complete change in character. A landscape that is productive of caribou pasturage turns into a useless mudhole, something that we are likely to see more and more as so-called development penetrates the tundra lands. Much redesign of the Alaskan oil pipeline had to be done for this reason.

OUTFLOWS OF WATER FROM THE SOIL

As in other storage zones the storage of water in a soil body provides an opportunity for partition among several modes of subsequent movement. During and after storms considerable lateral movement of water takes place in the upper layers of the soil on slopes. During much of the time between storms, water is moved upward by plants to their foliage; some water drains downward to depths beyond the reach of plant roots. Some of this internal movement takes place as vapor, most of it as liquid, as will be discussed later.

As indicated in the chapter on infiltration, the soil environment for water in motion is complicated. One panel felt that "much needs to be learned before we can use the basic attributes of a soil to soundly predict the nature of the pore system and its facility for water transmission under varying degrees of saturation" (Soil Science Society of America, 1966).

A writer on hydrology says that "the zone of aeration is the 'no man's land' of hydrology. It has received some attention from the groundwater hydrologist, who is interested in how water travels through it both downward to and upward from the water table; from the meteorologist and the surface-water engineer, who are interested in its . . . modification of runoff; from the soil scientist, who is interested in it mainly as a reservoir of moisture for plant growth; and from the sanitary engineer, who is interested in it as a transporter of and a temporary or permanent trap for various contaminants. . ." (Hackett, 1966). These questions of water storage and movement apply to infiltration and soil-moisture storage, and are likely to receive more attention as the renovating powers of the soil are increasingly employed to treat urban waste waters. In order to predict the complex chemical and biological processes that decide "the fate of waste-water constituents," to use the expression of Bouwer and Chaney (1974, p.

135), it will be necessary to understand soil–water relationships much better. These constituents—mass fluxes carried by infiltrated water—include suspended solids and dissolved salts, organic carbon, bacteria and viruses, nitrogen and phosphorus, fluorine, boron, and metals of all kinds, each differently acted on within the moist soil body. What happens in the soil of these special "land-treatment" ecosystems and of all ecosystems is of vital importance if the downward outflow of water is to be usable rather than harmful to the groundwater body that receives it and if the upward flows through plant systems are to maintain biological production. Rates and volumes of these outflows are subjects of following chapters.

REFERENCES

Albrecht, J. C. (1971). A climate model of agricultural productivity in the Missouri River basin, *Publ. Climatol.* **24,** No. 2, 107 pp.

Baker, D. G. (1972). Prediction of spring runoff, *Water Resources Res.* **8,** 966–972.

Barton, M. (1974). New concepts in snow surveying to meet expanding needs. *In* "Interdisciplinary Symposium on Advanced Concepts and Techniques in the Study of Snow and Ice Resources (Proceedings)," pp. 39–46, Nat. Acad. Sciences, Washington.

Bassett, J. R. (1964). Tree growth as affected by soil moisture availability, *Proc. Soil Sci. Am.* **28,** 436–438.

Bodman, G. B., and Colman, E. A. (1944). Moisture and energy conditions during downward entry of water into soils, *Proc. Soil Sci. Soc. Am.* **8,** 116–122.

Boulding, K. (1964). The economist and the engineer: economic dynamics of water resource development. *In* "Economics and Public Policy in Water Resource Development" (S. C. Smith and E. H. Castle, eds.), pp. 82–92. Iowa State Univ. Press, Ames, Iowa.

Bouwer, H., and Chaney, R. L. (1974). Land treatment of wastewater, *Adv. Agron.* **26,** 133–176.

Budyko, M. I. (1974). "Climate and Life" (D. H. Miller, ed.). Academic Press, New York, 508 pp. (*English transl.* of "Klimat i Zhizn' " Gidromet. Izdat., Leningrad, 1971, 472 pp.)

Chang, J.-H. (1963). The role of climatology in the Hawaiian sugar-cane industry: an example of applied agricultural climatology in the Tropics, *Pac. Sci.* **17,** 379–397.

Day, F. P. Jr., and Monk, C. D. (1974). Vegetation patterns on a southern Appalachian watershed, *Ecology* **55,** 1064–1074.

Dickinson, W. T., and Whiteley, H. (1973). Watershed areas contributing to runoff. *In* "Results of Research on Representative and Experimental Basins," *Vol. 1*, pp. 12–26. Internat. Assoc. Hydrol. Sci–Unesco, Paris.

Dreibelbis, F. R. (1963). Land use and soil type effects on the soil moisture regimen in lysimeters and small watersheds, *Soil Sci. Soc. Am. Proc.* **27,** 455–460.

Ekern, P. C. (1967). Soil moisture and soil temperature changes with the use of black vapor-barrier mulch and their influence on pineapple (*Ananas comosus* (L.) Merr.) growth in Hawaii, *Proc. Soil Sci. Soc. Am.* **31,** 270–275.

Fitzpatrick, E. A., Slatyer, R. O., and Krishnan, A. I. (1967). Incidence and duration of periods of plant growth in central Australia as estimated from climatic data, *Agric. Meteorol.* **4,** 389–404.

Hackett, O. M. (1966). Ground-water research in the United States, U. S. Geol. Surv. Circ. 527, 8 pp.

Hershfield, D. M. (1974). The frequency of freeze-thaw cycles, *J. Appl. Meteorol.* **13,** 348–354.

Horton, R. E. (1937). Determination of infiltration-capacity for large drainage-basins, *Trans. Am. Geophys. Un.* **18,** 371–385.

Hylckama, T. E. A. van (1956). The water balance of the earth, *Publ. Climatol.* **9** (2), 59–117.

Kulik, M. S., and Sinel'shchikov, V. V. (1966). "Lektsii po Sel'skokhoziaistvennoi Meteorologii." Gidrometeorol. Izdat., Leningrad, 340 pp., illus.

Leeper, G. W. (1964). "Introduction to Soil Science," 4th ed. Melbourne Univ. Press, Melbourne, 253 pp.

Manogaran, C. (1973). Economic feasibility of irrigating Southern pines, *Water Resources Res.* **9,** 1485–1496.

Marshall, T. J. (1959). Relations between water and soil, Commonwealth Agric. Bureaux Tech. Commun. 50, Farnham Royal, England.

Mather, J. R. (1974). "Climatology: Fundamentals and Applications." McGraw-Hill, New York, 412 pp.

McCalla, T. M., and Army, T. J. (1961). Stubble mulch farming, *Adv. Agron.* **13,** 125–196.

Moskvin, Yu. P. (1974). Investigations of the thawing of the active soil layer in the permafrost zone, *Sov. Hydrol.* **5,** 323–328.

Palmer, W. C. (1965). Meteorological Drought, Weather Bureau Res. Paper 45, 58 pp.

Palmer, W. C. (1968). Keeping track of crop moisture conditions, nationwide: the new crop moisture index, *Weatherwise* **21,** 156–161.

Pereira, H. C. (1973). "Land Use and Water Resources in Temperate and Tropical Climates." Cambridge Univ. Press, London and New York, 246 pp.

Rasmusson, E. M. (1968). Atmospheric water vapor transport and the water balance of North America. II. Large-scale balance investigations, *Mon. Weather Rev.* **96,** 720–734.

Rauner, Iu. L. (1965). O gidrologicheskoi roli lesa, *Izv. Akad. Nauk SSSR Ser. Geog.* No. 4, 40–53.

Richards, L. A. (1960). Advances in soil physics, *Proc. Int. Congr. Soil Sci., 7th* **1,** 67–79.

Rickard, D. S. (1968). Climate, pasture production and irrigation, *Proc. New Zealand Grassland Assoc.* **30,** 81–93.

Rickard, D. S., and Fitzgerald, P. D. (1969). The estimation and occurrence of agricultural drought, *J. Hydrol. (N. Z.) 7,* No. 2, 11–16.

Rode, A. A. (1969). "Osnovy Ucheniia o Pochvennoi Vlage, Tom II. Metody Izucheniia Vodnogo Rezhima Pochv." Gidrometeorologischeskow Izdatel'stvo, Leningrad, 287 pp.

Soil Science Society of America Water Resources Committee, C. H. Wadleigh, chairman. (1966). Soil characteristics in the hydrologic continuum, *Proc. Soil Sci. Soc. Am.* **30,** 418–421.

Staple, W. J. (1964). Dryland agriculture and water conservation. *In* Research on Water: A Symposium on Problems and Progress, pp. 15–30. Soil Sci. Soc. Am. (ASA Spec. Publ. 4), Madison, Wisconsin.

Stearns, S. R. (1966). Permafrost (Perennially Frozen Ground). U. S. Cold Regions Res. Eng. Lab., Hanover, New Hampshire, Cold Regions Sci. Eng. Pt. I, Sect. A-2, 77 pp.

Stoeckeler, J. H., and Curtis, W. R. (1960). Soil moisture regime in southwestern Wisconsin as affected by aspect and forest type, *J. Forestry* **58,** 892–896.

Thornthwaite, C. W. (1945). *In* Report of Committee on Evaporation and Transpiration, H. C. Wilm, chairman, *Trans. Am. Geophys. Un.* **26** (Pt. V), 683–693.

Thornthwaite, C. W. (1948). An approach toward a rational classification of climate, *Geog. Rev.* **38,** 55–94.

Thornthwaite, C. W., and Mather, J. R. (1955). The water balance, *Publ. Climatol.* **8,** No. 1, 104 pp.

Walker, J. M. (1965). Distribution of crops with respect to mean potential soil moisture deficit at the end of August, *Meteorol. Magazine* **94,** 47–51.

Ward, H. S., van Bavel, C. H. M., Cope, J. T. Jr., Ware, L. M., Bouwer, H. (1959). Agricultural drought in Alabama. Alabama Exp. Sta. Bull. 316, 53 pp.

Wisconsin Statistical Reporting Service (1972). Weekly Crop Weather Report, 10 October 1972.

Wisconsin Statistical Reporting Service (1973). Weekly Crop and Weather Reports.

WMO (1968). Practical Soil Moisture Problems in Agriculture, Tech. Note 97, 69 pp.

WMO (1974). "Methods for Water Balance Computations, An International Guide for Research and Practice" (A. A. Sokolov and T. G. Chapman, eds.). Unesco Press, Paris, 127 pp.

Chapter X

EVAPORATION FROM WET SURFACES

This and the next three chapters deal with water moving in a direction opposite to the downward movement described in the preceding chapters, that is, movement up into the atmosphere from systems at the earth's surface. This is a flux in which the critical processes are (1) to get the water molecules into the mobile gaseous state, and (2) to remove them from the evaporating surface, a process that is rather complicated for vegetation, and which is therefore discussed separately. The availability of water to be evaporated might be a third factor, and the process of evaporation from drying surfaces also is deferred to a later chapter.

Vaporization of water requires more energy than does vaporization of most other substances. The latent heat of vaporization is about 2500 kJ kg^{-1} of water if in liquid form, about 2800 if in solid form. In this chapter we look at environments in which liquid water is freely available and from which the removal of water molecules is relatively uncomplicated, leaving the supplying of energy as the primary consideration.

DETERMINING EVAPORATION RATES

When energy loosens the bonds that hold water molecules together in the liquid or solid state, two things happen: ice melts, water vaporizes. The first transformation is highly visible, and has been familiar to man for so long that our ancestral Indo-European language had words for melting and thawing. The second transformation, not visible at all, is expressed by a coined word, "evaporation," that is not recorded earlier than the year 1545. Although evaporation goes on in front of our eyes when we look at a forest or a pond, the process is not well understood nor easily or accurately measured.

To minimize these problems of determination, they can be approached in three ways: as the supply of water to the site where vaporization takes place, as the removal of vapor from this site, and as the rate at which energy is converted at the site into latent heat. Where does this energy come from?

The Energetics of Evaporation

The simple energy-budget equation is balanced in the form:

$+R_A$ radiant energy absorbed by the evaporating surface, including both short-wave and long-wave forms; also called "radiant-energy intake" (see Chapter XI)

$-R_E$ radiant energy emitted by the evaporating surface in accordance with the Stefan–Boltzmann relation

$\pm S$ heat exchange with layers below the evaporating surface

$\pm H$ exchange of sensible heat with the air

$-E$ conversion of energy by the vaporization process into latent heat

———

0

Here + means energy gained by the evaporating surface and − means energy lost from it. The fluxes of sensible heat (S and H) might be directed either toward or away from the evaporating surface under different conditions of water or air temperature distribution; their direction is indicated by + or −. We will examine these energy sources in the order given.

Radiant Energy Two forms of radiation impinge on ecosystems and water bodies at the earth's surface; for convenience they are called "short wave" and "long wave." Solar radiation falls in the short wavelengths (shorter than about 4 μm) and radiation emitted by gases and clouds in the atmosphere is called long wave (wavelengths longer than about 4 μm). Part of each flux is reflected by the earth's surface, although in the case of water surfaces the fractions are generally less than 0.05; the rest penetrates the surface and is progressively absorbed with distance traveled.

The average rates at which short-wave and long-wave radiation is absorbed in water bodies are large. For example, during summer in central North America they are about 200 W m^{-2} of short wave and 350 of long wave. These amounts of energy supply correspond to the daily vaporization of approximately 7 and 12 kg of water m^{-2} (depths of 7 and 12 mm day^{-1}). Both forms of radiation vary through the day, the familiar on-and-off cycling of solar radiation (600 W m^{-2} or more in the middle of the day, zero during the night) being particularly striking. Long wave is more steady in both the daily and annual cycles.

Not all of the energy from absorbed radiation can be converted into evaporation or other nonradiative energy conversions. The price of existence as a wet surface is to maintain a surface temperature of at least 0°C, and this implies the emission of long-wave radiation at a rate of 315 W m^{-2} from the surface.

Summing all the gains by absorption of short-wave and long-wave radiation and subtracting the 315 W m^{-2} just noted, we have the accompanying tabulation as typical of day and night in summer and winter in the interior of North America (in W m^{-2}).

Typical day	Absorption of solar radiation	Absorption of long-wave radiation from the atmosphere	Emission of radiation by surface[a]	Sum
Midday in summer	600	370	−415	+555
Night in summer	0	330	−415	−85
Midday in winter	130	290	−315	+105
Night in winter	0	270	−315	−45

[a] Taking the surface temperature as 20°C in summer, 0°C in winter.

Clearly the sum of all the fluxes of radiation is a highly variable source of energy; it is large in the long summer days, with a net loss every night. While these variations are more or less predictable, except for the effect of passing cloud decks, they nevertheless impose a pattern on the other, or nonradiative fluxes in the energy-budget expression cited earlier. By definition, the sum of the nonradiative fluxes, S, H, and E, must be equal and opposite to the sums of the radiative fluxes given in the accompanying tabulation.

Substrate-Heat Flux One alternative to radiant energy as a source is the substrate, that is, the layers of soil or water underlying the surface. The soil–heat flux is directed toward the active surface more than half the time, i.e., during the hours of later afternoon and darkness. In some cases it supports a small rate of evaporation from the soil itself.

Heat moves upward from the deeper layers of a water body to its surface during the months of fall and winter. Low intensity of radiation in these months has little inhibiting effect on evaporation, which proceeds vigorously until the time when ice forms on the lake surface, if it does. The enormous amounts of heat stored in deep water bodies make it possible for them to evaporate much more water than do adjacent land surfaces.

Sensible-Heat Flux The second alternative source of energy for evaporation is the atmosphere itself, which thus is simultaneously humidified and cooled by turbulent exchange between it and the underlying wet surface. In this situation the surface must be cooler than the air and also moister.

A simple example is the wet-bulb thermometer, in which a cotton wick keeps the bulb wet while it is strongly ventilated by being swung. If the air is quite dry, evaporation from the wick removes heat from the bulb, which has only a small amount of "substrate" heat storage in its tiny volume and also is shielded from any excess of radiant energy. The energy budget of the wet bulb thus contains only two items, the outward flux of latent heat E and the inward flux of sensible heat H. These offset each other exactly. The drier the air, the greater are both fluxes since the bulb temperature sinks to a level much lower than the air temperature.*

In natural situations the other fluxes at the surface of the earth are usually not zero, so the sensible-heat flux seldom becomes the sole source of energy for evaporation.

Combined Sources of Energy for Evaporation The energy sources that support vaporization differ from place to place and time to time. Each can act singly, but it is more likely that combinations of them provide the energy that is converted into latent heat.

In a Milwaukee November we see the grass still green, and deduce that it is continuing its life processes, including transpiration, although at a low rate, powered by the slanting rays of the low sun. It is not as obvious, perhaps, that at the same time Lake Michigan also is vaporizing water and at a much higher rate. Its rapid evaporation is supported less by solar energy than by the heat put into storage in its deeper layers a few months earlier and now being drawn upon in fall. Grass evaporation, having little access to stored energy and depending largely on current radiant energy, which is supplied at a slow rate, proceeds slowly at this season when lake evaporation is continuing at a daily level of 3–4 kg m^{-2} as a result of its good source of heat.

When a surface is energized from large nonatmospheric heat sources, either by absorbing radiant energy R or receiving a large flow

* This difference is utilized as a means of determining the dryness of the air—the psychrometric method.

of heat from beneath S, the two heat fluxes from it into the atmsophere find themselves in competition. Usually the latent-heat flux prevails over that of sensible heat, if the surface is wet. (If it is dry, there is, of course, little or no evaporation or latent-heat conversion taking place.)

The ratio between these two competing flows of heat, both transported by turbulent convection in the atmosphere, is called the "Bowen ratio," named for the oceanographer I.S. Bowen (1926). Usually the sensible-heat flux is much the smaller of the two, and the ratio is of the order of 0.1–0.3. However, by destabilizing the air this flux increases its diffusivity for water vapor, and the same vapor flux is moved at a smaller gradient, which allows a lower temperature at the evaporating surface. In a study of cooling ponds, those unique ecosystems built to move heat from thermal power plants into the environment, Ryan *et al.* (1974) found that thermal convection made possible the removal of an added flux of 500 W m^{-2} of dump heat from a 4-km^2 pond, with a rise of surface temperature above the 2-m air temperature of only 9°C. The rate of evaporation, 14 mm day^{-1}, is one of the highest from any system at the surface of the earth.

Surface Temperature and Vapor Pressure

Each energy flux, including evaporation, is determined in part by the temperature of the active surface generating it. Long-wave radiation is emitted in accordance with the fourth power of the absolute temperature of the surface. The flux of sensible heat moves away from the surface in proportion to the temperature gradient from the surface into the adjacent air. The heat exchange between the surface and substrate depends on the corresponding temperature gradient below the surface.

The latent-heat flux associated with evaporation also follows a gradient, that of specific humidity or vapor pressure. The surface value determining this gradient is a single-valued function of the surface temperature, expressed as the Clausius–Clapeyron relation (for two physicists of the nineteenth century) in the accompanying tabulation from the Smithsonian Institution (1966). The accelerating trend of specific humdity with increasing temperature (shown in the last column) means that during a rise in surface temperature the gradient driving the latent-heat flux usually grows faster than do the gradients driving the heat fluxes into the atmosphere and soil. Therefore the conversion of heat in evaporation tends to preempt most of the energy.

Surface temperature (°C)	Vapor pressure (mb)	Specific humidity at sea-level pressure (gm water kg^{-1} air or parts per 1000)	Rate of change in specific humidity for a 1° change in surface temperature (g kg^{-1} deg^{-1})
−20	1.0	0.6	0.1
0	6.1	3.7	0.3
10	12.3	7.4	0.5
20	23.4	14.6	0.8
25	31.7	19.8	1.2
30	42.4	26.3	1.5
32	47.6	29.7	1.7
35	56.2	35.2	1.9

Linacre (1964) and Priestley (1966) have shown that over a wet surface this preeminence of the latent-heat flux becomes virtually complete at a surface temperature of about 34°C. The water surface, continuing to humidify the air, has no further power to warm it. Thus 34°C (307K) represents a maximum temperature for any wet surface, regardless of how it might be loaded by radiant energy.

Verification of this fact is seen in the absence of ocean areas where water temperatures exceed 34°C, or where temperatures in oceanic air exceed 34°C (unless air has moved from hotter surfaces of dry land). Dry surfaces, on the other hand, are easily heated to high temperatures under strong radiation loading, and in turn heat the overlying air to temperatures higher than 34°C. The difference between a hot dry desert surface and the relatively cool foliage of an irrigated field in an oasis demonstrates that evaporation can remove large amounts of heat.

Surface temperature is at the same time a cause and a consequence of energy fluxes. It varies in response to changes in the rate at which the surface takes in energy and the ease with which it can get rid of this energy load. Corresponding variations in the surface vapor pressure produce changes in the rate of evaporation.

Uncertainties about Evaporation

The central problem in talking about evaporation is the scarcity of direct measurement. Only one method can be classed as direct measurement; all others depend (a) on statistical correlations of other fluxes to the vapor flux, (b) on sets of measurements in the energy-budget framework just described, (c) on sets of measurements in the water-budget framework, or (d) Dalton's evaporation law.

Statistical correlations between the vertical flux of vapor and the fluxes of momentum, heat, carbon dioxide or other substances are not precise because of our ignorance of the nature of turbulent transport in the atmosphere. The mechanisms of this transport do not necessarily work the same for different physical quantities, i.e., for heat as for water vapor.

In conditions of neutral thermal stability in the lower layers of the atmosphere and strong air movement, the turbulent transport mechanisms are most alike. In stable conditions marked by inversion, or in very unstable conditions this similarity cannot be counted on.

Measurements of energy fluxes to the evaporating surface of a lake are most accurate over long time intervals, for over a short time the flow of heat out of the lower layers of water is impossible to determine. The energy-budget method is most often applied to periods of a month or so, over which the cooling of the lower layers can be measured.

The water-budget framework also is most useful over long periods of time, since the movement of water into or out of the lake banks cannot be evaluated over short times. Over any length of time, however, the water-budget method is limited by the lack of measurements of precipitation falling into the lake. Because the lake energy budget often changes atmospheric stability over the lake from the stability conditions over land and because stability affects precipitation processes, it is risky to consider that lake rainfall is correctly represented by rain gages on nearby land sites. (See Chapter V.)

Evaporation from water surfaces is never limited by a shortage of the evaporating substance, therefore variations are associated with energy supply or with the vapor gradient from surface to air. Factors in evaporation can be considered from their connection with either supplying energy or removing the vapor molecules.

The case of evaporation from snow, which we examined in an earlier chapter, illustrates this circumstance. In general, evaporation from snow is limited by a lack of energy supply, as in the Arctic night. Often, in situations of apparent high energy level, as in advection of warm marine air over a snow cover, the vapor gradient is toward rather than away from the snow cover, and no evaporation occurs. Let us now proceed to examine evaporation from water substance in a quite different physical state, i.e., bulk water in lake basins and other aquatic ecosystems.

EVAPORATION FROM DEEP WATER BODIES

All water bodies except the most shallow are distinguished from other ecosystems by an extremely large capacity to accept heat into storage and subsequently to give it back. The availability of heat at the water surface to support evaporation depends less on short-time fluctuations in radiation than on the stirring of the water or the air. We need to look at this factor of heat storage in the water.

The Role of Heat Storage

Lake Ontario This lake has received considerable study, some of it connected with the importance of its level to the operation of the St. Lawrence Seaway, in hydroelectric production, and as a hazard to riparian property. In one study of its water budget, a heat budget was developed (Table I) that illustrates the annual cycle. Like the mountain snow covers described earlier, evaporation is absent in spring. Even a contribution of sensible heat from air warmed over surrounding lands in April and May does little to speed evaporation. Most of the heat from the warm air, like that from absorbed solar radiation at this season, simply goes to warm the deep water of the lake. The surface warms up slowly, and generates little outward movement of water vapor.

During the cooling phase of the annual cycle, which lasts from August until February or March, the situation is reversed as heat stored in the depths of the lake in earlier months now comes to the surface. Here it makes a large addition to the waning surplus in the radiation account. The total heat available to go into the atmosphere is now very great, and supports rapid evaporation during the dark winter. In fact, evaporation continues through nine months of the year, and is large in six months (September–February). Its energy support in September and early October comes chiefly from radiation, the rest of the time it comes from subsurface heat, which also buffers the thermal environment of the fish species in this aquatic ecosystem.

The Influence of Lake Depth The depth of a water body affects heat storage in the water because with sufficient stirring by wind—up to a depth of 50 m or more—a large mass of water is brought into thermal communication with the evaporating surface. The great heat-storage capacity keeps the surface cool and evaporation slow in the spring and summer and maintains a warm surface and rapid evaporation in the fall and winter. In May, for example, the latent-heat flux (dashed curve, in Fig. X-1a) from a 5-m deep lake or reservoir is 120 W m^{-2}, or about 4 mm day^{-1}, whereas from a 20-m lake it is only 30 W m^{-2}, a quarter as much. The deep lake is still taking more heat into storage in May than the shallow, and allowing much less to escape upward into the air. The converse occurs in fall.

TABLE I

Water and Energy Budget of Lake Ontario[a]

Parameter	J	F	M	A	M	J	J	A	S	O	N	D	J
Net income of energy from radiation exchanges[b]	+2	+35	+85	+140	+185	+200	+200	+160	+110	+50	+10	−5	+2
Exchange of heat between lake surface and deep water[b]	+250	+160	−15	−165	−245	−190	−130	−70	0	+65	+145	+200	+250
Net transport of heat in inflow and outflow rivers and snow melting[b]	−6	−5	−3	0	+1	0	0	−1	−1	0	−2	−5	−6
Energy available at lake surface[b]	+246	+192	+67	−25	−59	+10	+70	+89	+109	+115	+153	+190	+246
Surface temperature[c]	+3	2	2	2	4	10	19	20	17	11	7	5	3
Surface vapor pressure[c]	8	7	7	7	8	12	22	23	19	13	10	9	8
Atmospheric vapor pressure[d]	5	4	6	8	10	14	20	20	15	12	7	5	5
Evaporation[e]	−3.2	−2.6	−1.4	+0.6	−0.3	−0.1	−1.6	−2.6	−3.6	−3.0	−3.0	−2.9	−3.2
Latent-heat flux[e]	−90	−75	−40	+15	−10	−5	−50	−75	−105	−85	−85	−85	−90
Sensible-heat flux as residual[e]	−155	−115	−25	+10	+70	−5	−20	−15	−5	−30	−70	−105	−155

[a] Units: energy fluxes in W m^{-2}, water fluxes in mm day^{-1}, vapor pressure in mb, temperatures in °C. Note: The older data, especially, are subject to change as findings of the International Field Year on the Great Lakes are published. Figures are rounded. Evaporation E is calculated by the Dalton equation $E = KV(e_s - e_a)$, in which V is wind speed over the lake, e_a is vapor pressure over the lake, e_s is vapor pressure at lake–surface temperature, and K is a coefficient (Richards and Irbe, 1969).

[b] Bruce and Rodgers (1962).

[c] Webb (1970).

[d] Richards and Fortin (1962).

[e] Richards and Irbe (1969).

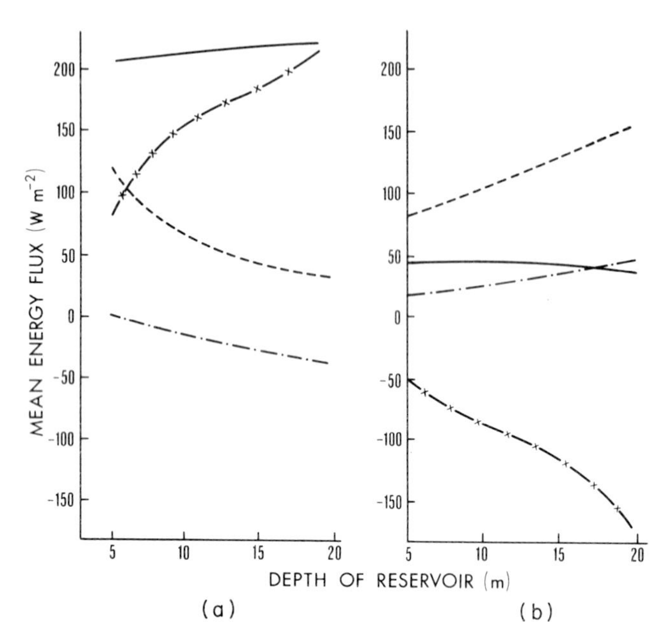

Fig. X-1. Energy fluxes in (a) May and (b) October in lakes and reservoirs of different mean depth (m). The net surplus of radiation of all wavelengths (solid line) is large in May at the surfaces of all the lakes. The change in heat content of the water (–×–×) is greatest in the deep lakes. In May, the larger lakes extract sensible heat (—·—·) from the air, while in October they give it up. Evaporation (– – –) decreases in deep lakes in May, increases in October (Nesina, 1967). From the standpoint of the water surface rather than the water body itself, the sign for change in heat content would be reversed, as it represents a removal of energy from the surface in May. The sign for latent heat flux would also be reversed.

The depth of the lake basin itself is, of course, not as critical as the depth of the layer in thermal communication with the lake surface. In a stratified water body under winter or summer conditions, the depth of the mixed layer above the thermocline is important. This is related to the fetch of wind on the lake. Arai (1966) shows that lake evaporation in spring and summer decreases with greater fetches F, which mix more cold water to the surface, the coefficient of diffusivity in the water being proportional to $F^{2/3}$ (Arai, 1972). He finds winter evaporation from deep mid-latitude lakes to increase proportionally to $F^{1/3}$.

Reservoirs If water from a deep storage reservoir in summer is taken from its middle or lower layers, which are usually colder than its surface layer,* the result is that less energy is removed in the discharge water than is brought into the reservoir by inflowing streams. Energy accumulates in the reservoir, its surface temperature rises, and evaporation increases. In Roosevelt Lake in Arizona this fashion of withdrawal probably increases annual evaporation 150 mm (Koberg, 1960).

In contrast, smaller reservoirs downstream on the same river, which are fed by cold water from Roosevelt Lake, experience less evaporation. For the system of reservoirs on the Salt River the total change was

7.0×10^6 m^3	more evaporation from Roosevelt Lake
2.6×10^6 m^3	less evaporation from the lower reservoirs
4.4×10^6 m^3	added loss of water by evaporation

If all reservoirs in the system are not under the same administration, a transfer of income equal to the value of the changed evaporation also takes place to the benefit of the downstream managers.

In warm dry regions where reservoirs must hold water over periods of drought, the radiation surplus is usually large, and adjacent lands generate streams of hot, dry air moving over the reservoirs. The result is that reservoir regulation of stream flow for irrigation or other purposes incurs excessive costs of water storage in the form of high evaporation.

This fact sets limits in physical terms, quite in addition to the economic costs of construction labor and materials, to the degree of regulation that is logical on any stream, regardless of the availability of subsidized low-interest funds. On the Colorado River, reservoir storage of 36×10^9 m^3 in 1959 provided an annual regulation of 7.8×10^9 m^3 at a cost of 1.0×10^9 m^3 lost by evaporation (Langbein, 1959). Net gain is 6.8×10^9 m^3 (7.8 less cost of 1.0).

Proposals to build more reservoirs to store a total of 97×10^9 m^3 of water would provide 9.8×10^9 m^3 of annual regulation, less 2.6×10^9 m^3 lost by evaporation. This is nearly three times the present amount of evaporation. A large expenditure on dam construction results in a net annual regulation of 7.2×10^9 m^3, not much more than the earlier 6.8 gain. Hydrologically, this gain over the present situation is insignificant. The toll of evaporation is a major obstacle in the way of reaching an artificially controlled water supply, in which the yearly

* In some cases, cold water is drawn to preserve downstream habitat for cold-water species of fish. In others, it is withdrawn for operating convenience and might cause damage to downstream water users, e.g., rice irrigators.

regime of natural input would be completely transformed into a yearly regime fitted to man's requirements.

The Role of Surface and Atmospheric Conditions

In Lake Ontario in fall, heat storage provides a large flow of heat to the surface, keeping it warm and its vapor pressure high. A large amount of heat is available for transfer from surface to atmosphere. The rate of transfer and the partition of the total between sensible heat and latent heat (as vapor) depends on two gradients from the surface into the air. These are the gradient of temperature, depending on the water-to-air temperature difference (3° in fall over Lake Ontario), and the gradient of vapor pressure, depending on the difference between surface vapor pressure (a function of surface temperature, as noted earlier) and atmospheric vapor pressure. The water-to-air difference in vapor pressure above Lake Ontario runs about 2 mb (Table I).

Estimates of the evaporation from a deep water body can be made from data on its surface temperature (and hence surface humidity q_s), atmospheric vapor pressure q, and wind speed u. Based on Dalton's law, a later form of the relation is the Shuleikin–Sverdrup equation

$$E = \chi \rho u (q_s - q),$$

in which χ is a coefficient and ρ is air density, the product $\chi\rho$ approximately equal to 2.5×10^{-6} g cm^{-3} (or 2.5 g m^{-3}) (Budyko, 1971, Eq. 2.88, p. 108).

Wind speed is important for two reasons: high winds accelerate the upward turbulent movement of vapor (and sensible heat) from the surface; and heat stored in deep water is brought to the surface by vertical mixing within the water body, brought about by wind stress on its surface.

If the second effect did not exist, the removal of sensible and latent heat from the surface would chill it, reducing both flows. As the Lake Ontario data show, however, the removal of as much as 200 W m^{-2} from the surface in fall reduces its temperature only slowly because the reservoir of subsurface heat prevents such chilling. Otherwise, cooling of the surface would result in a lowering of vapor pressure to a point at which evaporation would cease.

Wind-generated turbulence brings dry, often warmer air down to the water surface and carries vapor up from it. Because the vapor pressure of the surface is maintained reasonably constant, these variations in atmospheric turbulence produce large short-term variations in evaporation.

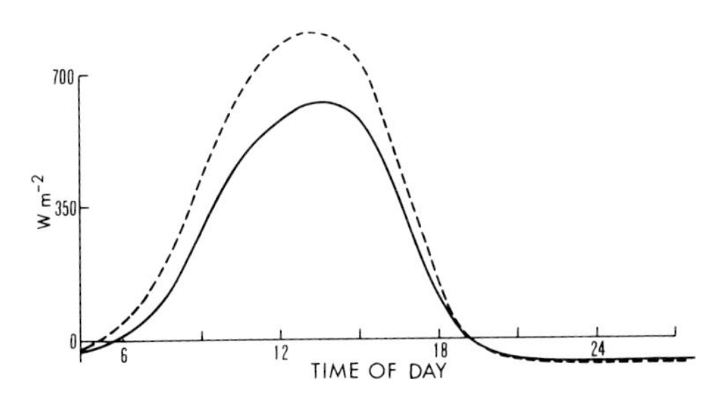

Fig. X-2. Diurnal variation of the net surplus (or deficit) of radiation of all wavelengths at the land surface (solid line) and the surface of Lake Sevan (dashed line) (Timofeev, 1960).

Regimes and Variations Detailed studies of heat budgets and evaporation at Lake Sevan in Armenia illustrate both regular and irregular variations in evaporation. Figure X-2 shows the variation in the net whole-spectrum radiation at the lake surface on an average day in summer, with the smaller values at an adjacent land surface for comparison (Timofeev, 1960). Figure X-3 shows the relatively constant regime of wind effect as indicated by eddy diffusivity K_1 (m^2 sec^{-1}) over the lake. It is generally less than the large daytime diffusivity over the heated land. Figure X-4 shows mean diurnal variation in lake evaporation, which peaks around 1600 at a rate of about 0.3 mm hr^{-1}.

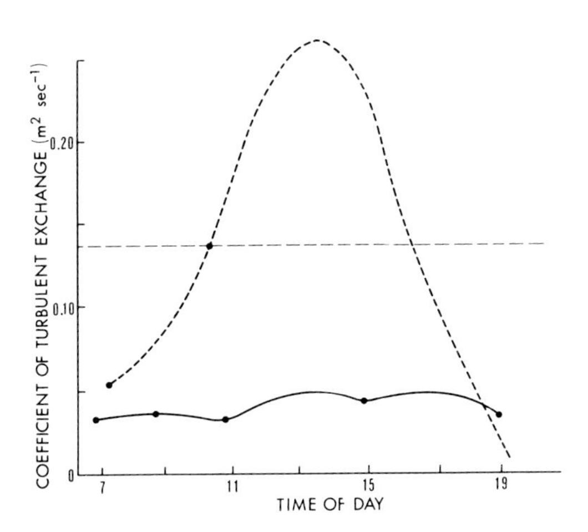

Fig. X-3. Diurnal variation of the coefficient of turbulent exchange in July over land (dashed line) and over Lake Sevan (solid line) (Timofeev, 1960).

It continues through the night, using stored heat, at a level about half that reached in its peak hours.

Changes from day to day are of interest (Fig. X-5). The fluctuations in the latent-heat and sensible-heat fluxes run roughly parallel, suggesting that both depend on turbulence and air temperature and dryness of the airstreams passing over the lake. They display little or no relation to fluctuations in the daily surplus of whole-spectrum radiation. Their independence of the radiant-energy input to the lake demonstrates the countering action of substrate storage of energy in sustaining a high level of evaporation that varies with atmospheric conditions affecting the removal of vapor from the warm water surface. Conditions that control vapor diffusion are more limiting than those that control energy availability.

Alteration of the Water Surface by Monolayers There are several long-chain alcohols that form a film one molecule thick on water surfaces. This doubles the resistance to passage of water vapor into the air. These films reduce evaporation appreciably if they can be maintained uncontaminated and intact on the water surface. Suppression of evaporation might, however, permit an increased surface temperature that has a counteracting effect.

The economics of a monolayer depend on the particular long-chain molecule chosen, its cost, the frequency of spreading, the size and wind exposure of the reservoir, and the value of the water saved. LaMer points out that many field experiments, cheaply and easily done, used impure chemicals and ineffective means of distributing them on the water surface (LaMer and Healy, 1965). Many so-called experiments were done with little preparatory study, e.g., laboratory research on the molecular mechanisms that produce the resistance that the would-be evaporating molecule of water must surmount.

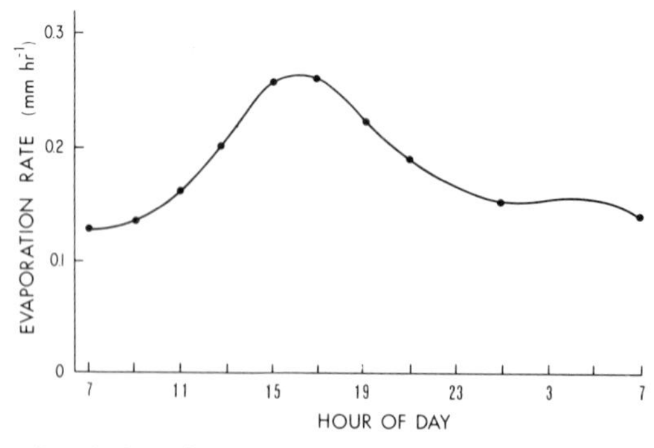

Fig. X-4. Diurnal variation of evaporation in July from Lake Sevan (Timofeev, 1960).

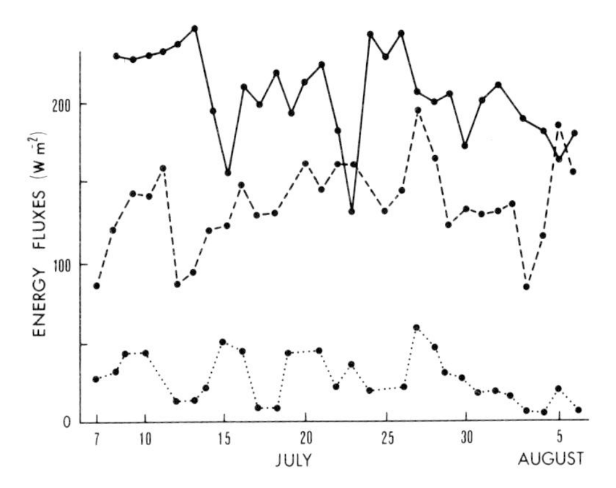

Fig. X-5. Interdiurnal variation of the net surplus of all-wave radiation (solid line) at the surface of Lake Sevan in 1956, with latent-heat flux (dashed line), and sensible-heat flux (dotted line) (Timofeev, 1960). Signs reversed on turbulent fluxes.

The molecular architecture, i.e., length of the chain, has considerable influence on this resistance, which varies from 0.7 sec cm^{-1} for chains with 14 carbon atoms to 6.0 sec cm^{-1} for those with 18 atoms. In the right situations, using the correct chemical formulations and the most efficient means of replenishing the monolayer slick as the wind pushes it aside, evaporation can be reduced by as much as half.

A city reservoir at Broken Hill, in interior Australia, achieved such a reduction. The cost per cubic meter of water saved was about one-half cent, competitive with water produced by other sources available to this city.

EVAPORATION FROM SHALLOW WATER BODIES

The amount of heat stored in a thin layer of water is not large. Insignificant in the annual regime, it exerts some effect on the diurnal regime. The role of heat storage is much smaller in evaporation from shallow water bodies than from deep ones.

Energy Considerations

When daily evaporation totals are desired, the changes in heat storage in the water from day to day can be combined with the daily radiant-energy surpluses to give the daily amounts of heat available at the water surface to supply the two atmospheric turbulent-heat fluxes. These amounts can then be divided according to the Bowen ratio method to give the separate latent-heat flux or evaporation rates. Such a method was developed by Cummings and Richardson (1927) soon after Bowen's work appeared. They applied it in reproducing daily evaporation amounts without having to resort to wind-speed data, at that time ubiquitous in evaporation research but very significant only with deep bodies of water.

Where wind plays no role in stirring a water body and expediting the flow of substrate energy to the surface, or does not carry sensible heat from dry hot land surfaces upwind, it was shown by these investigators to be irrelevant to the process of evaporation. In other words, Cummings and Richardson demonstrated that where air movement does not affect the substrate- or sensible-heat fluxes, it has nothing to do* with evaporation as an energy-conversion process at the surface of shallow water.†

The profile of absorption of solar radiation in an aquatic ecosystem is changed by submersed vegetation. For example, an algal bloom in a 1-m-deep pond of reclaimed waste water (another new kind of ecosystem!) at Phoenix increased the extinction coefficient and concentrated absorption in the top layer of the system, isolating it from the lower layers and widening its diurnal regime of temperature (Idso and Foster, 1974). Midday surface temperature reached 36°C (still lower than the temperature of the desert air). The vapor-pressure difference rose to 38 mb, 0.25 larger than it was on days without blooms, suggesting that evaporation increased. In other aquatic ecosystems, evaporation tends to increase by 0.2–0.3 with submersed vegetation (WMO, 1974, p. 46), although the amount depends on the specific radiation conditions. Emergent vegetation might produce quite different effects, depending on its albedo and aerodynamic properties.

Evaporation Pans

The use of shallow containers of water to derive an estimate of evaporation from deep water bodies has many weaknesses. One is

* The small amount of energy converted from the kinetic energy of the airstream to thermal energy at the water surface seldom exceeds a few watts per square meter.

† Wind has remained fixed in our thinking about evaporation, largely perhaps, because we tend to draw an analogy between our own bodies and natural surfaces. This analogy, however, is spurious. Our skin temperature is held relatively constant at 32°C by heat-circulation and heat-generation processes in its substrate, the layers of flesh beneath it. When we get out of a swimming pool and stand dripping wet in the wind we keenly feel the loss of latent heat. Instead of letting the skin temperature drop until income and outgo of energy come into balance, the mechanisms of the human body attempt to maintain it near 32°C. We operate more like a deep lake than like a pond or a field of plants.

that less short-wave radiation is absorbed in shallow than in deep water. Another is the small amount of heat storage, a third is the changed aerodynamics of the pan, and a fourth is the weak development of any boundary layer or vapor blanket.

An evaporation pan is a traditional device that cannot be called either a meteorologic or a hydrologic instrument, because what it measures cannot be defined in meteorologic or hydrologic terms. Pans were deployed in the nineteenth century weather-observing networks of several countries at a time when scientific understanding of the evaporation process was weaker than now, and when it was thought that the atmosphere had some mysterious quality, called its "evaporating power."

To talk about some "power" of the atmosphere to cause evaporation is to talk in anthropomorphisms. McIlroy (1957) suggests why the idea was popular. If an evaporative power could be discovered, it should be exerted on such devices as pans of water as well as on lakes and croplands. Once you measured its effect on a pan, you would automatically know its effect on natural landscapes. "More recently, however," McIlroy says, "the importance of the difference between various evaporating systems have been increasingly realized." The diversity of nature is here manifested as a diversity of evaporating systems and of the energy they make available to their wetted surfaces.

This history helps explain the acceptance of devices that have no physical basis. Even to indicate evaporation from other shallow bodies of water, pans are defective because of their small size, their poor aerodynamic shape factor (they protrude 25 cm above the ground surface), and their lack of thermal contact with the ground. These discrepancies in environment between pans and real water bodies prevent the analyst from developing genuine physical correspondences, a circumstance recognized by Horton 60 years ago: "The land-exposed evaporation pan appears to be about the poorest device humanly contrivable for the purpose of determining the evaporation losses from broad water surfaces" (Horton, 1917).

In some weather networks containers of water are sunk flush with the ground surface and avoid the aerodynamic problems of a pan raised above the surface. Figure X-6 (Konstantinov, 1968, p. 274) shows the diurnal and annual regimes of evaporation from a sunken basin of water, area 20 m^2, and a lake. The daily curves show some resemblance to one another but the annual curves do not. While evaporation from this 20-m^2 basin displays a symmetrical rise and decline, that from the lake rises steeply to a late and high peak.

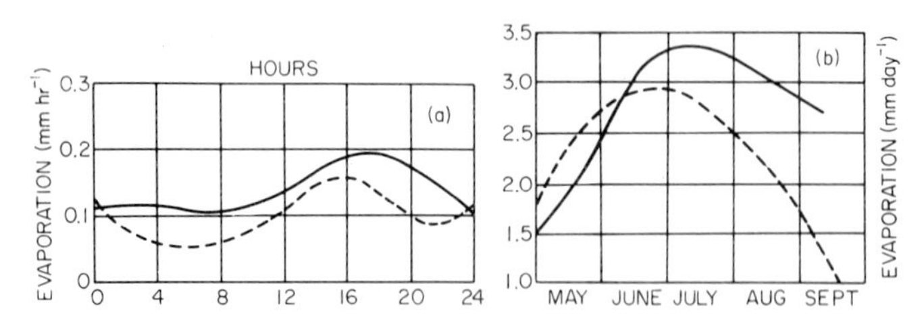

Fig. X-6. (a) Diurnal and (b) annual marches of evaporation from water surfaces: a lake (solid line) and a tank of surface area 20 m^2 (dashed line) (Konstantinov, 1968).

Diurnal Regimes

Some idea of the diurnal regime of evaporation from shallow water comes from flush-mounted weighing lysimeters 1.5 m deep, on which a 2-cm layer of water was ponded, in the middle of a large flooded field near Phoenix (Fritschen and van Bavel, 1963). In springtime conditions of relatively cool air and strong sunshine (360 W m^{-2}) the thin layer of water was quickly heated to a temperature (32°C at noon) much higher than that of the air (21°C at noon), and gave off considerable sensible heat (Fig. X-7). The Bowen ratio was about 0.5

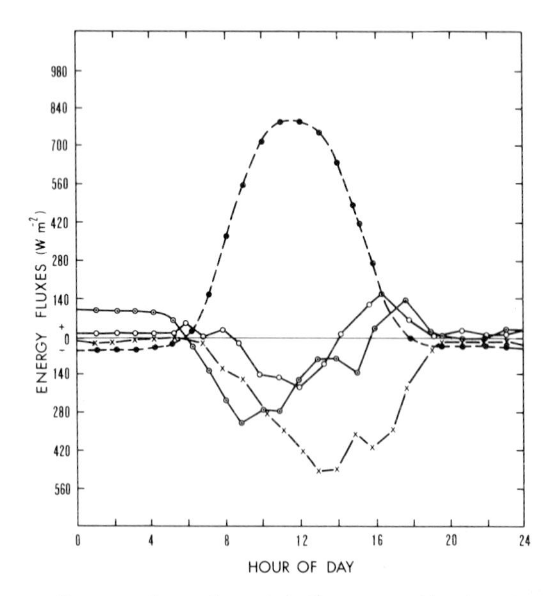

Fig. X-7. Energy fluxes at the surface of shallow water (depth = 2 cm) at Phoenix on 25 April 1961. Net surplus of all-wave radiation (●—●) mean value of 205 W m^{-2}; substrate heat flux (⊙—⊙); sensible-heat flux (○—○); latent-heat flux (×—×) mean value of 170 W m^{-2} (Fritschen and van Bavel, 1963).

during the middle of the day, i.e., the sensible-heat flux was 0.5 of the latent-heat flux.

The large flow of sensible heat tended to reduce the rate of latent-heat flux and evaporation, as did also, in the morning hours, a large flux of heat into the substrate, that is, into the water and the underlying soil of the lysimeter. The noon and the whole-day budgets ($W\ m^{-2}$) are shown in the accompanying tabulation (data from Fritschen and van Bavel, 1962, 1963). Heat stored in the substrate, in the soil as much as in the shallow water, was an important factor, as the large amount stored in the morning returned at a rapid rate during the night. It supported a small sensible-heat flux, met a moderate nocturnal deficit in the radiation budget, and provided for continuation of evaporation. The nocturnal rate of evaporation was about 0.15 of the rate at noon. The buffering action of the substrate-heat storage, although less than what it would be in a deep body of water, prolonged the period of evaporation at the cost of a reduction in its daytime rate.

Parameter	25 April 1961	
	1200	Mean over 24-hr day
Net surplus of whole-spectrum radiation R	+745	+205
Substrate-heat flux S	−225	+2
Sensible-heat flux (as residual) H	−175	−37
Latent-heat flux (lysimeter measure) E	−345	−170
Balance	0	0

EVAPORATION FROM A WET SOIL SURFACE

Evaporation from a shallow body of water filled with rock particles of all sizes, that is to say, a saturated soil in which the big pores are full of water, presents contrasts to evaporation from a water body of unimpeded internal circulation. The prevention of movement within the water by the rock particles means that most of the flow of heat to the surface takes place by molecular conduction, which is slow. Stored heat within the body can therefore not easily be exploited to support evaporation.

Measurements at Aspendale, Australia (McIlroy and Angus, 1964), shown in Table II, from soil in lysimeters of 1.1-m depth, indicate yearly evaporation from a soil-filled lysimeter that was watered at frequent intervals. The total, 940 mm, was 0.86 of that from a water-filled lysimeter.

The lower evaporation from the soil-filled lysimeter probably indicates a somewhat smaller radiation surplus than at the water-filled lysimeter. In addition, in the periods between sprinkler irrigations the soil might have dried superficially so that at times of peak radiation, water was not supplied fast enough to utilize all the radiant energy. In the water-filled lysimeter, on the other hand, much of the peak radiation went into storage that was drawn on to support evaporation by night as well as by day. Later measurements at soil-filled lysimeters surrounded by bare soil rather than grass give somewhat larger figures (Dilley and Shepherd, 1972), but the regime remains different from that of evaporation from a water body.

Large expanses of continuously wet soil are uncommon in nature, because if energy supplies are adequate the surface will soon be colonized by plants (unless biochemically inhibited). In cultural landscapes, in contrast, bare soil is found between the periods of crop-plant occupation.

For example, in Wisconsin corn fields do not lose their bareness until May. Radiant energy during May is strong and on the first day or two after a rain evaporation from the wet soil is rapid. At times it is so vigorous that when the vapor emanating from the fields condenses in the cool air it forms bodies of steam fog so dense as to block visibility when they drift across roads. In the steppe zone of Russia a great deal of valuable moisture evaporates during the "10 to 15 days between

TABLE II

Evaporation from Lysimeters at Aspendale Near Melbourne (38S)[a]

Lysimeter	J	F	M	A	M	J	J	A	S	O	N	D	J
Soil	5	4	3	2	1	1	1	1	2	3	4	4	5
Water	7	5	3	2	1	1	1	1	2	3	4	5	7

Sum (mm):		
	Soil	940
	Water	1095
	Rainfall	630

[a] Units: mm day^{-1}. Source: McIlroy and Angus (1964).

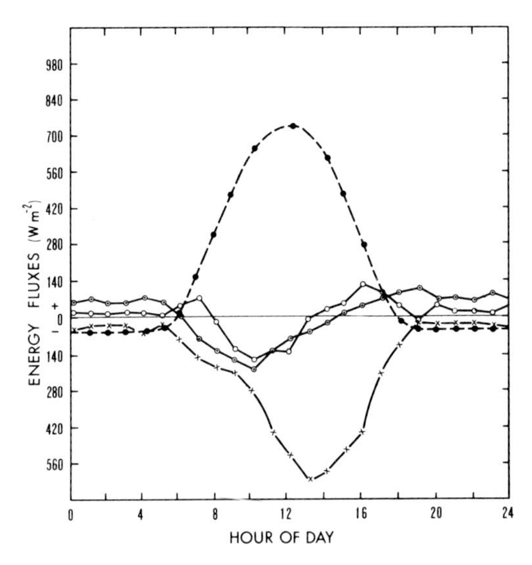

Fig. X-8. Energy fluxes at a wet soil surface at Phoenix, 29 April 1961. Net surplus of all-wave radiation (●—●), mean value of 197 W m^{-2}; substrate-heat flux (⊙—⊙); sensible-heat flux (○—○); latent-heat flux (×—×)), mean of 201 W m^{-2} (from Fritschen and van Bavel, 1962).

snowmelt and sowing"—so-called "unproductive" evaporation (L'vovich, 1973, p. 68).

Diurnal Regimes

A lysimeter study in Arizona (Fritschen and van Bavel, 1962) gives the daily regime of the energy fluxes at a wet soil surface (Fig. X-8). Here the latent-heat flux took all the radiation surplus and more in addition, when the day was totalled. In the morning, however, it removed little more heat than did the substrate- and sensible-heat fluxes. Both these fluxes returned heat to the surface in the late afternoon, aiding evaporation at that time. They also provided a small energy supply through the night.

A close association between the temperature of the evaporating soil surface and the rate of evaporation was found in a later experiment here (Conaway and van Bavel, 1967). Careful radiometric measurements of the surface temperature showed it varying, on a day in late spring, from 10°C at night to 32°C at midday, a level it held for hours. The associated rate of evaporation varied from 0.1 mm hr^{-1} at night to 0.7 at midday, for a daily total of almost 7 mm. More important, the rates of evaporation fit a Dalton-type (John Dalton, 1766–1844) formula

Evaporation = (Wind function) (Vapor-pressure difference between surface and air).

Correct measurement of the surface temperature, not ordinarily easy to do for soil or vegetation surfaces, determines the value of surface vapor pressure to be entered in the equation above.*

As crop plants begin to shade the soil, it gets less energy and evaporation diminishes. However, soil evaporation continues to contribute a quarter to a half of the total amount of vapor leaving the soil–plant system, a so-called "unproductive" share because it is not associated with biological productivity.

EVAPORATION

In examining evaporation from deep and shallow bodies of water, and from wet soil surfaces, we have been getting closer to water in terrestrial ecosystems. In the following chapter we will consider such systems in conditions of abundant water supply so that, as in the foregoing chapter, our attention is primarily focused on the supply of energy to the system or the removal of vapor from it.

REFERENCES

Arai, T. (1966). Evaporation from lakes in the temperate climate. *J. Agric. Meteorol.* **22,** 51–57.

Arai, T. (1972). Characteristics of thermal structure of the natural lake analyzed from the global viewpoint, *Geog. Rev. Jpn.* **45,** 601–615 (Engl. resume 612–615).

Bowen, I.S. (1926). The ratio of heat losses by conduction and by evaporation from any water surface. *Phys. Rev.* **27,** 779 – 787.

Bruce, J. P., and Rodgers, G. K. (1962). Water balance of the Great Lakes system. *In* "Great Lakes Basin" (H. J. Pincus, ed.). Washington: Amer. Assoc. Advance. Science, *Publ.* **71,** 41–69.

Budyko, M. I. (1971). "Klimat i Zhizn'." Gidromet. Izdat. Leningrad, 472 pp. (*English transl.*: "Climate and Life." Academic Press, New York, 1974, 508 pp.)

Conaway, J., and van Bavel, C. H. M. (1967). Evaporation from a wet soil surface calculated from radiometrically determined surface temperatures, *J. Appl. Meteorol.* **6,** 650–655.

Cummings, N. W., and Richardson, B. (1927). Evaporation from lakes, *Phys. Rev.* **30,** 527–534.

Dilley, A. C., and Shepherd, W. (1972). Potential evaporation from pasture and potatoes at Aspendale, *Agric. Meteorol.* **10,** 283–300.

* This case is one of the few in which this straightforward formula is used, for in most environments it has proven difficult to obtain a sufficiently accurate idea of surface temperature. Development of remote sensing devices may help fill this gap, though in vegetation evaporation might not occur precisely at the same place that is sensed by the radiometer.

Fritschen, L. J., and van Bavel, C. H. M. (1962). Energy balance components of evaporating surfaces in arid lands, *J. Geophys. Res.* **67,** 5179–5185.
Fritschen, L. J., and van Bavel, C. H. M. (1963). Evaporation from shallow water and related micrometeorological parameters, *J. Appl. Meteorol.* **2,** 407–411.
Horton, R. E. (1917). A new evaporation formula developed, *Engrg. News-Record* **78,** 196–199.
Idso, S. B., and Foster, J. M. (1974). Light and temperature relations in a small desert pond as influenced by phytoplanktonic density variations, *Water Resources Res.* **10,** 129–132.
Koberg, G. E. (1960). Effect on evaporation of releases from reservoirs on Salt River, Arizona, *Int. Assoc. Sci. Hydrol. Bull.* **19,** 37–44.
Konstantinov, A. R. (1968). "Isparenie v Prirode," 2nd ed. Gidrometeorolog. Izdat., Leningrad, 532 pp.
LaMer, V. K., and Healy, T. W. (1965). Evaporation of water: its retardation by monolayers, *Science* **148,** 36–42.
Langbein, W. B. (1959). Water Yield and Reservoir Storage in the United States. U. S. Geol. Surv. Circ. 409, 5 pp.
Linacre, E. T. (1964). A note on a feature of leaf and air temperature, *Agric. Meteorol.* **1,** 66–72.
L'vovich, M. I. (1973). "The World's Water: Today and Tomorrow." Mir Publ., Moscow, 213 pp. (*English transl.* by L. Stoklitsky of "Vodnye Resursy Budushchego." Izdat. Prosveshchenie, Moskva, 1972).
McIlroy, I. C. (1957). The measurement of natural evaporation, *Austral. J. Agric. Sci.* **23,** 4–17.
McIlroy, I. C., and Angus, D. E. (1964). Grass, water and soil evaporation at Aspendale, *Agric. Meteorol.* **1,** 201–224.
Nesina, L. V. (1967). Computation of the components of the thermal balance of planned reservoirs, *Sov. Hydrol.* **1967,** 183–189.
Priestley, C. H. B. (1966). Limitation of temperature by evaporation in hot climates. *Agric. Meteorol.* **3,** 241–246.
Richards, T. L., and Fortin, J. P. (1962). An evaluation of the land–lake vapor pressure relationship for the Great Lakes, *Univ. Michigan Inst. Sci. Technol. Great Lakes Res. Div. Publ. 9,* pp. 103–110.
Richards, T. L., and Irbe, J. G. (1969). Estimates of monthly evaporation losses from the Great Lakes 1950 to 1968 based on the mass transfer technique, *Proc. Conf. Great Lakes Res., 12th,* pp. 469–487.
Ryan, P. J., Harleman, D. R. F., and Stolzenbach, K. D. (1974). Surface heat loss from cooling ponds, *Water Resources Res.* **10,** 930–938.
Smithsonian Institution (1966). "Smithsonian Meteorological Tables" (R. J. List, ed.). 6th ed. Smithsonian Institution Publ. 4014, Washington, D.C., 527 pp.
Timofeev, M. P. (1960). "The Heat Balance of Water Bodies and Methods of Determining Evaporation (from Them)." U. S. Weather Bur. Transl. by G. S. Mitchell and N. A. Stepanova, Washington, D.C., 28 pp. [Teplovoi balans vodoemov i metody opredeleniia ispareniia. *In* "Sovremennye Problemy Meteorologii Prizemnogo Sloia Vozdukha" (M. I. Budyko, ed.). Leningrad, 1958, 43–60.]
Webb, M. S. (1970). Monthly mean surface temperature for Lake Ontario as determined by aerial survey, *Water Resources Res.* **6,** 943–956.
WMO (1974). "Methods for Water Balance Computations," A. A. Sokolov and T. G. Chapman (eds.). Unesco Press, Paris. 127 pp.

Chapter XI

EVAPORATION FROM WELL-WATERED ECOSYSTEMS

The foliage of ecosystems rooted in moist soil forms an environment in which much water is vaporized. As described in an earlier chapter, water of intercepted rain and snow evaporates readily from foliage. In addition, evaporation occurs within leaves at the wet walls of leaf cells, as an inevitable concommitant of their openness to the inflow of CO_2 during photosynthesis.

In classing well-watered soil–vegetation systems* along with other wet surfaces we specify that the soil hold water with minimum tension—$\frac{1}{3}$ atm or less—and that internal resistances to movement of water within the plants be small, as they usually are (Bonner, 1959). We also specify good hydraulic contact between soil and plant roots and favorable environments for root uptake of water. The case of vegetation in drying soil is deferred to the following chapter.

An ecosystem in sufficiently moist soil can be treated as a type of surface cover in which energy and atmospheric conditions, rather than water supply, are the controlling factors in the rate of vaporization of water. These are the same conditions that control vaporization at lake and snow surfaces, but vegetation is somewhat different when contrasted with these other surface covers. Like a snow cover, it can, for instance, draw on only little heat storage in its substrate to support evaporation; unlike snow cover, it absorbs a large fraction of the sunshine that falls on it, and its surface temperature can rise far above the 0°C that is limiting for snow and that restricts the possible vapor pressure at a snow surface.

* Such systems are not uncommon, being found where the spacing of rains is no more than a few days, in lands of high groundwater, and in artificially irrigated lands, including urban green areas.

In comparison with a water body, a vegetation–soil ecosystem has small heat storage, as good absorption of solar radiation, and a better linkage with the atmosphere by means of convective exchanges because it is rougher. These likenesses and differences suggest a little about the process of evaporation from ecosystems, the energy sources that power it and the external conditions that govern it.

TRANSPIRATION OF WATER FROM LEAVES

Transpiration plays a role in the operations of vegetation that is somewhat more than just a leakage of water that happens to occur when cells are opened to receive CO_2 from the environment. It represents an "energy subsidy" (Odum, 1971, p. 43) to the ecosystem from outside. This subsidy powers the intake of nutrients from the soil and thus is a mechanism for nutrient conservation in the soil–vegetation system (Odum, 1971, p. 95). Also, by cooling the leaves it helps to "reduce the respiratory heat loss (i.e., the 'disorder pump-out') necessary to maintain the biological structure" (Odum, 1971, p. 43). While this book tends to deal more with the circulations of water outside plant tissue in ecosystems—delivery of rain, interception, infiltration, and so on—we cannot forget the important role of water within the tissues of the plants in the system.

While transpiration is a part of the functioning of life processes of plants, it is also at the same time a physical process. The usual number of kilojoules of energy are converted to latent heat, drawing on one or another source of energy, when a certain mass of water is vaporized; water has to be delivered to the site where vaporization can take place; and the vapor has to be carried away.

Resistance to the Diffusion of Water Vapor

One characteristic of evaporation in an ecosystem not found in other evaporating systems is that the wet cell walls, the sites of vaporization, are not directly exposed to the free air, which is the ultimate sink for the vapor molecules. They open instead on intercellular spaces within the leaf, which interpose an additional resistance to molecular diffusion and tend to slow down the vaporization process.

The resistance of vapor moving from leaf to air is as much as 10^7 times the resistance to movement in 1-m length of stem, and 10^6 times the resistance to movement of water into a root. This internal resistance is added to the resistance located between leaf outer surface and air, which depends on wind speed and other factors.

Photosynthesizing cells in leaves must extract CO_2 from a medium, the atmosphere, in which it exists at a very low concentration. This extraction requires an extensive area of wet cell walls to which CO_2 molecules may diffuse in the intercellular air spaces and through which, dissolved in water, they can move into the cells. A square meter of crop or forest land supports several square meters of leaf surface area.* This area is multiplied by the reticulation of airways and interstices between cells of the leaf mesophyll, a net of air-filled spaces opening from each stoma.

Movement of CO_2 inward through these spaces to the moist cell walls, and movement of water vapor in the opposite direction, takes place by diffusion. Resistance to diffusion r_l is a kind of inverse conductivity term, and can be so entered in the gradient equation for transpiration E as

$$E = (\rho_l - \rho_{\mathrm{a}})/(r_l + r_{\mathrm{a}}),$$

where ρ_l and ρ_{a} indicate specific humidity or water vapor concentration at the cell wall and in the air, respectively (Gates and Hanks, 1967). The resistance terms are expressed in units of seconds per centimeter, r_l being that within the leaf while r_{a} represents resistance to diffusion of vapor from the leaf surface into the free air.

The resistance functions provide a means of separately analyzing the path of vapor diffusing in the turbulent atmosphere (air resistance index, which is a reciprocal of the turbulent diffusivity of the airstream) and the path of vapor diffusing through the interstices of leaves from the wet cell walls (stomatal resistance), or "external" and "internal" resistances, respectively. (Resistances are used here, rather than conductivities, for the same reason they are used in figuring thermal insulation of a house wall made of several materials of different conductivity, namely, that they can be combined by simple addition.)

EVAPOTRANSPIRATION FROM PLANT COMMUNITIES

Several approximations enter into the generalization that well-watered vegetation acts as a wet surface. In broad terms, however, the generalization is useful and includes a process, evaporation from the soil, that is difficult to separate from transpiration. It is linked with an

* The ratio between leaf surface area and ground area is called the "leaf area index." It is a useful parameter of a plant canopy, indicating differences among species and changing as a plant goes through its life cycle.

energy concept, potential evapotranspiration, which is the rate of moisture conversion obtaining at a vegetation-covered surface having these idealized characteristics:

(1) Plants short and densely spaced, growing actively, having access to unlimited soil moisture;
(2) Vegetation surface uniform, covering the soil, and infinite in extent.

This characterization lays stress on plant foliage *en masse* and hence on the features common to different species of plants, i.e., the physical exchanges of energy and moisture that occur at the surfaces of all leaves. It reduces emphasis on the differences between species, internal resistance or other changes under biological control, and soil effects, in order to concentrate on the physical process of vaporization.

Uniformity of cover is essential, as is completeness of the coverage of soil. The characteristic of infinite extent tends to reduce local differences that might result from proximity to a boundary. The most basic characteristic, of course, is that of unlimited soil moisture.

Sources of Energy

Under the conditions stated for potential evapotranspiration, what forms of energy are important? What are their sources?

(1) During the hours of daylight, when stomata are open and transpiration is in progress, little energy is likely to come from the *substrate*, which at these hours is taking in energy—not giving it out. Moreover, the air space contained and immobilized beneath a mantle of vegetation tends to insulate the soil from the outer active surface, which is now elevated from the soil surface to the upper part of the vegetation canopy. Little heat is transported across this space and the substrate plays only a minor role in the energetics of the situation, quite in contrast with the water bodies considered in the preceding chapter.

(2) Small amounts of energy are released by respiration in plants, and by decomposition of organic matter in the soil. This heat, not readily transported to the upper part of the plant canopy, has minor importance.

(3) Sensible-heat flux H from air to vegetation is minimized as a source of energy to support evapotranspiration by the criterion, stated earlier, of infinite extent. Air moving over such an infinite plane sooner or later comes to equilibrium with the underlying

surface, and any movement of sensible heat during the hours of radiation surplus and active evapotranspiration is upward.

(4) The sources that remain as important sources of energy for evapotranspiration are the fluxes of radiation. Potential evapotranspiration therefore represents a rate of energy conversion that approaches the summation of all the radiation fluxes, short wave and long wave, incoming and outgoing R. All fluxes are substantial during the hours when evapotranspiration is in progress.

The radiation flux that is most important, because its variations are echoed in the regimes of the other fluxes, is the incoming short-wave flux from the sun. This flux also contains the particular wavelengths that have photochemical efficacy.

The daily regimes in photosynthesis, i.e., operation with open stomata, and in thermal energy supply, are both aspects of the daily regime of solar radiation. Except in certain night-active epiphytes, they produce parallel regimes in solar radiation and evapotranspiration. A nondimensional representation of the typical daily regime is shown in Fig. XI-1 (Fleming, 1970, Fig. 3), for clear days. Note the slight lag after solar noon.

The general success of approximating methods in estimating evapotranspiration from well-watered vegetation cover indicates the central

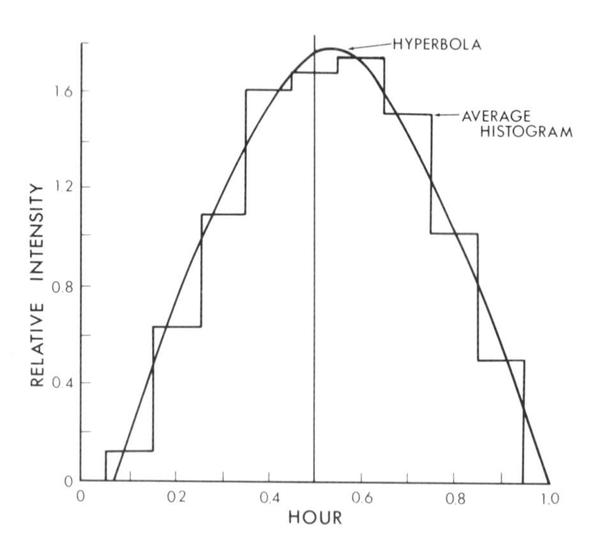

Fig. XI-1. Nondimensional form of diurnal regime of solar radiation on clear days (Fleming, 1970).

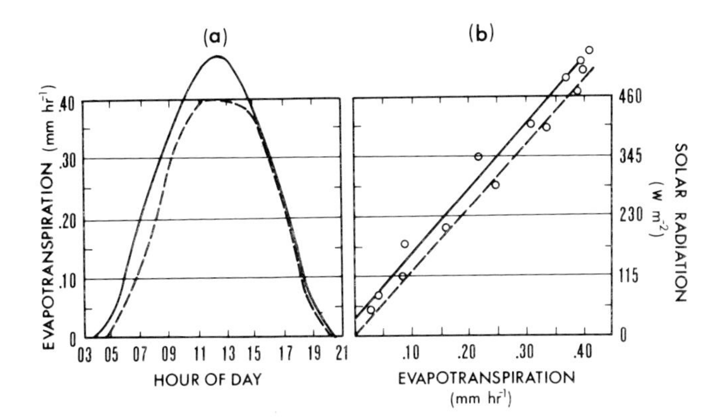

Fig. XI-2. (a) Diurnal marches of solar radiation (solid line) and evapotranspiration from meadow (dashed line). (b) Relation between hourly solar radiation and hourly evapotranspiration from meadow during the day. Before noon, solid line; after noon, dashed line (Kozlov, 1959).

role of radiative energy in evaporation from this type of surface cover. This role is well expressed in the daily cycle.

Radiation and Evapotranspiration in the Diurnal Cycle

The relative importance of radiant energy and air temperature as factors in the daily regime of evapotranspiration from a meadow is shown by comparing Figs. XI-2 and XI-3 (Kozlov, 1959). In Fig. XI-2

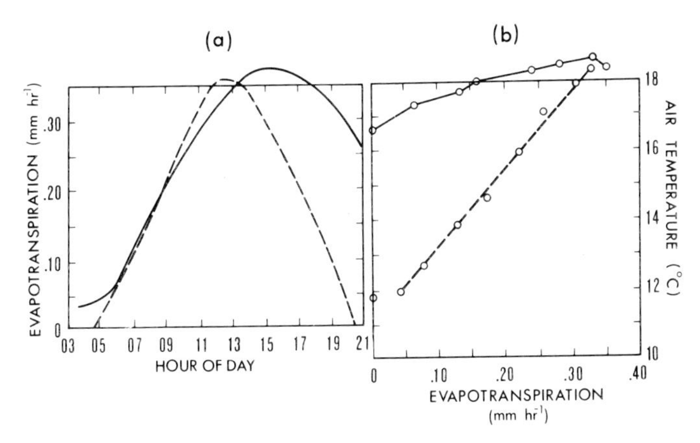

Fig. XI-3. (a) Diurnal marches of air temperature (solid line) and evapotranspiration from meadow (dashed line). Note how evapotranspiration decreases in the afternoon although the air stays warm. (b) Relation between hourly values of air temperature (ordinate) and hourly evapotranspiration. Before noon, dashed line; after noon, solid line (Kozlov, 1959).

the daily regime of short-wave radiation is seen to have about the same shape as the regime of evapotranspiration, except that radiation starts an hour earlier and stays about an hour ahead of the curve of evapotranspiration, illustrating a kind of warming-up process. Fig. XI-3 stands in contrast. While the rising limb of the daily march of air temperature coincides with the rising limb of the evapotranspiration curve fairly well, there is no resemblance between the two curves after noon. Evapotranspiration begins to decline at a time when air temperature is still climbing. It continues to decline through the afternoon hours, while air temperature is still as high as it was at noon; and it reaches zero while the air is as warm as it had been at 1100.

When evapotranspiration at each hour is plotted against the corresponding value of solar radiation or air temperature, we obtain the graphs in Figs. XI-2b and XI-3b. In such graphs, it is not uncommon to find that the rising and falling hours plot along separate lines to form a hysteresis loop, but the loops in these two graphs differ from one another. In the graph of radiation versus evapotranspiration the width of the loop is equivalent to about 0.04 mm hr^{-1}, that is, evapotranspiration for the same value of solar radiation is 0.04 mm hr^{-1} greater in the afternoon than in the morning. Since little heat is going into the soil in the afternoon, this difference is reasonable. In contrast, the loop in the graph of evapotranspiration versus air temperature (Fig. XI-3) is much wider. For example, at an air temperature of 17°C at 1100 in the morning, the evapotranspiration rate is 0.25 mm hr^{-1}, but at the same temperature in the afternoon (reached at 1930), it is only 0.04 mm hr^{-1}. In energy-flux terms, the width of the hysteresis loops is 150 W m^{-2} for the air temperature relation and only 35 W m^{-2} for solar radiation.

Clearly, the air-temperature regime is a poor predictor of the evapotranspiration regime. Much of whatever association might be found between them results from the circumstance that both are responses to the forcing function of solar radiation; but each makes its response to this external function in its own way. Evapotranspiration is more directly related, as we have seen, to variations in incoming solar energy than to the dependent environmental parameter of air temperature.

The Net Surplus of Whole-Spectrum Radiation Instead of using incoming solar radiation alone, it is common to include information on the energy transferred to and from the vegetation surface by the other fluxes of radiation. Vegetation reflects a fraction of the incoming solar radiation; it absorbs long-wave radiation emitted by the air and clouds, and in accordance with its regime of leaf temperature it emits long-wave radiation also.

The summation of these several fluxes—called the net surplus of whole-spectrum radiation, or sometimes just "net radiation"—represents all the radiative transfers of energy R. By definition it must therefore be the same as the sum of all the nonradiative transfers. Of these, the conversion of energy in vaporizing water E is by far the largest in the situation under study here; the exchanges of heat with the soil G and air H are minimized by the conditions set under the definition of potential evapotranspiration, as noted earlier.

There is, therefore, a tendency for many investigators to equate evapotranspiration with the net surplus of whole-spectrum radiation R, a quantity that is easily although not always accurately measured. Often only the daylight period, or the somewhat shorter period of radiative surplus, is examined. Rouse (1970) found high correlations of radiation surplus with two-week sums of evapotranspiration at sites on Mt. St. Hilaire, near Montreal.

Another way to look at this situation is to begin by writing the whole energy-budget equation

$$R_{S\downarrow} + R_{L\downarrow} + R_{SR\uparrow} + R_{LR\uparrow} + R_{LE\uparrow} + E + H + G = 0,$$

in which subscripts refer to short-wave (S) and long-wave (L) radiation, and subscripts R and E to reflected and emitted radiation, respectively. Subscripted arrows represent the direction; in all terms a flux toward the ecosystem surface is taken as +, a flux away from it as −. Applying this statement to a wet surface of great extent and small capacity to store heat, terms H (sensible-heat flux) and G (soil-heat flux) drop out, leaving

$$R_{S\downarrow} + R_{L\downarrow} + R_{SR\uparrow} + R_{LR\uparrow} + R_{LE\uparrow} + E = 0.$$

Rearranging terms, we can group on the right side all the energy fluxes that are independent of the temperature of the ecosystem surface, thus

$$R_{LE\uparrow} + E = -(R_{S\downarrow} + R_{SR\uparrow} + R_{L\downarrow} + R_{LR\uparrow}) = -R_A.$$

The new group on the right is equivalent to the fractions of short-wave and long-wave radiation absorbed by the ecosystem, and can be regarded as a loading, or, more positively, as an intake of radiant energy R_A* (Miller, 1972). As this energy intake changes, so does surface temperature. In response to changes in surface temperature,

* As referred to in the opening pages of Chapter X.

the two left-hand fluxes, both temperature dependent, also must change, E in response to surface vapor pressure, $R_{LE\uparrow}$ in accordance with the Stefan–Boltzmann fourth-power law.

Finally,

$$E = -R_A - R_{LE\uparrow}.$$

The right-hand side is equivalent to net radiation, but expresses the fact that a temperature-independent and a temperature-dependent component are combined in it.

In some situations it is not desirable to neglect the soil-heat flux G, especially if the study is restricted to daytime hours when G is prevailingly downward. It may be measured, or simply estimated as 0.1–0.2 of the net radiation surplus. Because this method still leaves out the sensible-heat flux, it is sometimes found to overestimate evapotranspiration (Davies and McCaughey, 1968) by as much as 1 mm day^{-1}.

The outflow of sensible heat from vegetation, as from a water surface, while minimized by the definition of potential evapotranspiration, might still be brought into the picture by the use of the Bowen ratio, a number that depends on the relative coolness and dryness of the air overlying the warm, moist vegetation mantle. In the postulated condition of ample soil-moisture supply, infinite areal extent, and uniformity, the fraction of the net surplus of whole-spectrum radiation converted into sensible heat is not large, about 0.1–0.2.

It can be included by using the Bowen ratio (beta) to partition the

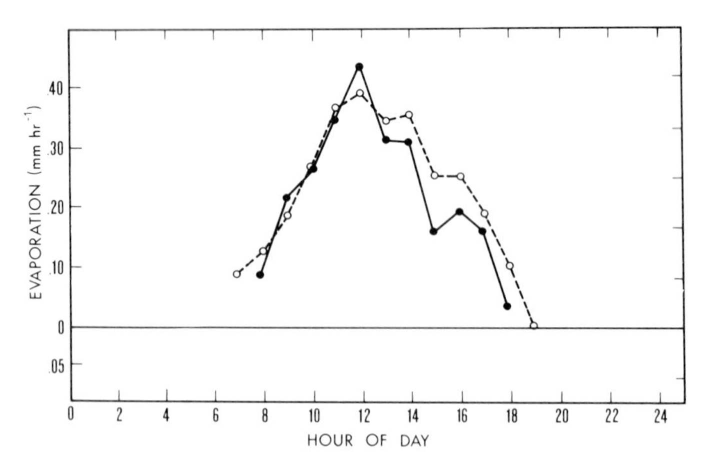

Fig. XI-4. Measurements of hourly evapotranspiration (mm hr^{-1}) from unirrigated Themeda grassland at Krawarree in the southern tablelands of New South Wales in early summer. Readings are made by the energy-partition evaporation recorder (solid line) and weighing lysimeter (dashed line) (McIlroy, 1971).

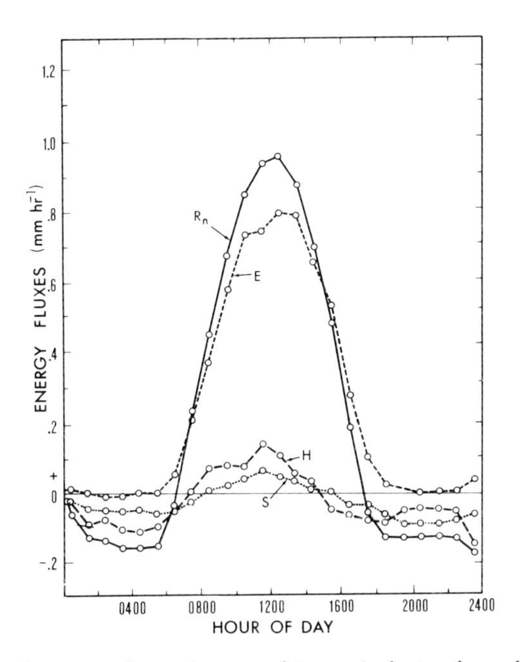

Fig. XI-5. Hourly energy fluxes (expressed in equivalents of mm hr^{-1} evaporation, i.e., 1 mm hr^{-1} = 700 W m^{-2}) at field of alfalfa and brome grass on Plainfield sand in Wisconsin, 4 September 1957. R_n is the net surplus (or deficit, at night) of all-wave radiation fluxes, E the evapotranspiration, H the sensible-heat flux (relatively small), and S the soil-heat flux (Tanner, 1960). Signs reversed for nonradiative fluxes. Reprinted with permission from the Soil Science Society of America.

value of R (or of $R - G$) into H and E, H being equal to beta E. In the right conditions the Bowen ratio can be computed from measurements of atmospheric temperature and vapor pressure at two heights above the evaporating surface. These measurements, made every few minutes, can be fed into a small field computer that calculates the ratio. McIlroy (1971) has carried this method one step farther, by hooking up the Bowen-ratio meter and computer to a recording net radiometer in a system called EPER (energy partition evaporation recorder). Figure XI-4 shows the record from this instrument over rough grazed pasture of Themeda grass at a CSIRO site at Krawarree, N.S.W. in early summer. The comparison is with lysimeter measurements of water loss from this soil–plant ecosystem, which is typical of much upland pastoral country in southern Australia. The peak rate of evaporation is reached at noon and is equivalent to 0.47 mm hr^{-1}. Good comparisons were also found with the sums of carefully measured R, G, and H in one of the few sites where H has actually been measured (Dilley, 1974, p. 23).

Figure XI-5 (Tanner, 1960) shows the tendency of the flux of latent heat from pasture in Wisconsin to approach the net surplus of the radiation fluxes. In this situation, soil-heat and sensible-heat fluxes are small.

The Combination Equation McIlroy's combination equation (1971) for potential evapotranspiration is a more exact statement. It shows how atmospheric dryness, expressed as the depression D below air temperature of a wet-bulb thermometer can be used to modify the basic relation to net radiation less soil-heat flux $(R - G)$:

$$(\text{Pot ET})(L) = (s/(s + \gamma))(R - G) + hD.$$

The first term on the right side of the equation varies with air temperature, and includes the change in the dependence of vapor pressure on temperature (s) and the psychrometric constant (γ).* A coefficient of heat transfer between air and foliage h is related to roughness of the vegetation canopy, i.e., the resistance to diffusion of vapor molecules and sensible heat.

Even this improved formula does not always exactly express the rate of evapotranspiration from well-watered vegetation. Experiments suggest that a degree of biological control is sometimes exerted (Shepherd, 1972), although its effect is small in annual totals.

Advective Factors

Wind Effect The reader might have noticed that no consideration of wind appears in the foregoing discussion of potential evapotranspiration. Wind speed has been found, in fact, to have little or no influence.

Potential evapotranspiration tends to approach the net surplus of whole-spectrum radiation, which is made up of radiation fluxes that are little affected by wind speed. Even in the combination equation, which brings in the factor of atmospheric dryness, it is unnecessary to introduce any index of air movement.

In this respect, potential evapotranspiration is not analogous to evaporation from a body of water. Stirring of a water body by wind action increases the movement of heat from lower layers to the surface, but stirring of a vegetation canopy by wind makes available little additional heat, assuming the air to be no warmer than the leaves.

An increase in wind speed increases the coefficient of turbulent exchange, but this is counterbalanced by a decrease in the leaf-to-air

* A dimensional conversion between vapor-pressure gradient and temperature gradient.

gradients of both temperature and vapor pressure. As more sensible heat is removed from the leaves they cool, and their vapor pressure decreases. The fluxes of latent and sensible heat remain about the same as before.

In conditions that do not meet the criterion of infinite extent of vegetation, however, the heat-supply situation is different. An airstream coming over a small area of transpiring vegetation does not immediately attain equilibrium with it but remains warmer or drier, or both, depending on the condition of the upwind area it came from. Instead of accepting a small flux of sensible heat from the underlying transpiring surface, as shown in the foregoing graph of Tanner's experiments, it may convey sensible heat to it. Advection of hot air accelerates evapotranspiration by providing additional energy to ecosystems situated in dry surroundings.

Local Transport of Heat A few years ago, when researchers were changing their thinking about wind as the all-powerful factor in evapotranspiration and were beginning to look to radiation to take its place in the scheme of things, they were occasionally disturbed by anomalous results from field experiments. Small plots of irrigated corn, for example, transpired at rates significantly higher than could be accounted for by radiant energy.

It was found that these plots were receiving appreciable amounts of sensible heat carried by the wind from unirrigated, hotter lands lying upwind. The fetch of the airstream over the irrigated crop was far too short to allow the airstream to come to equilibrium with the underlying moist surface. Moreover, if the irrigated plants were tall and widely spaced, projecting above the surrounding surfaces, even greater heat transfer from the hot air was found to occur.

These additions of a nonradiative heat flux to the existing surplus of radiant energy were thus identified. The first was named the "oasis" effect, and the second the "clothesline" effect. The oasis effect in particular has considerable interest to us on a micro- and mesoclimatic scale. Table I suggests the size of the nonradiative energy intake by a transpiring stand of tall grass (1-m height) in Arizona under two conditions (Bavel *et al.*, 1963).

(1) On 23 July the lysimeter grass was in the midst of grass of the same height and density. It was transpiring actively, obviously with the aid of some supplementary heat from the warm (42°C maximum) air. This supplement averaged approximately 100 W m^{-2} over the 0600–1800 period, being especially strong in the late afternoon.

TABLE I

Latent-Heat Flux from Sudangrass, 23 and 26 July 1962 (Phoenix)

Parameter	Period (hr)							
	06	08	10	12	14	16	18	Mean 06–18
Net surplus[b] of whole-spectrum radiation	±0	+150	+450	+615	+580	+385	+55	+370
Latent-heat flux:								
(1) Lysimeter within the crop stand (23 July)	−45	−220	−495	−650	−725	−570	−300	−470
(2) Lysimeter plants isolated (26 July)	−90	−335	−600	−875	−1170	−1020	−530	−720

[a] Units: W m^{-2}. Source: Bavel *et al.* (1963).
[b] Mean of both days.

(2) On the other day, after the surrounding grass had been cut, transpiration rates rose abruptly. In the 12-hr period the supply of advected heat was nearly as large as the net surplus of energy in the radiation budget. The grass, better ventilated than on the first day, was extracting great quantities of heat from the desert air.

Riparian vegetation in the desert extracts large amounts of sensible heat from air moving from adjacent dry slopes, and may evaporate as much as 1500 mm yr^{-1}. Efforts to reduce this evaporation by clearing the tamarisk or cottonwoods are often short lived, as the sites are soon reoccupied (Horton, 1972).

This experience suggests that sites favorably situated for the meeting of energy and water seem destined to become occupied by some biological system. If water is accessible, the transpiration from such vegetation remains large even when half the transpiring surface area is removed by thinning or pruning (Hylckama, 1970). The loss of foliage is quickly made good (see Plate 20).

Regional Scale Transport of Heat Even where irrigated vegetation is surrounded by a buffer zone many meters wide, heat is still extracted from warm air. For example, measurements of evapotranspiration from lysimeters along the coast of Queensland are smaller after general rains than before (Tweedie, 1966). The rains moisten large areas over which the air is then less strongly heated, so that general air movement brings less sensible heat to the lysimeters.

Plate 20. Riparian vegetation on the floodplain of the River Murray in South Australia. The dry sandy soils (note blowouts) of this Mallee country contrast with the moisture available to flood-plain vegetation (September 1966).

In Table II the evapotranspiration from a grass-covered lysimeter is shown to be larger in all seasons than from the surface of water (McIlroy and Angus, 1964). The difference suggests a favorable budget of radiation at the grass surface, and may also indicate the ability of the large leaf area of the grass to extract sensible heat from air flowing

TABLE II

Evaporation from Lysimeters at Aspendale, near Melbourne (38S)[a]

Lysimeter	J	F	M	A	M	J	J	A	S	O	N	D	J	F
Grass[b]	8	6	4	3	1	1	1	1	2	4	5	6	8	6
Water	7	5	3	2	1	1	1	1	2	3	4	5	7	5
Sum (mm): Grass	1310													
Water	1095													

[a] Units: mm day^{-1}. Source: McIlroy and Angus (1964).

[b] Later measurements on a sward of somewhat modified species composition are 4, 5, 5, and 4 mm day^{-1} in November–February (Dilley and Shepherd, 1972).

over the irrigated experimental site. Supplementation was least on days when the air at the site had come off a water surface. It was intermediate on days when the stream had a fetch over land surface that had received rain within the preceding three days. It was greatest in summer on days when the air had a fetch over dry land. In later studies, after the site had become surrounded by subdivision tracts, this addition increased still more (Dilley and Shepherd, 1972).

McIlroy and Angus do not believe, however, that all of this supplemental contribution of heat came from land surfaces immediately up-wind, but rather ascribe the source of some of the heat to large-scale subsidence in the frequently anticyclonic flow over southern Australia.

A similar regional-scale movement by heat to transpiring vegetation has been noted in southern Africa (Thompson and Boyce, 1967). Here, it appears to increase the evaporation rate by about 0.1, in a large-scale oasis effect.

Biological Factors

In the real world a number of changes are needed in the concept of potential evapotranspiration. These can often be added without losing the useful generality of the idea, and bring it closer in touch with the fact that ecosystems are biological constructs though they must also obey the laws of physics.

Stomatal Resistance Physicists studying evapotranspiration, at first reluctant to recognize the nonphysical factors in the situation, have found that plants sometimes exert a degree of control on the process. Plant canopies are not "passive evaporating surfaces" but as Lee (1968) says, "if plants are indeed 'wicks,' the plant scientist must argue that they are wicks of a unique kind. They are wicks with varying hydraulic conductivities. They are wicks coated with an epidermis that changes its permeability to gaseous exchange diurnally and seasonally and shows characteristic variations with plant species and type." *

Closing of stomata under heavy energy loading, reducing the rate of transpiration by about 0.2, was in fact found in careful lysimeter

* This overdependence on physical laws to explain all aspects of an ecosystem phenomenon does not mean that such laws should not be considered. However, an overemphasis on them as sole predictors of plant behavior might be considered an overcompensation to earlier neglect. Leighly (1937) long ago pointed out that botanists had not been successful in their research on water stress because they did not formulate it in correct physical terms.

measurements at Aspendale (Shepherd, 1972). Diffusion resistance r_l, usually 3–5 sec cm^{-1}, increased in these days by up to 2 sec cm^{-1}, leaf water content declined, and evaporation was reduced. This reduction over the year, however, was not considered to exceed 0.01–0.03 of the total evapotranspiration.

The idea of "external" and "internal" resistances is applicable whether a separate leaf or a whole ecosystem is under study. In the latter case the value of r_l (also called "surface diffusion resistance") is about the same, but r_a now expresses the ventilation of the whole canopy in dependence on its openness and also on wind speed. Some representative values over a growing season (sec cm^{-1}) are given in the accompanying tabulation to show the relative role that each resistance plays at different phases of the growing season (Fig. XI-6). There is a clear relation of r_a to the height (and roughness coefficient) character-

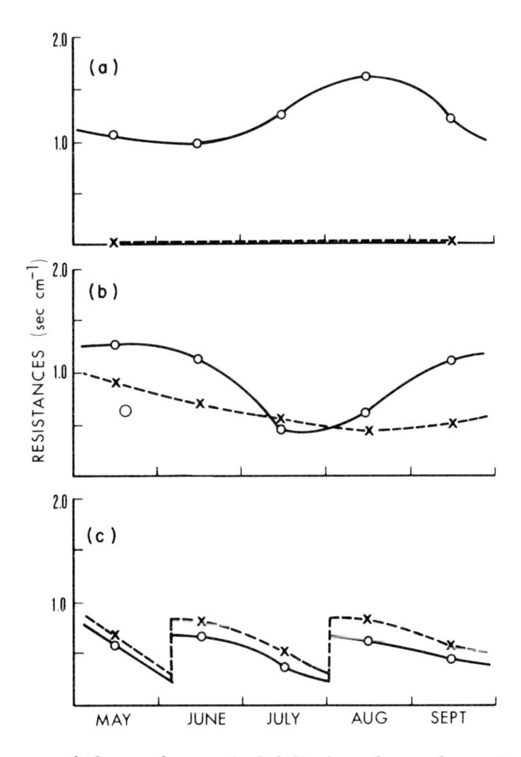

Fig. XI-6. Change of the surface r_s (solid line) and aerodynamic r_a (dashed line) resistances to water-vapor diffusion from three types of vegetation through the growing season: (a) pines, (b) potatoes, (c) alfalfa. Changes in foliage area are evident in the curves (b) (c) (hay cut in early June and early August) (Szeicz *et al.*, 1969).

istic of each ecosystem, greater height bringing stronger ventilation. It was suggested that the large r_l in pines compensates for the very small r_a, since in the evaporation equation their effects are added to one another.

Ecosystem[a]	r_l	r_a
Alfalfa	0.3–0.6, depending on height Both changing after each cutting	0.3, slightly less than r_l
Potatoes	0.4–1.2, lower values for complete ground cover	0.5–1.0, decreasing through the season
Pines	1.0–1.5	0.03, much less than r_l

[a] Source: Szeicz *et al.* (1969).

Phases of Plant Development Our original idea of potential evapotranspiration refers to a uniform green cover rooted in moist soil, but few, if any, plants retain an amount of foliage that is constant with time; most go through stages of growth, which are characterized by

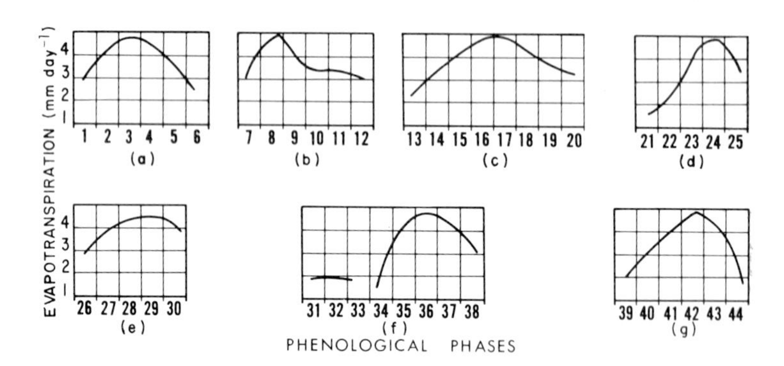

Fig. XI-7. Daily rates of evapotranspiration during the life cycles of seven crops: (a) flax, (b) clover, (c) spring wheat, (d) sorghum, (e) alfalfa, (f) winter wheat, (g) corn. Phenological phases: 1—germination, 2—start of stem growth, 3—formation of inflorescences, 4—flowering, 5—green maturing, 6—golden maturity, 7—germination, 8—formation of lateral shoots, 9—formation of inflorescences, 10—flowering, 11—midphase, 12—withering of tops, 13—germination, 14—third leaf, 15—tillering, 16—stooling, 17—heading, 18—flowering, 19—milky ripeness, 20—waxy ripeness, 21—germination, 22—third leaf, 23—tillering, 24—stooling, 25—silking, 26—resumption of growth, 27—formation of lateral shoots, 28—stem growth, 29—formation of inflorescences, 30—sprouting, 31—germination, 32—third leaf, 33—tillering, 34—shooting, 35—heading, 36—flowering, 37—soft ripeness, 38—full ripeness, 39—germination, 40—seventh leaf, 41—tasseling, 42—flowering, 43—soft maturity, 44—full maturity (Konstantinov, 1968, p. 414).

extension (or atrophy) of leaf area, growth in height, changes in reflectivity, increase in root intake surfaces, and so on. These phases are particularly marked in annual crops and those grown for roots or seeds rather than biomass, because transpiration is primarily a vegetative function rather than one connected with the processes by which plants transform biomass into metabolites. At the height of the growing season, evapotranspiration rates approach or equal potential evapotranspiration if water is available. For the design of systems for sprinkler irrigation Pillsbury (1968, p. 74) puts these peak rates in the range of 4–8 mm day^{-1}, depending on radiation and advective heat inputs to a climatic region.

Figure XI-7 presents daily rates of evapotranspiration from seven crops under optimum conditions of soil moisture (Konstantinov, 1968, p. 414). All curves rise to a peak, then decline, but the time of the peak varies with the crop. In clover, part (b), it is reached early, since a full vegetation cover is established early; with sorghum, part (d) it comes late in the life cycle. These changes are implicit in the annual regime of potential evapotranspiration from vegetation. Maintaining fully moist conditions in a crop by irrigation depends on knowledge of its growth phases; research was done on irrigated soybeans in Kansas to determine when water applications are most effective (Brady *et al.*, 1974).

Analogous data in Fig. XI-8 (Rauner, 1972, p. 139) show relative evapotranspiration from deciduous forest (curve 1), also in central Russia. A later start is displayed (curve 2) by an opening in this forest-steppe region, while curve 3 shows the relatively brief activity of grass on the open steppe.

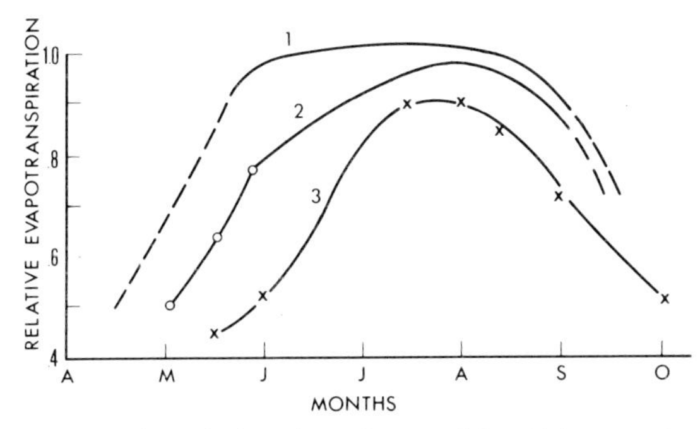

Fig. XI-8. Variation in relative (actual–potential ratio) evapotranspiration from different ecosystems in the central forest-steppe region through the growing season. Curve 1—Deciduous forest; curve 2—forest opening with native meadow-steppe vegetation, curve 3—open steppe (from Rauner, 1972, p. 139).

Species Effects on Evapotranspiration Not only do plants vary with time in characteristics pertinent to potential evapotranspiration, like ground cover, but there also exist differences that consistently pertain to species. In many cases these differences lie in the internal resistance to vapor diffusion, which was discussed earlier; in other cases they relate to a different tempo of life cycle.

Sometimes they are more subtle. In the watered Aspendale lysimeter, the original grass was gradually displaced by kikuyu grass. When it formed half the sward, potential evapotranspiration seemed to be down by about 0.2. There was "little change in apparent roughness" of the canopy (Dilley and Shepherd, 1972), but perhaps an "increase in stomatal resistance or enhanced sensitivity to the slightest of water stresses."

Roughness differences are important, however, in many situations. Sibbons (1962) shows that the greater roughness of crop plants provides more energy for evapotranspiration, as contrasted with mowed or grazed grassland.

Comparisons of potatoes with grass at Aspendale reveal a neat correspondence of the species difference in potential evapotranspiration with the growth of leaf-area index, height, and ground cover of the potatoes. At six weeks after emergence, the potatoes were vaporizing 0.3–0.4 more water than the grass (Sibbons, 1962). This difference is ascribed to greater absorption of radiation and the aerodynamically rougher surface.

Surface roughness is a canopy parameter that varies over nearly three orders of magnitude among plant species. Woody vegetation in particular develops a high roughness, as we saw earlier in describing the impaction of raindrops and snowflakes on forest.

The Vertical Dimension of Ecosystems and Its Influence on Evapotranspiration Evapotranspiration from such deep canopies as those of forest has been studied less than that from grassland or cropland, due in part to logistical difficulties in positioning instruments. Other reasons include the inhomogeneity of the rough upper surface, and the relative magnitude of boundary effects between forest stands and adjacent areas of lower vegetation. Yet a strong throughflow of water to be transpired seems to be a factor in the great height of trees; a throttling of this stream has a stunting effect on growth (Odum, 1971, p. 376).

Perhaps also, the fact that trees are bigger than man has led us to approach a forest at *our* level, i.e., ground level. Yet this is the level of the stand that is thermodynamically least important. As Geiger pointed out long ago, referring to many days on a platform above a forest stand in southern Germany, it is in the canopy that the most vigorous energy conversions occur, as well as most of the exchanges of matter (O_2, water vapor, CO_2) between the forest and its atmospheric environment. We have been looking at the logs in a forest, not the zone where the action is!

A deep canopy zone has several characteristics that differentiate it from low vegetation. Discussion of these may help us better understand many aspects of evapotranspiration.

(1) In a deep zone, the absorption of radiant energy is spread over a layer that is several meters, not a few centimeters, thick. Albedo is correspondingly low because incoming solar radiation is trapped by multiple reflections (Miller, 1955, pp. 94–104).

(2) Emission of long-wave radiation takes place from a considerable thickness of canopy, in which few surfaces are likely to be overheated and hence to emit strongly. The outflow of energy as long-wave radiation tends to be reduced.

(3) The canopy is clearly separated from the ground, with its capacity to store heat, by a trunk space in which air motion usually is slow. Little energy is transported to or from the ground across this zone in a net sense.

(4) The entire canopy zone contains about the same leaf area (three to five times the ground area of the forest stand) as lower vegetation. Rauner (1968) shows that this means a low density of biomass per unit volume of crown space, e.g., 0.5 $m^2\ m^{-3}$ (or 0.005 $cm^2\ cm^{-3}$).

(5) Both this low spatial density and its elevated position make for good ventilation. The atmospheric resistance to movement of vapor r_a from the leaves is generally small, 0.1 sec cm^{-1} in some cases, a mere fraction of the stomatal resistance. As a result, any change in stomatal opening has a marked influence on the summed resistance to diffusion of water vapor. General height has been shown related to roughness by Lettau and others, and roughness to small values of atmospheric resistance (Szeicz *et al.*, 1969).

(6) The rough nature of the upper surface of the canopy also is helpful in extracting sensible heat from a deep layer of warm air. This large roughness dimension of forest, evident even in an 8-m Douglas fir plantation, strengthens the convective term in the combination equation so that evapotranspiration does not follow short-term changes in the radiation fluxes (McNaughton and Black, 1973).

These characteristics all influence the potential evapotranspiration from forest. In comparison with a cover of low vegetation, the forest often displays a higher rate of evapotranspiration and this superiority is increased in warm air advection (Rauner, 1960, 1963, 1965).

The deep zone of interaction in forest harbors some interesting spatial distributions. For example, from what level does the vaporized water come? Figure XI-9 shows profiles of leaf area (at left) and the vertical flux of water vapor (at right) out of an 11-m aspen stand near Kursk (Rauner, 1972, p. 98). Evaporation from the ground adds about 0.06 mm hr^{-1} to the upward stream of vapor, and the dense undergrowth (note the leaf area at 0.8 of the depth of the stand, i.e., 3 m from the ground) adds about 0.09 mm hr^{-1} more. The rest of the vapor flux that emerges from the top of this ecosystem comes from the aspen leaves. The total flux is 0.56 mm hr^{-1}, or, represented in energetic terms, −395 W m^{-2}, a substantial share of the available energy (see Plate 21).

Measurements in a pine plantation near Canberra, Australia, (foliage top 5.5 m above the ground, depth of canopy 5.0 m) (Denmead, 1964) show (Figs. XI-10 and XI-11) the vertical profiles of the fluxes of radiant energy, sensible heat, and latent heat in a thick canopy. In the period 1125–1205 on 20 May 1963, for instance, the upward latent-heat flux at the level 3.2 m from the top was −45 W m^{-2}, but the surplus in the whole-spectrum radiation budget at this level was only +35 W

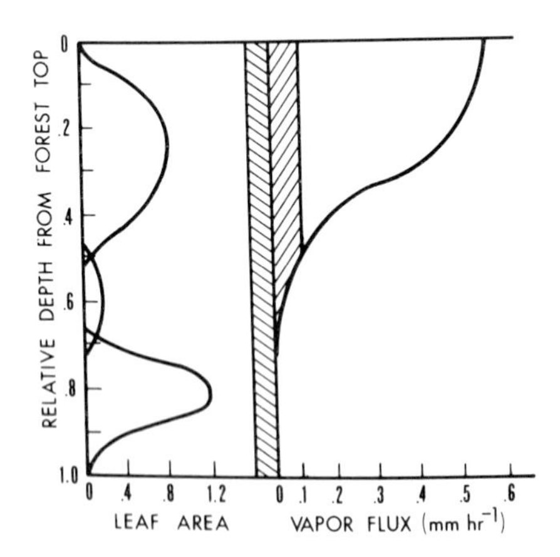

Fig. XI-9. Vertical profiles through a closed aspen ecosystem of leaf area (left graph) and upward flux of water vapor (right graph). Note that depth into the stand rather than height is shown as ordinate. Cross-hatched bars portray the flux from soil evaporation and understory transpiration (from Rauner, 1972, p. 98).

Plate 21. Humidity profile observations in heat-budget study in aspen near Kursk. Assmann aspirated thermometers at heights of 6, 8, 12, 14, and 16 m are read precisely by a small telescope (Anan'ev) and recorded by Rauner (in beret) in August 1969.

m^{-2}. The additional transpiration was supported by a downward sensible-heat flux of +10 W m^{-2}.

This situation is not uncommon in forest, because the radiation-absorbing layer is generally well elevated, while transpiration is distributed throughout the green area. In pines in the Sierra Nevada, for instance, the level 10 m above the snow was consistently several degrees warmer than lower levels (Miller, 1955, p. 110), although the stand was quite open.

The accompanying tabulation shows how the flows of radiant energy change as we descend from the sunlit upper levels of the forest, and also the changing partition of this energy by wavelengths. The sensible-heat flux reverses from an upward direction (−), indicating

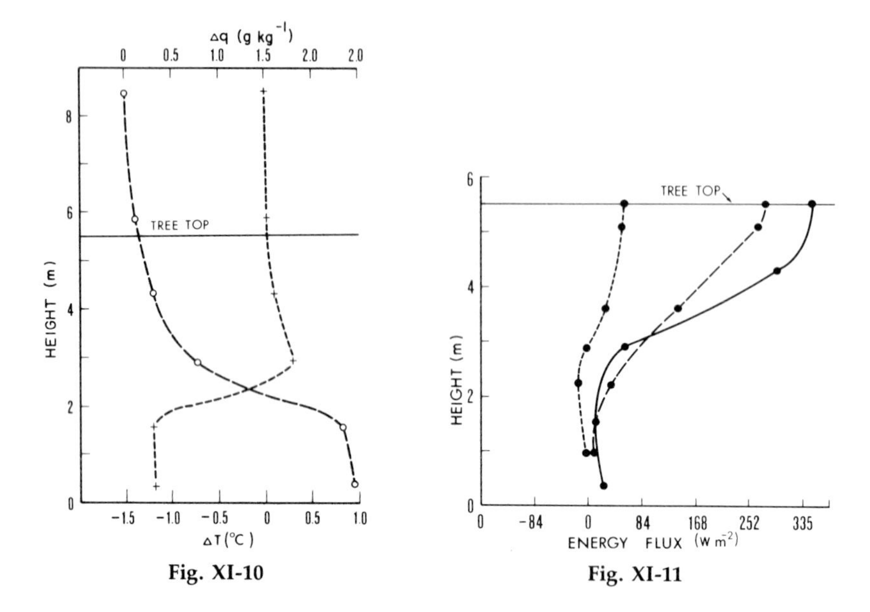

Fig. XI-10. Vertical gradients of temperature (– – –) and specific humidity (— — —) in pine forest at 1145, 20 May 1963 (Denmead, 1964). Note the coldness of the lower canopy and the trunk space, and the high humidity of this layer. Values are referenced to those in the air above the canopy.

Fig. XI-11. Vertical profiles of energy fluxes: net surplus of radiation (—); latent-heat flux (— — —); sensible-heat flux (– – –) in pine forest at 1145, 20 May 1963. Positive values indicate upward turbulent energy fluxes (Denmead, 1964).

that it is competing with the latent-heat flux, to a downward direction (+) about 2.5 m down. At the dark 3.2-m level, it is helping to provide heat for evaporation.

Level, measured downward from crown top (m)		Energy fluxes[a] (W m^{-2}) Net surplus of whole-spectrum radiation	Sensible heat	Latent heat
Top	0	+350	−70	−280
	1.5	+240	−60	−180
	2.5	+90	±0	−90
	3.2	+35	+12	−47
Near ground	4.5	+25	±0	−25

[a] Denmead, 1964.

The vapor flux that emerged from the top surface of the canopy at midday was equivalent to 0.4 mm hr^{-1} of water. When, later in the day a warmer airstream moved above the forest, producing an inversion, evapotranspiration drew its support both from radiation and the downward flux of sensible heat to the forest from the free air.

Vertical translocation of sensible heat to lower parts of the canopy allows leaves through a depth of several meters to take part in transpiration. A model of transfer processes in vegetation (Philip, 1964) shows different vertical distributions of the sources or sinks of sensible heat, radiation, and latent heat, depending on the physical structure of the ecosystem.

Moreover, another smoothing effect operates, on a smaller scale (Knoerr, 1967); sunlit leaves on a plant are sources of sensible heat that moves to the shaded leaves. Transpiration from the shade leaves is supported to a degree by this microscale movement of heat, increasing the total evaporation from the system.

A model has been developed to describe the processes in the depths of a transpiring ecosystem (Waggoner and Reifsnyder, 1968) and to reproduce the fluxes of heat within it, as well as the vertical profiles of the resulting conditions of air temperature and humidity. The model requires information on the rate at which radiation and wind speed are extinguished downward into the canopy, which depends on such physiognomic parameters as the leaf-area index at each level and the depth of foliage. The investigators obtained satisfactory reproduction of conditions in clover and barley. A further development of the model, applied to Denmead's (1964) observations in the pine plantation (Waggoner *et al.*, 1969), satisfactorily reproduced the vertical distributions of the energy fluxes and the microclimatic profiles of air motion, temperature, and humidity. Transpiration takes place in the lower layers of the forest canopy more strongly "than the absorption of radiation would appear to indicate."

By this model it is possible to run experiments that simulate changes in vegetation characteristics or in the initial conditions of radiation or atmosphere. For example, an increase in surface area of foliage by 0.5 results in an increase in transpiration by about 0.2. Increased penetration of wind into the canopy causes a small increase in transpiration. Doubling the net surplus of whole-spectrum radiation increases transpiration by about 0.3.

EMPIRICAL PATTERNS OF POTENTIAL EVAPOTRANSPIRATION

Observations

Potential evapotranspiration can be determined for shallow-rooted plants growing in lysimeters developed by Thornthwaite and others, and those used at Aspendale by McIlroy, to which water is applied frequently enough to keep micropores in the soil filled. Readings made at Thornthwaite's laboratory in New Jersey are shown in 10-day means in Fig. XI-12, in terms of their equivalent in energy terms (W m^{-2}). Table III shows monthly means at Seabrook and other sites, to present differences due to latitude and season.

Available data indicate that potential evapotranspiration often exceeds rainfall in climates generally regarded as "humid." At Aspendale the yearly excess is 500 mm, at Townsville 600.

Annual Regime If monthly values of evapotranspiration from well-watered grass are plotted (McIlroy and Angus, 1964) against net whole-spectrum radiation, a hysteresis loop is formed; the relation of potential evapotranspiration to radiation surplus differs in spring and fall. At a given radiation surplus in the fall, there is about 1 mm day^{-1} more evaporation than in the spring.

About a quarter of this discrepancy is probably due to the annual cycle of heat storage in the soil. On spring days the cold soil takes in heat that might otherwise support evapotranspiration, and also limits water uptake by plant roots. The balance of the discrepancy probably represents greater regional advection of sensible heat in fall than in spring, and a more extensively developed leaf area. Even so, the lag of evapotranspiration after the date of the radiation maximum is less than the well-known lag of air temperature, which is about one month. As a result, annual regimes of evapotranspiration and air temperature are not in phase with one another, any more than they are in the diurnal cycle, discussed earlier in this chapter.

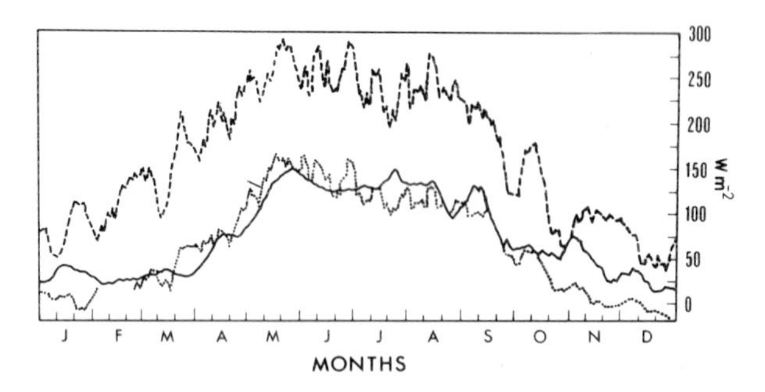

Fig. XI-12. Annual march of solar radiation (– – –), net surplus of all-wave radiation (···), and potential evapotranspiration (—) at the Thornthwaite Laboratory of Climatology in New Jersey in 1956 (Mather, 1974, p. 62).

TABLE III *Potential Evapotranspiration from Grass or Grass-Clover Sward*[a]

Location	Month												Year
	J	F	M	A	M	J	J	A	S	O	N	D	
Seabrook, N.J.[b]													
4 yr, 39N lat.													
Mean	0.8	0.6	1.1	2.2	3.7	3.9	4.8	4.2	3.3	2.3	1.5	0.7	890
Std. error	0.2	0.2	0.3	0.1	0.2	0.2	0.3	0.2	0.3	0.2	0.3	0.1	
Raleigh, N.C.[b]													
3 yr, 36N lat.													
Mean	—	—	—	—	—	4.3	4.4	3.6	2.9	1.3	—	—	—
Std. Error	—	—	—	—	—	0.2	0.3	0.1	0.2	—	—	—	
Davis, Calif.[c]													
2 yr, 39N lat.													
Mean	0.8	1.8	2.8	4.0	5.2	7.1	7.0	5.7	4.2	2.9	1.5	0.8	—
Std. error	—	—	—	0.3	0.2	0.1	0.1	0.1	0.3	—	—	—	
Hong Kong[d]													
2 yr, 22N lat.													
Mean	2.3	2.6	2.8	3.2	4.0	4.0	4.3	3.7	4.2	4.5	3.7	2.7	1280
Std. error	0.1	0.2	0.2	0.3	0.3	0.4	0.3	0.3	0.4	0.1	0.2	0.1	
Aspendale, Vic.[e]													
3 yr, 38S lat.													
Mean	7.8	5.6	4.2	2.9	1.3	1.0	0.9	1.4	2.3	4.1	5.2	6.0	1310
Townsville, Qld.[f] (Kalamia)													
1 yr, 19S lat.													
Mean	6.4	4.5	5.2	4.3	4.0	4.3	5.3	6.8	5.3	6.8	8.7	11.0	1850

[a] Units: mm day^{-1}.
[b] Bavel (1961).
[c] Tanner (1967), for April–September, and personal communication for October–March.
[d] Ramage (1959).
[e] McIlroy and Angus (1964).
[f] Tweedie (1956).

At the same air temperature in fall, evapotranspiration is less than in spring, primarily because at a given air temperature the radiation supply in fall is smaller. For example, air temperatures at Aspendale in November and April are about the same, but the radiation surplus in November (spring) is more than double that in April; potential evapotranspiration is almost double.

Insofar as the monthly rates of evaporation are affected by the ground coverage or growth phases of ecosystems, their annual regimes were described earlier in the discussion of biological factors.

Geographic Distribution

These data on potential evapotranspiration derived from a few series of lysimeter measurements have been supplemented by calculated values of potential evapotranspiration. There are many ways to make these calculations, e.g., those developed in the Glavnaia Geofizicheskaia Observatoriia in Leningrad by Budyko (1974, pp. 335–346), Zubenok, and others in this large research group.* They continually are being improved as more is found out about the physical and biological processes at work. For our purposes here, it will suffice to look at some of the patterns that result when one or another of these methods is applied.

At Midsummer The smallest values in June—less than 20 mm for the month—are found at the shores of the Arctic Ocean, which are swept by cold maritime air streams. Tundra vegetation loses a large amount of sensible heat into this cold air at the expense of evapotranspiration. Lands farther from this cold ocean have values of evapotranspiration of 75–100 mm, more in accord with their supply of radiant energy.

In most agricultural regions of the western Soviet Union and Europe, potential evapotranspiration ranges between 120 and 150 mm during June, i.e., 4–5 mm day^{-1}. The agricultural regions of eastern North America have potential evapotranspiration rates in summer of 4–6 mm day^{-1}. In oases, this quantity ranges up to 8 mm day^{-1}, aided no doubt by hot air off the adjacent desert.

* The agreement between these methods for computing potential evapotranspiration and those of Thornthwaite (1931, 1948; Mather, 1974, pp. 66–67), which is commonly used in North America, appears to be closest in summer in the United States and the Soviet Union, and in equatorial climates (Zubenok, 1974). Thornthwaite (1960) recognized that the dependence of his method on air temperature was a limitation that he hoped subsequently to remove; this dependence seems to explain areas of lack of agreement.

Annual Total Excluding the ice sheets, the lowest amounts are found on the northern shores of North America and Eurasia. They are smaller than 200 mm.

A marked latitudinal gradient, which was not seen on the map for June, appears in the annual map. In Europe, potential evapotranspiration decreases from 1300 mm or more in lands near the Mediterranean to 500 mm in Baltic lands, indicating the longer summers and the mild winters of the south.

In North America, potential evapotranspiration on the coastal plains of the Gulf of Mexico exceeds 1500 mm. From this there is a decrease northward to 1000 mm in the corn belt and 700 mm in the dairy belt, and to still smaller values in the boreal forest and tundra.

EVAPORATION DIFFERS WITH ECOSYSTEMS

Well-watered vegetation evaporates differently than water bodies do. It has less access to heat stored below the surface, and the effect of wind on evaporation is different and usually much smaller. Both systems have ample water supply, but different sources of energy and different coupling with the atmosphere, so their amounts and rhythms of evaporation are in contrast.

We have now examined evaporation from the surfaces of several forms in which water substance exists in nature—snow, water bodies, shallow water, wet soil, and well-watered vegetation of varying thickness and other properties. In each type different factors are found to influence evaporation.

Heat storage in the substrate plays a large role in some cases, none at all in others. Wind has an effect on evaporation from water bodies, little on snow, and minor in vegetation. Sensible-heat flux supports evaporation under some conditions, but competes with it under others. The radiation fluxes that make up the surplus of net whole-spectrum radiation play varying roles, direct and immediate in the case of vegetation.

None of these wet surfaces can reliably be equated with another. Snow, plants, water bodies, and soil—each evaporates water in its own way. A further complication can now be introduced: systems experiencing a shrinking supply of water.

REFERENCES

Bavel, C. H. M. van (1961). Lysimeter measurements of evapotranspiration rates in the eastern United States, *Proc. Soil Sci. Soc. Am.* **25,** 138–141.

Bavel, C. H. M. van, Fritschen, L. J., and Reeves, W. E. (1963). Transpiration by sudangrass as an externally controlled process, *Science* **141,** 269–270.

Bonner, J. (1959). Water transport, *Science* **129,** 447–450.

Brady, R. A., Stone, L. R., Nickell, C. D., and Powers, W. L. (1974). Water conservation through proper timing of soybean irrigation, *J. Soil Water Cons.* **29,** 266–268.

Budyko, M. I. (1974). "Climate and Life" (English transl. ed. by D. H. Miller). Academic Press, New York, 508 pp.

Davies, J. A., and McCaughey, J. H. (1968). Potential evapotranspiration at Simcoe, southern Ontario, *Arch. Met. Geoph. Biokl.* **B16,** 391–417.

Denmead, O. T. (1964). Evaporation sources and apparent diffusivities in a forest canopy, *J. Appl. Meteorol.* **3,** 383–389.

Dilley, A. C. (1974). An energy partition evaporation recorder, Austral. Commonw. Sci. Indus. Res. Organ., Div. Atm. Phys. Tech. Paper 24, 25 pp.

Dilley, A. C., and Shepherd, W. (1972). Potential evaporation from pasture and potatoes at Aspendale, *Agric. Meteorol.* **10,** 283–300.

Fleming, P. M. (1970). A diurnal distribution function for daily evaporation. *Water Resources Res.* **6,** 937–942.

Gates, D. M., and Hanks, R. J. (1967). Plant factors affecting evapotranspiration. *In* "Irrigation of Agricultural Lands" (R. M. Hagan *et al.,* eds.). pp. 505–521. Am. Soc. Agronomy, Madison.

Horton, J. S. (1972). Management problems in phreatophyte and riparian zones, *J. Soil Water Cons.* **27,** 57–61.

Hylckama, T. E. A. van (1970). Water use by salt cedar, *Water Resources Res.* **6,** 728–735.

Knoerr, K. R. (1967). Contrasts in energy balances between individual leaves and vegetated surfaces. *In* "Forest Hydrology" (W. E. Sopper and H. W. Lull, eds.), pp. 391–401. Pergamon, Oxford.

Konstantinov, A. R. (1968). "Isparenie v Prirode," 2d ed. Gidrometeorolog. Izdat., Leningrad, 532 pp.

Kozlov, M. P. (1959). Sutochnyi khod summarnogo ispareniia s luga i ego sviaz' s sutochnym khodom meteorologicheskikh elementov. Vses. Gidrolog. S'ezd, 3d, 1957, *Trudy* **3,** 166–173.

Lee, R. (1968). Reply, *Water Resources Res.* **4,** 667–669.

Leighly, J. (1937). A note on evaporation, *Ecology* **18,** 180–198.

Mather, J. R. (1974). "Climatology: Fundamentals and Applications." McGraw-Hill, New York, 412 pp.

McIlroy, I. C. (1971). An instrument for continuous recording of natural evaporation, *Agric. Meteorol.* **9,** 93–100.

McIlroy, I. C., and Angus, D. E. (1964). Grass, water and soil evaporation at Aspendale, *Agric. Meteorol.* **1,** 201–224.

McNaughton, K. G., and Black, T. A. (1973). A study of evapotranspiration from a Douglas fir forest using the energy balance approach, *Water Resources Res.* **9,** 1579–1590.

Miller, D. H. (1955). "Snow Cover and Climate in the Sierra Nevada, California." Univ. Calif. Press, Publ. Geog. 11, Berkeley, California, 218 pp.

Miller, D. H. (1972). On the variations of radiant-energy intake over time with some notes on the responses of evaptranspiration and other energy fluxes as functions of the temperature of the surface of the earth, *Publ. Climatol.* **25**(3), 47–67 (Thornthwaite Memorial Volume II).

Odum, E. P. (1971). "Fundamentals of Ecology," 3d ed. Saunders, Philadelphia, Pennsylvania, 574 pp.

Philip, J. R. (1964). Sources and transfer processes in the air layers occupied by vegetation, *J. Appl. Meteorol.* **3,** 390–395.

Pillsbury, A. F. (1968). "Sprinkler Irrigation," FAO Agric. Devel. Paper 88, Rome. 179 pp.

Ramage, C. S. (1959). Evapotranspiration in Hong Kong: a second report, *Pac. Sci.* **13,** 81–87.

Rauner, Iu. L. (1960). Teplovoi balans lesa, *Izv. Akad. Nauk SSSR Ser. Geograf.* No. 1, 49–59.

Rauner, Iu. L. (1963). Izmerenie teplo- i vlagoobmena mezhdu lesom i atmosferoi pod vliianiem okruzhaiushchikh territorii, *Izv. Akad. Nauk SSSR Ser. Geograf.* No. 1, 15–28.

Rauner, Iu. L. (1965). O gidrologicheskoi roli lesa, *Izv. Akad. Nauk SSSR Ser. Geograf.* No. 4, 40–53.

Rauner, Iu. L. (1968). Biometricheskie pokazateli lesnoi rastitel'nosti v sviazi s izucheniem ee radiatsionnogo rezhima. *In* "Aktinometriia i Optika Atmosfery" (V. K. Pyldmaa, ed.), pp. 335–342. Izdat. Valgus, Tallinn.

Rauner, Iu. L. (1972). "Teplovoi Balans Rastitel'nogo Pokrova." Gidrometeoizdat, Leningrad, 210 pp.

Rouse, W. R. (1970). Relation between radiant energy supply and evapotranspiration from sloping terrain: an example, *Can. Geog.* **14,** 27–37.

Shepherd, W. (1972). Some evidence of stomatal restriction of evaporation from well-watered plant canopies, *Water Resources Res.* **8,** 1092–1095.

Sibbons, J. L. H. (1962). A contribution to the study of potential evapotranspiration, *Geog. Ann.* **44,** 279–292.

Szeicz, G., Endrödi, G., and Tajchman, S. (1969). Aerodynamic and surface factors in evaporation, *Water Resources Res.* **5,** 380–394.

Tanner, C. B. (1960). Energy balance approach to evapotranspiration from crops, *Soil Sci. Soc. Am. Proc.* **24,** 1–9.

Tanner, C. B. (1967). Measurement of evapotranspiration, *In* "Irrigation of Agricultural Lands" (R. M. Hagan *et al.,* eds.), pp. 534–574. Am. Soc. Agronomy, Madison.

Thompson, G. D., and Boyce, J. P. (1967). Daily measurements of potential evapotranspiration from fully canopied sugar cane, *Agric. Meteorol.* **4,** 267–279.

Thornthwaite, C. W. (1931). The climates of North America according to a new classification, *Geog. Rev.* **21,** 633–655.

Thornthwaite, C. W. (1948). An approach toward a rational classification of climate, *Geog. Rev.* **38,** 55–94.

Thornthwaite, C. W. (1960). Personal communication.

Tweedie, A. D. (1956). The measurement of water need in Queensland, *Austral. Geog.* **6,** 34–39.

Tweedie, A. D. (1966). Personal communication.

Waggoner, P. E., Furnival, G. M., Reifsnyder, W. E. (1969). Simulation of the microclimate in a forest, *Forest Sci.* **15,** 37–45.

Waggoner, P. E., and Reifsnyder, W. E. (1968). Simulation of the temperature, humidity and evaporation profiles in a leaf canopy, *J. Appl. Meteorol.* **7,** 400–409.

Zubenok, L. I. (1974). Evaporation deficit under various climatic conditions on land, *Sov. Hydrol.* **1974,** 251–257.

Chapter XII

EVAPORATION FROM DRYING ECOSYSTEMS

While many surfaces in nature, such as those of water and snow, are always moist, most terrestrial systems dry out during periods between rains. The uncertain spacing of rainstorms, emphasized in the first chapters, means that drying periods of unpredictable length are inherent in the environment of most terrestrial ecosystems. During these periods the decline in stored soil moisture tends to produce a decline in the evaporation rate, although the specific connection is not always a precise one.

BARE SOIL SURFACES

A wet bare soil surface given a moderate energy supply rapidly dries out. Water in the top layers of the soil body is exhausted and water from deeper layers is supplied slowly; surface evaporation soon diminishes to a very small rate. A deep tank filled with clay loam, irrigated so as to be wetted nearly to its bottom, was found to lose 30 mm of water during the first month, but only 23 mm more during the rest of the first year and only 90 mm in all, during more than four years (Veihmeyer and Brooks, 1954). At the end of the 1547-day experiment most of the water in the soil was still there. The mean rates of evaporation were 1 mm day^{-1} during the first month, 0.15 mm day^{-1} during the first year, and 0.06 mm day^{-1} over the whole period. In contrast, a similar tank planted with two crops of vetch evaporated 655 mm in 280 days (an average of 2.3 mm day^{-1}). Evaporation from the bare soil in the first year was less than one-tenth of this amount, and was a still smaller fraction in succeeding years.

It is clear that over distances more than a few tenths of a meter, water, except in roots, moves very slowly in the soil. Movement of water out of the soil body as vapor depends therefore not only on the energy supply, as has been true with the systems described in the two preceding chapters, but also on how fast water can move from its initial location in the soil body to the level where both water and energy are available and vaporization can take place. This zone of water and energy convergence is initially at the surface, but shifts as the soil body dries. This shift is illustrated in the successive phases of the evaporation process.

The Phases of Evaporation from Soil Bodies

Philip (1957, 1964) shows that evaporation from bare soil progresses through three phases:

(1) While the soil surface is wet, evaporation proceeds as from a saturated surface. It responds chiefly to the local supply of energy, that is, the radiation surplus and such atmospheric factors as the profiles of temperature and humidity. In this phase both energy and water are available in the surface layer.

(2) As the moisture content of the top layers of the soil θ declines, the hydraulic conductivity or permeability of the soil K, which is a complex function of moisture content, also declines. Water under the same pressure head moves more slowly than before, because there is a reduced cross section for it to move through. At the same time, the tension forces holding the water in the micropores and in films on the soil particles (capillary pressure or moisture suction ψ) are increased, for they also are functions of soil-moisture content, as noted in the chapter on soil-moisture intake. In this phase, evaporation from the soil surface depends only on the vertical distribution of moisture in the soil. If moisture lost from the top layers is not readily replaced from below, evaporation slows down. Atmospheric conditions and energy input have a diminished effect on evaporation, not being able to affect the supply of moisture to the layer where energy can vaporize it.

(3) With still smaller moisture content and hence less hydraulic conductivity K, the movement of liquid water comes nearly to a standstill. Outflow of vapor from the surface is supplied mainly by the upward movement of vapor through the soil. Vapor diffuses in the soil in response to the gradient of vapor pressure, which means that it moves from warm levels to cooler ones. The temperature profile of the soil is therefore of primary importance to evaporation in this phase.

In winter, when the upper layers of soil are cooler than the deeper ones, vapor moves upward to the surface and out of the soil body. In summer, on the other hand, the upper layers are warm and the lower ones relatively cool, so the flux of vapor is directed downward. Evaporation from the surface is small or altogether lacking. This counter-seasonal regime resembles the regime of evaporation from deep bodies of water, described earlier.

In this third phase of evaporation from the soil, radiation and atmospheric factors play a role that contrasts with their role in the case of a saturated surface, phase (1). The energy that can support vaporization of water is distributed vertically like the distribution of temperature in the soil column. In any case, the rate of energy supply is small because the zone where liquid water is located is far below the irradiated surface where the supply of energy is greatest. Water is effectually separated from a substantial source of energy, and evaporation is correspondingly small.

In speaking of these three phases in desiccation of a soil body, Philip (1957) comments that it "appears futile to seek to relate evaporation from soils with dry surfaces to the 'evaporating power of the air,' as represented by evaporation from water surfaces of 'wet' soils." We saw earlier that the notion of "evaporating power of the air" was a chimera in explaining evaporation from wet surfaces; it is just as inadequate for drying soil bodies. It is clearly necessary to examine the physics of the system as a whole.

Energy Fluxes and Temperature at the Soil Surface

Effects of Changes in the Energy fluxes Lysimeter measurements of evaporation in early stages of a drying soil (Fritschen and Bavel, 1962) and the associated heat fluxes are shown in Table I for the wet soil described earlier, and on later days (Fig. XII-1). As the soil surface dried, the increase in its reflectivity and temperature resulted in a net surplus of radiant energy that was smaller by about 0.15 than when the surface was wet. The drier surface lost less heat by evaporation, and its higher temperature forced more heat to move from it into deeper soil layers S and into the air H. Associated with these three shifts in heat flow, the rate of conversion of heat by vaporization decreased by almost half.

The rate at which water in the soil body moved upward toward an evaporating surface was measured (Rose, 1968) for movement in both the liquid and the vapor state. Liquid-water flow in the soil was at about 40 mm $sec^{-1} \times 10^{-6}$ on the first day after the soil was wetted (or about 4 mm day^{-1}), and at half or less of this rate on the third and fourth days. The flux of water in vapor form was nearly zero at night,

TABLE I

Exchanges of Energy at a Drying Soil Surface, Phoenix, Arizona (24-hr average fluxes of energy)[a]

Parameter	Date		
	29 April	30 April	2 May
State of surface	Wet	Moist	Dry
Surface temperature (°C)	30	32	38
Net surplus of energy in all radiation exchanges R	+196	+193	+160
Heat exchange with soil body S	+4	−5	−16
Heat exchange with the air H	+1	−15	−34
Heat of vaporization E	−201	−173	−110
Equivalent evaporation (mm)	7.0	6.0	3.8

[a] Units: W m^{-2}. Source: Fritschen and Bavel (1962).

and by day proceeded at somewhat higher speeds than the flux of liquid water—but it was directed downward, away from the heated upper layer. The vapor flow out through the surface was obviously supplied by the liquid flux, not the vapor. Evaporation at night indicated a continued upward liquid-water flow in the soil body at a low rate.

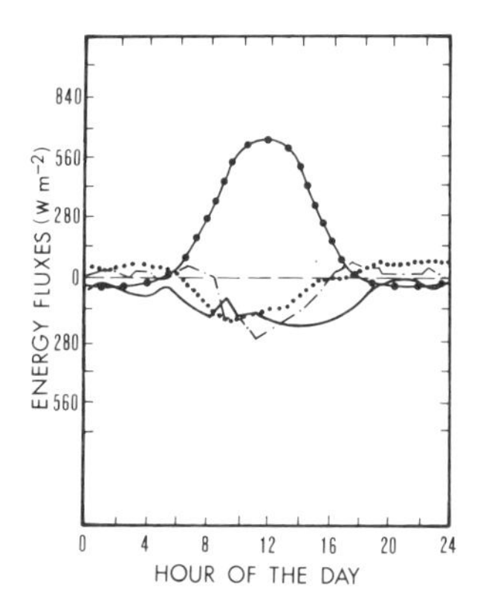

Fig. XII-1. Diurnal course of the energy fluxes (—•—• net all-wave radiation, — latent-heat flux, —·—·sensible heat flux, ··· soil-heat flux) at the dry surface of soil five days after it was irrigated (Fritschen and Bavel, 1962).

Because most of the water for evaporation must come from the top layers of the soil mass, the amount of evaporation is not much greater from soil deeply wetted by a big rain than from soil merely superficially dampened by a small one. Over a period of time, total evaporation depends less on infiltration amounts than on the number of times per season that the surface is wetted and subsequently has access to enough energy to evaporate the small depth of water within its reach. As was true with water intercepted on litter and leaf surfaces (Chapter VI), total evaporation is determined chiefly by the number of rains per season or the duration of surface wetness. Bare soil behaves much like city pavements and roofs, which draw upon no subsurface store of water and evaporate only the 1 mm or less that is caught in depression storage and the film remaining after rain. In these cases of superficial evaporation energy is assumed available and is evident in the temperature of the surface.

Surface Temperature Estimates of evaporation from a drying soil, as well as from the wet soil we discussed in a preceding chapter, can be made if surface temperature is known. This implies that the effect on evaporation of the progressive drying of the soil surface, to a moisture content as low as 0.005, is taken into account in the increase in surface temperature.

Fuchs and Tanner (1967), in an experiment on sandy soil in Wisconsin, measured a maximum surface temperature of 44°C, which was 16°C higher than air temperature and also much higher than the surface temperature (32°C) at the wet soil surface in the Arizona experiment described earlier. If we consider that some level around 32°C represents the upper temperature limit at an extensive, freely evaporating surface, it is clear that drying out the top 1–2 cm of a sandy soil will be associated with a substantial rise in the temperature of this layer and the temperature at its surface. Since the layer *is* dry, however, such a temperature rise does not produce more evaporation but rather less.

The decrease of the evaporation rate is given by the expression

$$(\rho c_p h/s)(e_0^* - e_z)$$

(Fuchs *et al.*, 1969) in which ρ is the density of air (kg m^{-3}); c_p specific heat of air (J kg^{-1} deg^{-1}); h transfer coefficient for heat and vapor from the surface to height z in the air (m sec^{-1}); s slope of the curve that relates saturation vapor pressure to surface temperature (mb deg^{-1}); e vapor pressure (mb); e_0^* saturated vapor pressure that corresponds to the temperature of the surface (mb); e_z vapor pressure in the air at height z (mb). As the surface heats, e_0^* increases and actual evaporation departs more and more from the potential rate. Thus, under steady radiant heating, for example, the progressive drying of a surface reduces evaporative cooling, the surface gets hotter, e_0^* rises, and actual evaporation becomes a small fraction of potential evaporation.

Midday rates of evaporation from the dry, sandy top layer in Wisconsin were only 0.06 mm hr^{-1}. The daily total was about 0.3 mm, an order less than the partly dried soil in the Arizona experiment.

This "combination equation is limited to surfaces such as bare soils" (Fuchs *et al.*, 1969), at which the turbulent exchanges of heat with the air are made at a well-defined interface. It then shows how evaporation from soil becomes a self-limited process.

In vegetation, in contrast, the zone filled with leaf surfaces at which the exchanges of sensible and latent heat take place is thick, and in it the idea of a "surface temperature is not well-defined" (Fuchs *et al.*, 1969). Let us now look at this more complicated system.

EVAPOTRANSPIRATION FROM A DRYING SOIL–VEGETATION SYSTEM

Evaporation is altered when plants occupy a soil body. Leaves grow over the soil surface, and roots occupy the soil body. Energy and water now come together at a new convergence zone, the leaf surfaces. These are supplied with energy from the sun and with water brought up from deep in the soil body.

Exposed to radiation and the atmosphere, the cells of plant leaves must contain, and also leak, water. Therefore they must be supplied with water, which is actively searched out in the soil by roots. Because the forces holding water in the small pores of the soil increase as soil moisture declines, roots find it increasingly hard to extract water during periods between rainstorms.

While movement of water in bare soil is difficult to formulate and measure physically, the location and movement of water in the soil–plant–atmosphere system is far more complex. Even within the plant, water movement forms one of the classical problems of plant physiology (Bonner, 1959). Dependent on the varying forces in drying soil and the regular and irregular regimes of heat supply and vapor removal at foliage surfaces, evapotranspiration from drying vegetation is a highly complicated process, although a common one. Bare soil tends to be quickly colonized by plants. A special example is the solid-waste disposal site, or landfill, which yields a leachate percolation

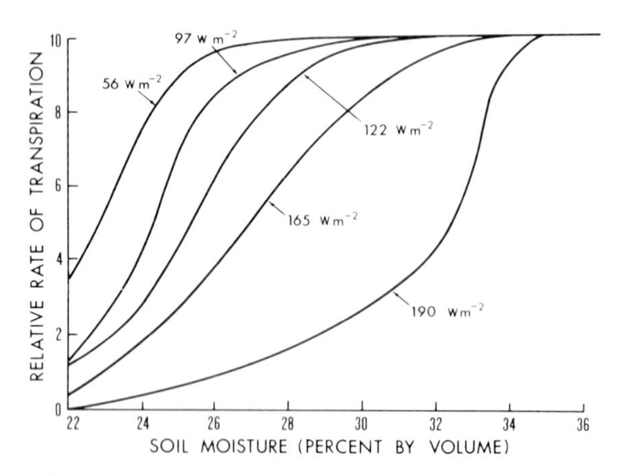

Fig. XII-2. Relative rate of transpiration in tenths as a function of volumetric soil moisture on days of different energy levels (potential evapotranspiration, but labeled in energy terms W m^{-2}). Field capacity of the soil was 36%, the 15-bar content 22% (Denmead and Shaw, 1962).

stream of complex and often harmful composition. Even if a landfill is sealed from the native groundwater body, the volume of leachate should be minimized, and this can be done by promoting evapotranspiration. Molz *et al.* (1974) reduced leachate volume by 300 mm yr^{-1}, or 0.7, by planting the landfill surface. Similar efforts are made on septic-tank effluent fields.

Different plants behave differently in the same conditions of soil and atmospheric forces. Behavior patterns such as closing stomata, offering resistance to diffusion of water from leaves, extending the absorbing roots, and tending to grow roots at different depths are species dependent, as we saw earlier.

Effects of Drying Soil on Ecosystem Evapotranspiration

Evapotranspiration is reduced by slow movement of soil water to root hairs as the soil dries. When this happens the tension gradients along which water moves must increase in order to keep water moving across the space between the soil particles and the root hairs. For this increase to develop, the diffusion pressure deficit in the leaves increases; there is less leaf turgor and the guard cells close the stomata. Transpiration declines.

If the energy supply is high, this change in the water stream moving from soil to root to leaf to air might take place even when the moisture content of the soil is high (Denmead and Shaw, 1962). "We expect transpiration rates to decline with decreasing soil moisture content and we expect that this decline will be evident at higher and higher soil moisture contents as the potential transpiration rate increases," i.e., on days of a large supply of energy. On days of great energy supply—say 200–300 W m^{-2}—transpiration falls short of the potential rate even when plants are rooted in a soil of moisture content close to field capacity (0.36 moisture content) (Fig. XII-2).

On days of low energy supply (about 100 W m^{-2}), however, Fig. XII-2 shows that transpiration proceeds at the potential rate even from plants rooted in soil having a small moisture content. In this case, moisture content is much less than field capacity. Only in very dry soil does transpiration on such low-energy days fall short of the potential rate.

The soil-moisture content at which actual transpiration begins to depart from the potential rate is identified by Denmead and Shaw (1962) as the point at which leaf turgor is lost. In a given soil type this point varies with energy supply, as Fig. XII-3 shows. In the soil illustrated, the following critical levels of soil moisture are found:

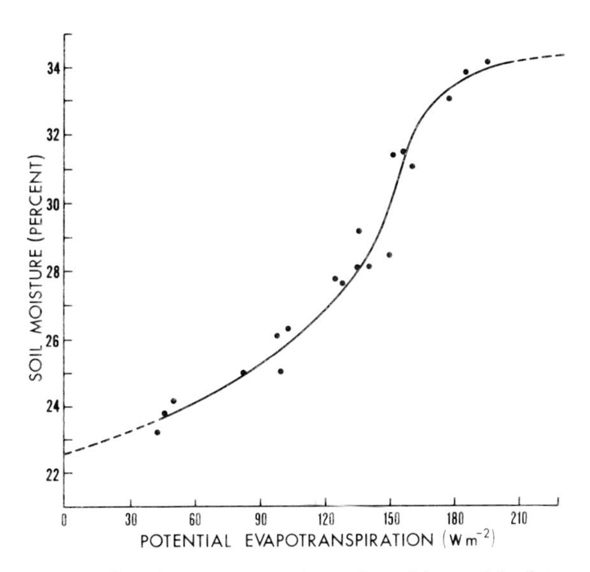

Fig. XII-3. Estimated soil moisture at the point of loss of leaf turgor on days of different energy levels or potential evapotranspiration. On days of high energy, leaf turgor is lost at relatively high levels of soil moisture (percent by volume) (Denmead and Shaw, 1962).

Soil moisture 0.36 (Field capacity). Soil suction 0.3 atm.

Soil moisture 0.34 Turgor-loss point on days when the potential evapotranspiration rate is 6 mm day^{-1} (energy conversion rate approximately 300 W m^{-2}).

Soil moisture 0.27 Turgor-loss point on days when potential evapotranspiration is 4 mm day^{-1}.

Soil moisture 0.23 Turgor-loss point on days when potential evapotranspiration is 1.4 mm day^{-1} (approximately 80 W m^{-2}). This is also the level of a soil suction of 15 bars.

As we saw in a preceding chapter, soil moisture in nature is not evenly distributed, and this affects transpiration. Calculations of evapotranspiration at 25 sites in a field of grass in Ontario revealed a large dispersion; over a 3-month period in which the average evapotranspiration at all sites was 252 mm, the standard deviation among sites was 33 mm, giving a coefficient of variation of 0.13. These differences resulted from differences not in energy supply or vapor removal but in "large soil moisture differences at the individual sites" (Rouse, 1970). Uncovering these differences required both the careful measurements of soil moisture itself, by two methods, and also the calculation of downward movement of water F beyond 1.8-m depth by use of a flux gradient relation:

$$F = -K_z \, \partial\phi/\partial z,$$

where K is a coefficient of transport related to the gradient of potential ϕ (Rouse and Wilson, 1972). Evaporation rates determined by this water budget of the soil were not acceptable over one-day periods, but were adequate for periods of several days if no rain fell and if at least six soil-moisture sites were measured.

Forests approach the potential evapotranspiration ideal more closely than does lower vegetation, considering the course of the whole growing season. Compared to annual crops, they are active longer in late summer and autumn; compared with most low ecosystems, they root more deeply and can keep on transpiring during dry spells in summer. Both their darker appearance (they absorb more solar radiation) and roughness (extracting more heat from the air) give forests a larger supply of available energy to support transpiration. As Rauner demonstrates (1972, p. 146), actual evapotranspiration from forests tends to approach potential evapotranspiration, except where dry seasons occur. In Douglas fir in southwestern Oregon differences between actual and potential transpiration indicate strong stomatal control in summer (Reed and Waring, 1974). Growth of the fir in spring is limited only by energy input, but in late summer by plant resistance. Over monthly periods it is related to the ratio of actual to potential transpiration.

Effects of Stomatal Resistance

When loss in leaf turgor brings about a narrowing of the stomata, the stomatal resistance to diffusion of water r_l increases. For example, when stomata narrow from a width of 5–1 μm, stomatal resistance increased in one experiment from 0.8 to 2.8 sec cm^{-1}, as shown in Table II. With no change in the external resistance to vapor diffusion r_a, the "total resistance nearly doubles, but transpiration decreases by only one fifth because the leaf becomes warmer and the water concentration difference increases by about half" (Waggoner and Zelitch, 1965), i.e., from 24 to 35 mb. In other words, the greater heating of the leaf counteracts some of the throttling effect of closing stomata by increasing the vapor-pressure difference between leaf and air.

In the usual flux versus gradient formula, the transpiration stream depends both on the difference between the concentration of water vapor at the cell walls and in the atmosphere, and on the resistance to

TABLE II

Leaf Transpiration in Quiet Air[a]

Parameter	Width of stomata (μm) 0	1	5
Resistance (sec cm^{-1})			
Stomatal r_l	∞	2.8	0.8
Atmospheric r_a	1.7	1.7	1.7
Total	∞	4.5	2.5
Temperature of leaf (°C)	44	35	31
Vapor pressure in leaf (mb)	91	56	45
Leaf–air temperature difference (°C)	14	5	1
Leaf–air vapor-pressure difference (mb)	70	35	24

[a] Wind speed 5 cm sec^{-1}, radiation surplus +410 W m^{-2}, air temperature 30°C, vapor pressure 21 mb, leaf width 5 cm. Data from Waggoner and Zelitch (1965).

diffusion of vapor from cell walls into the free atmosphere. The reduction in transpiration by a fifth puts more of the heat-removal burden on the sensible flux and long-wave radiation, which can handle the load if the leaf gets warmer (as is shown in Table II).

In stronger air movement (Table III), narrowing of stomata from 5 to 1 μm increases the stomatal resistance r_l, as in the previous example. The effect is enhanced on total resistance, which now is chiefly composed of the stomatal component and contains only a small contribution r_a due to atmospheric resistance in the moving air. Increase in stomatal resistance results in a large decrease in transpiration, which falls to half its former rate.

Energy Considerations

As a transpiring ecosystem runs through its store of available water, as soil moisture gets harder to extract from the tiny pores, and as stomatal resistance to vapor diffusion increases, the vaporization of water at cell walls in the leaves decreases. Given the same supply of energy to the leaves in the sunny weather between rainstorms, there must be a shift in its division; less can be converted into the latent heat of vaporization, more must be converted into sensible heat and transferred into the air H.* This is represented as an increase in the Bowen ratio ($\beta = H/E$), as H grows and E diminishes.

Some of the methods described in the preceding chapter to determine rates of evapotranspiration from well-watered ecosystems can also be utilized in the circumstances of drying soil. One of these is the Bowen ratio method, in which the ratio is approximated by calculation from the gradients of temperature T and specific humidity q in the air over the ecosystem, as

$$\beta = c_p(\Delta T + \Gamma\ \Delta z)/L\ \Delta q,$$

where c_p (specific heat of air at constant pressure = 1 J g^{-1} deg^{-1}), and L (latent heat of vaporization of water = approximately 2500 J g^{-1}) serve as coefficients that convert temperature and humidity readings into units of heat and hence make the ratio nondimensional. The term $\Gamma\ \Delta z$ is a minor adjustment that corrects for the adiabatic cooling of ascending air with change of height z.

* A small increase will also occur in the subsurface movement of heat S, as seen in the Arizona energy budget for drying soil (Table I), although less than where soil is bare. Warmer foliage also radiates away more energy; this shift is also of minor significance.

TABLE III

Leaf Transpiration in Moving Air[a]

Parameter	Width of stomata (μm)		
	0	1	5
Resistance (sec cm^{-1})			
Stomatal	∞	2.8	0.8
Atmospheric	0.2	0.2	0.2
Total	∞	3.0	1.0
Temperature of leaf (°C)	34	31	28
Vapor pressure in leaf (mb)	53	45	38
Leaf–air temperature difference (°C)	+4	+1	−2
Leaf–air vapor-pressure difference (mb)	32	24	17

[a] Wind speed 223 cm sec^{-1}, radiation surplus +410 W m^{-2}, air temperature 30°C, vapor pressure 21 mb, leaf width 5 cm. Data from Waggoner and Zelitch (1965).

This ratio is then applied, as noted earlier, to divide the net available heat at the surface, i.e., radiation surplus less subsurface heat intake. In dry conditions Δq is small, since both crop and air are dry, but ΔT is large, because the crop surface gets very hot. Instead of a few tenths of a unit as over well-watered surfaces, the ratio reaches values an order of magnitude greater; a value of 2, for instance, indicates that twice as much heat is removed from the ecosystem in sensible form as in latent. Values of the Bowen ratio were determined from micrometeorological measurements in a four-year study (Denmead and McIlroy, 1970) of a lysimeter in which wheat was growing, and agreed satisfactorily with those found from measurements of R, S, and E in the lysimeter. This Australian experiment confirms the validity of the meteorologically determined ratio when it is applied to divide available heat at the surface of a crop often cultivated in dry climates.

In a Canadian study (Wilson and Rouse, 1972) a variation of the combination model for evapotranspiration mentioned earlier ($E = (R - S)[s/(s + \gamma)]$) was found also to work well on moderately dry days. The partition of available energy ($R - S$) is still represented by the term $s/(s + \gamma)$, until the soil gets too dry. Budyko (1971, p. 94) notes that this formula is useful during periods when the major components of the heat budget are "not very small," i.e., "mainly daytime conditions in the warm season"—the time when evaporation is greatest.

Environmental monitoring of ecological niches in such areas as high tundra in Alaska can be helped by simulation modeling of the energy budget and its latent-heat component. Data from each ecosystem—soil depth, diffusivity and heat capacity, site slope and orientation, and surface albedo and roughness—are combined with atmospheric inputs to yield evapotranspiration (Brazel and Outcalt, 1973). The diurnal regimes of evapotranspiration were reconstituted at three sites in an Alaskan pass; daytime mean fluxes were 220 W m^{-2} from a southeast sandy clay slope, 150 from a sandy gravel flat, and 30 from a northwest basalt talus slope. Peak rates occurred at 1000, 1230, and 1600 hr, respectively, following the moving sun.

Methods of employing the Bowen ratio between the two turbulent heat fluxes involve an assumption that turbulence above the surface transports sensible and latent heat equally, a topic of controversy for many years. Dyck (1972) concludes that "the energy budget method is fundamentally sound. More precise results await the theoretical determination of the relation between diffusivities for water vapour and heat under non-adiabatic conditions. . . ."

Determination of Evaporation Rates

The water-vapor flux from ecosystems has seldom been measured directly. Only short observation runs above such low systems as grass have been made and these in special conditions of the mesoscale environment that maximize its homogeneity with respect to albedo, roughness, moistness, and so on.

In spite of the practical importance of evaporation to such surface scientists as agronomists, soils men, and engineers, as well as to atmospheric scientists, serious problems of instrumentation remain. Evaporation has been saddled with nineteenth century measuring devices; these are based on an inadequate understanding of the physics of the vapor flow, which they do not measure directly, and are sometimes hampered by even older misconceptions. We saw in Chapter X that pans do not measure even the evaporation from aquatic ecosystems; still less are they appropriate for terrestrial ecosystems. As Thornthwaite and Holzman (1939) say, they "do not supply actual information on water-losses from drainage areas."

Empirical Methods Formulas for estimating evapotranspiration from conventional simple meteorological observations (readings of temperature and humidity of the air at 1.5 m, and of air motion at 10 m) are numerous, being based on the idea that air temperature and humidity often are related to energy fluxes at the evaporating surface. Indeed, the regimes of these atmospheric conditions often are consequences of the regimes of the fluxes, but not always immediate or precise consequences. Gaps in this general relation place an inherently low limit on the accuracy of air-temperature methods of estimating evapotranspiration.

These gaps can be narrowed by the use of factors that yield estimates of *surface* temperature, or other means (Konstantinov, 1968, Chapter 3) to extract the greatest amount of information out of more sophisticated meteorological records, including gradients and stability indexes. These adjustments were developed by Soviet investigators,who had access to data from a nationwide network of surface temperature and radiation fluxes at plots where the meteorological measurements are taken.

Without such auxiliary data, it is difficult to work out usable procedures for going from conventional atmospheric data to evapotranspiration. It would be pointless to reproduce here the innumerable empirical formulas for evapotranspiration that have been compiled, without data on the fluxes and temperature at the active surface. Such empiricisms are confined to the climatic area where they were derived, and are highly perishable in transit to other areas. The reader who needs to resort to an empiricism should refer to its original form so that he can be informed of all the limitations entailed in its correct use.

Geophysical Methods Water-budget accounting which indirectly determines evapotranspiration as the residual in the budget equation has long been applied in lysimeters and plots, and in small and (presumably) tight drainage basins. The larger the system the longer the time interval needed to obtain accurate results; the larger systems may yield only one value per year of record. Errors in measuring the other fluxes in the water budget are not necessarily negligible, as we have seen in earlier chapters—and, as in any residuals approach, they tend to be incorporated in the value calculated for evapotranspiration.

Aerodynamic research of the 1930s was applied to this geophysical problem by Thornthwaite and Holzman (1939, 1940) and subsequent workers. Even with increasing understanding of atmospheric turbulence and the effects of ecosystem roughness and atmospheric stability on vapor flux, however, application to real conditions remains difficult in terrains that are mosaics of many individual ecosystems that contrast in roughness and moistness. Another indirect approach to evapotranspiration, through the energy budget, may also have a residuals problem. As in the aerodynamics methods, boundary-effect problems are often encountered.

Because of the past lack of direct measurement "evapotranspiration research has become the patron of all kinds of study of the heat and water balances of the earth's surface and of such relevant surface characteristics as surface temperature, roughness, albedo, and moisture content" (Miller, 1965, p. 205). Now, however, direct measurements by the eddy-correlation method are beginning to be made of the upward stream of vapor above certain types of evaporating surfaces. Most of these have been made in special, experimental situations, and here and there on a regular basis.

This method requires sensors (psychrometer for concentration of vapor and vertical anemometer for its upward or downward movement) that "must operate with very low inertia. High demands are made on the sensitivity and efficiency of the devices for measurement and analysis" (Dyck, 1972). To move from special experimental situations into a greater range of local research studies in which "more robust but less accurate devices" can be checked out and calibrated, Dyck recommends further development work on the equipment for direct measurements.

Pending wider adoption of the eddy-correlation method, we have to make do with soil lysimeter measurements, application of aerodynamics and mass-transfer formulas in the lower layer of air, and energy-balance methods. These last have potential for precision that is partly sacrificed if neither turbulent heat flux (H or E) is measured separately and only their sum is partitioned by Bowen ratio or other relations mentioned earlier. From these several methods, however, it is possible to find sets of evaporation determinations made in conditions likely to minimize inherent errors, or sites where two independent methods yield similar values of evaporation rates and rhythms. Some of these are used here to illustrate typical time and spatial distributions of this important water transfer.

Influence of Ecosystems and Terrain The geophysical approaches to evaporation must also deal with the spatial interactions among ecosystems. Although uniform in the properties that govern their coupling with sun and atmosphere, ecosystems are spatially limited, and many are too small for good atmospheric measurements. Contrasts in albedo, moisture, or roughness among neighboring ecosystems produce mesoscale linkages of vapor and sensible heat that make it difficult to measure the vapor flux rising out of a specific ecosystem.

It is usually accepted that while the accuracy of evaporation formulas is good for water surfaces over intervals of time long enough to permit determining changes in heat storage in the water, it is poor over short periods of time for land surfaces. This is especially true if the terrain is heterogeneous in roughness and moistness, and, of course, is true if it differs from the region in which the applied empirical coefficients were derived. The "significance of the varying form and properties" (Sibbons, 1962) of the surface of the earth has as yet hardly been taken into account in studies of evapotranspiration. Characteristics of soil, vegetation, and relief of small areas affect evaporation estimates. Budyko (1971, p. 105) concludes that "the task of determining evaporation from limited areas in terrain where the earth's surface has a heterogeneous structure may be very complicated. Difficulties associated with its solution are at present still not fully overcome."

While current methods give, however, little reliable information about the microscale or mesoscale patterns in time or space, they have more utility for work at larger scales. Indeed, lacking a network of local measurements of evaporation, we have no other way of deriving information about regional evaporation than by use of estimating methods. Annual values can be determined within 0.05–0.10 (Budyko, 1971, p. 128).

VARIATIONS IN EVAPOTRANSPIRATION OVER TIME

Periods of Airstream Dominance

The rate of actual evapotranspiration from a drying soil–vegetation system follows the supply of energy, the supply of liquid water, and the atmospheric removal of water vapor. With respect to the last-named factor, it has been found that the rates of evapotranspiration into different airstreams are quite different. Measuring turbulent vapor diffusion from grassland near Washington in summer (on a farm later swallowed up by the Pentagon), Thornthwaite and Holzman (1940) found these rates of evapotranspiration:

Into continental air (22 days)	2.0 mm day^{-1}
Into maritime air (25 days)	1.2 mm day^{-1}

The continental airstream is characteristically drier, and usually is hotter and sunnier.

Evaporation into unusually cold air, say on the shores of a cold ocean or lake, is reduced because the large temperature gradient from surface to air favors a large flux of sensible heat. This situation is expressed in a large value of the Bowen ratio.

Periods of Deficient Soil Moisture

Moisture Stress on Iowa Corn Days or groups of days when deficient soil moisture affects the growth and transpiration of vegetation can be identified as important features of the regime of evapotranspiration. The soil-moisture percentage at which leaf turgor is lost, a function of the rate of supply of energy, as has been noted (Fig. XII-3), seems to represent "the lower limit of availability of soil water for dry matter accumulation" (Denmead and Shaw, 1962). In other words, on days when the soil moisture is less than the turgor-loss point, "the plant virtually ceases to assimilate carbon dioxide" (Denmead and Shaw, 1962). In a sense the soil–vegetation ecosystem goes into a standby status.

Determination of the turgor-loss point on each day, for comparison with the concurrent soil moisture, provides a means of identifying days when the plant is under moisture stress so great as to affect its assimilating function. In Fig. XII-4 (Dale and Shaw, 1965) daily values of soil moisture in the root zone of corn in two different years are

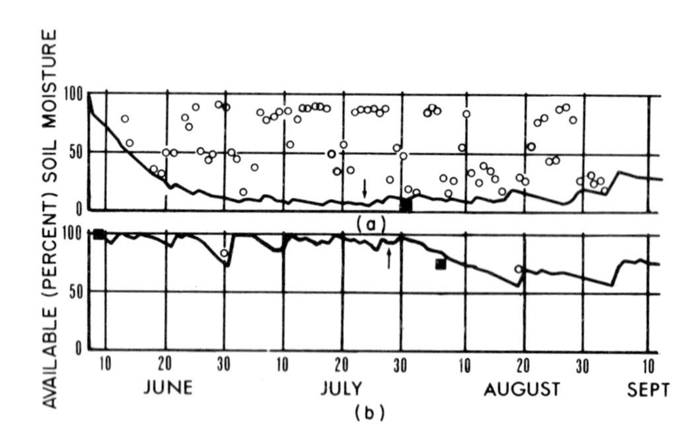

Fig. XII-4. Available soil moisture in the summers of (a) 1956 and (b) 1958 in root zone of corn at Ames, Iowa on well-drained soils that hold 230 mm at field capacity in the top 1.5-m depth. Circles represent the soil moisture at which under the daily energy level leaf turgor would be lost and assimilation cease, plotted only when greater than the soil-moisture level existing and so representing a condition of moisture stress. Squares represent measured values of soil moisture (from Dale and Shaw, 1965).

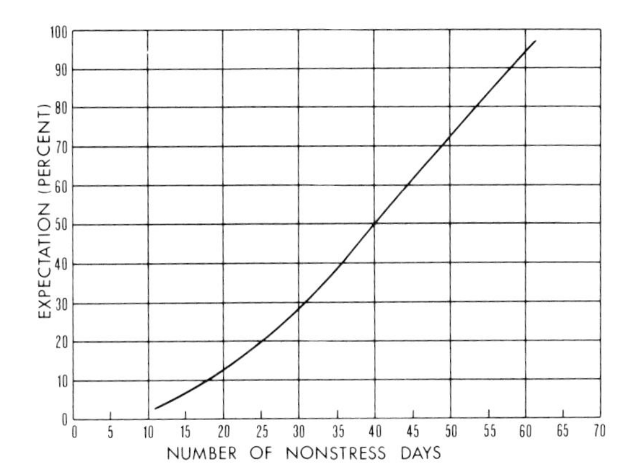

Fig. XII-5. The percentage chance of experiencing the indicated number or fewer of nonstress days for the moisture status of corn on Nicollet silt loam at Ames, Iowa in the 63 days from six weeks before silking to three weeks after. Field capacity is 230 mm available; period of analysis is 1933–1962 (Dale and Shaw, 1965).

shown by the solid line and the occurrence of days of moisture stress by the dots. In 1958, for example, the soil was sufficiently moist for corn during most of the growing season. Only two days of moisture stress occurred (30 June and 19 August). It can be considered that this corn field generated water vapor at a rate close to the rate of potential evapotranspiration. In 1956, in contrast, the soil dried out in mid-June and did not recover during the rest of the summer. Moisture content remained less than 20% of the available root-zone capacity. Moisture stress days were unrelenting; 58 such days occurred in the 63-day period believed to be critical to the yield of the corn plant (42 days before to 21 days after silking). These two years represent extremes in the recorded climatic history of Iowa corn. Other summers plot in between these in the frequency distribution shown in Fig. XII-5. In half the summers it can be expected that 23 stress days or more will occur in the critical 63-day period in this part of the Midwest.

Relation of Soil Moisture with Evapotranspiration in Ohio From extensive measurements of grass evapotranspiration Mustonen and McGuinness (1967) developed equations for daily and monthly values in the growing season at Coshocton, Ohio, which show the relative weight of moisture factors (Plate 22). For daily rates of evapotranspiration they find

$$\mathrm{ET_D} = 0.70(\mathrm{Pot\ ET})^{1.0}(\mathrm{soil\ moisture})^{0.25}.$$

Plate 22. Three lysimeters at the Coshocton experiment station, Ohio, in early spring. The first two are drainage types, the third weighing. Note the use of several rain gages to measure accurately the input of water to the lysimeter vegetation (April 1965).

For monthly values they find

$$ET_M = 0.60(\text{Pot ET})^{1.0}(\text{soil moisture})^{0.25}(\text{rainfall} + 10 \text{ mm})^{0.125}.$$

In both equations the first term, potential evapotranspiration, can be regarded as energy supply. It has a somewhat higher weighting in day-to-day variations than month-to-month.

Moisture supply in the monthly expression is expressed by a soil-moisture term and a rainfall term to a small power. The rainfall term might be regarded as an index to the duration of a water film on the foliage, permitting rapid evaporation, as described in the chapter on intercepted rain.

The more important indication of water availability is the measurement of soil moisture, taken to the $\frac{1}{4}$ power. A decrease in soil moisture is not fully felt in evapotranspiration.

Phases of Plant Growth

In drying soil, the actual evapotranspiration rates depart from the potential rate as water becomes more difficult to obtain, but departures from the potential occur for other reasons also. Internal translocations and transformations of matter in the latter part of a plant's growth cycle become more important, and photosynthesizing and the transpiration associated with it diminishes in these stages.

Cutting grass reduces its area of transpiring foliage, hence the evapotranspiration rate. From the long series of measurements in Ohio (Mustonen and McGuinness, 1967), the effect of two summer cuttings of hay are shown in Fig. XII-6 in terms of the computed rates of evapotranspiration for the existing conditions of water and energy availability as they fluctuated from day to day. The points are averages over 15 yr. The first cutting, in June, reduced the rate of evapotranspiration to about 0.3 of what it had been just earlier. Recovery took several weeks. The second cutting, in August, usually smaller than the June cutting, reduced evapotranspiration to about 0.5 of the rate just preceding. Recovery after this second cutting was slower than after the first.

Evapotranspiration in relation to the growth phase of grassland in Australia is presented in Table IV from one of the few sets of direct measurements made of this flux. In the eddy-correlation method the speed and humidity of upward flow in gusts is recorded, as is the speed and humidity of downward gusts, and the net sum over a period of $\frac{1}{2}$ or 1 hr measures the net upward vapor flux from the grass. The site at Kerang provided the long fetch over smooth land necessary for this measurement technique. A large share of the total of the nonradiative energy fluxes is taken by latent heat when the grass is green and transpiring, and when it is tall. Roughness is an obvious

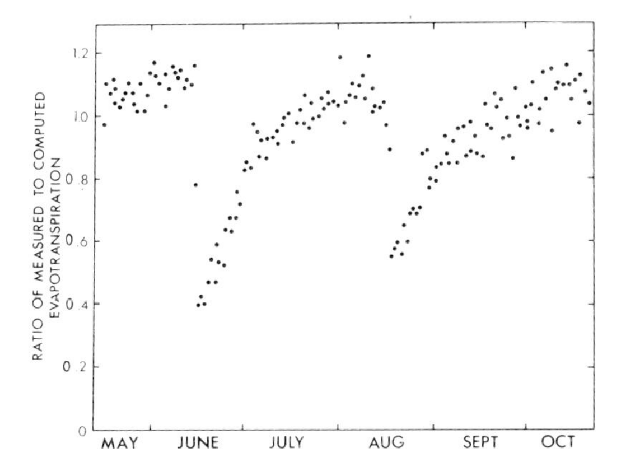

Fig. XII-6. Effect of cutting hay on the ratio of measured evapotranspiration to that computed for daily conditions of available water and energy, taken as long-term averages (Mustonen and McGuinness, 1967).

TABLE IV

Latent-Heat Flux over Grass at Kerang, Australia[a]

Series	Condition of grass	10–12	12–14	14–16	Mean (1000–1600) Flux	As fraction of all nonradiative energy fluxes	Vapor flux (mm hr^{-1})
A[b]	Dry, short	−70	−90	−45	−68	0.25	0.10
B	Green, short (5 cm)	−150	−125	−135	−137	0.51	0.21
C	Green, taller (5–10 cm)	−90	−195	−135	−140	0.60	0.22
D	Green, tall (10 cm)	−240	−185	−115	−180	0.80	0.26

[a] Units: W m^{-2}. Source: Data in Swinbank and Dyer (1968). Observations made in different seasons.

[b] By difference; Series B, C, and D were measured.

factor in evapotranspiration from ecosystems subjected to heat advected from drier regions, and varies with season.

Growth stages in annual crops have a more marked effect than in grass cover. During the growing season, annuals go through their life cycle: bare ground, emerging seedlings, increasing foliage, greater height and roughness, complete ground cover, perhaps a period of special sensitivity to water stress at the time of flowering, then declining leaf area, drying leaves, collapse, and death. A grass cover serves as a base line against which these life-cycle changes can be seen.

At Coshocton, for instance, where established meadow had about the same evapotranspiration rates as permanent grass, its replacement crop when the land was prepared for corn (in May) or wheat (in October) produced much smaller rates. Later, however, in the mid-growth phases of corn (July) and wheat (fall and spring), rates of evapotranspiration were higher than from grass. Corn in the flowering stage in a hot summer transpires twice as much water as does meadow, but in a cool summer only a third more (Konstantinov, 1968, p. 386). Earlier and later its water demands are more like those of meadow; over the whole season they might total less.

Corn, a brief-summer crop, has the greatest rate of water demand, reaching 6 mm day^{-1}. In most years both heat and water are readily available at the same time. Wheat transpires slowly through the winter and reaches rates equal to those from corn before it is harvested in June. Evapotranspiration from the meadow averages 2.3 mm day^{-1} over the year. Its peak rates are similar to those of the other crops in the rotation. (See Table V.)

Combining all crops of the rotation, the evapotranspiration regime rises to a peak of 5 mm day^{-1} from a winter low of 1 mm day^{-1}. The ratio of the summer half-year to the winter half-year is 3:1, which evidences a considerable degree of seasonality.

The Seasonal Regime of Actual Evapotranspiration

The degree to which actual evapotranspiration departs from the potential rate, i.e., the degree to which vegetation is kept from responding to the seasonal forcing function of the energy supply because of limitations in its water supply and the demands of its biological growth phases, is seen in Fig. XII-7. The station, L'vov in the Ukraine, displays a regime of zero potential evapotranspiration in winter, a sharp rise during the spring, and a slower rise from May to the maximum in July, which is 4 mm day^{-1}. The decline is more uniform than the rise, and a little slower. This curve represents the energy function.

TABLE V

Evapotranspiration from a Rotation Cycle on Lysimeters Growing Corn, Wheat, and Meadow (Grass and Alfalfa) under Improved Practice at Coshocton, Ohio, Mean 1944–1958[a]

Crop	J	F	M	A	M	J	J	A	S	O	N	D
Corn	—	—	—	—	3	4	6	4	2			
Wheat										1	1	1
Wheat	1	1	2	3	5	4						
Grass							3	4	2	2	1	1
Grass	1	1	2	2	5	4	4	3	3	2	1	1
Grass	1	1	2	3	5	5	4	3	3	2	1	1
Grass	1	1	2	3	—	—	—	—	—	—	—	—
Average over whole rotation	1	1	2	3	4	5	5	4	2	1	1	1

Totals:			
Corn	581 mm in 150 days.	Mean (mm day^{-1})	3.8
Wheat	562 mm in 273 days.	Mean (mm day^{-1})	2.1
Grass	2335 mm in 1034 days.	Mean (mm day^{-1})	2.3
Whole rotation	868 mm in 1 year.	Mean (mm day^{-1})	2.4

[a] Units: mm day^{-1}. Source: Data from Harrold *et al.* (1962, Table 35).

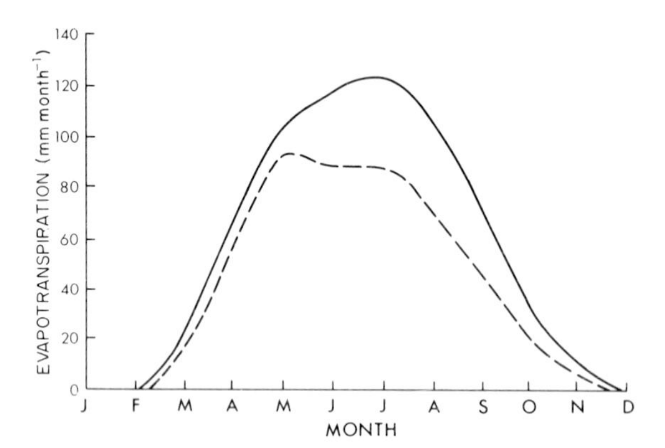

Fig. XII-7. Annual regimes of potential (solid line) and actual (dashed line) evapotranspiration at L'vov (Zubenok, 1965).

At first the response of the vegetation follows it faithfully, although actual rates in March and April are 10–15% less than potential. This suggests slow establishment of new growth in the grass cover, and also spring droughts in occasional years.

After the first half of May the actual curve increasingly departs from the potential. In some summers the soil-water reserves from winter get depleted to the extent that transpiration is entirely dependent upon the vagaries of the intervals between replenishing rains. In July and August actual evapotranspiration falls short of potential by one third.

The End of Summer Changes in evapotranspiration at the end of summer are illustrated in three observation periods in August 1953 in a large interdisciplinary observation program in Nebraska (Lettau and Davidson, 1957). Table VI shows the diurnal variation in periods II, V, and VI, a series marked by decreasing soil moisture. At this time most of the grass was dying or already dead, and the latent-heat flux from grass and soil claimed only a third of the total energy transported from the grassland by turbulence. Table VII shows the link between soil climate and evaporation. Between periods II and V the drop in evaporation is particularly large, from more than half the total turbulent-heat transport in period II to only 0.2 of it in period V. Moisture stress is not uncommon in interior North America in late summer, because potential evapotranspiration is large and soil moisture may be depleted, as we saw in Chapter IX.

Other Dry Seasons Shortfalls of evapotranspiration do not always occur in late summer and fall. In Denmark, for instance, a shortfall severe enough to reduce grass growth substantially (and with it the

TABLE VI

Latent-Heat Flux at Drying Prairie Surface near O'Neill, Nebraska at Three Observation Periods in 1953[a]

Period	Time					
	0635	0835	1035	1235	1435	1635
II (13–14 August), soil moisture 5.9% at −10 cm, 4.1 g in 0–45-cm layer						
Latent-heat flux (W m^{-2})	0	−70	−230	−230	−218	−128
Fraction of the sum of turbulent fluxes	0	0.5	0.6	0.5	0.6	0.7
Vapor flux (mm hr^{-1})	0	0.10	0.32	0.32	0.31	0.19
V (25 August), soil moisture 4.2% at −10 cm, 2.6 g in 0–45-cm layer						
Latent-heat flux (W m^{-2})	+12	−35	−70	−60	−104	0
Fraction of the sum of turbulent fluxes	0.3	0.2	0.2	0.1	0.3	0
Vapor flux (mm hr^{-1})	+0.02	−0.05	−0.1	−0.08	−0.15	0
VI (31 August), soil moisture 3.2% at −10 cm, 1.8 g in 0–45-cm layer						
Latent-heat flux (W m^{-2})	−45	−35	−92	−23	−35	−60
Fraction of the sum of turbulent fluxes	1.0	0.2	0.2	0.04	0.1	0.2
Vapor flux (mm hr^{-1})	−0.07	−0.05	−0.13	−0.03	−0.05	−0.08

[a] Source: Lettau and Davidson (1957, Vol. II).

TABLE VII

Latent-Heat Flux at Drying Grassland, O'Neill, Nebraska[a]

Period	Dates August 1953	Soil moisture at −10 cm (%)	Depth in 0–45-cm layer (mm)	Latent-heat flux (0600–1800)		
				$W\ m^{-2}$	mm water	Fraction of total turbulent transport
II	13–14	5.9	41	150	2.5	0.55
V	25	4.2	26	43	0.7	0.20
VI	31	3.2	18	48	0.8	0.15

[a] Source: from Lettau and Davidson (1957, Vol. II).

yield of butter, cheese, etc.) comes in early summer, reaching a peak deficit of 30 mm in July (Aslyng, 1960 and elsewhere.) Concern about this "hidden" drought led to a large water and energy research program at the experimental farm of the Agricultural University that has produced some of the longest homogeneous records of geophysical fluxes to be found anywhere.

In the Soviet Far East, shortfalls are largest in spring (Zubenok, 1974), when they average 40 mm in May, and recede with the summer inflow of vapor in the East Asia monsoon. At the end of the cloudy rainy summer, evaporation approaches its potential value, as may be generally true in lands dominated by the Asiatic monsoons.

In latitudes where rain comes with the intertropical convergence, such as Northern Australia, the "winter" (season when the noon sun is farthest from the zenith) is the period of large shortfalls in evapotranspiration and almost no biomass production—a time when few cattle make weight gains and many do not survive. Winter in higher latitudes cannot, in contrast, be regarded as a season of moisture shortage regardless of the amount of soil-moisture storage, because its low energy level means a low value of potential evapotranspiration.

Annual Sums of Actual Evapotranspiration

The shortfall in moisture utilization when actual evapotranspiration is less than potential evapotranspiration provides a practical index to the moisture health of an ecosystem over a season or year. This index is sometimes applied in the form of a difference, or sometimes a ratio. The difference is that between actual and potential evapotranspiration, i.e., the kilograms per square meter of water that, if it had been available at the right time, would have been evaporated by the ecosystem; this number serves as the basis for estimating irrigation requirements. The ratio is that of actual to potential evapotranspiration, i.e., 0.60 in the median year at Blue Canyon, in the California Sierra (Muller, 1972).

Both the difference and the ratio forms are useful in analyzing crop yields (Mather, 1974, pp. 207–214). A small difference or shortfall, or a high ratio, indicate that moisture stress was generally small or absent, although of course such other environmental factors as nutrient availability also affect crop yields. One general depiction of the relation between the actual–potential evapotranspiration ratio and the actual–potential yield ratio is shown in Fig. XII-8 (Chang *et al.*, 1963). This graph further illustrates the idea that moisture stress late in the growing season reduces yield less than stress coming early, during the phase of active vegetative growth when high transpiration is a concommitant of rapid photosynthesis.

The ratio of actual–potential evapotranspiration varies from one crop year to another, and its variation gives an index of one factor in the uncertainty of agricultural yields that is so important in a hungry

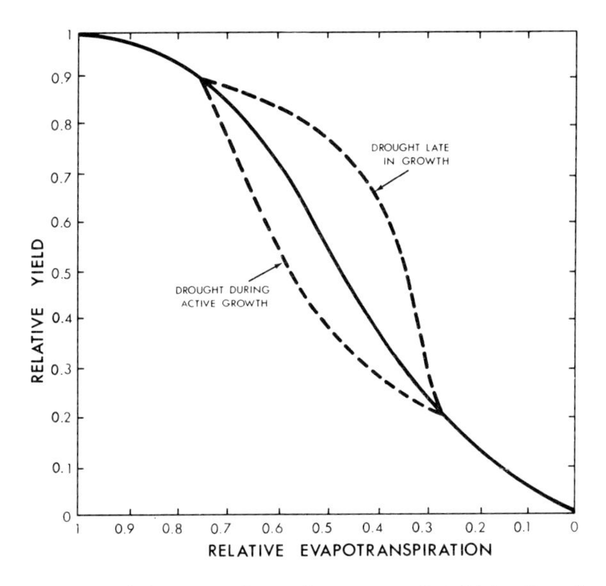

Fig. XII-8. Relation of the ratio of actual to potential yield (ordinate) to the ratio of actual–potential evapotranspiration (abscissa) (Chang *et al.*, 1963). Reproduced from Agronomy Journal, Vol. 55, pp. 450–453, 1963.

world. Muller (1972) calculated these ratios for each year from 1941 through 1970 at a number of places, and some features of the resulting frequency distribution can be summarized in the accompanying tabulation (see Fig. XII-9). The dry summer of the Sierra means that in the median year actual evapotranspiration is only 0.60 of the potential, so that there is a shortfall averaging 0.40 of the potential. The ratio of actual to potential is even smaller in the dry, hot climate of south Texas, and in the driest decile it is very low indeed. In Florida, on the other hand, summer rainfall, in phase with the regime of potential evapotranspiration, is fairly reliable in most years; the median value of the ratio is high and the range in expectation is more like that in California, as shown by the smaller difference between the 0.1 and 0.9 deciles. (See Fig. XII-5 earlier for a similar expression of moisture stress or shortfall in Iowa, emphasizing a critical period rather than the entire year.)

Station	Wettest decile of years	Median	Driest decile of years	Range between 0.1 and 0.9 deciles
Blue Canyon, California	0.68	0.60	0.47	0.21
Brownsville, Texas	0.70	0.50	0.35	0.35
Miami, Florida	0.97	0.92	0.78	0.21

In irrigated lands, the object of the land manager is to raise the ratio of actual to potential evapotranspiration as high as possible without going beyond the point of diminishing returns or incurring excessive operating or labor costs from the too frequent application of water. If such costs are not limiting, however, as is usually true with lawns of city householders, actual evapotranspiration probably approaches potential even more closely because watering is almost a cultural obligation and supports vigorous transpiration and growth. Not all urban green areas remain green during dry weather, because large parks may meet cost limitations that reduce the amount of water applied.

The city is a mix of ecosystems that have different evaporation characteristics such as pavements and parking lots, grass that goes unwatered in a dry spell, street and park trees, and lawns on which water is lavished. Considering the addition of vapor from combustion and industrial processes, the city is indeed a complex evaporating system; not surprisingly, we have no water budget for a city. Nevertheless, the fraction of green area in the urban landscape has been found to be important in determining the intensity (Rauner and

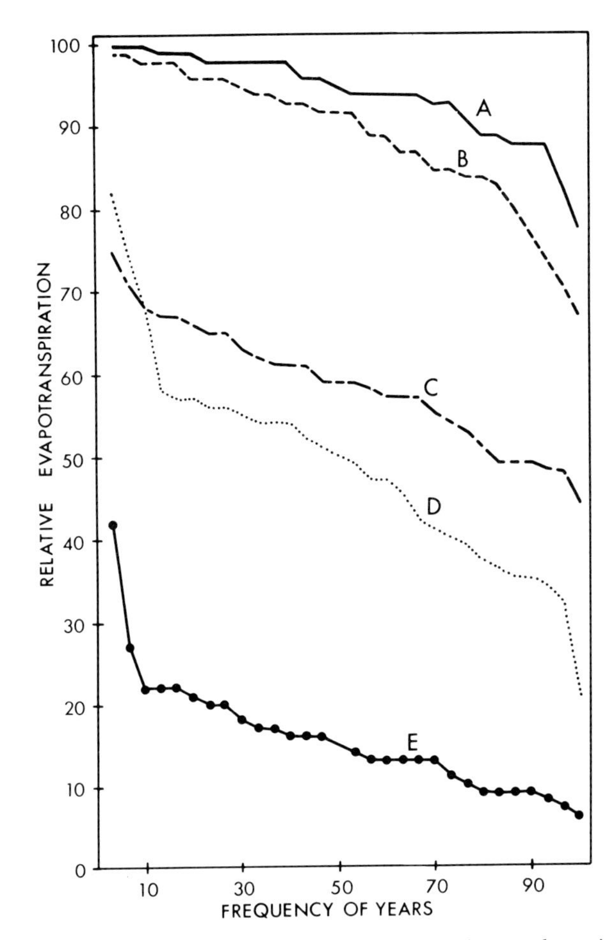

Fig. XII-9. Cumulated frequency of distribution of the yearly ratios of actual–potential evapotranspiration for A = Morgan City; B = Miami; C = Blue Canyon; D = Brownsville; E = Phoenix. At Morgan City, Louisiana, moisture shortfalls occur in few years; the median ratio is 0.95. At Phoenix, on the other hand, the median ratio is 0.15, actual evapotranspiration being a small fraction of potential (from Muller, 1972).

Chernavskaia, 1972) and the shape (Outcalt, 1972) of the urban heat island, because evaporating ecosystems give off little sensible heat and contribute little to the heat island. On the contrary, being well tended and well watered, they yield perhaps as much vapor and biomass as any other ecosystem does.

While the potential productivity of ecosystems is connected with radiant energy through photosynthesis and associated phenomena as described by Budyko (1974, p. 419), actual productivity often falls short for lack of available moisture. Budyko (1974, p. 425) shows how actual productivity Π is related to both the net surplus of radiant energy and a radiative index of dryness. Because both Π and annual evaporation E can be stated as functions of this index and both reflect gas transports through leaf stomata, they are associated with one another. This association is implicit in many discussions of biological yield and has been formulated explicitly by many investigators. One such formulation is

$$\Pi = 3 - 3 \exp[-0.00097(E - 20)],$$

with Π and E both in kg m^{-2} (Lieth and Box, 1972). With evaporation of 300 mm yr^{-1}, for instance, net biological productivity is 0.7 kg m^{-2}, or 7 tons ha^{-1}.

The relation of grain yield to evapotranspiration does not pass through the coordinate origin but has an intercept that for corn is about 300 mm, an amount which may include "unproductive" evaporation from the soil (Hillel and Guron, 1973). These investigators concluded that any moisture stress may reduce corn production and recommend that the root zone be kept moist for maximum production and also maximum efficiency of water use (production per unit volume of water). Each added millimeter of water is important in dryland farming.

Longer Variations in Time

Other measurements at Coshocton show an interesting comparison between improved practices of land management and those prevailing in the region when the experiments started a generation ago (Plate 23). In summer the change in management practices led to greater infiltration and storage of water in the soil. Evapotranspiration from the corn benefitted from the available moisture to increase by 50 mm, or 0.11. The increase for meadow was 120 mm, or 0.33. Entry of more water into the soil means that moisture is available to plants for a longer time after each storm. Days of soil-moisture stress are fewer.

Actual evaporation from terrestrial systems can, of course, be increased by several kinds of human action. Shelterbelts reduce turbulent diffusion over the enclosed fields, reduce potential evaporation, improve the state of soil moisture, and increase actual evaporation somewhat, as Budyko (1974, pp. 473–479) explains from theoretical reasoning. This was confirmed by extensive field work in one of the regions of the world that has had extensive modification of its surface into this physically interesting mix of low and tall ecosystems, southern European Russia.

Plate 23. Small basin at the Coshocton experiment station, with lysimeters in the background. The nearest gage measures subsurface flow.

The effect of irrigation on actual evaporation is obvious. Less obvious, perhaps, is the fact that irrigated ecosystems absorb more solar radiation and lose less long-wave radiation so that they have a general effect on the heat budget of the earth as a whole. This effect has been evaluated (Budyko, 1974, pp. 465–471) and is "not negligibly small."

Like new chemical compounds, new kinds of ecosystems are also originated by human ingenuity and activity. These include many that display unusual evapotranspiration regimes, like cooling and waste-water ponds (Chapter X), landfill sites, described earlier in this chapter, septic-tank fields, and of course cities themselves. While evapotranspiration from these new systems is no easier to measure or calculate than that from natural ecosystems, some information can be gained by modelling, as we saw for cities.

Simulation models for evapotranspiration also give information on percolation or water surplus in terms of vegetation characteristics. When verified, such models can be used to estimate effects of vegetation conversions. For example, a model done in the Eastern Deciduous Forest Biome Project of the International Biological Program at Coweeta Hydrologic Laboratory (Swift *et al.*, 1975) shows that conversion to pines would reduce water yield about 200 mm, and clearcutting would increase it 360 mm annually.

LARGE-SCALE PATTERNS

For large-scale determination of evaporation, Priestley and Taylor (1972) advocate using bulk aerodynamic methods, as were used in the Lake Ontario studies described in an earlier chapter, for sea surfaces, and energy-budget methods for land surfaces. The Bowen ratio is usually involved in such methods and the authors note that over saturated land surfaces it is smaller at high temperatures than low, being about 0.1 over 25°C surfaces and approaching zero as the surface temperature approaches the critical value of 34°C mentioned earlier. For drying lands the authors rely on a variety of experimental findings to recommend that evaporation rates calculated for saturated-land conditions be assumed to obtain until some specific sum of water has been evaporated and then linearly decreased to a zero rate after 50 mm more water has been evaporated. This is a simple scheme but is reasonable in view of the complex meso- and microscale distributions in the mosaic of ecosystems that cover land surfaces.

Seasonal Regimes over Large Regions

Eastern North America Maps in Budyko's "Atlas Teplovogo Balansa" (1963), based on such methods, illustrate seasonal variation of evapotranspiration over regions of North America and the adjacent seas (Table VIII). In June the distribution with latitude is more uniform than in other seasons. From the Gulf of Mexico to Labrador, the change is only from 90 to 60 mm month^{-1}. The transect is quite different in December. Very low values in the north indicate a dormant winter landscape, usually snow covered and cold, so that its vapor pressure is very low—only a few millibars. Evapotranspiration continues in the south, though only at a third of its summer rates. In the Gulf itself, on the other hand, winter is the most active season of

TABLE VIII

Evapotranspiration in Eastern North America[a]

Location	Dec	Mar	June	Sept	Dec	Year
Gulf of Mexico	190	160	90	140	190	1830
Gulf Coast to Ohio River	40	70	90	75	40	800
Ohio River to Great Lakes	15	50	85	60	15	620
Lakes–St. Lawrence	10	20	75	50	10	450
Labrador–Ungava	~5	~10	60	40	5	320

[a] Units: mm month^{-1}. Source: Budyko (1963).

the year, with evaporation averaging more than 6 mm day^{-1}, exemplifying the effect of heat stored in a water body in producing a counter-seasonal annual cycle of evaporation that supplies much of the vapor inflow into the continent (Chapter II) in the season when it is accumulating moisture.

From the whole of eastern North America, from the Rockies to the Atlantic, research on vapor-flux convergence (Rasmusson, 1971) shows the amount of water added to the atmosphere from the underlying fields and forests. This large-scale averaging of the upward transfer of vapor shows the following regime through the year (in mm day^{-1}):

N	D	J	F	M	A	M	J	J	A	S	O	N	D
1.0	0.7	0.7	0.4	0.8	1.3	2.0	2.7	<u>3.0</u>	2.4	1.7	1.3	1.0	0.7

The summer maximum is five times as large as the wintertime rates.

The winter minimum comes rather late, in February, and the maximum comes in early July. The rise from low to high rates of evaporation takes place in less than five months of the annual cycle. The decline is more drawn out; evapotranspiration continues moderately strong well into the fall.

The fact that the rate in September is double the March rate probably reflects heat stored in the soil and the fact that soil–vegetation systems are established and in operation. Besides, rainfall in September is heavier than in March (hurricane effect over much of the area under consideration).

Australia The maps in the "Atlas Teplovogo Balansa" (Budyko, 1963) make it possible for us to develop figures for areally averaged evapotranspiration from Australian regions of some size and uniformity, which show contrasting regimes (Table IX) at lower latitudes than

TABLE IX

Evapotranspiration in Australia[a]

Location	June	Sept.	Dec.	March	June	Year
Center	13	14	20	<u>30</u>	13	250
North	13	8	75	<u>100</u>	13	580
East Coast	28	37	68	<u>75</u>	28	630
Southeast	20	42	<u>50</u>	35	20	470
Tasmania	13	32	<u>58</u>	32	13	400
Southwest	30	<u>50</u>	20	17	30	370

[a] Units: mm month^{-1}. Source: Budyko (1963).

North America. Evapotranspiration from the immense center of the continent—more than half the total area—is low (0.5 mm day^{-1}) in winter and spring, somewhat higher at the summer solstice, and reaches the year's peak in fall (about 1 mm day^{-1}). The forcing function here is not the energy regime but the water regime, i.e., the summer rains and their aftermath. In any single year, the uncertain advent of these rains may produce a quite different pattern of evapotranspiration, best described as a month or two months of frantic biological activity following the unpredictable rains. (Remember July is *not* midsummer.) The North and the East Coast also have more evapotranspiration in autumn than in summer. The East Coast, however, gets winter rains and evapotranspiration is active in winter and spring as well as in autumn. The higher-latitude regions of the Southeast and Tasmania display a summer maximum deriving from energy supply. Their winter minimum is about the same size as the dry-season minimum in the Center. Southwestern Australia is unique, with a definite rise of evapotranspiration in winter (mild and wet) and a peak in spring. Evapotranspiration in the dry summer here is as small as in the dry Center. Yearly totals are small in the Center (250 mm), and greatest (600 mm) in the tropical North and along the East Coast. In contrast, yearly evaporation from the oceans surrounding Australia is 1500–2000 mm.

Annual Totals over Large Regions

As a water-conversion process that requires energy, evapotranspiration represents the bringing together in space and time of supplies of both energy and water. Water is delivered episodically to the earth's surface, in contrast with the steadier supply of energy in a given season. Therefore the critical feature of the climate is how often rains come to replenish the water budget of ecosystems and the critical feature of the terrain is its capacity to store water in periods between rains.

Annual amounts of evapotranspiration are greatest where the soil or a water body offers a large capacity to store water. The opportunity for convergence of water and energy is thereby prolonged, and the total amount of water evaporated is larger.

These general considerations are evident in the fact that evapotranspiration from the land usually is less than from the sea, other considerations being equal, because the period of convergence of water and energy is generally shorter at land surfaces. Long ago John Dalton recognized that evaporation from the land must be less than rainfall (Tuan, 1968, p. 128) and that it is less in winter than in summer.

A more sophisticated statement of these considerations is provided in a concise graph by Budyko (1971, p. 311) shown in Fig. XII-10. Yearly precipitation is the abscissa and yearly evapotranspiration the ordinate. Yearly heat supply is introduced by a set of curves on the face of the graph that represent different levels of the net surplus of radiant energy of all wavelengths. Thus regions of low rainfall show relatively little difference in evapotranspiration for different levels of radiation surplus. Nearly all the rain evaporates and supplying more energy cannot have much further effect. In regions of sufficient rainfall, say more than 500 mm, differences in energy begin to produce differences in evapotranspiration.

Relation with the Radiative Index of Dryness For yearly amounts, a comparison of the water supply and the energy supply values that bring actual and potential evapotranspiration into balance can be expressed in the fraction R/LP, where R is the net surplus of radiant energy, which tends, as we saw in the preceding chapter, to approximate the rate of potential evapotranspiration, P the rainfall, as an approximation of water supply to the soil body, and L the latent heat of vaporization that puts both R and P into the same dimensional units. Called the "radiative index of dryness" by Budyko (1974, p. 322), this ratio serves to generalize one aspect of geographical zonality (Budyko, p. 346), vegetation productivity (p. 426), and also annual evapotranspiration (p. 324) by the relation

$$E = P - P \exp(-RL/P),$$

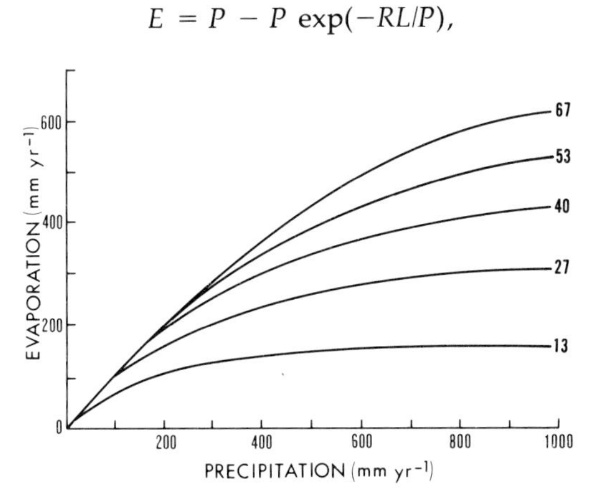

Fig. XII-10. Relations of annual evapotranspiration to rainfall at different levels of the net surplus of whole-spectrum radiation (W m^{-2}) (Budyko, 1971, p. 311).

which states that actual evapotranspiration E falls short of rainfall P by an amount that is equal to $P \exp(-R/LP)$, or by $P/e^{R/LP}$. At small values of the radiative index of dryness the denominator of this last term is small, and evapotranspiration falls far short of rainfall, the consequence being a wet landscape like the tundra. At large values of the radiative index of dryness, the term $P/e^{R/LP}$ is small, and evapotranspiration tends to approach rainfall. When the radiative index of dryness lies in the central part of its range, around a value of 1, then energy supply balances water supply, and the efficient forest ecosystem is favored. A rough correspondence of the index with structure of ecosystems is:

Index <0.3	Tundra ecosystems
0.3–1.0	Forest of various types depending on R
1.0–2.0	Steppe of different types
2.0–3.0	Semidesert of different types
>3	Desert of various types, depending on R (e.g., Arctic, high-mountain, mid-latitude interior, or low-latitude types of desert).

In North America, values of the radiative index of dryness (called the "dryness ratio" by Hare) range from 0.5 in Newfoundland and on the coast of British Columbia and Washington, to more than 3 in the arid southwestern United States. They tend to approach 1 over most of

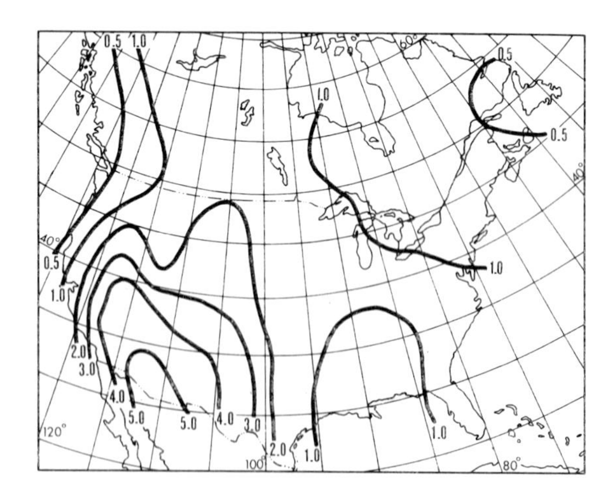

Fig. XII-11. Annual dryness ratio (net surplus of whole-spectrum radiation to the energy required to evaporate the annual precipitation) over North America (Hare, 1972).

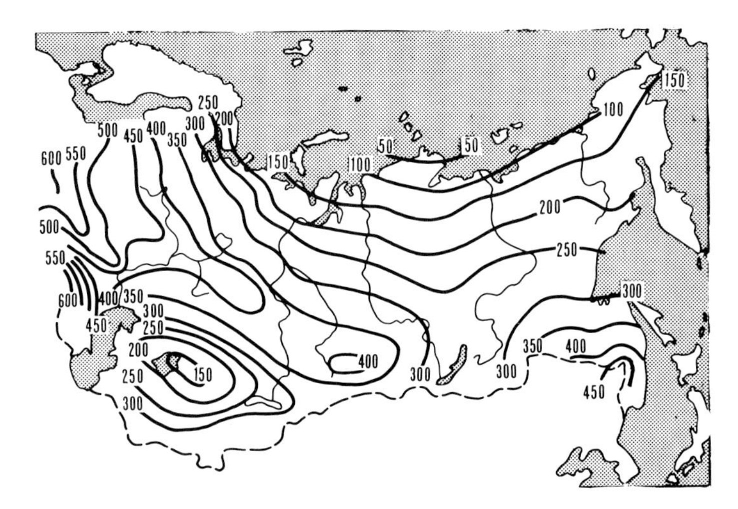

Fig. XII-12. Average annual evapotranspiration (mm) from standard meteorological observation surfaces in the Soviet Union (Konstantinov, 1968, p. 443).

the eastern United States and Canada (Fig. XII-11) (Hare, 1972). In spite of the equivalence of *R* as energy supply and *LP* as energy conversion in evaporation, there are also episodic or seasonal imbalances; "phase differences between the annual cycles of radiation and precipitation allow considerable runoff," i.e., short-term excesses of water as well as short-term shortfalls. This out-of-phase situation is characteristic of California; the season of heavy rain and snow comes at the energy ebb-tide of the year, with resulting floods. The hot, sunny, and dry summer is a long period of moisture shortfall to the degree of severe drought. Even in a climate in which annual *R* equals annual *LP* exactly, surplus water occurs at some times of the year, and periods of moisture shortfall, ecosystem stress, and low biological production at other times.

Areal Patterns The geographic distribution of annual evapotranspiration follows logically from the general considerations just discussed. Figures XII-12 and XII-13 are maps of the distribution of annual evapotranspiration in the Soviet Union, the first being for the standard observing situation and the second for forest (Rauner, 1972). In any climatic region forest ecosystems experience a more complete or longer-lasting convergence of energy with water and support more evapotranspiration than low vegetation.

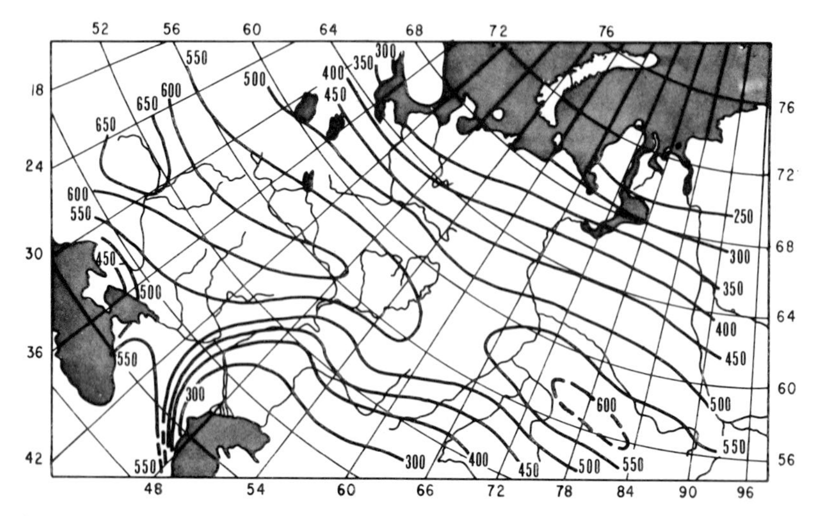

Fig. XII-13. Average annual evapotranspiration (mm) from forest land of the western part of the Soviet Union (Rauner, 1972, p. 166).

Evapotranspiration averages (mm) of vegetation zones in the European part of the Soviet Union are:

Tundra	200
Taiga	280
Mixed forest	365
Forest-steppe	300
Semidesert	200

(L'vovich, 1973, p. 91). More energy is available at lower latitudes, but less water. More can be done by land management to conserve water than energy, hence the emphasis placed by L'vovich on such practices, especially in the high-energy forest-steppe zone.

Mean evapotranspiration from all the continents is about 400 mm yr^{-1}, corresponding with the figure for the middle-latitude continental areas. It is only one-third the mean evaporation from the ocean surface, which is from 1200 to 1300 mm.

Means by continents (mm) display an amazing uniformity:

Europe and Asia	375
Africa	430
North America	385
Australia	370

with only one exception: South America, 750 mm. Wide in the

equatorial latitudes, its well-watered, well-energized forests produce a latent-heat flux double that of most other continents.

THE ERA OF EVAPORATION

If the period beginning in the 1930s is called the "Era of Infiltration," as we saw in Chapter VIII, a later period can be called the "Era of Evaporation" (or "Evapotranspiration," to use a word coined by Thornthwaite). The infiltration theory gave a workable technique for reconstituting events during a flood-producing storm and focused attention on the state of readiness of ecosystems to handle water delivered by a storm. Such information was particularly needed for developing flood forecasting procedures as well as in hydrograph analysis of past storms, which was carried out to understand how a drainage basin reacts to heavy rain. In order to determine "the rate at which field moisture-deficiency develops" (Thornthwaite and Holzman, 1939) it became clear that evaporation research had to be put on solid physical grounds.

Moreover, as flood and rain hydrology overcame some of the most baffling problems of events happening during rainfall, the interstorm periods began to receive attention formerly given only to the spasmodic episodes of storms. Evaporation was the key to understanding water demands of ecosystems in drainage basins that might generate floods, in grasslands and grain regions incurring losses in biological production due to water shortage, and in irrigation districts. A rethinking of evaporation centered on the aerodynamics and the energy requirements of evaporating ecosystems during the long periods between storms.

This question had been under study by three of the most original minds in hydrometeorology, one of them at least impelled by concern about drought as he experienced it. In 1948 all three men published important works (Budyko, 1948; Penman, 1948; Thornthwaite, 1948). Their approaches were rapidly applied to problems of irrigation-system design, drought (and hidden drought) and lost production in humid-climate agriculture, evaporation loss from reservoirs and even stock ponds in the West, effects of shelterbelts on grain production, as well as the engineering problem of assessing the moisture status of a drainage basin on the eve of a storm. While difficulties remain, especially in the lack of direct measurements of this moisture flux, the basic physics is better understood as a result of the research since 1948.

The regime of evapotranspiration through the year represents a response to the two forcing regimes of moisture supply and energy supply. Where these are not in phase, the season of evapotranspiration may be brief. In most parts of the world, soil-moisture storage suffers depletion at random times during the energy-rich season. This results in the occurrence of days of moisture stress, often in long series, in which biological production is seriously inhibited.

The flow of vapor from the surface into the atmosphere is elusive and varies greatly with physical and biological characteristics of the surface, substrate, and atmosphere. The examples given in this chapter illustrate some conditions affecting evapotranspiration from bare soil and vegetation surfaces in drying states, and show how great is the variety and how incessant the changeability of the factors that determine this conversion and flux of water.

REFERENCES

Aslyng, H. C. (1960). Evaporation and radiation heat balance at the soil surface, *Arch. Meteorol. Geophys. Biokl.* **B 10,** 359–375.

Bonner, J. (1959). Water transport, *Science* **129,** 447–450.

Brazel, A. J., and Outcalt, S. I. (1973). The observation and simulation of diurnal evaporation contrast in an Alaskan alpine pass, *J. Appl. Meteorol.* **12,** 1134–1143.

Budyko, M. I. (1948). "Isparenie v Estestvennykh Usloviiakh," Gidromet. Izdat., Leningrad.

Budyko, M. I. (ed.) (1963). "Atlas Teplovogo Balansa Zemnogo Shara." Mezhduvedomstvennyi Geofizicheskaia Komitet pri Prezidium Akademii Nauk SSSR, Moscow. Glavnaia Geofizicheskaia Observatoriia im A. I. Voeikova, Rezul'taty Mezhdunarodnogo Geofizicheskogo Goda, 69 pl.

Budyko, M. I. (1971). "Klimat i Zhizn'." Gidromet. Izdat., Leningrad, 472 pp.

Budyko, M. I. (1974). "Climate and Life" (*English transl.* ed. by D. H. Miller.). Academic Press, New York, 508 pp.

Chang, J.-H., Campbell, R. B., and Robinson, F. E. (1963). On the relationship between water and sugar cane yield in Hawaii, *Agron. J.* **55,** 450–453.

Dale, R. F., and Shaw, R. H. (1965). The climatology of soil moisture, atmospheric evaporative demand, and resulting moisture stress days for corn at Ames, Iowa, *J. Appl. Meteorol.* **4,** 661–667.

Denmead, O. T., and McIlroy, I. C. (1970). Measurements of non-potential evaporation from wheat, *Agric. Meteorol.* **7,** 285–302.

Denmead, O. T., and Shaw, R. H. (1962). Availability of soil water to plants as affected by soil moisture content and meteorological conditions, *Agron. J.* **45,** 385–390.

Dyck, S. (1972). Report [of Committee] on evaporation, *Int. Assoc. Hydrol. Sci. Bull.* **17,** 51–59.

Fritschen, L. J., and Bavel, C. H. M. van (1962). Energy balance components of evaporating surfaces in arid lands, *J. Geophys. Res.* **67,** 5179–5185.

Fuchs, M., and Tanner, C. B. (1967). Evaporation from a drying soil, *J. Appl. Meteorol.* **6,** 852–857.

Fuchs, M., Tanner, C. B., Thurtell, G. W., and Black, T. A. (1969). Evaporation from drying surfaces by the combination method, *Agron. J.* **61,** 22–26.

Hare, F. K. (1972). The observed annual water balance over North America south of 60°N, and inferred convective heat exchange, *Publs. Climatol.* **25** (3), 7–15 + 9 p. (C. W. Thornthwaite Memorial Volume II).

Harrold, L. L., Brakensiek, D. L., McGuinness, J. L., Amerman, C. R., and Dreibelbis, F. R. (1962). Influences of Land Use and Treatment on the Hydrology of Small Watersheds at Coshocton, Ohio, 1938–1957. U. S. Dept. Agric., Tech. Bull. 1256, 194 pp.

Hillel, D., and Guron, Y. (1973). Relation between evapotranspiration rate and maize yield, *Water Resources Res.* **9,** 743–748.

Konstantinov, A. R. (1968). "Isparenie v Prirode," 2nd ed. Gidrometeorolog. Izdat., Leningrad, 532 pp.

Lettau, H. H., and Davidson, B. (1957). "Exploring the Atmosphere's First Mile," 2 vols. Pergamon, Oxford.

Lieth, H., and Box, E. (1972). Evapotranspiration and primary productivity: C. W. Thornthwaite memorial model, *Publ. Climatol.* **25** (3), 37–46 + 16 pl. (C. W. Thornthwaite Memorial Volume II).

L'vovich, M. I. (1973). "The World's Water: Today and Tomorrow." Mir Publ., Moscow, 213 pp. (Transl. by L. Stoklitsky of Vodnye Resursy Budushchego. Izdat. Provesh-chenie, Moskva, 1972).

Mather, J. R. (1974). "Climatology: Fundamentals and Applications." McGraw-Hill, New York, 412 pp.

Miller, D. H. (1965). The heat and water budget of the earth's surface, *Advan. Geophys.* **11,** 175–302.

Molz, F. J., Van Fleet, S. R., and Browning, V. D. (1974). Transpiration drying of sanitary landfills, *Ground Water* **12,** 394–397.

Muller, R. A. (1972). Application of Thornthwaite water balance components for regional environmental inventory, *Publ. Climatol.* **25** (2), 28–33 (Thornthwaite Memorial Volume I).

Mustonen, S. E., and McGuinness, J. L. (1967). Lysimeter and watershed evapotranspiration, *Water Resources Res.* **3,** 989–996.

Outcalt, S. I. (1972). A reconnaissance experiment in mapping and modeling the effect of land use on urban thermal regimes, *J. Appl. Meteorol.* **11,** 1369–1373.

Penman, H. L. (1948). Natural evaporation from open water, bare soil and grass, *Proc. Roy. Soc. London* **(A) 193,** 120–145.

Philip, J. R. (1957). Evaporation, and moisture and heat fields in the soil, *J. Meteorol.* **14,** 354–366.

Philip, J. R. (1964). The gain, transfer and loss of soil-water. *In* "Water Resources Use and Management." pp. 257–275. Melbourne Univ. Press, Melbourne, Australia.

Priestley, C. H. B., and Taylor, R. J. (1972). On the assessment of surface heat flux and evaporation using large-scale parameters, *Mon. Weather Rev.* **100,** 81–92.

Rasmusson, E. M. (1971). A study of the hydrology of eastern North America using atmospheric vapor flux data, *Mon. Weather Rev.* **99,** 119–135.

Rauner, Iu.L. (1972). "Teplovoi Balans Rastitel'nogo Pokrova." Gidrometeoizdat., Leningrad, 210 pp.

Rauner, Iu. L., and Chernavskaia, M. M. (1972). Teplovoi balans goroda i vliianie gorodskogo ozeleneniia na temperaturnyi rezhim, *Izves. Akad. Nauk SSSR, Ser. Geograf.* No. 5, 46–53.

Reed, K. L., and Waring, R. H. (1974). Coupling of environment to plant response: A simulation model of transpiration, *Ecology* **55,** 62–72.

Rose, C. W. (1968). Evaporation from bare soil under high radiation conditions, *Int. Congr. Soil Sci., 9th, Trans.* **1,** 57–66.

Rouse, W. R. (1970). Effects of soil water movement on actual evapotranspiration estimated from the soil moisture budget, *Can. J. Soil Sci.* **50,** 409–417.

Rouse, W. R., and Wilson, R. G. (1972). A test of the potential accuracy of the water-budget approach to estimating evapotranspiration, *Agric. Meteorol.* **9,** 421–446.

Sibbons, J. L. H. (1962). A contribution to the study of potential evapotranspiration, *Geog. Ann.* **44,** 279–292.

Swift, L. W., Jr., Swank, W. T., Mankin, J. B., Luxmoore, R. J., and Goldstein, R. A. (1975). Simulation of evapotranspiration and drainage from mature and clear-cut deciduous forests and young pine plantation, *Water Resources Res.* **11,** 667–673.

Swinbank, W. C., and Dyer, A. J. (1968). Micrometeorological Expeditions 1962–1964. Commonwealth Sci. Indus. Res. Organ., Div. Meteorol. Phys., Australia, Tech. Paper 17, 48 pp.

Thornthwaite, C. W. (1948). An approach toward a rational classification of climate, *Geog. Rev.* **38,** 55–94.

Thornthwaite, C. W., and Holzman, B. (1939). The role of evaporation in the hydrologic cycle, *Trans. Am. Geophys. Un.* **20,** 680–685.

Thornthwaite, C. W., and Holzman, B. (1940). A year of evaporation from a natural land-surface, *Trans. Am. Geophys. Un.* **21,** 510–511.

Tuan, Yi-fu (1968). "The Hydrologic Cycle and the Wisdom of God: A Theme in Geoteleology." Univ. Toronto Press, Toronto, 160 pp.

Veihmeyer, F. J., and Brooks, F. A. (1954). Measurements of cumulative evaporation from bare soil. *Trans. Am. Geophys. Un.* **35,** 601–607.

Waggoner, P. E., and Zelitch, I. (1965). Transpiration and the stomata of leaves, *Science* **150,** 1413–1420.

Wilson, R. G., and Rouse, W. R. (1972). Moisture and temperature limits of the equilibrium evapotranspiration model, *J. Appl. Meteorol.* **11,** 436–442.

Zubenok, L. I. (1965). World maps of evaporativity, *Sov. Hydrol.* 274–289. (*English transl.* by S. Molansky from *Tr. Glav. Geofiz. Obs.* **179,** 144–160, 1965).

Zubenok, L. I. (1974). Evaporation deficit under various climatic conditions on land, *Sov. Hydrol.* **1974,** 251–257.

Chapter XIII

WATER IN THE LOCAL AIR

Moistening of the atmosphere by evaporation from the underlying surface proceeds quickly and invisibly every day when energy and water are available. As we saw in the preceding chapter, this moistening process goes forward at daily rates up to 3–4 kg m^{-2}. The surface is the source of the water vapor that is mixed through the earth's atmosphere, a constituent that has great significance in the boundary layer.

The boundary layer of the atmosphere is that lowest kilometer or so most affected by the conditions at the earth's surface. The lower part of this layer, which might be called the "local air," can be considered to be a part of a regional landscape, trapped in the concavities of the surface and closely associated in its temperature, movement, and moisture with what goes on at the surface. The local air is the immediate environment of terrestrial ecosystems, permeating their foliage canopies and participating in their moisture balances. Water in this zone, in either invisible or visible form, reflects processes at these ecosystems and in turn plays an important role as a factor in their environment.

WATER VAPOR IN THE LOCAL AIR

Vapor moving upward from the source at the earth's surface would, in time, diffuse evenly through the entire atmosphere like CO_2 except for the fact that it is intermittently removed by condensation in the middle levels of the atmosphere. The interplay between the surface as sole source of vapor and the cold traps or vapor sinks represented by ascending columns or sheets of air in the atmosphere produces a more or less equilibrium balance that is represented by a vertical decrease in vapor concentration. The profile of humidity decreases upward in an exponential decay curve $\log e_h = \log e_0 - h/6.5$, in which e represents vapor pressure at sea level (subscript 0) and at any height h (in kilometers) (Hann, 1897, p. 279). The removal mechanisms operate throughout the troposphere but with decreasing effect at the higher altitudes where they find less vapor to condense on the average. Atmospheric vapor concentration is greatest in the local air that bathes the ecosystems on the earth's surface.

Table I shows values of vapor pressure (in millibars) at typical wet surfaces in nature, and the sea-level specific humidity of moist air in contact with these surfaces (as grams of vapor per kilogram of air). The dryness of cold high-latitude surfaces and of the air above them form one extreme. The other extreme is found in the high humidities of equatorial oceans or rainforest and the air in contact with them, in which vapor concentration is more than two orders of magnitude greater than in the high latitudes, as the surfaces approach the 32–34°C temperature limit discussed in Chapter X.

Regimes

Diurnal Regimes The diurnal regime of vapor pressure in the air near the evaporating surfaces depicts the response of a storage function in the local air, which buffers moisture input from evapotranspiration during daylight hours to a varying outflow, the upward mixing of vapor. Over many land surfaces nocturnal inversions that

TABLE I

Vapor Pressure and Specific Humidity at Selected Surfaces[a]

Surface type	Typical surface temperature (°C)	Corresponding humidities: Vapor pressure (mb)	Corresponding humidities: Specific humidity ($g\ kg^{-1}$)
Snow surface of ice cap	−40	0.13	0.1
Mid-latitude snow field in winter	−20	1	1
Melting snow	0	6	4
Mid-latitude water or vegetation	+15	17	11
Low-latitude ocean or forest	+25	32	20
Under intense radiation	+33	50	32

[a] Source: Smithsonian Tables (1966).

persist into the forenoon hours limit upward mixing and a maximum in the daily vapor pressure cycle occurs in the late morning.

This often is followed by a minimum at the time when upward mixing of vapor out of the local air is most vigorous. A second maximum may follow, when turbulent mixing diminishes in the evening. The primary minimum, at night, indicates the lack of input from the surface. Sometimes, in fact, a net downward flow of vapor takes place out of the air to the surface, and the vapor pressure in the local air declines still more.

The vapor concentration in the local air over water surfaces, which generate a fairly constant upward flow of vapor, exhibits a different diurnal regime. This regime is primarily influenced by the variations in upward mixing of vapor out of the local air into the free atmosphere. The maximum vapor pressure therefore comes at the time of weakest vertical mixing, near sunrise, and the lowest at the time of strongest mixing.

Annual Regime The dependence of atmospheric vapor pressure on the rate of evaporation from the underlying surface influences its variation through the year. Over Lake Ontario, for example, it varies from 4.5 in late winter to 20 mb in late summer. The effect of the underlying surface is further demonstrated by the fact that during the months of active evaporation from the lake, the atmospheric vapor pressure over it is substantially greater than over the land adjacent (Richards and Fortin, 1962). The ratio of lake-to-land values shown over the annual cycle in Fig. XIII-1 is largest—about 1.3—in midwin-

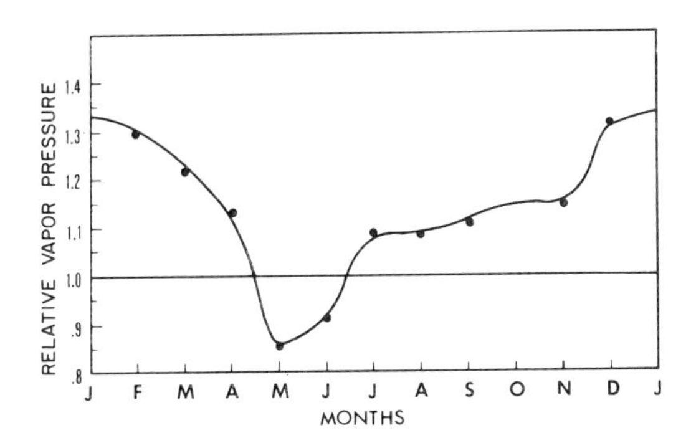

Fig. XIII-1. Annual regime of the ratio between vapor pressure over Lake Ontario to that over adjacent land surfaces (Richards and Fortin, 1962).

ter, when evaporation is weak from the frozen land but active from the lake. The ratio during May and June, on the other hand, indicates the coldness of the lake surface relative to the surrounding lands; vapor pressure is lower over the lake than over the land.

Annual regimes of vapor pressure at four places in Australia are given in Table II. Between winter and summer at Sydney, vapor pressure doubles, a response more to the change in evaporation from land than to the smaller change in sea-surface temperature and evaporation. The author's experience with Sydney humidity in February and March, however, confirms the high vapor pressures then, which are associated with the continued warmth of the near-shore waters. Days of easterly (on-shore) winds are especially sultry.

The change between winter and summer at Darwin is large, 13 mb. The whole annual cycle is found at a higher level than at the other stations; this indicates the warmth of the Timor Sea. The large change through the year indicates the effect of a small change in a high sea-surface temperature as well as the occurrence of dry air in winter (June and July) moving from the interior of the continent. In the Northern Hemisphere continents, such movement of dry air also occurs in winter, but it is much drier than in Australia.

The World Pattern

The air over the cold snow surfaces of the northern continents in winter is particularly dry. Over great areas, vapor pressure is less than 2 mb. These are source regions of very dry air, which moves south and dominates many winter days in the midwestern United States.

The contrasting surfaces of moist warm areas of the equatorial latitudes average higher than 25 mb in vapor pressure. Latitudinal means of vapor pressure over the land are shown in Table III, which

TABLE II

Vapor Pressure in Australia[a]

Location	J	F	M	A	M	J	J	A	S	O	N	D	Mean
Darwin, 12S	31	31	31	27	22	18	18	21	25	28	29	30	26.0
Brisbane, 27S	21	22	20	17	14	12	11	11	13	15	18	20	16.4
Sydney, 34S	18	19	18	15	12	10	10	10	11	13	15	17	13.9
Hobart, 43S	9	12	11	10	9	8	8	8	8	9	9	10	9.3

[a] Units: mb. Source: Commonwealth of Australia Yearbook (1965).

TABLE III

Vapor Pressure in the Air near the Surface of the Continents[a]

Latitude	Jan	July
90N	<1	3
60	2	12
40	5	18
30	8	20
20	14	23
10	22	28
Equator	28	28
10S	28	25
20	22	15
30	16	10
40	14	8
60	—	—
90	4	1

[a] Units: mb. Source: Kessler (1968, p. 14).

shows that the zone of high humidity extends from 20S to 10N latitude.

Figure XIII-2, the world pattern in July (Landsberg, 1964), displays large areas where vapor pressure exceeds 30 mb. The Caribbean Sea, the Gulf of Mexico, and their coastal lands are well known sources of

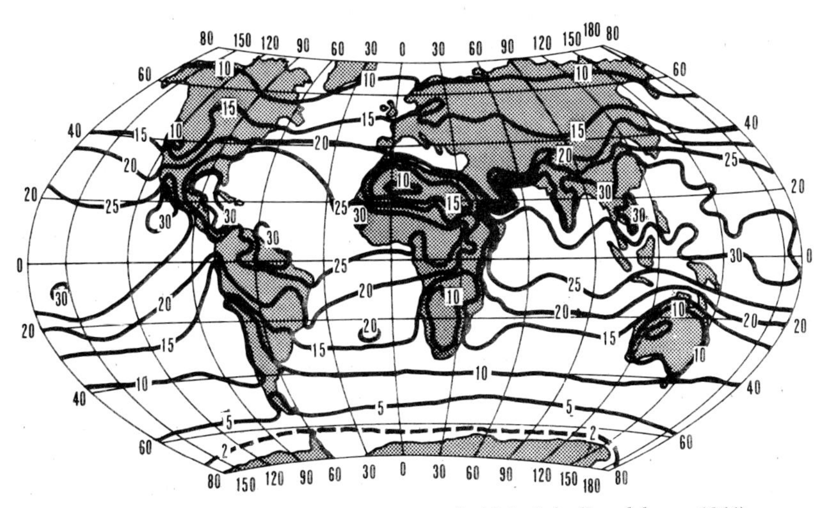

Fig. XIII-2. Average vapor pressure (mb) in July (Landsberg, 1964).

humid air streams. Days of sweltering humidity in central North America occur during southerly airflow, which on its traverse across the evaporating fields and forests of the lower Mississippi Valley gains even more moisture.

It is significant that, when we look for the sources of the rain and snow that fall on central and eastern North America, we see over the western Pacific Ocean a vapor pressure that lies between 12 and 15 mb during most of the year. Over the Arctic Ocean we see a vapor pressure lower than 10 mb. In contrast, over the Gulf of Mexico we see a vapor pressure of 20 mb in winter and 30 mb in summer. Which body of water is most likely to provide moisture for the central plains of the continent?

Humidity As a Component of the Environment

Vapor content is affected by the source of the airstream, but this quality changes as the air moves over drier or moister land. Characteristics of the boundary layer and even more of the local air reflect conditions at the directly underlying surface. Movement of the local air is braked over rough ecosystem canopies and its warmth, CO_2, and water-vapor content are modified by the fluxes of sensible heat, CO_2, and vapor from the ecosystems that make up its porous lower boundary.

Over an ecosystem surface having "infinite" extent, as was hypothesized in the definition of potential evapotranspiration in an earlier chapter, the partnership between ecosystem and local air tends to be dominated by the ecosystem. The rhythms of warmth, moisture, and so on, in the local air are forced by the rhythms of the corresponding fluxes at the underlying surface.

In reality, this hypothetical situation is rare. Most terrestrial ecosystems are limited in size, often covering an area of only a few hectares, 100 m or so across. The air that fills their foliage volumes is still dominated by exchanges of water and energy at the leaf surfaces, as evidenced by the vertical profiles of humidity or temperature within a forest stand; but above these systems the main body of the local air moves on across the countryside, floating above a mosaic of many contrasting ecosystems and carrying, for example, the moisture it acquires from one system over a neighboring system.

For this reason, the expressions for evapotranspiration discussed previously include terms for atmospheric humidity. The Bowen ratio can be approximated by measuring the differences in temperature and moisture between an ecosystem and the overlying air. The gradient of moisture is important in the evaporation process and depends on the concentration of vapor within the storage zone represented by the local air.

One effect of high specific humidity is, by reducing transpiration, to slow the flow of nutrients brought into a plant in the stream of water moving from soil to leaves. This decrease in nutrient intake might well account for the stunted growth of trees in perennially cloudy, foggy zones of some mountains (Odum, 1971, p. 376).

The local air, not being precisely defined as to thickness, forms a reservoir of indefinite size for water vapor. We can obtain a rough idea of its storage capacity from the following: Assume it to be 1 km deep; it then contains 1300 kg of air in a column of 1-m^2 cross-sectional area. Over an ice cap, at a specific humidity of 0.1 g kg^{-1} (see Table I), this column contains about 130 g of vapor m^{-2} of surface. Over an equatorial forest, at a specific humidity of 20 g kg^{-1} (Table I again), it contains 26 kg of vapor. Obviously, the storage of water in this zone can vary tremendously. Suppose we take a typical mid-latitude instance of 10 kg of vapor stored m^{-2} of area. This value, 10 kg m^{-2}, is of the same general size as the mass of water delivered to the surface in a day of moderate rain (10 kg m^{-2} day^{-1}), or the amount evaporated from a corn field in two or three days of hot summer weather.

Conditions of high moisture concentration slow down the further transfer of water from the ecosystems into the air, as quantitatively shown in the various evaporation formulas. Dry air, on the other hand, accepts vapor from below avidly, and brings about high evaporation rates best exemplified where desert air invades an oasis. The cool air we feel as we drive from the desert onto a road between irrigated alfalfa fields is a consequence of the fact that available energy at the alfalfa system is channeled almost entirely into evaporation, leaving little or none to warm the air. The moisture stored or not stored in the local air thus is an important factor in the environment of an ecosystem, affecting its rate of evaporation and indirectly its openness to assimilate CO_2 from the air.

Gentilli (1955) investigated the generally accepted correspondence of dew-point temperature (a common measure of vapor pressure or specific humidity) with the minimum air temperature reached during the preceding night (a common forecast variable). The correspondence in the Plains States and eastward is good, especially in seasons when nights are warmer than about −10°C. It demonstrates the important role played by water in the atmosphere. Water vapor effectively transfers atmospheric energy to the ground by radiation, and maintains ecosystems in an equable environment during the stress period of the night. This energetic bond from vapor in the local air to moisture at the surface means that vapor concentration influences snow melting, freeze-thaw cycles, evaporation, and dew formation. This income of 30–50 W m^{-2} in long-wave flux density is particularly important in the small energy budgets of night hours, when the warmth it brings accelerates plant respiration and reduces net photosynthetic productivity of ecosystems, and adds to human heat stress in urban systems.

Humidity in the Environment of Man Moisture in the local air also affects man when it reaches high values. The isarithm of 20 mb in Fig. XIII-2 is of interest because this value is generally accepted as an index of sultriness in the human environment. In Asia this line takes in most of China and Japan in July; it covers the eastern Mediterranean and its southern shores. In North America it takes in the eastern part of the continent to a latitude of 40N. The zonal averages of Table III show that the 20-mb area stretches from 15S to 30N latitude and encompasses a major fraction of the world.

Air conditioning, the American term for artificial cooling and drying of summer air, has become one of the major consumers of electric power in this country. A rough indication of the desire for it is given by a combined index* of air temperature and dew point. Dew point is uniquely related to vapor pressure and is weighted about a third as heavily as air temperature in the formulation.† Table IV shows the frequency distribution of the units of this index at Baltimore in midsummer, when evapotranspiration in eastern North America is most active. At index values above 75 units about half the people are uncomfortable. These values occur 0.82 of the time in the afternoon and even at night 0.09 of the time. At index values above 80 units nearly everyone is uncomfortable; such values occur 0.50 of the afternoon hours at Baltimore, in the open air. In the city, and especially in buildings, the frequency of occurrence is larger.

The physiological effects of moisture in the atmosphere are not well understood, beyond the obvious fact that high vapor pressure suppresses evaporative cooling of the heat-stressed human body. One panel (Sargent *et al.*, 1967) also notes that "there is need to know the

* At first called the "discomfort index" (Thom, 1956), this was later euphemized to "temperature–humidity index."

† A similar index, called "effective temperature" (Landsberg, 1969, p. 54) is affected about the same by an increase of 6 mb in vapor pressure alone (= 4 g kg^{-1} increase in specific humidity) as by a 6°C rise in air temperature alone.

TABLE IV

Frequency Distribution of the Discomfort Index (or Temperature–Humidity Index) at Baltimore in July[a]

Scale units	Night 0000–0500	Afternoon 1200–1700
<60	5	0
60–64	22	0
65–69	21	1
70–74	43	17
75–79	9	32
80–84	0	38
85–89	0	12
	100	100

[a] Units: percent of occurrences. Source: Thom (1956).

chronic effects of exposure to very low humidity, for man now spends so much of his life in artificial atmospheres that may be exceedingly dry." As midwesterners know, heating polar air at a vapor pressure of only 2–4 mb without humidifying it results in very low relative humidity in the home and severe respiratory problems.

VISIBLE FORMS OF WATER IN THE LOCAL AIR

Most atmospheric water is in the vapor state, but sometimes a myriad of tiny droplets or ice crystals also are present. Although these total only a small mass of water substance, they are conspicuous in the landscape and form an important part of the environment of ecosystems at the surface.

Amounts and Significance

A typical figure for the liquid-water content of a cloud is 0.5 g m^{-3}. This means only 0.5 g of water per 1300 g of air, or, in terms of specific humidity, only 0.4 g kg^{-1}. Compare this figure with ordinary vapor-content values of 10 g kg^{-1} in the middle latitudes (Table I), or more than 20 g kg^{-1} in the low latitudes. Yet clouds and fog are prominent features of the environment of ecosystems.

The significance of the small amounts of condensed water in clouds or fogs stems from the finely divided state. Each tiny crystal or droplet (10^{-6} of the size of a raindrop) refracts and reflects light. A cloud of droplets reduces vision through the atmosphere to almost zero, cuts off the direct beam of the sun, and becomes a source of diffused shortwave radiation and of long-wave radiation to the ecosystems below. A cloud deck passing across the sky causes a rapid change in the flow of radiant energy to an ecosystem. The total input of energy changes, usually decreasing, and its spectral composition shifts to favor both the chemically and biologically effective blue wavelengths and far infrared wavelengths that have chiefly a heating effect. At the same time, the change from a direct beam to a diffused source for the shortwave radiation brings about deeper penetration of light into the ecosystem, benefiting photosynthesis. Water droplets and ice particles in the local air thus have important effects on the functioning of ecosystems, affecting their water balances.

Radiation Fog

One mode of formation of visible water particles in the local air results from condensation in place of vapor already present in the air. On nights of weak vertical mixing, radiation cooling is concentrated in a shallow layer of air, and if initial humidity were high, some of the vapor in this layer would condense as radiation fog. Radiative cooling of the top surface of the fog now goes on more rapidly after droplets have formed, and supports further condensation of vapor, strengthening the development of the fog.

Radiation fog has many hours in which to develop during winter nights. It continues well into the next morning, perhaps even lasting through the day if it is thick enough to prevent solar heating of the ground.

Cold air that has drained into topographic depressions is often the medium in which vapor easily condenses into fog. Inversions above the cold air reduce upward mixing of the droplets and confine vapor, cold air, and fog droplets with their associated pollutants within the basins (see Plate 24). Ecosystems in such sites are subjected to prolonged attacks by such pollutants as sulfur dioxide, often in the form of droplets of dilute sulfuric acid. They are experiencing "one of the most common causes of the accumulation of pollutants to obnoxious concentrations for long periods" (Scorer, 1968, p. 31) that human activity produces.

In periods of weak general movement in the atmosphere, such fog

Plate 24. Patches of radiation fog in Morioka, northern Japan, in winter. Some downslope movement also is taking place (December 1966).

tends to perpetuate itself because it screens the wet, cold soil from solar heating. The author recalls a winter of frequent radiation fog at airports on the Colorado Plateau, a region in which fog is seldom forecast. Early winter rains saturated the soil and were followed by a long period of stagnant air circulation, in which water repeatedly circulated between the wet soil and the foggy local air. Closed airports disrupted the short-hop air operations of the day (1943–1944), trapping almost all of one airline's planes at Albuquerque. The lifting of the fog was forecast correctly after the high level of soil moisture was taken into account.

The tectonic valley of the Rhine between the Vosges and the Black Forest often fills with cold, damp air in periods of anticyclonic stagnation. One such period in 1972 and 1973 lasted for 32 days continuously (Weischet, 1973). The sun can barely be discerned, visibility at ground level is low, and heavy frost forms on trees. Meanwhile, above the fog clear skies reign. Skiers go up into the Black Forest, only 0.5 km higher in altitude, and enjoy solar warmth, an air temperature around 5°C higher, and a distant view of the Alps. Indeed an abrupt transition to be caused by a thimbleful of water! The contrast increases as pollutants continue to accumulate in the valley fog.

Plate 25. Visible-channel scanning by DMSP satellite 7529 R showing radiation fog filling the lowlands of Puget Sound, the Willamette Valley, and the Columbia Basin on 18 December 1975. Two thousand holiday travelers were immobilized for several days at the Seattle–Tacoma airport alone, and a like number waited at other airports to get into Seattle. Meanwhile, AMTRAK trains arrived on schedule.

The humid conditions here can be expected to worsen still more after new power-plant cooling towers are built and further humidify the hapless valley's air. The wet plumes from these gigantic evaporating devices* represent an impact on the local air that needs study in many parts of the industrial world.

The valleys of the western US under winter anticyclones often fill with long-lived bodies of radiation fog. Plate 25 is a satellite photograph taken at a time of anticyclonic dominance in December 1975. The upper, visible surfaces of the fog bodies are uniformly white but the ecosystems in the valleys—and people, too—live in a gray, windless, clammy world of weak light and slowed biological activity.

The amount of liquid water that shows so clearly in these photographs from space is very small. If we take the fog bodies as being 200 m deep and having a liquid-water content of 0.5 g m^{-3}, we compute a

* The source strength of such a system approaches 10^5 tons of vapor per day.

content of 0.1-kg liquid water in a 1 square meter atmospheric column. In terms of a layer of liquid water, this is only 0.1 mm thick. If it fell as rain, it would barely be measurable.

Such a film is two orders smaller in mass than the amount of water present in the atmospheric column as invisible, transparent water vapor. Yet the fog shown in the photograph covers most of the homes and working places in this part of the world, and all its airports.

Water Fluxes from the Underlying Surface

Ice Fog Interior basins of Alaska are so cold in winter that vapor begins to condense when it reaches a level of about 0.1 g of vapor kg^{-1} of air, that is, only 100 parts per million (ppm). If nuclei are absent, the critical temperature for condensation is −40°C; if they are present, as they are in basins inhabited by man, −30°C is sufficient chilling for condensation to occur.

The products of condensation are ice crystals. In bulk these form ice fog, which remains trapped in the basin until a warm-up occurs and it evaporates. It represents a special kind of storage of water. How long does it take for vapor flow from the underlying surface to fill this storage?

Cities provide the nuclei that hasten condensation and also the water itself. The vapor flux from fuel combustion at a fort near Fairbanks is 3.6×10^5 kg day^{-1} (Wilson, 1969, p. 89). How large a volume will this single vapor flux saturate? At a vapor fraction in the air of 0.1 g kg^{-1}, the mass of the air associated with the ice fog is

$$(3.6 \times 10^5)/10^{-4} = 3.6 \times 10^9 \text{ kg}.$$

At a density of 1.4 kg m^{-3}, this associated air takes up a volume of

$$(3.5 \times 10^9) \text{ kg}/1.4 \text{ kg m}^{-3} = 2.5 \times 10^9 \text{ m}^3 = 2.5 \text{ km}^3.$$

The area of the basin containing Fairbanks and the fort is 75 km^2. How deep will the layer of ice fog be?

$$2.5 \text{ km}^3/75 \text{ km}^2 = 0.033 \text{ km} = 33 \text{ m}.$$

One day's vapor flux from just one of several human sources of vapor thus can form a fog 33 m deep. Obviously it is not difficult to generate enough vapor to produce a thick layer of fog in a climate in which the saturation vapor pressure is so low. Such ice fogs perpetuate themselves and trap pollutants as well as vapor. It is no surprise that even small settlements in the Arctic are considered as polluted as any on earth. The native ecosystems, lacking any such vapor-generating capability in winter, can hardly be blamed for the fog.

*Steam Fog** This type of fog forms from an extremely rapid flux of vapor (100 g m^{-2} hr^{-1} or more) from a warm, wet surface in the presence of moderate vertical mixing. In fall or early winter it is a visible flux of water rising as wisps of steam from warm leaf surfaces (Plate 26), and especially from water surfaces.

When cold air moves onto the lake surface it immediately becomes saturated. Condensation of the vapor poured into it from below becomes visible as clouds of water droplets, like the condensing cloud that forms just beyond the spout of a teakettle. Steam fog on Lake Michigan is well described by Church (1945), from observations on many crossings on the car ferries for a lake energy-budget study.

Vapor and fragments of the fog are mixed aloft in the unstable air. The layer will persist if the upward mixing of water out of the fog layer is not as rapid as the input of vapor from the lake. Like other cases of water storage, this can be interpreted in terms of vapor flux convergence; more water enters the layer than leaves it (Miller, 1946). This means that

$$A_l(\partial q/\partial z)_l > A_t\ (\partial q/\partial z)_t,$$

where subscripts l and t represent lower and upper planes in the air near the water, z height, q specific humidity, and A the exchange coefficient, which is small at the lower plane and larger at the upper. This expression can be restated as

$$(\partial q/\partial z)_l/(\partial q/\partial z)_t > A_t/A_l.$$

The ratio of the humidity gradients at the lower and upper planes must exceed the inverse ratio of the exchange coefficients if water is to accumulate in the layer between heights l and t, and form fog.

Steam fog occurs when the polar airstream is colder than −15 to −20°C, at lake temperatures near freezing. Such low temperatures represent a southward transport of cold air chilled over high-latitude snow surfaces, which does not occur before early winter.

How much vapor is required to saturate a layer 10 m deep over the lake at a temperature of −15°C? The required specific humidity is larger than in the cold basin of the ice–fog example; it is 1.2 g of vapor kg^{-1} of air. A 10-m column of 1-m^2 cross section contains 12 kg of air, so the vapor requirement for saturation is 14 g of vapor m^{-2} of lake surface. This represents 0.014 mm of water, a relatively tiny amount at

* Although other kinds of fog occur, they are less the products of the interaction between ecosystems and local air than ice, steam, and radiation fogs.

Plate 26. Steam fog being mixed aloft from wet forest in the Otway Ranges, Victoria, about noon on a midwinter day.

a surface where daily evaporation is of the order of 3–4 mm (cf. Lake Ontario evaporation figures in Chapter X). In one particular outbreak, on 10 January 1968, the rate was 8 mm day^{-1} (Phillips, 1972), representing a strong source of vapor for fog and cloud formation.

Lake Clouds in Winter Over the large lakes of the North American Middle West, steam fog forms in every stream of Canadian air. If the wind is strong the fog is lifted by turbulence into low stratocumulus clouds that cover most of the lake and may extend many kilometers onto the lee shore.

Over Lake Michigan (Plate 27), as elsewhere, steam fog forms extensive layers. It is a visible form of water in the local air in every polar outbreak. In turbulent air the layers form stratus decks from which convective towers rise in sharp contrast with the cloud-free air of the upwind land areas. From the sunny, paralyzingly cold conditions at the upwind shore they look like a boiling fluid, an apt likeness to the actual physical process. The sight is an unseasonable recalling of the towers and expanding bulges of summer thunderheads, and may on occasion include columns of condensate rising 500 m from the layer of steam fog to the cloud deck and resembling waterspouts (Lyons and Pease, 1972).

Plate 27. Steaming water (Lake Michigan near Milwaukee) and convective clouds in an Arctic airstream (30 January 1971 at −20°C).

Snow Showers Convective activity that breaks through low-level inversions in the cold air and grows upward often produces snow showers. These occur at times over lakes as small as Lake Mendota (area 40 km^2). Humidification of the thin layer of air involved can be accomplished in a short over-water traverse where evaporation is so rapid.

The small convection cells reach the level of ice-crystal formation relatively early and produce snow in light flurries. These return to the lake a fraction of the water that evaporated from it only a short time before.

Small snow showers continue as they drift away from the lake. They drop small amounts of snow on the downwind shores, for example, on the campus of the University of Wisconsin that borders Lake Mendota.

These cells become large and vigorous over the Great Lakes. When they come ashore over lowland highways and cities several may follow the same line of airstream convergence and dump heavy snow on them. Few winters pass without newspaper front-page pictures of a city such as Buffalo immobilized by 40–50 cm of snow that fell in one day. Syracuse, on the lake plain south of Lake Ontario, is the snowiest big city of North America, and the expensively engineered thruways are closed for 50–100 km after these snow showers.

While the snow so strategically concentrated on one city or freeway in depths sufficient to put it out of commission is locally great, the overall transfer of water from lake to land in this fashion is not large. The contribution of lake-effect snowfall to the water budget of the whole land area draining into Lake Erie, for instance, is only 0.06 (Webb and Phillips, 1973). This illustrates, as did the fog cases also, what effect a relatively small mass of water can have in certain circumstances.

Rain Showers

During settled weather in summer, some rain showers represent the visible part of a cycling of water within the local environment. Such showers are a visible result of evapotranspiration from the local surface. Much of the water in these showers is derived from the rapid vapor flux from the underlying moist, warm surface, perhaps as recently as the morning of the same day. The high vapor content of air in summer gives these convective cells more energy than is released in the snow showers just discussed.

The vapor feeding such a shower system is collected from an area of the order of a few square kilometers into columns of the late-morning cumulus clouds. By midafternoon one of these convection cells has become successful in organizing vapor collection from a much wider area. Its neighbors are starved out, while it goes on into the thunderstorm stage. "A convective thunderstorm typical of the southwestern United States requires a region of atmosphere covering 1000 or more square miles of land to develop and sustain its action" (Workman, 1962).

The rain from such a storm represents water concentrated from several hours of evaporation at a surface several thousand square kilometers in area, which might total several million cubic meters of water. Part of this is condensed, as we saw in an early chapter, and formed into a shaft of heavy rain. The storm moves slowly across the landscape, distributing water to it. As a water-processing system, the storm has rearranged water in the surface-atmosphere system, concentrating it many-fold and bestowing it on a small part of the original area of contributing ecosystems.

For the sake of continuity, it should be noted that most of the vapor in the local air eventually is mixed throughout the depth of the lower atmosphere; even the local thunderstorm of the preceding example leaves a good deal of water behind it at high altitudes in the air (much of it in its great anvil of cirrus). Away from the earth, the vapor is easily carried great distances before its turn comes to encounter a precipitating cell, as was described in the discussion of vapor transport in Chapter II.

Over a large region like one of the oceans or a well-watered continent in summer, a certain proportion of the locally generated vapor is condensed and precipitated within the boundaries of the region. Such precipitation of local vapor accounts for approximately 0.1 of a total summer rainfall in large regions like the European part of the Soviet Union (Zhakov, 1966; Budyko, 1974, p. 240), depending on the strength of the upper circulation and the frequency of precipitating storm systems. By raising the humidity of the below-cloud air layer, it increases the probability of downwind rainfall, e.g., over Illinois as a result of expanded irrigation in Nebraska. Whereas at 50% relative humidity the chance of precipitation in specific conditions is only 0.05, at 90% relative humidity it is 0.21 (Drozdov, 1974).

Even over large regions, then, most of the vapor in the local air is ultimately carried away. A small fraction is recycled locally by mesoscale processes like snow or rain showers, and some returns by direct condensation on the underlying surface.

CONDENSATION OF VAPOR ON THE UNDERLYING SURFACE

Not all the water input to ecosystems comes from convective local storms or from storms feeding on long-distance transport of vapor, which were described in early chapters. A small downward flux of water vapor occurs when conditions favor a downward gradient in vapor concentration, and vapor condenses on the earth's surface and on its ecosystem surfaces. In many ways this process is quite different from condensation in storms in the middle atmosphere, even though the same 2500 kJ of energy is released when each kilogram of water condenses. Other aspects of the energetics are quite different and the process is slower and more subtle. We speak here of dew and frost.

Dew is a common, highly visible manifestation of condensation in ecosystems, and has received much biological attention. Interest has centered on absorption of water through the leaves and other possible mechanisms through which water might move from the atmosphere to the biosphere without traversing the tortuous paths through canopy and soil that we have been following in this book. The consensus is that the role of dew in ecosystem water supply is marginal, but fascinating.

For example, ragweed appears to have mechanisms for channeling dew it collects into stemflow that moistens its own roots and gives it some advantage "in early plant succession within old-field ecosystems" (Shure and Lewis, 1973), an environment conducive to heavy condensation of dew. It might also help translocate nutrients from leaves to soil and roots, thus playing a role in nutrient recycling.

The amounts of such stemflow are, however, not very large—about 0.05 mm per night (Shure and Lewis, 1973). In a climate where daily rates of evapotranspiration reach 4–5 mm, the place of dew in the total water budget of the ecosystem is minor. The possibility that it might play a larger role in dry periods has, however, always loomed large. Its high visibility has created a folklore about its water-budget importance that has been abetted by faulty measurement methods.

Dew-measuring instruments suffered from much the same misconceptions as did evaporation pans, i.e., the belief in some "dew-producing power" in the atmosphere. Exposures and thermodynamic characteristics of dew meters were unrepresentative of natural surfaces, and measurements were not physically explainable (Slatyer and McIlroy, 1961). Much agricultural writing on dew still exaggerates its quantity, sometimes putting it in the same class as rain as an input of water to croplands. Only when energy-budget analyses were applied to this phenomenon in the mid-1930s, was its correct magnitude discovered.

The Energetics of Dewfall*

When the latent heat of condensation (or vaporization) is put into the context of an energy budget, it is obvious that when dew is formed substantial amounts of energy are released and have to be disposed of somehow.† This fact was known by the seventeenth century, but the question remained: "What was the source of the cold necessary for the condensation of the vapour?" (Middleton, 1966, p. 178).

By the end of the eighteenth century, experimenters had verified this need of a cold sink but could not decide among five or more possible explanations for the sources of the cold. "Never can a natural

* The word "dewfall" is a relic of the old controversy whether dew "rose" or "fell" (Middleton, 1966, p. 178). That is, did the vapor "come from the earth or from the air." Even accepting the idea that it came from the air, at the beginning of the 19th century "there was not even agreement on whether dew 'fell' or was condensed out of the air" (Middleton, 1966, p. 186).

† As indicated in the earlier chapter on atmospheric storms and condensation, this question of removal of the latent heat of condensation appears in the cloud environment. Here, however, it is disposed of by being converted into the energy needed to expand a rising parcel of air—an effective heat-removal mechanism.

phenomenon have been more utterly in need of an organizing hypothesis. Yet by this time the materials for an explanation were at hand in the discovery of invisible radiant heat [long-wave radiation], the existence of which had been suspected for more than a century, but was not established beyond a doubt until about 1774" (Middleton, 1966, p. 185).

The connection was made by Wells, in 1814, who "had an extremely clear view of what is now called the radiation balance" (Middleton, 1966, p. 189). Accurate measurements of long-wave radiation, however, were lacking for another century. The rate of formation of dew continued to be grossly exaggerated until energy budgets were constructed by such investigators as Hofmann (1956) and Monteith (1957).

If the energy released at the surface by condensation of vapor is not removed from the scene, condensation cannot continue. Energy accumulates, the surface is warmed, and its vapor pressure rises to a level at which it negates the gradient along which vapor had been moving toward the surface. What processes might remove energy from the surface? As Wells noted in 1814, visible dew or frost is formed only in conditions in which the major energy-removing process is radiation.

The Radiation Deficit At terrestrial temperatures radiation removes 300–400 W m^{-2} from terrain surfaces. This rate of removal would appear to make room for a large input of the latent heat of condensation. Atmospheric vapor also is a radiator, however, and along with such other minor constituents of the atmosphere as CO_2, ensures that the underlying surface receives 200–300 W m^{-2} of energy as a downward flux of long-wave radiation. (Please refer to the discussion of the energy budget of snow cover in Chapter VII.)

If the atmospheric vapor condenses into cloud droplets, the downward flux of radiation becomes greater, and the net deficit of radiant energy at the surface becomes very small. Figure XIII-3 (Hofmann, 1956) shows the immediate response of the radiation deficit at the earth's surface to a break in a cloud deck. It also shows the immediate reduction of the deficit and the cessation of dew formation as soon as the cloud deck returns and radiates some 50 W m^{-2} of energy to the surface being observed. In fact, the dew that had condensed during the cloudless period gradually evaporated afterward. Even when short-wave radiation is out of the picture at night, the net deficit of radiation seldom exceeds 60–90 W m^{-2}. This value sets the upper limit on the rate of formation of dew.*

* In mass terms, 60 W m^{-2} represents the condensation of water at a rate of 0.09 mm hr^{-1}.

Other Contenders for Radiative Cooling Part of the radiation deficit at the surface is ordinarily met from other sources, such as the soil body. Monteith's (1957) measurements of soil-heat flux on clear nights show that this energy source often approaches, sometimes exceeds, the magnitude of the radiation deficit, thereby eliminating much chance for formation of dew.

The soil-heat flux is especially likely to be large during nights in clear weather, because in such weather the daytime intake of heat is large. Furthermore, if the air is humid enough to be brought to the condensation point ("dewpoint temperature") by a few hours of radiant surface cooling, it is likely that the soil that had supplied water for transpiration during the preceding day is also moist. Moist soil has high thermal conductivity, which favors a nocturnal movement of soil heat to the radiating surface.

Dew cannot form under a completely calm atmosphere.* There must be enough air motion to create mechanical turbulence that will carry vapor downward to the cold surface. Such turbulence, however, also transports sensible heat downward.

At low temperatures and a saturated atmosphere, the amount of sensible heat transported is somewhat greater than the amount of latent heat. At high temperatures it is usually less (Monteith, 1957), but is is always present as a competitor with latent heat to meet the radiative cooling at the surface. The process of condensation of vapor is never the sole heat source free to fill the deficit of the radiation budget.

Rates of Formation For this reason the rate of dew formation is ordinarily much less than the limiting rate based on radiative cooling, approximately 0.1 mm hr^{-1}, as noted earlier. Hofmann (1956) feels that in Central European conditions the maximum amount during a whole night of downward vapor flux, which is 0.5–0.7 mm, is seldom approached. On the nights studied by Monteith (1957), the rate of vapor flux was from 0.01 to 0.015 mm hr^{-1}.†

* This fact probably accounts in part for the discrepancy between the fact that dew forms less frequently on foliage than a downward gradient of vapor pressure occurs. Such gradients occur 0.28 of the time at Argonne, Illinois (Moses *et al.*, 1967); Geiger (1961, p. 111) shows them as being even more frequent. Vapor cannot, however, move downward along these gradients rapidly enough to form a visible deposit of dew if mechanical turbulence is weak.

† It is possible that in some conditions a downward flux of vapor ends, not in vapor condensing in visible form on leaves but in vapor absorbed in the soil. This seems likely in windy, humid weather, especially at rather high ambient temperatures. It has been noted in some areas (Prohaska, 1968) where sand may be slightly moistened at night to a depth of several centimeters.

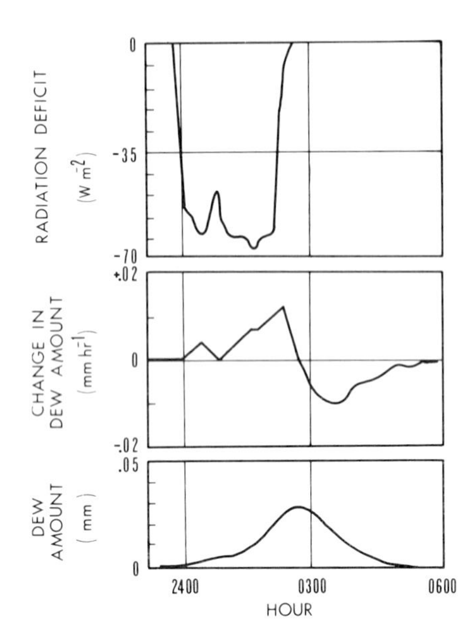

Fig. XIII-3. Formation of dew during several hours of clear sky on the night of 16–17 November 1953, and subsequent evaporation of it when clouds reappeared. During cloudless hours downward long-wave radiation was small and the net exchange reached a deficit of nearly 70 W m^{-2}. Reappearance of clouds increased the flux of downward long-wave radiation and the dew evaporated (Hofmann, 1956).

Perhaps on a favorable night 0.15 mm would be usual; this amount of water, if it fell as rain, would just be measurable in present gages. So much for dewy legend. The sharp pencils of energy-budget analysts have deflated it to a minor item in the energy budget and a negligible one in the water budget.

Dew, along with water of distillation* on leaves, however, represents a change in a basic characteristic of the earth's surface—its moistness. This external water forms a receptive surface for fungus spores impacted on leaves by atmospheric eddies. Many kinds of crop

* It is necessary to distinguish dew formation from distillation, which is a movement of vapor from warm, wet soil to the foliage above, where it condenses. This occurred in Monteith's (1957) experiments on nights of calm, when mechanical turbulence is weak and brings no vapor to the foliage from above. Distillation makes visible a small, closed cycle of evaporation and condensation within a soil–plant system when it is sufficiently isolated from outside influences. Though no more than a redistribution of water in the soil–plant system, distillation can result in very wet grass that will thoroughly drench a walker's shoes and pant legs. It might also serve to recycle nutrients dissolved from the leaves.

blight start when leaves are wet for long periods at the appropriate temperatures. Geiger (1961, p. 340) notes that one experimenter found dew wetness to occur on only 18 hr of a fall period under the canopy of a beech forest, but on 334 hr in a site under open sky. In the open site, less radiant heat is received at night than under the trees, dew formed more often, and remained longer. Plants in the open site thus are more vulnerable to attack by invading fungus diseases. (See also Chapter VI.)

Frost When condensation takes place at a temperature below 0°C, the deposit takes the feathery form of frost rather than that of dewdrops (Plate 28), but the energy relations are the same. The transfer rate from atmosphere to surface is limited for the same reason,

Plate 28. Frost on stalks of dead plants near the Dreisam River in Freiburg, Germany. The deposits on stalks of 2–3-mm diameter are about 10-mm diameter (December 1971).

although we seldom think of frost in its water-supply aspect, probably because, unlike dew, it occurs in a season when there is no shortage of water.

Frost is more conspicuous than dew because of its high reflectivity. When it forms on the branches of a bare tree it becomes one of the more spectacular forms that water can take in our environment. If damp stagnant air from which frost is deposited fills a basin for days at a time, as is true of the Rhine graben in southern Germany, at temperatures close below 0°C, frost forms profusely every night. It is deposited thickly on surfaces cut off from soil heat storage, and lasts hours in the short winter days.

The impaction of supercooled droplets on vegetation changes the process described in Chapter VI because the droplets freeze into a sheath. This is denser than hoarfrost, which grows molecule by molecule. Loading by frozen fog on the aerial parts of ecosystems (Plate 29) or the electric wires of human systems is more important ecologically than frost but not as destructive as freezing rain.

Plate 29. Results of impaction of supercooled droplets of fog on trees in the Rhine trench near Freiburg in December 1971, when fog laden cold air filled the valley for days at a time. This illustrates one mode of delivery of water substance to ecosystems at the earth's surface, as well as a locale of storage of water that later will melt and drip to the ground (photo by W. Weischet).

Condensation of Vapor on Surfaces Cooled Nonradiatively

On Lakes Where a cold substrate can remove heat rapidly from a water surface, atmospheric vapor may condense on it. Such condensation is not visible and not directly measurable, but can be deduced from vapor-pressure gradients. It is not limited to the rate of heat removal by the radiation exchange, as is dew or frost, but can reach large rates.

It is common in spring, when cold lake water underlies streams of air warmed and humidified by adjacent land surfaces, which at this season are likely to be bare, wet, and steaming. In Richards and Fortin (1962), the vapor pressure at the surface of Lake Ontario is shown to be 7–8 mb in these months, whereas bare fields on the lake plain during a sunny day, in contrast, reach vapor pressures of 12–15 mb. Warm, humid land air moving out over the lake brings both latent and sensible heat. These are transported downward to the water by the turbulent diffusivity of the air stream. In one case over Lake Ontario (23 April 1968) the rate of condensation was calculated to be 1.5 mm day^{-1} (Phillips, 1972). If the parallel downward transport of sensible heat cools the air to the dew point, the zone of condensation broadens from the lake surface and fog forms in the lower layers of the air stream.

On Snow and Ice A melting snow or ice surface has, as mentioned earlier, an insatiable capacity for heat. No amount supplied to it by radiation or the sensible-heat flux can raise its surface temperature, nor compete directly with heat that might be supplied by condensation. The only limit on the rate of condensation of vapor is the capacity of the turbulent transport mechanism to operate in conditions of great atmospheric stability, and this depends on wind speed, wind shear, and terrain roughness. The surface of an ice cap is not rough, but the snow fields of mountain glaciers or of mid-latitude mountains lie in dissected, sometimes forested, landscapes that produce mechanical turbulence in moving air.

The examples of snow melting as a result of condensation of vapor on it in advective weather, given in an earlier chapter, give rates of water accretion that are of the order of 1 mm day^{-1}. This process is clearly much more vigorous than dew formation, although sporadic in occurrence.

Condensation on nonmelting snow or ice also occurs, but is limited by the same energetic factors that limit it at foliage or other surfaces. The radiation deficit provides the major heat sink, but because this sink is partially filled by the downward sensible-heat flux and the substrate-heat flux, not all of it is available to accommodate the released heat of condensation.

Condensation of frost on these cold surfaces plays a role in their energy budgets, especially in the long polar nights of Greenland and Antarctica, but in these places the deficit of long-wave radiation is small anyway. The total contribution by the downward latent-heat flux, speculative at best, probably is less than 3–4 W m^{-2} during a winter month (Miller, 1956). This rate is equivalent to about 0.1 mm day^{-1} in mass of water. Low rates that continue through weeks of darkness add to moderate totals, however. The results are seen as deposits of feathery frost on top of the snow, reported at a frequency of 0.3 or more on Greenland (Wegener, 1939).

The Return Vapor Flow from Atmosphere to Surface

The downward vapor flux reaches high rates only at cold lake (or ocean) surfaces or at melting snow and ice. At other surfaces the means of removing the heat liberated by dew or frost condensation are limited. The latent-heat flux must compete with other potential sources to get its share of these limited amounts of chilling. As a result, the heat released by condensing vapor does not seem likely to exceed 5 W m^{-2} in the middle latitudes. Formation of dew does not exceed 0.1–0.2 mm per night.

On many nights the downward flux of vapor is entirely absent. If the wind dies down, if the soil is warm, if the air is dry, if the atmosphere or clouds radiate too strongly to the earth, if the wind is strong—dew or frost do not form, as our own observation tells us.

In contrast to the hundreds of millimeters in the yearly upward flux of water vapor from the surface of the earth, we think of surface condensation in terms of only 10 or 20. This downward flux plays an interesting but minor role in the hydrodynamics of terrestrial ecosystems.

Only small amounts of water are stored in the local air. Sometimes this water is in highly visible form and has substantial effects on the energy budgets of ecosystems at the earth's surface and, hence, small effects on their water budgets. More often, the storage of invisible water vapor in the local air affects the hydrodynamics of underlying ecosystems by reducing the rates of evaporation from them. Return flows of water from local air storage to the surface, except as entrained up into the strong vertical motion of the storm cells described in Chapter II, are small. Such deliveries of water to ecosystems can never give rise to the massive fluxes of water that take part in the processes of infiltration into the soil of ecosystems, evaporation from them, or deep percolation downward from them.

REFERENCES

Australia, Commonwealth Bureau of Census and Statistics (1965). "Official Year Book of the Commonwealth of Australia," No. 51-1965. Commonwealth Govt. Printer, Canberra, 1354 pp.

Budyko, M. I. (1974). "Climate and Life" (English transl. ed. by D. H. Miller). Academic Press, New York, 508 pp.

Church, P. E. (1945). Steam-fog over Lake Michigan in winter, *Trans. Am. Geophys. Un.* **26,** 353–357.

Drozdov, O. A. (1974). Zakonomernosti vlagooborota v atmosfere i vozmozhnosti dolgosrochnogo predskazaniia ego izmenenii. *In* "Vlagooborot v Prirode i ego Rol' v Formirovanii Resursov Presnykh Vod" (G. P. Kalinin, ed.), pp. 70–83, Stroiizdat, Moscow.

Geiger, R. (1961). "Das Klima der bodennahen Luftschicht," 4te Aufl. Die Wissenschaft, F. Vieweg, Braunschweig, Bd.78. 646 pp. illus. (Translated as "The Climate Near the Ground." Harvard Univ. Press, Cambridge, Massachusetts, 1965).

Gentilli, J. (1955). Estimating dew point from minimum temperature. *Bull. Am. Meteorol. Soc.* **36,** 128–131.

Hann, J. (1897). "Handbuch der Klimatologie," 2te Aufl., I Band: Allgemeine Klimatologie. Engelhorn, Stuttgart, 404 pp.

Hofmann, G. (1956). Verdunstung und Tau als Gleider des Wärmehaushalts, *Planta* **47,** 303–322.

Kessler, A. (1968). Globalbilanzen von Klimaelementen. Ein Beitrag zur allgemeinen Klimatologie der Erde. Hannover: Hochschule Inst. Meteorologie Klimatologie, *Ber.* **3,** 141 pp.

Landsberg, H. E. (1964). Die mittlere Wasserdampfverteilung auf der Erde. *Meteorol. Rundschau* **17,** 102–103.

Landsberg, H. E. (1969). "Weather and Health. An Introduction to Biometeorology," Anchor Books, Sci. Stud. Ser. 59. Doubleday, Garden City New York. 148 pp.

Lyons, W. A., and Pease, S. R. (1972). Picture of the month—"steam devils" over Lake Michigan during a January Arctic outbreak, *Mon. Weather Rev.* **100,** 235–237.

Middleton, W. E. K. (1966). "A History of the Theories of Rain and Other Forms of Precipitation." Franklin Watts, New York, 223 pp.

Miller, D. H. (1946). Discussion of P. E. Church's "Steam fog over Lake Michigan in winter," *Trans. Am. Geophys. Un.* **27,** 575–577.

Miller, D. H. (1956). The influence of snow cover on the local climate of Greenland, *J. Meteorol.* **13,** 112–120.

Monteith, J. L. (1957). Dew, *Q. J. R. Meteorol. Soc.* **83,** 322–341.

Moses, H., and Bogner, M. A. (1967). Fifteen-Year Climatological Summary, January 1, 1950–December 31, 1964. Argonne National Laboratory, DuPage County, Argonne, Illinois, 671 pp (ANL-7084).

Odum, E. P. (1971). "Fundamentals of Ecology," 3d Ed. Saunders, Philadelphia, Pennsylvania, 574 pp.

Phillips, D. W. (1972). Modification of surface air over Lake Ontario in winter, *Mon. Weather Rev.* **100,** 662–670.

Prohaska, F. (1968). Personal communication.
Richards, T. L., and Fortin, J. P. (1962). An evaluation of the land–lake vapor pressure relationship for the Great Lakes. Univ. Michigan Inst. Sci. Technol., Great Lakes Res. Div. Publ. 9, pp. 103–110.
Sargent, F. II *et al.* (1967). Biometeorology today and tomorrow: principal findings of study group on bioclimatology, *Bull. Am. Meteorol. Soc.* **48,** 378–392.
Scorer, R. S. (1968). "Air Pollution." Pergamon Press, Oxford, 151 pp.
Shure, D. J., and Lewis, A. J. (1973). Dew formation and stem flow on common ragweed *(Ambrosia artemisiifolia), Ecology* **54,** 1152–1155.
Slatyer, R. O., and McIlroy, I. C. (1961). "Practical Microclimatology, with Special Reference to the Water Factor in Soil–Plant–Atmosphere Relationships." UNESCO, Australia Comm. Sci. Indus. Res. Organiz.
Smithsonian Institution (1966). "Smithsonian Meteorological Tables" (R. J. List, ed.), 6th ed. Smithsonian Inst. Publ. 4014, Washington, D.C., 527 pp.
Thom, E. C. (1956). Measuring the need for air conditioning, *Air Cond. Heat. Vent.* **53** (8), 65–70, August.
Webb, M. S., and Phillips, D. W. (1973). An estimate of the role of lake effect snowstorms in the hydrology of the Lake Erie basin, *Water Resources Res.* **9,** 103–117.
Wegener, K. (1939). Ergänzungen für Eismitte, *Wiss. Ergeb. deut. Grönland-Exped. A. Wegener 1929 und 1930/1931* **4**(pt 2), 87–136.
Weischet, W. (1973). Personal communication.
Wilson, C. (1969). Climatology of the Cold Regions. Northern Hemisphere II, Cold Regions Res. Engr. Lab., Hanover, New Hampshire, Cold Regions Sci. Eng. Monogr. I, Sec. A-3b, 158 pp.
Workman, E. J. (1962). The problem of weather modification, *Science* **138,** 407–412.
Zhakov, S. I. (1966). "Proiskhozhdenie Osadkov v Teploe Vremia Goda." Gidrometeor. Izdat., Leningrad, 251 pp.

Chapter XIV

PERCOLATION FROM ECOSYSTEMS

When a layer of soil is saturated by heavy rain, meltwater, or irrigation water, its larger pores, as well as the small ones, fill up; they begin to transmit water to lower layers of the soil body, from which it drains still deeper. When the input of water ceases, downward movement continues to draw water out of the larger pores. As they empty movement continues through the smaller pores, but at a slackening rate because the hydraulic conductivity has decreased. As the large pores fill with air they no longer transmit water; gravity gives way to capillary forces, which can pull water in any direction, i.e., wherever a drier soil exists to create a gradient of capillary potential.

Movement that continues as the soil becomes unsaturated presents a difficult measurement problem for the scientist. His work would be easier if he could assume that a known depletion of soil moisture at a specific depth represented only withdrawal of water by plant roots. As it is, he has to consider that a part of the depletion might represent continued drainage of the soil.

In his broad perspective of water dynamics, Horton (1933) proposed "field-moisture capacity" to show the volume to be filled before excess water "percolates downward to the water-table. Thus all the water that can be utilized by vegetation is retained in the soil and only the excess is rejected," or passed on through the soil.

SHALLOW PERCOLATION

Downward movement of water within the soil body itself can be illustrated by measurements after rains at a shallow depth (−10 cm) in a pine forest in Germany (Brechtel, 1965) (Table I). Delivery of water at

TABLE I

Water Budget of the 0–10-cm Layer of the Soil in Pine Forest of Varying Density in Rains of 4.6 mm day^{-1} Mean Intensity[a]

Factors	Forest Density Expressed in Relative Illumination				
	0.5	0.6	0.7	0.8	1.0 (open site)
Catch at forest floor (1206 gages)	2.9	3.4	3.9	4.5	4.6
Standard deviation	0.32	0.37	0.32	0.18	—
Percolation below 10 cm (600 gages)	0.5	0.6	0.8	1.1	0.3
Standard deviation	0.19	0.22	0.26	0.30	—

[a] Units: mm day^{-1}. Source: From Brechtel (1965).

the forest floor ranged from 2.9 to 4.6 mm per rainy day, depending on forest-canopy density; the mean rate of movement through the −10-cm level ranged from 0.3 to 1.1 mm day^{-1}. This flow exhibited a local variability, expressed as a standard deviation, equal to 0.2–0.3 mm day^{-1}. The variability was so great that the investigator had to use 600 small lysimeters to obtain a reliable sample.

Of interest also is the small flux of water past the −10-cm level in unforested soil, only a third of that in the forest. This results from rapid removal of water from the 0–10-cm layer by evapotranspiration, so that space was likely to be available to store rainwater within the layer. The dense structure of the soil in the open was another factor.

Shallow percolation prevails in the flat grass ecosystems of the Great Plains, where the water-budget surplus is usually small. Employing Thornthwaite's bookkeeping of soil moisture (Chapter IX) to compute the water surplus, Arkley (1963) found that percolate fluxes were closely associated with the depths at which a carbonate horizon formed in the soil profile. As one goes west across the Plains, percolation decreases and the carbonate layer is found nearer the surface. In these flat systems, water surplus is generally manifested as percolation, but on slopes, especially if forested, much shallow percolate moves off-site as subsurface runoff, which will be described in Chapter XVI.

DEEPER PERCOLATION

Measurements to the depth of 1 m show the water fluxes in a thicker layer, the root zone of chaparral in the foothills of the Sierra Nevada in

California during the water year 1939–1940 (Table II). These data are selected from a long study of the disposition of rainfall that was made by Rowe and Colman (1951). A major outflow of water in this budget is by way of deep percolation downward from and out of the main body of soil.

Variation with Time

Seasonal Incidence Rowe and Colman recognized three divisions of the year:

(a) The wetting season, when the root zone was building up its storage from water in the progressive downward movement of the wetting fronts of infiltration in successive storms; little or no water percolated to greater depths until the small pores were replenished. This process required 115 days in this particular year. Under a similar regime of fall and winter rain in South Australia, but with less deeply rooted vegetation, Holmes and Colville (1970a) found "a delay of 60–100 days between the effective beginning of the rains and the first drainage out of the lysimeters."

(b) The percolation season, when heavy rains kept soil moisture high (in spite of some evapotranspiration in the mild winter climate) and sporadically filled the large pores with water, which subsequently drained downward out of the soil body.

(c) The drying season, after the heavy rains of winter had ceased to keep the large pores in the soil filled. As a result, little or no downward flow occurred.

TABLE II

Percolation and Evapotranspiration in Woodland Chaparral at North Fork, California, 1939–1940[a]

Parameter	Wetting season 20 Sept 1939–3 Jan 1940	Percolation season 4 Jan–4 Apr 1940	Drying season 5 Apr–1 Oct 1940
Infiltration into soil	+290	+720	+10
Evapotranspiration	−110	−130	−160
Percolation	0	−590	0
Change in soil-moisture charge	−180	0	+150

[a] Units: mm. Source: Data from Rowe and Colman (1951). Note: The sign convention is with reference to the ground surface. Movement of water from the surface into root-zone storage in fall and early winter is therefore negative. From the soil's standpoint, the reverse would be true.

While water was pulled upward out of the root zone to be evaporated or transpired in each season, downward percolation occurred chiefly during season (b) when storage capacity in the root zone was full. After heavy rains ended in April, evapotranspiration began to draw down soil-moisture storage and also asserted first claim on new water from the few small rains. There was no chance for a surplus of water in the root zone. The large pores remained empty; percolation to deeper layers ceased. The massive though intermittent percolation flux expresses the fact that in Californian ecosystems the regimes of water and energy inputs are out of phase (Chapter XII). Even moderate rainfall in the low-energy season can generate a large surplus, which in forest ecosystems becomes percolation.

Travel time of percolate to groundwater is often a matter of months. In sandstone in England, it is 8 months for a distance of 34 m (Chadwick *et al.*, 1974).

Intermittent Occurrence It is likely, as was true here, that the percolation process is intermittent almost everywhere. In some years it might be confined to a few months, and in others might not occur at all. On the Great Plains, for example, no percolation below the root zone occurs in most years.

Farther north, in the Canadian prairies, the events when water percolates to the groundwater body "are isolated in both time and space" (Freeze and Banner, 1970). They occur only during the spring snow melting season (but even then much of the meltwater does not reach groundwater) and in occasional heavy (50–100 mm) rains in summer, but "only under favorable conditions of water-table depth, soil type, and antecedent soil moisture content" (Fig. XIV-1).

The importance of moisture status of the soil cannot be overemphasized. Unless it is satisfied, little or no water percolates deeper. Of the water that enters the soil of North America, 0.82 remains there to be extracted by plant roots; only 0.18 penetrates farther (L'vovich, 1972). Worldwide, 0.86 remains in the soil and 0.14 percolates to groundwater.

Even where significant average amounts of water percolate from the root zone, as at North Fork, the changes from year to year may be large. In the three years preceding the year cited in Table II, percolation was 680, 1080, and 210 mm.* As Slatyer and Mabbutt (1964) point out for groundwater bodies in arid lands, those "fed directly by

* The mean year-to-year change in the period used here was 550 mm and is nearly as large as the 4-year mean percolate flow, 640 mm. The predictability of this water flux is obviously not very good.

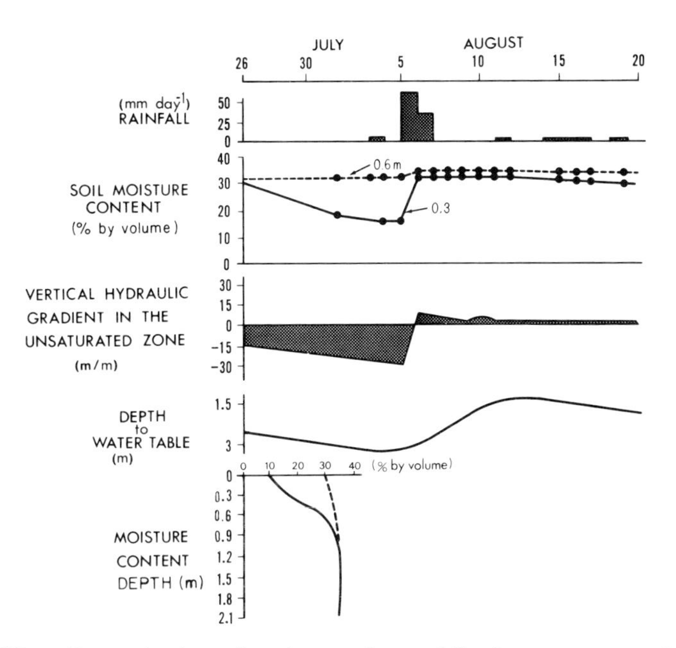

Fig. XIV-1. Events in the soil and groundwater following a summer rain on the prairies. Note the increase in moisture at 30-cm depth, the change in the vertical hydraulic gradient in the unsaturated zone, and the rise in the water table (from Freeze and Banner, 1970). The bottom graph shows the profile of moisture before (solid line) and after (dashed line) the rain of 5 August.

rainfall have high-quality water, but are commonly associated with a low ratio of storage to flow, and hence with considerable water-table fluctuations.''

Year-to-year variability in percolation is also illustrated by nine years of data from the Coshocton lysimeters (Dreibelbis, 1957), where the standard deviation of the annual totals of percolation from a grass–brome cover was 130 mm. This is about 0.75 of the mean percolation.

This high variability can be compared with that of annual evapotranspiration, 60 mm, relative to mean evapotranspiration of 605 mm. Not only is the absolute variation greater by twice for percolation, but also the relative variation (compared to the 9-year mean) is greater—seven times as much. Evapotranspiration, taking the first cut at the incoming water, is relatively stable from year to year; percolation, getting the leftovers, fluctuates wildly. This illustrates the common situation that a storage in some places in an ecosystem may cause increased variability in one of the outflows from it.

The episodic nature of percolation at Coshocton is shown in Fig. XIV-2, which comes from a water-budget modeling study (England and Coates, 1971). While the figure shows variation among years, due in part to the regime of demand by the crop in that particular year of the four-year rotation, the general tendency for most percolation to take place in winter and spring is evident.

Influence of Ecosystems

Surface Condition and Root Depth The effect of soil condition was investigated in California (Table III) by means of lysimeters filled with a homogeneously mixed sandy clay loam, in which pines and chaparral shrubs had grown for ten years and grass for five. Moisture storage of the soil was 0.18 by weight (and 0.28 by volume) at field capacity; it was 0.07 by weight at the wilting point (Zinke, 1960), so about 360 mm of water was available for evapotranspiration if the soil was thoroughly wetted by winter rains.

This did not happen every winter, however. As indicated in the table, no percolation occurred from lysimeters occupied by woody vegetation in the winter of 1956 to 1957. In the next winter, a rainy one, about 100 mm percolated. The grass, rooted only in the upper layers of the soil, allowed percolation to occur in both years.

Some differences in percolation under different ecosystems go back to the branching process at the soil surface, that is, the partition of rainwater between infiltration and surface runoff. The bare-soil lysimeter took in little water through the soil surface, so there was little to be divided at the subsequent branching between soil-moisture take-up and percolation; percolation got none.

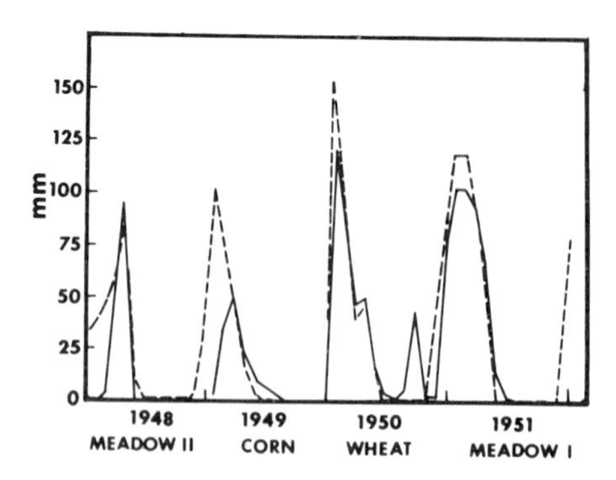

Fig. XIV-2. Percolation at Coshocton, Ohio, as measured in a lysimeter (solid line) over a four-year crop rotation and as calculated by a mathematical model of hydrologic processes at the U. S. Department of Agriculture Hydrographic Laboratory (England and Coates, 1971). Courtesy of the American Water Resources Association.

TABLE III

Percolation Yield from Different Species of Vegetation in Lysimeters in a Dry and in a Wet Year, San Dimas Experimental Forest, California[a]

Parameter	1956–1957 following 4 dry years	1957–1958
Rainfall	515 (200 mm below the mean)	1230 (near the record)
Pinus coulteri (3-m ht.)	0	160
Quercus dumosa, Adenostoma fasciculatum, Eriogonum fasciculatum (average of 3 lysimeters)	0	80
Native bunch grasses	65	280
Bare soil	0	20

[a] Units: mm yr^{-1}. Source: Sinclair and Patric (1960). Depth of lysimeters 1.8 m; surface areas 20 m^2.

The local nature of percolation, reflecting differences in soil and rock permeability as well as in ecosystems, is emphasized by results of a study on the Canadian prairies (Freeze and Banner, 1970). Much meltwater, if generated before the ground has thawed, does not percolate to groundwater, and most rains do not either. Percolation is focused on depressions in the land surface; other areas might never experience percolation events.

Other differences in percolation follow from differences in rooting depth, hence the capacity of the soil volume that is periodically emptied of water. When a shallow root zone is emptied by transpiration, less capacity exists to be filled from the next rain than when a deep root zone is emptied. A smaller uptake of rainwater in the soil leaves more free to drain downward. Percolation from the lysimeters with shallow-rooted grass was appreciably larger than from those in which pine and chaparral ecosystems occupy a deep soil volume.

Greater percolation from shallow-rooted than deep-rooted grass occurred in a long series of measurements in lysimeters containing undisturbed silt loam at Coshocton, Ohio (Table IV). The mean rate of percolation was greatest—approaching 1.5 mm day^{-1}—in late winter, declining thereafter except in years of heavy spring and summer rains. In autumn percolation was rare from the shallow-rooted grass, and nil from meadow and corn–wheat–meadow–meadow rotation.

Some investigations in South Australia showed that useful ground-

TABLE IV

Percolation of Water from Lysimeters at Coshocton, Ohio, Mean 1947–1955[a]

Factors	Season of rare percolation (Aug–Nov)	Percolation occasional years (Dec)	Percolation heavy (Jan–Apr)	Percolation declining (May–Jul)	Yearly sum (mm)
Muskingum silt loam					
Poverty grass (shallow roots)	0.2	0.6	1.5	0.8	295
Meadow (deep roots)	0.0	0.0	1.0	0.4	165
Keene silt loam					
Rotation of corn–wheat–meadow under prevailing agrcultural practice	0.0	0.5	1.5	0.3	215
Under improved agricultural practice	0.0	0.3	1.3	0.2	180

[a] Units: mm day^{-1}. Source: Harrold *et al.* (1962).

water was to be found only under belts of grassland, the "adjoining areas of mallee (*Eucalyptus*) scrub yielding only small supplies" (Slatyer and Mabbutt, 1964). Holmes and Colville (1970a,b) found that pine plantations in the southeastern corner of South Australia percolate very little water, the regional water table being supplied mostly from the grasslands.

Strip-cutting of jack pine in Michigan reduced evapotranspiration for several years and produced an increase in percolation. As determined by using two kinds of measurements of the groundwater body, the increase in percolation was 70 mm yr^{-1} (Urie, 1971). Following the third year after cutting, as vegetation reoccupied the vacant sites, percolation diminished. In such studies, however, it should be noted that the forest floor and soil remain intact; hydrologic processes in them probably continue to favor percolation, which is also enhanced by the reduction in competing transpiration. In a yield model employing functions of terrain, soils, and geology (Jarboe and Haan, 1974), the maximum daily percolation can be calculated from data in the regression equation to be about 0.3 mm more under forest than under low vegetation.

Vegetation Zones Generalized values of annual percolation to groundwater from ecosystems of the major zonal vegetation regions of the European part of the Soviet Union show that large excess water amounts occur where energy for evaporation is weakest; tundra ecosystems must pass on a large fraction of the infiltrated water because their soil reservoirs so rarely are emptied by transpiration, as shown in the accompanying tabulation (L'vovich, 1973, p. 91). Moderately large annual volumes of percolation from the forest ecosystems, in spite of their efficiency in converting available energy into evaporation, reflect several factors that tend to conserve water. (1) Snow melting releases much of the year's total water supply at a season when energy for evaporation is lacking; (2) atmospheric processes deliver more rain and snow to the forest than to the low ecosystems; and (3) the forest floor is open for rapid infiltration. Grassland ecosystems percolate far less water to groundwater bodies since their water budget is small.

Zone	Percolation (mm yr^{-1})	Ratio of percolation to infiltration
Tundra	90	0.30
Taiga	70	0.20
permafrost taiga	25	0.10
Mixed forest	55	0.13
Forest-steppe	15	0.04
Steppe	3	0.01
Semidesert	1	<0.01

The situations of water-filled small pores that permit significant percolation to occur tend to be limited in time and space. Pores may remain full when transpiration is slow, as in a low-energy season, or when rains are closely spaced or snow cover melts day after day at rates of 20–30 mm day^{-1}. In a spatial sense, pores may remain full longer in sites where stemflow or some other feature of vegetation or terrrain concentrates water onto a small area. In either case percolation is intermittent in time or discontinuous in space; it is even more episodic than rain.

MASS TRANSPORTS BY PERCOLATING WATER

Percolating water transports other forms of matter. Some, like chlorides, enter the soil from the atmosphere or in irrigation water. Some, like carbon dioxide and bicarbonates, are picked up by the water from soil compounds and are important for continued rock weathering. Percolation is important in the release of nutrients from the geological reservoir into the exchangeable pool of nutrients that ecosystems can draw on.

In Rural Landscapes

Water moving through the soil and upper layers of rock in the Sierra Nevada quickly picks up soluble material. The content of totally dissolved solids in the snow is about 5 ppm (Feth *et al.*, 1964), while in water that has traveled through the soil it is about 30; the pickup of silica was most rapid, amounting to about half the gain, or 5–10 g m^{-2}. There also was a large increase in the HCO_3^- radical.

Adsorption of metal ions of some salts onto the surfaces of clay particles and organic matter holds them from being swept downward wholesale by the percolate water. When they are present, however, in amounts greater than that for which base-exchange capacity of the soil particles can provide space, they are leached out when water moves downward.

The mass flux of dissolved solids in the percolation stream varies from place to place. In the upper Colorado basin it averages about 3 g m^{-2} yr^{-1} (Iorns *et al.*, 1965) in natural landscapes. Where percolation comes from irrigated land, however, the mass flux is far greater, about 100 g m^{-2} yr^{-1}. From irrigated lands of the Gunnison River basin, it is half again larger still, indicating a considerable leaching of nutrients as a result of the application of irrigation water. Some of the material originates in the accelerated breakdown of the original soil under abnormally heavy flows of water; some represents, no doubt, fertilizers added by the irrigators. It is an obvious burden to possible later users of the percolated water, much of which eventually gets into the river and is carried down to California and Arizona. The question is who will pay for cleaning the water—the upstream users who loaded it or the downstream people who need to use it again (McGaughey, 1971).

In Mechanized, Chemicalized Farming If, in irrigation agriculture, most of the applied water leaves the site by evaporation and transpiration, its freight of solutes remains behind. In surprisingly few years a salt concentration can build up in the root zone, where its osmotic pressure renders it more difficult for roots to extract water from the pores. The plants begin to experience virtual drought. Downward movement of water in the soil body is necessary to prevent salt accumulation, but the salt does not disappear; it is carried along with the percolating groundwater body and to the streams.

In recent years many other geochemical circulations have come to be associated with the movement of water below the root zone. "Agriculture is being accused of being the great polluter of water supplies in irrigated areas. For example, a major problem has developed in the San Joaquin Valley of California concerning the disposition of waste waters that are believed to contain residual nitrogen and other nutrients leached from agricultural lands" (Soil Science Society of America, 1966).

A large literature on the interactions of percolate with soil particles and organisms confirms the expectation that in so complex a medium as a soil body, great variation exists. Chemicals and heavy metals little studied in the past are carried in percolate, increasing our ignorance of what it does in the soil or what the soil can do for it.

Fertilization of sugar cane in Oahu has increased the nitrate content of the basal groundwater body eight times (Mink, 1962a). Cheap energy has made great tonnages of nitrogen fertilizers available, not only in irrigation farming, where the ammonia tanks sit along the canals, trickling nutrients into the irrigation water, but also in humid land farming. Table V shows a mass budget of dairy country in the northeastern United States (Frink, 1969) for which the basic information came from a program in farm accounting. Management data give

TABLE V

Mass Budgets of Plant–Nutrient Elements in Dairying Region of New York–New England[a]

Factors	N	P	K
Input of Nutrients:			
In rainfall	4	0	0
Fixed by vegetation	90	0	0
In fertilizer	50	22	42
In concentrate feeds	81	18	20
Sum	225	40	62
Output of Nutrients:			
In milk	31	5	8
In meat	5	1	1
Volatilized and carried off in atmosphere	70	0	0
Sum	106	6	9
Net leakage from cow–grass system into soil	119	34	53

[a] Units: kg cow^{-1} yr^{-1}. Source: Frink (1969).

a cow density of 1.2 $hectare^{-1}$, and annual expenditures of about $40 per cow for fertilizer for forage and $200 for concentrate feeds.

Local recycling of the fertilizers applied (nitrogen, phosphorus, and potassium) takes place in grass, manure, and soil, but some may leak out of this local system. While most of the phosphorus and potassium seem to be taken by the soil and remain in the soil–plant–cow cycle, this is not true of the nitrogen, which is not as easily held in the system.

The calculated movement of nitrogen from the soil zone of a farm is at least 13 g m^{-2}, mostly in the percolation stream. When this is expressed as the concentration of NO_3^- in groundwater, it exceeds the quality standards set by the Public Health Service. The saving feature is that farms do not occupy all the land, so the regional average is somewhat less. In the Middle West, the lavish use of fertilizers may have more serious consequences (Commoner, 1971, pp. 81–93), by entering city water supplies. Native ecosystems and their clean percolate are less common here than in New England.

Barnyard nitrates percolating to small bodies of groundwater in northern Wisconsin over the years have contaminated them to the point that the water causes cows and women drinking it to miscarry (Tanner, 1974). Feed lots, which have become important in North American agriculture, process massive quantities of nitrogen because animal populations per hectare are so much greater than in the older low-density farming, and can generate very large mass transports of nitrogen to groundwater.

Growth of the practice of highway salting has increased the percolation of salt-bearing water into shallow groundwater along the roads. In New Hampshire, for example, salt is regularly applied to about 6000 km of highway and shows up in about 200 more wells each year. The rate of application is about 15 tons km^{-1}, or 15 kg m^{-1} of road. If a strip of say 100 m from the highway is affected, the unit-area application is thus 15,000 g/200 m^2 or 75 g m^{-2} yr^{-1}. This is several times the rate at which salt is delivered in rain or snow (Dowst, 1967).

When waste water is applied to agricultural ecosystems at rates not much greater than evapotranspiration, e.g., 20–100 mm weekly, it can be regarded as a kind of irrigation and as in other forms of irrigation attention must be paid to the passing through of solutes. Nitrogen applications, for example, should not be much in excess of the nitrogen requirements of the crop (Bouwer and Chaney, 1974, p. 147). Low-rate applications of waste water in arid regions are likely to lead to excess salt inputs to groundwater (Bouwer, 1974), and should be carefully considered before adoption.

Urban Percolates

The foregoing illustrations of leachates in rural environments are part of the larger picture of lavish expenditures of energy and material in an urbanized technology. The reduction of infiltrated water in modern cities with their large impervious areas and extended sewer networks, means that urban areas no longer load the earth immediately beneath them with contaminants. Rather, they now distribute them into the terrain more widely.

For example, the disposal of solid wastes by methods that do not involve recycling, or incineration and resulting air pollution, may lead instead to relatively uncontrolled introduction of materials into the ground. Landfill disposal sites have been found to generate all kinds of leachates, including entirely new compounds formed by unexpected chemical reactions within the disposal site or in the earth into which it drains.

One report notes that there are "thousands of distinct organic substances presently known. Of these, a substantial and variable number are found in water bodies. Detection and identification of minute, but possibly harmful, amounts of these is currently an immensely difficult, expensive procedure. Indeed, the vast majority of such trace pollutants occurring in public water supplies are not identified. Moreover, the chronic physiological effects of minute amounts of chemical substances are almost completely unknown" (Wolman, 1962). The number of these substances has certainly not diminished since Wolman wrote these words.

Return to the Soil On the other hand, if provision is made to use the renovating power of the soil, organic urban wastes need not be contaminating. In an experimental application of waste-water effluent from a sewage-treatment plant in Pennsylvania to forest land, about 90% of the applied water became percolate to groundwater (Sopper, 1968). A small fraction apparently was retained in the soil at periods when soil-moisture storage was low, and supported additional transpiration by the forest. The purpose of the experiment was, however, less to irrigate the forest than to renovate water having a high concentration of nutrients not easily removed by ordinary sewage-treatment methods.

The forest soil and its associated forms of life are, under the right conditions, effective at removing excess nutrients such as soluble compounds of phosphorus and nitrogen. Sopper suggests, however, that an observed downturn in the efficiency of renovation might indicate that the ecosystem's recycling mechanism, involving leaf-fall and root uptake of nutrients, might be reaching the limit of its capacity to absorb additional nutrients.

In this study percolation rates are about 3–6mm day^{-1}, and the system operates only in the warm season. The yearly total of artificial percolation is of the order of 500–1000 mm.

High-rate applications of urban waste water employ rates of 500 to as much as 3000 mm $week^{-1}$, and clearly aim at renovating the water rather than simply disposing of it via evapotranspiration, as with low-rate systems. This renovated water may be pumped for recreational or industrial use, or allowed to recharge the groundwater body (Bouwer and Chaney, 1974). The degree of renovation depends on how completely the soil processes dispose of its various burdens, some toxic solutes not being readily removed from the water. Sites for such heavy percolation need to be selected carefully, and operations need to be monitored (Walker, 1974).

Urban effluents processed only through home septic tanks that feed large percolation fluxes represent a more primitive, yet widespread, form of mass transport downward from the soil body. This condition is so familiar that little discussion is necessary here except to note that the water flux itself, quite aside from its power to transport nutrients, bacteria, and viruses, is not negligible in size. A typical backyard percolation field of approximately 100-m^2 area can generate a downward flux of the order of 5–10 kg m^{-2} day^{-1}. This is no problem where downward drainage is good, but where it is not it can create a considerable swamping problem. Single properties of land that contain both water supply and waste-water disposal components of a household system are predicted to increase by 5 million (Bernhart, 1973), and for health reasons it is desirable to minimize interaction between these components. (As we were taught in country school, keep the well away from the barnyard!) Transfer of contaminants and nutrients can be reduced if the septic-tank evaporation bed is large and microbially active enough to permit no percolation to occur. A bed of 0.4-m depth needs to be about 200 m^2 in area (Bernhart, 1973).

SIGNIFICANCE OF PERCOLATION

In Ohio a change in the amount of percolating water probably has only minor importance, but where groundwater is valuable and percolation is small, any change might be important. In the basin of the Goulburn River in New South Wales, for example, the annual percolation is between 5 and 10 mm yr^{-1} (Chapman, 1963). In terms of the total yield from the river basin (28 mm yr^{-1}), percolation represents an important fraction. It would be even more important in more porous terrains.

The extensive Gambier Plain in South Australia is underlain by a limestone aquifer that is a potential resource of great value in this summer-dry climate, not only to its own region but also to the whole state. Recharge to this reservoir comes almost entirely from grassland ecosystems on the Plain, although the rate is not heavy—65–85 mm yr^{-1} (Holmes and Colville, 1970a). The forest ecosystems, which are being extended as new areas, are planted to *Pinus radiata* for timber and pulp, evaporate almost all their winter rainfall of 600 mm by early summer, and percolate no water to the groundwater body (Holmes and Colville, 1970b). These investigations show the significance of even relatively small amounts of percolation, which is vulnerable to further conversion of grass ecosystems to forest, and limits the potential for irrigation in the region (Holmes and Colville, 1970a).

In a contrasting situation of high percolation, an investigation was made in the upper basin of Kipapa Stream on Oahu (Mink, 1962b). In this basin some subsurface water moves at shallow depths into the stream channels, but more of it percolates downward into the basal aquifer that underlies large areas of Oahu. Rainfall in Kipapa basin is heavy (about 4800 mm yr^{-1} at the maximum station), and a major part of it is converted via percolation to the recharge of this economically priceless basal aquifer. Deep percolation, as much as 1500 mm annually, is perhaps as high as anywhere in the world.

Other places with heavy percolation are found under stream channels. These are particularly important in semi-desert regions, where most of the land surface contributes only sporadically to percolation because only in occasional years is the root zone brought to a stage of saturation permitting gravitational flow. Percolation opportunity beneath stream channels is, of course, prolonged because of the continual saturated state of the ground.

Augmented Percolation

Percolation of water from irrigation canals and ditches is a similar process that may dispose of a third or more of the water brought into an irrigation district. Because excessive percolation from ditches and irrigated fields may create groundwater problems, attempts are made to control it by lining canals and metering the amounts of water applied to field furrows.

Where water is cheap it often is applied without much thought; it might be applied up to the limit of a legal water right whether needed or not. The resulting large flow is likely to raise the water table into the aerated layer of the soil, causing the swamping problem that is nearly ubiquitous wherever man irrigates. It usually is not until after swamping has started that drainage is provided for.

Water is usually applied to crops in excess of the rate of potential evapotranspiration in order to assure a net downward movement of water and dissolved salts. This flow represents an artificial percolation, including an associated mass flux of salt.

Some irrigation districts will encourage percolation to a groundwater body that later may become a source of water for pumped irrigation; ditches and canals on the alluvial fan of the Kings River near Fresno, California are commonly kept full of water all winter to increase percolation. Some fields have water ponded on them for this same purpose.

Artificial Recharge The situation just described merges into one in which percolation is increased by building spreading grounds near stream channels on alluvial fans, and ponds on tighter valley soils. Artificial recharge has been carried on for many years in California.

In recent years it has been done not only with local flood water, but also with recycled waste water and with water imported into southern California at times when full aqueduct capacity is not needed for current supplies. Thornthwaite, in about 1952, was one of the first to use sprinkler irrigation in a natural ecosystem, recharging waste water from a vegetable freezing plant in New Jersey. Polluted river water has been recharged in England at an application rate of 5400 mm yr^{-1} (Chadwick *et al.*, 1974). This massive flux apparently did not alter the balance of species in the lowland heath ecosystem that was irrigated, and as measured at 1.8-m depth the water improved in quality in most respects.

As water becomes less a free good in our economic thinking, artificial recharge will increase. Furthermore, more thought will be given to forestalling activities that might have a deleterious effect on natural percolation. For example, urbanization around some metropolitan areas may be prohibited by zoning provisions where percolation to a valuable aquifer is recognized as important. Examples are found near Chicago and on Long Island. Zoning laws were upheld to prevent location of a new factory and attendant urbanization in an area 30 km west of Milwaukee known to be vital for intake of water into an important sandstone aquifer. On the other hand, a recharge area recognized as important to an aquifer used by several towns near

Chicago went unprotected. Authorities were unable to find a way to continue the forest ecosystem in this function (Platt, 1972, pp. 107–129).

PERCOLATION AND RECHARGE

World averages of percolation should be mentioned for comparison with the fluxes being cited. Estimates by L'vovich (1972) are 90 mm yr^{-1}, half the Coshocton figure. The estimate for North America, 84 mm, contrasts with that for Australia, 7 mm. In a dry country, where saturating volumes of water seldom enter the soil, almost no water flows out of it as percolate. As a result, bodies of saturated soil and rock, i.e., groundwater, are scattered and small in such areas. A valuable kind of water resource is absent. On the other hand, groundwater aquifers are extensive and abundant in lands of sufficient percolation and represent a valuable form of water in the local terrain, to be described in the next chapter.

REFERENCES

Arkley, R. J. (1963). Calculation of carbonate and water movement in soil from climatic data, *Soil Sci.* **96,** 239–248.

Bernhart, A. P. (1973). Protection of water-supply wells from contamination by wastewater. *Ground Water* **11**(No. 3), 9–15.

Bouwer, H. (1974). Design and operation of land treatment systems for minimum contamination of ground water, *Ground Water* **12**, 140–147.

Bouwer, H., and Chaney, R. L. (1974). Land treatment of wastewater, *Advances Agron.* **26**, 133–176.

Brechtel, H. M. (1965). Metodische Beiträge zur Erfassung der Wechselwirkung zwischen Wald und Wasser, *Forstarchiv* **35**, 229–241.

Chadwick, M. J., Edworthy, K. J., Rush, D., and Williams, P. J. (1974). Ecosystem irrigation as a means of groundwater recharge and water quality improvement, *J. Appl. Ecol.* **11,** 231–247.

Chapman, T. G. (1963). Rainfall–runoff relations in the upper Goulburn catchment, N.S.W., *Inst. Eng. Austral. Civil Eng. Trans. CES* **1,** 25–35.

Commoner, B. (1971). "The Closing Circle. Nature, Man, and Technology."Knopf, New York, 326 pp.

Dowst, R. B. (1967). Highway chloride applications and their effects upon water supplies. *J. New England Water Works Assoc.* **81,** 63–67.

Dreibelbis, F. R. (1957). Average monthly percolation from the Coshocton monolith lysimeters. *J. Soil Water Conserv.* **12,** 85–86.

England, C. B., and Coates, M. J. (1971). Component testing within a comprehensive watershed model, *Water Resources Bull.* **7**, 420–427.

Feth, J. H., Roberson, C. E., and Polzer, W. L. (1964). Sources of Mineral Constituents in Water from Granitic Rocks, Sierra Nevada, California and Nevada, U. S. Geol. Surv., Water-Supply Paper 1535-I, Washington, D. C., 70 pp.

Freeze, R. A., and Banner, J. (1970). The mechanism of natural groundwater recharge and discharge. 2. Laboratory column experiments and field measurements, *Water Resources Res.* **6,** 138–155.

Frink, C. R. (1969). Water pollution potential estimated from farm nutrient budgets, *Agron. J.,* **61,** 550–553.

Harrold, L. L., Brakensiek, D. L., McGuinness, J. L., Amerman, C. R., and Dreibelbis, F. R. (1962). Influence of Land Use and Treatment on the Hydrology of Small Watersheds at Coshocton, Ohio, 1938–1947, U. S. Dept. Agric., Tech. Bull. 1256, 194 pp.

Holmes, J. W., and Colville, J. S. (1970a). Grassland hydrology in a karstic region of southern Australia, *J. Hydrol. (Amst.)* **10,** 38–58.

Holmes, J. W., and Colville, J. S. (1970b). Forest hydrology in a karstic region of southern Australia, *J. Hydrol. (Amst.)* **10,** 59–74.

Horton, R. E. (1933). The relation of hydrology to the botanical sciences, *Trans. Am. Geophys. Un.* **16,** 23–25.

Iorns, W. V., Hembree, C. H., and Oakland, G. L. (1965). Water Resources of the Upper Colorado River Basin—Technical Report. U.S. Geol. Surv. Prof. Paper 441, 370 pp + pl.

Jarboe, J. E., and Haan, C. T. (1974). Calibrating a water yield model for small ungaged watersheds, *Water Resources Res.* **10,** 256–262.

L'vovich, M. I. (1972). "Vodnye Resursy Budushchego." Izdat. Proveshchenie, Moscow.

L'vovich, M. I. (1973). "The World's Water: Today and Tomorrow." Mir Publ. Moscow, 213 pp. (Transl. by L. Stoklitsky of Vodnye Resursy Budushchego. Izdat. Proveshchenie, Moscow, 1972).

McGaughey, P. H. (1971). Waste water reclamation—urban and agricultural. *In* "California Water" (D. Seckler, ed.), pp. 161–173. Univ. California Press, Berkeley.

Mink, J. F. (1962a). Excessive irrigation and the soils and ground water of Oahu, Hawaii, *Science* **135**, 672–673.

Mink, J. F. (1962b). Rainfall and runoff in the leeward Koolau Mountains, Oahu, Hawaii, *Pac. Sci.* **16**, 147–159.

Platt, R. H. (1972). The Open Space Decision Process: Spatial Allocation of Costs and Benefits. Univ. of Chicago Geog. Dept. Res. Paper 142, Chicago, 489 pp.

Rowe, P. B., and Colman, E. A. (1951). Disposition of Rainfall in Two Mountain Areas of California. U. S. Dept. Agric. Tech. Bul. 1048, 83 pp.

Sinclair, J. D., and Patric, J. H. (1960). The San Dimas disturbed soil lysimeters, *Int. Assoc. Sci. Hydrol. Publ.* **49,** 116–125.

Slatyer, R. O., and Mabbutt, J. A. (1964). Hydrology of arid and semiarid regions. *In* "Handbook of Applied Hydrology" (V. T. Chow, ed.), Chapter 24, pp. 1–46. McGraw-Hill, New York.

Soil Science Society of America. Water Resources Committee, C. H. Wadleigh, chairman (1966). Soil science in relation to water resources development: I. Watershed protection and flood abatement, *Proc. Soil Sci. Soc. Am.* **30,** 421–424.

Sopper, W. E. (1968). Waste water renovation for reuse: key to optimum use of water resources, *Water Res.* **2,** 471–480.

Tanner, C. B. (1974). Personal communication.

Urie, D. H. (1971). Estimated groundwater yield following strip cutting in pine plantations, *Water Resources Res.* **7,** 1497–1510.

Walker, W. H. (1974). Monitoring toxic chemical pollution from land disposal sites in humid regions, *Ground Water* **12,** 213–218.
Wolman, A. (1962). Water Resources. A Report to the Committee on Natural Resources of the National Academy of Sciences—National Research Council. U. S. Nat. Acad. Sci., Publ. 1000-B, 35 pp.
Zinke, P. J. (1960). The influence of a stand of *Pinus coulteri* on the soil moisture regime of a large San Dimas lysimeter in southern California, *Int. Assoc. Sci. Hydrol. Publ.* **49,** 126–138.

Chapter XV

GROUNDWATER AND ITS OUTFLOWS INTO LOCAL ECOSYSTEMS

THE ENVIRONMENTS OF GROUNDWATER

Water percolates downward below the root zone until it reaches impermeable layers, where it accumulates in groundwater bodies that saturate porous rock layers or unconsolidated materials that underlie many terrains of the world. Like other storages, these bodies can be analyzed in terms of inputs, outputs, and resulting changes in the mass of water stored.

Groundwater bodies differ greatly in the amount of water they store. The porosity of rock ranges from 0.2 to 0.4 in gravel and sand, 0.1 in sandstone, and less than 0.1 in other consolidated rocks.

The term "specific yield" is an analog to "available moisture" that was used when we discussed moisture in the soil. The volume of water that can be removed from an aquifer by gravitational forces, i.e., excluding the water retained in small pores, lies in the range of 0.1–0.2 (Todd, 1959, p. 23). The immense aquifer of the alluvial fill of the Sacramento Valley, 10^4 km^2 in area, has a mean specific yield of 0.07 and a volume above the 65-m depth of 48×10^9 m^3. This volume is comparable with the total volume of surface storage in Shasta Lake and all the other reservoirs of the valley put together.

The environment for groundwater varies a great deal from place to place since the geology is complex. For example, in the Hawaiian Islands many individual bodies of groundwater exist at different elevations and under different pressures, although near one another. The largest float upon sea water; smaller bodies are perched on impermeable layers of soil sandwiched in the layers of basalt or are held in compartments formed by vertical dikes. Often the underground structure so isolates them that separate water budgets must be cast for each one. In California alone, 500 groundwater basins are objects of investigation and usually management (Peters, 1972).

Stallman (1964) points out that "the distribution of water might be viewed as dependent on either supply and/or environmental factors." For example, surface runoff is a direct function of supply, that is, the input of rainfall, mediated through relatively small storages in the interception zone and the surface-detention zone of an ecosystem. Groundwater, on the other hand, "is so greatly dependent on environment that groundwater investigations are almost completely devoted to an interpretation and description of environmental factors alone," i.e., the geologic structure of the rock below the ecosystems at the earth's surface. The spatial differentiation in groundwater environments does not necessarily display the same pattern as do the ecosystems at the surface, except where seepage and recharge areas in shallow groundwater are prominent or where such intrusive systems as landfills have been established. With increasing depth these spatial differences tend to diminish and more uniform aquifer conditions underlie the mosaic of surface ecosystems.

Two writers on groundwater outline the opportunities—indeed, the urgencies—for research on the "kaleidoscopic variety of environments in which ground water occurs, and of modifications made by man" (Thomas and Leopold, 1964). Emphasis on the environment indicates that often it is the *capacity*, not solely the water as such, that may be most important. In this respect the reservoir serves much as do the underground spaces in the Middle West in which natural gas is stored during the summer for use the following winter. Essential qualities are porosity and means of rapid input and output. "Underground space" is now recognized as a resource having value in itself, which is enhanced where it provides a secure, clean environment.

Some underground reservoirs are managed like surface reservoirs, water being withdrawn in dry seasons or dry years and replaced in seasons of abundance. Both input and output would be artificially accelerated under integrated administrative control, often by the agency that also manages the surface reservoirs. For example, an irrigation district that coordinates well pumping also provides for artificial recharge, as described earlier. In such joint use the underground reservoir is likely to be utilized for long-term storage, i.e., from year to year, while the surface reservoir, more rapidly filled and emptied, serves shorter-term balancing purposes.

This management technique, however, becomes unworkable if drawdown of fresh water allows the inflow of contaminants from adjacent rock layers. Saline water in deeper layers or from the ocean, or contaminants from industry or cities, can render the space useless for further freshwater storage. Dozens of coastal aquifers exist in environments that are threatened by incursion of salt water if they should be drawn down too far (American Society of Civil Engineers, 1969).

GROUNDWATER RECHARGE

Percolation downward from the soil almost always can be looked at from the converse point of view, by which it becomes a replenishment to the groundwater body, or recharge. The exception occurs when the ground water is being destroyed and its water table is receding into the depths faster than the percolation stream can follow it, as is the case near Phoenix.

Groundwater storage is the balance between incoming water, which results from percolation that is either widespread or localized, and various kinds of outflow. We have discussed the intermittent character of percolation, and now should note how the size of storage affects the response of the groundwater body to this episodic inflow. This relation can be expressed as the ratio of stored water to the amount of water annually flowing into storage. Such a ratio, with the dimension of time (mm divided by mm yr^{-1}), is small where inflow is large relative to the size of the body.

In small, rather tight basins it may be of the order of 0.01–0.05 yr. In the drainage basin of the Goulburn River in New South Wales, the storage-to-flow ratio, 1.05, represents a combination of one year's equivalent of flow stored in the consolidated rocks of the whole basin, plus 0.05 year's equivalent flow stored in river alluvium, which probably is the principal area of recharge (Chapman, 1964). In the adjacent Hunter River basin, an area with extensive areas of alluvium and pervious upland rock formations, the ratio is about 5 yr (McMahon, 1964).

Ratios smaller than about 50 yr indicate that groundwater storage is not great enough to absorb large fluctuations in the input when rainfall and percolation change, as a group of wet years gives way to a group of drier ones. Such bodies are vulnerable to fluctuations in percolation. Larger ratios suggest "inherited" groundwater rather than "cyclic" (Nace, 1969), an example being the water in the Ogalalla formation in west Texas, which has been 13,000 yr in transit. Management of inherited water is difficult because demands can easily come to exceed inputs. When these deposits have mined out the overlying ecosystems (largely cotton fields in this case) that have been irrigated from them will be confronted with change.

A problem with large amounts of water in storage is that changes are difficult to detect and calculation of inputs and yield are subject to large error. Even in the small basin of Aquitaine, in southern France, 25–30 millennia are required for water to pass through, and Schoeller (1959) points out that during these millennia great changes might have taken place in the pattern of recharge to the basin by percolation.

THE VOLUME OF STORED UNDERGROUND WATER

The volume of water in groundwater storage is determined by the specific yield fraction and the depth of the saturated layer. The lower edge of this layer usually is indefinite, as the small fissures or fractures pinch out with depth.

Another lower level is analogous to the economic depth of mining used in estimating coal reserves, and is set by the energy and dollar costs of raising water to the ground surface. Water at greater depths may be regarded under present conditions as unrecoverable. This depth often is assumed as 60 m, but obviously varies with the local energy costs of pumping (e.g., extension of natural gas pipelines into eastern Colorado has led to a large increase in pumping for irrigation), as well as the local value of pumped water. In many parts of the southwestern United States, pumping lifts are as great as 200 m. One wonders how much this definition of lower depth will change as energy becomes more costly in the United States.

Volume of groundwater can be compared with the vapor mass in the atmosphere, or with the liquid water in lakes and streams at the earth's surface. In the United States about 100 times as much water (8000 kg m^{-2}) is estimated to exist below the ground as in rivers and lakes on its surface (75 kg m^{-2}), although much of the groundwater is below the economically accessible limit mentioned above.*

Vapor in the atmosphere averages about 20 kg m^{-2} over the United States, 1/400 of the volume that exists, nearly immobile, beneath the ground. Vapor in the local air, the zone overlying surface ecosystems, is 5–10 kg m^{-2}. While this mass is important in its mobility and its exchanges with ecosystem water (Chapter XIII) it is small in comparison with the tons per square meter below surface ecosystems.

* Volume of groundwater in Australia is estimated at 1500 kg m^{-2}, river and lake water at 5 kg m^{-2} (Chapman, 1964).

The Water Table

The upper boundary of the groundwater body is more definite than the lower. It is the familiar water table. Except in geologic environments of very high permeability this surface is not level but follows the ups and downs of the land surface above it, from which it receives percolation. In one small area of Sweden, for example, the level of the water table was 3–4 m below the ground surface "irrespective of the height of the terrain" (Gustafsson, 1968), which varied from an elevation of 4 m above sea level to 38 m. Contour maps of the water table are a recommended means of inventorying aquifers and are useful for validating models of groundwater budgets (Peters, 1972).

The water table can be seen at the bottom of wells. Large galleries just above it, excavated for skimming wells taking water from the basal aquifer of Oahu, present a rather impressive sight to tourists. It also is visible in landscapes pocked with lake basins, as it forms the actual surface of some of the lakes. Such "spring ponds" have particular value as fish habitats because of their coldness.

Annual variation in the water table, i.e., in groundwater storage volume, indicates the season when most percolation occurs. The specific yield fraction determines the change in elevation. For instance, if the input from percolation during the winter of 1939–1940 from the chaparral site at North Fork, California, was 590 mm, movement into a groundwater environment with 0.07 specific yield would raise the water table by 590/0.07 = 8430 mm = 8.4 m.

In some terrains the highest level of the water table comes above parts of the land surface, forming temporary ponds or marshes. Recession of the water table subsequently allows these water bodies to dry up.

If groundwater is confined by an upper impermeable layer, there is no free water table, and for purposes of observation its place is taken by the equivalent pressure (piezometric) surface. Short-term and seasonal changes in pressure are not usually as marked as in a free water surface. This is partly because a confined water body does not receive brief pulses of percolating water, but rather a slow steady inflow, perhaps by leakage from other aquifers.

Changes in the volume of water may be buffered by changes in the aquifer volume itself. It will recover spatial volume when the water volume increases if its behavior under heavy loading from above is elastic. If, on the other hand, its behavior is plastic, the structure collapses when dewatered and permanent loss of storage capacity ensues.

Long-Term Changes in Volume

Slower changes also take place in the volume of groundwater bodies. For example, the amount of percolation from the overlying layers may decrease with the paving of the surface or the replacement of cesspools or septic tanks by sewers. Mining of groundwater deposits inherited from the past inevitably reduces their volume.

These slow changes are evident in several ways, depending on the environment of the groundwater body.

(1) If it is confined under pressure, there is a loss in head; if early wells were artesian, they cease to flow.*

(2) If the groundwater body is confined laterally in its own basin, any decrease in its volume is evident as a drop in the water table. In a basin near Phoenix, Arizona, for instance, the water table has been receding for several decades at a rate of one to several meters per year.

(3) If the groundwater body is in a basin that it shares with ocean or other mineralized water, the decrease in volume of fresh water is evident as a shift in the position of the interface between the two waters of differing density. Sea water encroaches into the aquifer space vacated by the shrinking freshwater volume.

Stability in Oahu Long-term balance in groundwater volume is exemplified in an aquifer in the island of Oahu. Long ago the nature of the aquifers that exist on this volcanic island was established, although neither the high-level bodies of water held by vertical dikes nor the basal lens of fresh water floating on salt water are common elsewhere in the world (Fig. XV-1) (U.S. Water Resources Council, 1968).

Concern has long been felt for the state of health of the basal lens in particular, which is under the ever-present threat of salt-water encroachment at its lower interface. Much attention has been given to how fast it responds to changes in inflow or outflow (Wentworth, 1947).

Invisible though it is, it is generally recognized as the principal resource in the water economy of the island (U.S. Water Resources Council, 1968, pp. 1–28), and is treated with the consideration this warrants. "The probability of overdraft and resultant salt water intrusion is directly related to increased land development for residential, industrial, and resort use" (U.S. Water Resources Council, 1968, p. 6-

* Though they frequently appear in textbook illustrations, flowing artesian wells are not common any more, the exploitation of groundwater being so widespread.

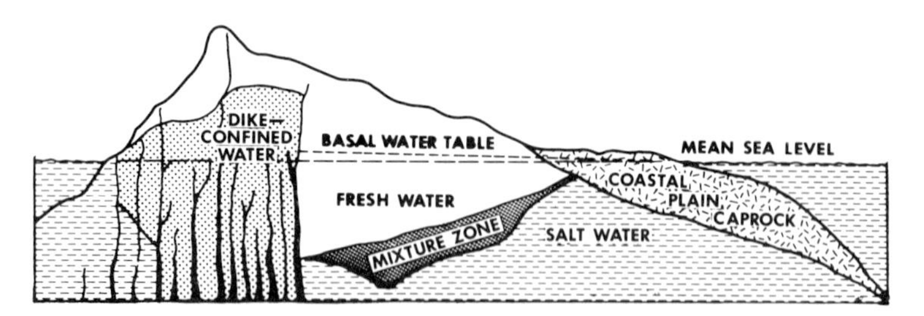

Fig. XV-1. Cross section of the geology and groundwater of Oahu (U.S. Water Resources Council, 1968).

19-5). As a result, the outlook is for "permanent operation of monitoring wells; continued studies to maximize safe production" (American Society of Civil Engineers, 1969).

In addition to the threat of salt-water encroachment, the basal aquifer has acquired other mass budgets. Nitrates in the percolate from fertilized cane fields increased nitrate concentration in the upper layer of the lens from 1 to 8 ppm. The concentration of silica has doubled, due to heavy percolation of irrigation water from the cane fields because the percolate speeds up the weathering of the rock mantle (Visher and Mink, 1964).

Waste in the Salt River Valley A long-term out-of-balance situation that contrasts with the well-managed basal freshwater lens under Oahu is illustrated by the groundwater body enclosed in the alluvial basin of the Salt River Valley in Arizona.

High groundwater was a problem that early irrigators in this valley, the Hohokam, had to contend with to about the fourteenth century. It was probably the reason their occupation of the region ended. When irrigation from the river resumed in the late nineteenth century, and was accelerated in the first decade of the twentieth century by the federally built Roosevelt Dam (Skibitzke *et al.*, 1961), swamping promptly showed up again. This led to local pumping to lower the water table, a practice that rapidly spread.

This body of water, like others in the Southwest, is now in decline and has been since at least 1923. During the forty years from 1923 to 1964, its shrinkage in volume is indicated by a drop in the water table that totals, on the average over the whole alluvial plain, 45 m (Anderson, 1968).

The situation is well known to both the irrigators and other groups in the state; the management of water is "a matter of primary interest to the general public and the various interest groups" (Mann, 1963, p.

67). This knowledge, however, has had little apparent effect, for the shrinkage continues unabated. Recent values of the average drop in the water table are 3 m yr^{-1} (U.S. Water Resources Council, 1968, p. 6-14-4).

It is interesting to look for possible reasons for the way people perceive this situation. One reason is its remoteness: far out of sight. Moreover, "recent economic growth, in Arizona at least, has not been stimulated by irrigation and has not been impeded by water scarcity" (National Research Council, Committee on Water, 1968, p. 15), so the problem is also remote economically. A more detailed economic analysis found that much of the pumped water makes little contribution to either the agricultural or the overall economy of the region (Kelso *et al.*, 1973, p. 27). "The problem is a 'man-problem' rather than a 'nature-problem'."

The legacy that hampers clear sight of the problem is the view of groundwater in past times (although Arizona is one of our newest states), when court cases and territorial legislation were setting the framework for the present day. Groundwater was looked on as having potential value chiefly for domestic wells and for watering livestock on the range—the windmill and tank combination so vital in settling the West. "There appeared to be little necessity of vesting it with a public character similar to surface water" (Mann, 1963, p. 43). Institutions created at that time hardly fit the current situation in which ground water outshadows surface water.

Another problem lies in a myth about great underground rivers, which influenced what courts wrote into their judgments. This myth is expressed in most nineteenth century textbooks in physical geography, with vivid descriptions of caves and streams in limestone karst terrain. Widely believed in many places, this myth has nothing to do with groundwater in the Salt River Valley. This ignorance of the environment of groundwater carries over into a present reluctance to support research on the geological environment of the basins in which water is found (Mann, 1963, p. 237).

A prediction of continued decline in the volume of this body of groundwater has been made from an analog modelling study. Over the period 1964–1974, the model indicates an average decline of 16 m in the water table. Over the period 1964–1984, it predicts a 32-m average drop, with a shift in areal distribution (Anderson, 1968). Such a decline is, again, part of the local perception of the situation. While some of the recent developers of groundwater "were interested in a permanent livelihood on the desert, many were of the 'suitcase' variety, willing to make an investment for short-term profits with full knowledge that the resource eventually would play out" (Mann, 1963, p. 51).

MASS BUDGETS ASSOCIATED WITH GROUNDWATER

Water in its underground environment is affected by the temperature and chemical composition of the rocks. The progressive mineralization of meltwater with an initially diverse but small content of dissolved matter was studied in the Sierra Nevada by Feth *et al.* (1964), using measurements of chemical composition of water that had spent varying lengths of time in the soil and rock of this granite batholith (Table I). Their data illustrate how water charged with the bicarbonate ion from carbon dioxide in the soil air vigorously dissolves minerals from the underlying rocks.

Mineral content of the water increases seven times (to 36 ppm) as it passes through the soil and shallow layers of the ground water en route to ephemeral springs or seeps. It doubles again in water that passes through deeper rock layers before coming out in perennial springs (75 ppm).

Water in thermal springs, which here mostly originated as snow or rain, shows further mineralization, to 410 ppm. It acquires both heat and salts in its long journey from the soil through the rocks and probably into deep fault zones. Chlorides, sulfates, and nitrates, which display only an initial or terminal increase, do not seem to be derived from the rocks but from the atmosphere or deep fault-zone water. Decaying litter is important for nitrates.

Water long buried underground often becomes highly mineralized. That in the Great Artesian Basin of Queensland, for example, is too salty for irrigating crops—fortunately for the conservative management it requires.

Movement of nitrate in percolate water into bodies of groundwater was noted in the preceding chapter, and many other examples of manmade contamination of ground water by rural and urban systems could be cited. Removal of pollutants is often slow and sometimes impossible, and this useful space remains contaminated. A mismanaged ecosystem at the overlying surface has left a permanent blot below ground. Walker (1974) notes that some techniques for reducing air or water pollution produce residues that managers want to dump somewhere, and that "the 'somewhere' undoubtedly will be a landfill, seepage lagoon, spray irrigation plot or some other ground-water recharge–disposal system," and is concerned over the time-bomb

TABLE I

Dissolved Solids in Meltwater after Varying Periods of Contact with Granite Rocks of the Sierra Nevada, California[a]

	Snow	Snowmelt runoff in streams	Base flow in streams	Ephemeral springs	Perennial springs	Thermal perennial springs
No. of samples	77	34	34	15	56	8
pH	—	7.0	7.0	6.2	6.8	8.6
HCO_3	2.9	14	21	20	57	180
SiO_2	0.16	8.1	12	16	25	51
Cations (Alkali)						
Na	0.46	1.5	2.7	3.0	6.0	122
K	0.32	0.7	1.0	1.1	1.6	2.6
Divalent						
Ca	0.40	2.4	4.4	3.1	10	13
Mg	0.17	0.6	0.9	0.7	1.7	1.0
Anions						
SO_4	0.95	1.3	2.0	1.0	2.4	59
Cl	0.50	0.4	1.3	0.5	1.1	59
NO_3	0.07	0.1	0.2	0.02	0.3	0.2
Total dissolved solids	4.7	22	35	36	75	410

[a] Units: ppm. Source: Feth *et al.* (1964).

aspects of buried toxic chemicals. Groundwater is in motion, and leachate plumes from landfills on Long Island have been measured as long as 3 km (Kimmel and Braids, 1974).

To avoid such irreversible changes in a valuable resource, the budget of mass inflows, storages, and outflows is commonly modeled. These budgets, if used properly within their limits (Evenson *et al.*, 1974), can aid the understanding of the mass budget of water and help the land planners and managers. Even if special bodies of groundwater are dedicated to the function of servicing urban waste water and the solid-waste landfills, these bodies must be monitored. Out-of-sight, out-of-mind cannot be the guide! Rather, it will be imperative to watch the volumes of these special water bodies, their spread or migration, and the concentrations both of the substances carried into them by landfill leachate and wastewater applications and of new chemical compounds resulting from the mixture of a great many reactive substances in aqueous solution. Mass budgets of each ion as well as of the water itself need to be cast for each of these groundwater entities.

LOCAL OUTFLOWS OF WATER FROM UNDERGROUND STORAGE

Groundwater bodies generally extend over greater areas than typical ecosystems do, and tend, like the local air discussed in a preceding chapter, to be a "common" element shared by many ecosystems. As water moves within a groundwater body it traverses distances greater than the diameter of the ecosystem from which it initially percolated; the characteristics of regimen or dissolved salts impressed upon it by the originating ecosystem are carried beneath other ecosystems and eventually mixed into a composite typifying the whole groundwater body.

This is, of course, one of the problems with leachates from solid-waste disposal sites or pathogens percolating from suburban wastes. Outflows from groundwater carry a mix of influences from the numerous ecosystems from which the aquifer was recharged. They begin to represent a movement of water away from the generating ecosystem, an off-site flow similar to those to be discussed in following chapters.

Outflows should be seen within the contexts of the hydrogeologic environment and the groundwater regime. "The hydrogeologic environment is defined by characteristics of the topography, geology, and climate. The groundwater regime is described by parameters such as the amount of water, the pattern of groundwater flow, the rate of discharge, the chemical composition of groundwater, the temperature of groundwater, and the variations of these properties" (Tóth, 1971). Outflows therefore appear as "springs, seepages, quicksand, soap holes, geysers, frost mounds, pingos, groundwater lakes and marshes, and certain near-surface and surface accumulations of salts, landslides, slumps, soil creep and gullying. . ." (Tóth, 1971), all of which shape the ecosystems in their vicinities.

Outflows may be restricted locally, and forced to occur at a distance from the site where percolation fed the groundwater body, a topic we defer to Chapter XVII.

Outflows that occur locally from groundwater can be classified in terms of the environment into which the water is discharged. (a) At linear depressions, the emerging groundwater is concentrated in volume sufficient for its potential energy to power its flow into a river system and away from the local area altogether. This route taken by the water-yield from the locality is discussed in a later chapter. (b) Outflow in small trickles into high-energy environments allows the absorption of radiant energy by wet soil or wet-soil plants to power the evaporation of most or all of the water. These sites provide conditions favoring water flux from the groundwater body into the local air, quite a different departure route than that taken by ground water that feeds into rivers.

Outflows into Vaporizing Environments

Several outflows from groundwater bodies exist in nature, and others have been added by man. The natural ones often bear an element of mystery, since their water comes from an invisible source. They represent a local cycling of water from the local rock formations to the local landscape and its ecosystems, occurring on a small spatial scale.

Small Seeps and Springs Springs have long been regarded as magical, if not actually sacred, each with its own goddess, and there is a friendly air about even a seep of water that supports a little patch of green in a dry landscape or an outflow that keeps only a few meters of channel moist.

The seeps of water at the foot of a forested slope represent a delayed near-surface flow following a storm. If they are sustained after a storm or snowmelting period ends, they represent outflow from a shallow groundwater body. Wet-weather seeps are outflows from groundwater bodies that have a longer life during the season when heavy percolation keeps the water table high, and persist for a month or more after such percolation ceases. In the geological view, where less permeable rocks force water to the ground surface, the area (and its ecosystems) becomes a seepage area (as in Fig. XV-2). In a storm, such an area is quick to generate off-site flow (Chapter XVI).

These small outflows from different groundwater bodies are common in the basin of the Central Sierra Snow Laboratory, both from glacial deposits and from the contact zone where permeable volcanic layers overlie the basal granite. "The atmospheric waters easily penetrate the porous volcanic rocks and, collecting on the bed-rock surface below them, reappear at points along the contact" (Lindgren, 1897). In late summer, long after the store of soil moisture has been exhausted, these seeps continue to trickle water out to small meadows, strips of willows and shrubs down the dry hill slopes, and along the upper

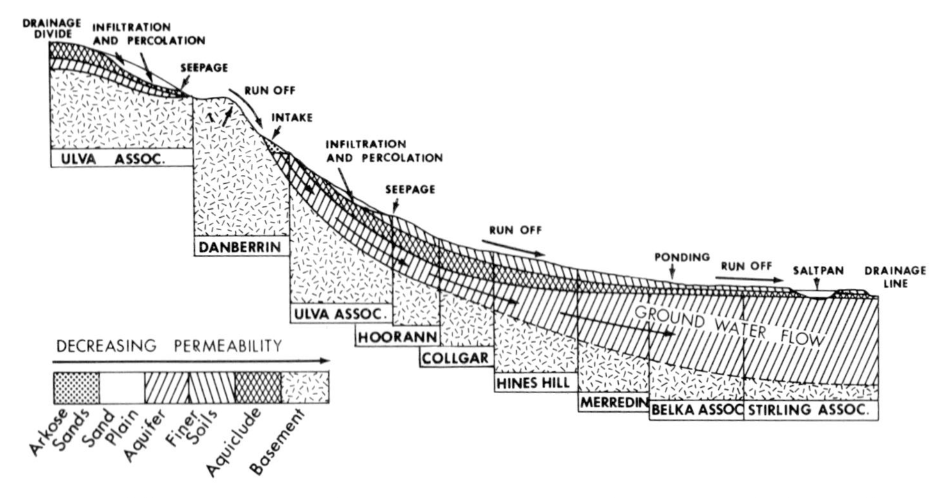

Fig. XV-2. Section several kilometers long across valley and upland of the Belka landscape of Western Australia, showing occurrence of groundwater and areas of intake, movement and discharge in the water–salt system. Vertical scale greatly exaggerated; this landscape is worn down and of extreme age (from Bettenay *et al.*, 1964).

reaches of phreatophyte-lined stream channels. The total volume of water, referred to the whole area of the basin, is perhaps 20 mm. By tapping these long-lived storages of water, some ecosystems in the basin continue transpiring late into summer and add substantially, as in the meadows grazed by sheep that have worked their way up the Sierra block from the Central Valley, to the total biological production of the area.

From deeper, more stable bodies of water come springs that flow all year with little change in volume or temperature. Their ability to carry the cold of winter over into summer makes them interesting and useful; we always tend to think of spring water as cold, and for years spring houses were built to use this coldness to preserve food. Outflows in springs and spring ponds form the habitat of distinctive ecosystems, in a steady thermal and chemical environment; some, like Silver Springs in Florida, are highly productive.

Hamlets in countrysides of agglomerated rural settlement and farmsteads in countrysides of dispersed settlement commonly located themselves at springs as reliable sources of water. Spring water is also biologically cleaner, as well as freer of sediment, than creek water.

> A flow would well up from the ground and water the whole surface of the earth.
>
> *Genesis* 2

Capillary Rise Less obvious than seeps are the areas in which the ground surface is near the water table. Capillary forces move water through small pores almost as easily upward as downward, and while the distances of movement are not great, they may be sufficient to bring water from a groundwater body into a zone where evaporation can occur.

If the ground is bare, capillary water evaporates readily from the wetted soil, sometimes leaving a memento of salts or alkali behind. If vegetation is present, its roots extract water from the capillary fringe and supply the transpiring leaves. In both situations a short vertical trip brings water from the ground-water supply into contact with the requisites for evaporation: energy supply plus atmospheric ventilation.

Field studies show that the rate of evapotranspiration based on capillary-fringe water depends on depth of the water table and the vegetation type, i.e., its rooting habit. The low parts of many tectonic basins in the western United States are occupied by distinctive ecosystems capable of exploiting this reliable source of water.

The height of the capillary rise above the freewater level is determined by pore sizes and other factors affecting surface tension. It can contribute moisture to the root zone and hence to the evapotranspiration flux if the water table is within about 0.7 m of the ground in fine sand, 1 m in loamy soil.

In a Swedish experiment (Fleetwood and Larsson, 1968), the isopleth of 0.45 soil moisture was found to follow the water table as it receded from a depth of 1.1–1.8 m. The 0.45 isopleth remained about 0.3 m above the water table. This capillary zone was exploited by one of the crops studied—rye—but not by the more shallow-rooted potatoes.

The figures differ with depth of rooting by different vegetation types. Alfalfa is known for its deep rooting and its ability to extract water from near the water table when the groundwater lies several meters below the ground surface.

Biological drafts upon groundwater occur in depressions in the terrain and support a considerable volume of evapotranspiration from vegetation cover. These favored places are apparent after a spell of dry weather when adjacent upland slopes have run out of soil moisture and their plants have dried up. When rains come these low areas, being already saturated, become sources of a quick flow of rainwater into the stream channels, as will be discussed in Chapter XVI.

The zone of capillary rise in sandy soils can be made to supply plant roots if it is held at a more or less constant level by controlling the level of the water table. Water-table level in such permeable soils can be sustained by running water *into* soil drain lines, from which it moves into the groundwater body and thence into the capillary fringe. This mode of water management is sometimes called "subirrigation." It was at one time peculiar to certain irrigated areas, but now appears to be feasible where tile lines have been put in place for purposes of drainage, e.g., in sandy, high-water-table soils of the Atlantic coastal plain (Kriz and Skaggs, 1973).

Phreatophytes Another outflow of groundwater is generated by plants whose roots can live in the saturated zone, or can survive fluctuations in the water table if they live deep in the capillary fringe. These plants move water almost directly from the groundwater body. Since groundwater is often most accessible in alluvial lands along streams, many riparian plants are phreatophytes that benefit from the steady and ample availability of water. These trees and shrubs form ecosystems that line streams and rivers in most parts of the world, living on the groundwater in the flood-plain alluvium and flourishing even when upland vegetation endures long rainless periods.

In climates with long periods of soil-moisture stress, gallery forests along the streams among dry, grassy uplands outline the drainage system with fidelity. In deserts, a soil that is dry most of the time coexists in the same landscape with a shallow body of riparian ground water exploited by trees or shrubs, a line of different ecosystems that make an attractive contrast in the landscape. (See Plate 17, in Chapter IX.)

Silting of river channels and reservoirs often provides an ideal environment for phreatophytes. Man's activities in the southwestern U.S. have resulted in great increases in the acreage covered by such vegetation as salt cedar.

Phreatophytes are not dependent on the zone of capillary rise, having roots that are accommodated to a low oxygen concentration in the aquifer environment itself. Such plants are able to exploit a source that is more generous and steady than is ordinary soil moisture, gaining a competitive advantage over upland plants that helps compensate the low-oxygen conditions their roots must live in. The vegetation of marshes and swamps, low parts of basins in the West, and low flood plain terraces and stream banks occupy thousands of square kilometers in the United States.

Since not all phreatophytes are useful economic plants, as is rice for example, sporadic campaigns to eradicate them as "water hogs" are mounted. In lands with a high water table, however, few plants can replace them, and a second look suggests that they do in fact create substantial benefits as habitat; cottonwood trees along mountain streams in Arizona and California are examples. Even the tamarisk is a productive ecosystem for much animal life (Horton, 1972). Ecosystems with heavy evapotranspiration (up to 1500 mm yr^{-1}) occupy only a small fraction of a flood plain; Horton and Campbell (1974, p. 7) estimate that the net result of attempts at conversion to different ecosystems would be less than 300 mm yr^{-1} saved.

Wetlands Fed by Groundwater If the land surface is pocked with basins, these often fill to become marshes or lakes that are generally connected with the groundwater body. Wetlands, for example, which "represent a surface exposure of the upper water table" (Southeastern Wisconsin Regional Planning Commission, 1970, p. 369), occupy 0.10 of the total basin of the Milwaukee River, lakes 0.01 more, and lands presently drained for farming 0.05, a total initial wetland area equal to 0.16 of the basin. Some of the existing wetlands are grass ecosystems, some shrub, and half are tree systems.

In some wetlands, or in some seasons of the year, water flows from the groundwater body into the wetland; in other cases, or at other seasons, the flow is from wetland unit into groundwater. Wetlands thus may act as discharge areas for the local groundwater system. If groundwater is an important resource the direction of flow is important. Sometimes close study is needed to determine which way it goes. Should wetlands be drained because they deplete the groundwater resource, or preserved because they supply it?

Several kinds of local outflow of groundwater are illustrated in Fig. XV-2, a section across an ancient, worn-down landscape in Western Australia, in which the groundwater has been studied because the deeper levels are very saline. Note the seepage areas at the foot of slopes, and where impervious basement rock or fine-textured soils protrude.

Capillary rise with evaporation from the soil surface or transpiration from salt-tolerant vegetation occurs in the valleys (Plate 30). Much valley land has been salted and swamped as a result of recent rise of the groundwater. Flow from groundwater into saline marshes (salt pans) occurs at the lowest part of the cross section. This salty groundwater was set in motion by increased percolation into it consequent on the replacement of forest by wheat on the uplands. In a nearby area, such clearing (Peck and Hurle, 1973) increased percola-

Plate 30. Former wheat land of the Belka land group (see Fig. XV-2) in Western Australia, destroyed by salting. Scarifications are attempts to aid regeneration of salt-tolerant shrubs as a replacement crop. Original woody vegetation that escaped being bulldozed for wheat agriculture is shown in the background.

tion by 45 mm yr^{-1}, a large increment to the 3–5 mm yr^{-1} prior to clearing, when most of the rainfall was held until transpired in the soil and therefore did not enter the saline groundwater or start it moving.

Mass Fluxes Associated with Local Movement of Groundwater

It is not uncommon for salts to be concentrated in the shallow groundwater even in terrain where the moisture in the root zone of the soil is relatively free of mineral content. They are then carried by it to one or another area of outflow where further concentration may take place.

Salts carried in capillary water are left in the soil when the plants extract the fresh water they transpire. In humid lands where upward capillary flow occurs only in the dry season, the salt flux is reversed during the season of more abundant soil water; then percolation flushes the salts downward and no long-term accumulation of salts builds up in the soil.

The salt flux in deserts, however, is directed upward most of the time, whether it derives from irrigation water or a high water table.

Eventually, often sooner rather than later, salts accumulate near the soil surface in such degree as to make life impossible for many crop plants. Irrigation districts in the Murray Valley of Australia began to experience salt problems when percolating cheap irrigation water brought the top of the groundwater body within 1.2 m of the land surface. "No forms of surface drainage or of soil treatment are then effective until the water table is lowered" (Pereira, 1973, p. 188).

"In arid and semi-arid areas, excessive salinity in the root layers of the soil is almost always the potential corollary to irrigation" (Ward, 1967, p. 232). Failure to consider this hydrologic fact of life has wrecked the fiscal status of many a hopefully undertaken irrigation project that was to make the desert bloom.

The concommitants of swamping and salting plague even the most expensive irrigation systems. They have to be corrected at great cost, or the project abandoned. Projects in middle Asia, in places where irrigation is centuries old, have suffered these ills, both on the Soviet (Gerasimov, 1968) and Chinese side (Wiens, 1967).

The still more ancient lands of Mesopotamia (Jacobsen and Adams, 1958) had problems of this kind long ago. The Indus and Nile Valleys are having them now. One might wonder if it is inevitable for every project to carry the seeds of its own destruction. A Texas writer speaks of dry-country irrigation in west Texas as an "old technology" with functional problems. These are inherent because its processes "fight against the way that nature has set things up" (Graves, 1971, p. 94), i.e., the way its ecosystems work, in comparison with, for example, crop ecosystems in the valleys of humid east Texas or marsh ecosystems in its fabulously productive estuaries. As in the Salt Valley irrigation area, west Texas is mining the only source of water—the Ogalalla formation—that is cheap enough (at present) to be afforded by the low-value crop systems on which it is being lavished.

An Australian soil scientist sums up the situation thus: "Under irrigation, soluble salt is a world-wide problem; there is usually a store in the soils of dry climates, and under the new wetter soil climate this salt is bound to move and may well come to the surface in low-lying or badly drained areas" (Leeper, 1970).

Minerals brought by the atmosphere are important in some Australian landscapes. Sea salt brought down in rainfall tends to accumulate in the ground water, often in great quantities.

A sudden increase in percolation resulting from lower evapotranspiration after woody, deep-rooted ecosystems are converted to shallow-rooted, short-lived cereal crops, upsets the local water budgets. This starts the water with its solutes in motion, and redistributes it in downslope areas. In Leeper's words again, "after the timber, with its year-round transpiration, has been cleared, the soil profiles become wetter, and salt water moves down slopes, eventually forming a water-table at the foot of the slope within capillary reach of the surface" (Leeper, 1970).

Vegetation in problem areas in Victoria (Downes, 1961) was converted from native forest and scrub to grazed perennial grasses. Under grazing pressure these changed to shallow-rooted annual grasses, and on many slopes to sparse over-grazed plant cover. The resulting decrease in on-site evaporation brought a corresponding increase in percolation, and movement of underground water and dissolved salt.

In the tablelands of New South Wales, Dijk (1969) identified different sources of salt deposits that appeared at the soil surface and eliminated vegetation over considerable areas. In no case was the source in the soil itself, which contained less than 100 ppm salt, but rather in different bodies of groundwater in the local terrain.

An early salt movement came from groundwater in the floors of the basins, a deposit of great age, yielding a "continued supply of saline water." Later, sporadic movements of salt issued in intermittent springs from bodies of groundwater perched on localized impermeable layers in the rock. Heavy rains wash the salt from these deposits to the surface, where it effloresces in white patches in the succeeding dry weather. Salt movement thus results from variations in rainfall and groundwater balance and outflows. These disturbances may be set in motion by any reduction in transpiration: greater areas of summer fallow, as in eastern Montana (Halvorson and Black, 1974) where saline seeps cover 300 km^2 and are increasing 10% annually; by changes in grazing cover; or by conversion of deeprooted shrub and forest ecosystems to grass in the Victorian and New South Wales cases discussed, or to wheat in Western Australia.

ARTIFICIAL OUTFLOWS FROM UNDERGROUND STORAGE

Shallow Groundwater in Cultural Landscapes

Small-scale outflows from groundwater bodies sometimes occur as a consequence of man's efforts to produce properly aerated soils for crop plants. This requires draining the big pores in the soil by giving gravity a little room to work in.

If shallow groundwater is so near the ground surface that it restricts the depth of root zone for plants not tolerating free water in the large soil pores, it presents man with a problem. He either can find economic use for phreatophytes or take measures to lower the water table so as to replace them with ecosystems of his own choosing. Lowering the groundwater body amounts to increasing the thickness of the soil layer that contains oxygen in its large pores and the water in its small ones, hence providing good environment for roots.

Drainage in Traditional Agriculture Deep furrows let free water drain out of ridges on which crop plants grow, a common practice in plow agriculture. In hoe agriculture it is a common practice to build up plant beds between field ditches in order to provide a deeper root zone above the saturated layer. The complex farming practices of the New Guinea highlands includes careful means of controlling water in the soil. Vast areas of the New World were shaped into raised fields (Denevan, 1970), often in seasonally inundated savanna regions, which were densely populated at the time. A broad range of raised fields is also found all through the Old World (Denevan and Turner, 1974).

In some of these complexes the ditches held water permanently and produced fish and other sources of protein. Often they also provide aquatic vegetation and mud that fertilize as well as elevate the built-up crop beds. Such lands in central Mexican basins are still highly productive (Wilken, 1969) after centuries of intensive cultivation; "techniques for handling saturated soils provide the key to successful land management." (See Plate 31.)

Drainage in Commercial Agriculture The practice of making the vital root–air–water layer thicker is of great antiquity, as we have seen. It increased further in the English agricultural revolution of the eighteenth century, when many methods of drainage were invented and experimented with. The increased application of mechanical power to ditching and tiling operations made possible further expansion up to the present time.

Great expanses of North America (e.g., the lake plains of northwestern Ohio and the glacial swamps of Wisconsin*) and of Europe are drained by networks of open ditches or tile drains (Plate 32). Tile drainage is invisible in the landscape, and this form of water management is not as conspicuous as is irrigation, yet the net results in the United States in terms of crop productivity are at least as important.

These networks ensure that the water table remains far enough

* Of the agricultural lands near the south end of Lake Michigan, 0.16 still have drainage problems. In contrast to this evidence of groundwater influence, only 0.04 are affected by surface-water flooding.

Plate 31. Farmer dredging aquatic vegetation from canal in chinampa near Puebla, Mexico, for use on plots behind him. The canals draw down the level of the groundwater table and assure an aerated soil body of adequate rooting depth. The border trees and shrubs maintain the land–water separation and probably also provide microclimatic benefits to the high-value crops.

below the ground surface, particularly in spring, that plants can establish and strongly develop their roots. Once the plant mechanisms that extract and transpire water are established and in operation they can usually keep soil water below the saturation point except in the heaviest rains.

Capacity of the drains has to match the expected inflow of water into the shallow groundwater body, depending on the change in storage, i.e., rise of water table, that can be tolerated. For "fair drainage" of soil in the north central United States, an outflow of 15 mm day^{-1} is planned; if "excellent drainage" is the objective, ditches must accept an outflow of 35 mm day^{-1} (Ogrosky and Mockus, 1964). These numbers express water budgets of the soil's big pores in a region where the 0.5-frequency rainfall intensity is 50 mm in 6 hr, 75 mm in a day.

Managing the groundwater level is also frequently used in wetland areas to control vegetation, flooding out certain ecosystems and encouraging those desired for wildlife habitat. Many sites are now known to be biologically more productive as marsh than as cropland, especially if they are located strategically for wildfowl mating or nesting. The biomass produced by many marshland ecosystems is as large as that of crops, and such systems serve as cover as well as food source; draining such lands might be counterproductive in terms of their gross value to society (see Plate 33).

In groundwater management, as in many other contacts of man with nature, it is necessary to determine and count all of the costs of a given course of action. Production of wildfowl, trees, meadow, or field crops are competing uses of the land. They require different management practices for shallow groundwater and exist in different institutional settings in our economy.

An example of recycling water within the same irrigated property, regarded as a managed system, is called a "return water system." It consists of a drain or a sump in the lowest part of the field, and a sump pump and pipeline that returns excess water into irrigation ditches or sprinklers. Looked at in one way, this is a means of withdrawing water from the big pores of the ground below the root zone and giving it a chance to be taken up into the small pores of the root zone. In another way, it moves water to a higher elevation and thus is a reverse kind of percolation flux. The solute content of water thus recycled tends to increase, although some of the costs imposed by higher salt content are offset by the nutrient value of the recirculated nitrate. This cycle within a single land holding resembles the situation

Plate 32. Lines of drainage tile ready to be laid to accelerate groundwater outflow from initially wet-prairie ecosystems in southwestern Minnesota (early June 1973).

Plate 33. Horicon Marsh, in southeastern Wisconsin, which, after a checkered history of other attempted uses, is now again a wildfowl wetland system. Among other birds, it provides several months of habitat for Canada geese in their hundreds of thousands, and bird-watching for visitors in like numbers (October 1970).

described in Chapter XIV of the suburban lot with its own well and septic outflow. Here, however, the land owner tries to strengthen the local cycling of water, not to break it.

Outflow of Groundwater into Wells

Another form of groundwater management involves more than using gravity; rather, it requires working against it by pumping water out of wells. When a hole is punched into a groundwater body and water pumped out of it, a slope is set up in the surrounding water table and water begins to flow into the hole. A well works by creating a cone of depression in the water table (or pressure surface) that will force water to move by gravity flow into a place from which it can be lifted to the ground surface. The flow into this well hole depends, of course, on features of the local geological environment that affect the transmission of water.

Because well water is generally applied to raise soil moisture in the overlying ecosystems, it reverses the percolation flux described in Chapter XIV. This integration of aquifer and ecosystem is most evident where the water table is near the surface. A manual form of irrigation still practiced in some valleys in Mexico ties the groundwater a few meters below a field with the crop ecosystem by human energy that lifts water 6–7 m and carries it 7–8 m to basins around the plants (Wilken, 1975). Fields are small, since the rate of application is only about 1 ton hr^{-1}, but the aquifer–ecosystem association is very close.

Withdrawal of water from underground storage by pumping is now a large outflow. Over the whole country it may represent 0.1–0.2 of the total outflow of water from groundwater bodies, or 10–20 kg m^{-2} of the total area each year. In local areas it is by far the dominant outflow, depending on the degree to which the groundwater body has changed, in man's perception of it, into a groundwater resource that can be most easily exploited by pumping.

Pumping from wells sometimes intercepts water that would otherwise be discharged in seeps or low areas from which it would evaporate, and produces little net change in volume of outflow from the given groundwater body. Sometimes it intercepts water that otherwise would flow out of the groundwater body into channels, thereby reducing off-site yield of water in rivers. In these areas, surface and groundwater have to be analyzed as one system to understand the way the exploitation of one outflow will affect the regime of the others.

In cases where the intent is to keep outflows in balance with recharge by percolation, an upper limit is set to pumping. Kayane (1971) suggests that annual percolation to groundwater in alluvial lands of Asia is unlikely to exceed 200 mm (cf. Coshocton, 130 mm). In climates where the deficit between potential evapotranspiration and actual evapotranspiration is larger than 200 mm, he feels that "groundwater cannot be considered a renewable resource for offsetting the annual water deficit."

Costs of Pumping Groundwater One advantage of wells is their place value. They often can be drilled just where the water is to be used, and so minimize subsequent transport costs. Because of the local nature of water bodies, the area influenced by pumping from wells may be fairly small, but concentrated. Locally heavy draft on some groundwater bodies with many wells means that the individual cones of depression tend to coalesce. This widespread lowering of the water table then generates a gravity-powered flow from outer areas of the aquifer, which can allow intrusion of water of low quality from other aquifers or from the ocean. Lowering the water level also increases the energy and financial costs of pumping.

Water is heavy for its value. Often the rate of pumping is limited by the sheer energy cost of overcoming gravity, even from shallow ground water. Clark (1967, p. 55) gives some monetary costs for different ways of lifting water (cents per cubic meter) as shown in the accompanying tabulation. "Primitive methods of raising water from wells by man and ox power prove to be barely remunerative even at high Indian prices for grain, at low Indian wages, and with the water table near the surface" (Clark, 1967, p. 94). The influence of cheap energy from fossil fuels is clear from the accompanying figures. As energy rises in price, effects can be expected on the amount of groundwater pumpage that is practical.

Method	Energy source	Capital cost	Operating cost	Total
Well sweep	Manpower	0.4	2.1	2.5
Persian wheel	Man and oxen	0.2	3.3	3.5
Small power pump	Fossil fuel	0.4	0.4	0.8

In discussing techniques that extend the range of recovery of water (Ackerman and Löf, 1959, p. 269), three of the four cases described were taken from groundwater exploitation. These are new techniques in drilling wells, in stimulating the flow of water into them, and in pump design. Each has had a major effect on pumped outflow.

Pumping as a new medium of outflow from a groundwater body takes its place in the system at the expense of former media of outflow or at the expense of stored water. In the Oahu and the Salt River Valley cases, discussed under the section on groundwater volume, rates of pumping over the entire areas (the island of Oahu, 1500 km^2; the Salt River Valley 9000 km^2) average about 400 mm yr^{-1} ("a very high rate" McGuinness, 1963, p. 290). At a typical specific-yield fraction of 0.2, this means the dewatering of 2 m of aquifer space per year, i.e., a 2-m drop in the water table if there is no recharge.

Oahu Pumped water in Oahu goes to irrigate sugar cane, at an application of 2000–3000 mm yr^{-1} (U.S. Water Resources Council, 1968), and to supply most of the urban use of water, which gets about a third. Future water supplies for widening urban land use will have to come from the basal aquifer; "the fiscal capacity and operational efficiency of the Honolulu Board of Water Supply provide reasonable assurance that emerging public water problems on Oahu will generally be solved" (U.S. Water Resources Council, 1968 p. 6-19-8).

All outflows from this basal lens are measured or estimated as precisely as possible. Outflows into the short streams are gaged;

leakage into the ocean and the mixing of fresh and salt water at the lower interface, especially under the influence of tidal action, has been studied for decades. Research elsewhere on this lens type of aquifer (it was first identified and studied in the sand dunes on the coast of the North Sea) has been transferred where possible to the Oahu case, with modifications to incorporate newer findings on the movement of water through the ground.

Good records of pumping have been kept for a long time and correlated with changes in the water table, which are closely watched. Pumps are immediately shut down in areas that begin to look touchy. "Leaky wells are controlled, pumping limited to amount of recharge, and locations chosen carefully" (American Society of Civil Engineers, 1969). The quality of the water is also carefully monitored.

Salt River Valley In contrast with this case of a groundwater body that is managed well by a self-reliant region and that will assure the region's prosperity, we turn to a different kind of management. In this example a body of ground water is being used up for rather small current gains to low-value crops.

Pump irrigation, once started as a mode of drainage, continued as a mode of water supply and soon came to overshadow the ditch-water system. Since the early 1920s, pumping outflow has continuously exceeded the rate at which the groundwater body is replenished (Anderson, 1968).

New developers continued to enter the state, extending agriculture beyond the limits of the total water supply. The availability of new sources of cheap energy in the state, of which a third is used to run pumps, has also accelerated the extension of pumping since the 1940s. "Overdraft was already substantial in large areas by 1948" (McGuinness, 1963), with the declines in ground-water storage noted earlier in this chapter.

The history of these events is reflected in the elevation of the water table shown in Fig. XV-3 (Harshbarger *et al.*, 1966, p. 34). It portrays the rise produced by percolation from surface-water irrigation beginning about the turn of the century, the slow decrease during the 1920s and 1930s, and the precipitous fall after big pumps were brought in by land developers after 1945. Comparison with the curve of pumping rate indicates that leveling off of the rate of pumping about 1953 did not stop the shrinkage of the groundwater body, since withdrawals continued in very large amounts.

The increase in pump lift has, to some degree, been offset by lower costs of lift per meter, so that the cost of applying a cubic meter of

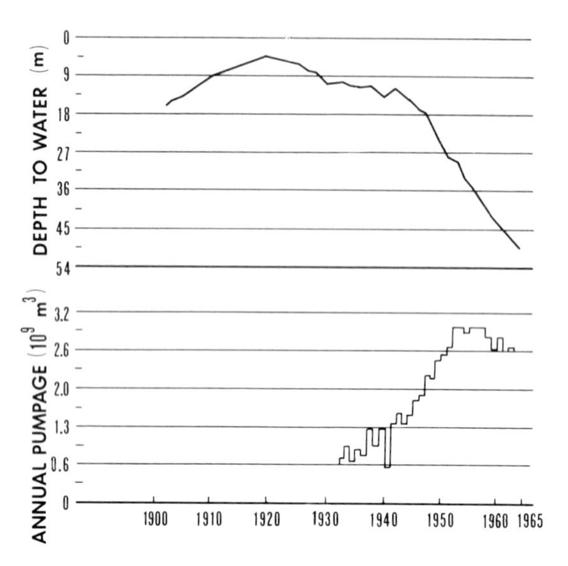

Fig. XV-3. History of pumpage of groundwater from the Salt River Valley and changes in depth to the water table (Harshbarger *et al.*, 1966).

water to cropland fell from 1930 to 1950, and since then (to 1967), remained about constant at 0.7¢ m^{-3} (Martin and Archer, 1971). "The current Arizona code restricts expansion of pumping facilities in critical areas, but does not limit the pumping rate as long as the owner's actions conform to the principle of reasonable use. Heavy overpumping persists in the critical areas" (U.S. Water Resources Council, 1968, p. 6-14-7).

Only 5% of the pumped water goes to livestock, people, and industry together; 95% is applied to cotton, alfalfa, sorghum, and a few minor specialty crops. The volume of water applied to low-value forage and feed crops "almost equals the estimated overdraft of groundwater" (National Research Council, Committee on Water, 1968, p. 73), but gives very low returns—about 1¢ personal income ton$-^{1}$ of water. In contrast, the smaller amounts of water applied to high-value crops gives a personal income of 5¢ ton^{-1}; water applied to mining and utilities sectors of the economy brings in about \$2 ton^{-1}, and that applied to manufacturing about \$60 ton^{-1}.

One reason for the excessive pumping of the groundwater might be its use in low-value applications. "Although agriculture uses 92 percent of all water consumed in Arizona, it directly contributes about 7 percent of the personal income in the state" (National Research

Council, Committee on Water, 1968 p. 72). The contrast would be greater if we had figures for the Phoenix basin alone.

One investigation suggests that "the best cure for a threatening water shortage is not necessarily more water; savings in water use, or transfer of water use to less-consumptive, higher-yield applications, or discovery of new techniques of water management may offer better solutions" (National Research Council, Committee on Water, 1968, p.100).

Such transfers of water to higher uses are now taking place in many areas of the West, and outflows from groundwater bodies by pumping are serving a higher economic purpose. "In Arizona . . . although groundwater withdrawals far exceed the annual recharge of aquifers, the 1948 groundwater code limits only the number of new wells drilled in certain 'critical areas,' not the quantity of water pumped from existing wells. . .(National Research Council, Committee on Water, 1968, p. 18).

Mann (1963) gives an account of the deadlocks in the state's political institutions, which might have been expected to take action to reduce the rate of withdrawal of water from this declining resource. "The decisions that should have been made in the 1940's in regard to ground water still had not been made in the early 1960's" (Mann, 1963, p. 256). This situation of water mining is similar (Graves, 1971) to the exhausting of the Ogalalla aquifer in the High Plains of western Texas.

GROUNDWATER BUDGETS

In this chapter we have looked at the underground environment of water, the inflow of percolating water into it, and some outflows from it. The budget idea is obvious for the groundwater reservoir. We can visualize the inflow by deep percolation during rainy seasons of the year, and have all seen outflows in springs and pumped wells. Visualizing also the closed nature of the system, with other modes of entry and exit of water being limited or absent, it is easy to think in budgeting terms. This is particularly relevant when we are considering cyclic rather than inherited groundwater, to use Nace's distinction.*

* Bodies of inherited water marked for mining or other forms of exploitation out of equilibrium with the rate of recharge are analyzed by more complex models which incorporate economics of pumping, alternate water sources, and alternate uses of water. In these cases, the pipeline analogy or more less equally balanced input and output is inappropriate (Nace, 1969).

Indeed the budget is the usual concept applied when man starts to develop or manage a body of groundwater, although difficulties are found in procuring accurate quantities for the inflows and outflows. Some bodies of groundwater also are difficult to delimit. Most groundwater bodies, being constantly in a state of changing equilibrium as inflows and outflows fluctuate, present a complex problem in evaluating how they respond to some change in conditions.

Nevertheless, a simplified restatement of the budget is useful. Deep percolation of water, when in excess in the soil, or from wetted stream channels or sites of artificial recharge, usually enters the groundwater body only during certain seasons of the year. Day-to-day changes in its rate are in general unknown. It usually carries freight in the form of solutes dissolved from the soil, salts or fertilizers, or urban wastes.

The consequent characteristics of the body of cyclic groundwater fed by percolate are thus its chemical or biological quality, its changing volume, and its temperature. These properties can be monitored more easily than the rates of inflow, and from these records we can deduce the health of the groundwater body.

One interest in the state of shallow bodies of groundwater lies in the level of the water table, whether it is too close to cultivated land surfaces or whether it is at the right level for wetland plants and habitat. Our other concern with the health of a groundwater body lies with its outflows: will springs dry up, will trout streams dwindle, is well water safe, will pumps have to be reset deeper in the wells, are we pumping too much fresh water and inviting the invasion of salt water? Some of these characteristics come under the heading of off-site yield of water, and are taken up in Chapter XVII.

REFERENCES

Ackerman, E. A., and Löf, G. O. G. (1959). "Technology in American Water Development." Johns Hopkins Press for Resources for the Future, Baltimore, Maryland, 710 pp.

American Society of Civil Engineers. Task Committee on Saltwater Intrusion (1969). Saltwater intrusion in the United States, *Am. Soc. Civil Eng. J. Hydrol. Div.* **95(Hy-5),** 1651–1669.

Anderson, T. S. (1968). Electrical Analog Analysis of Ground-Water Depletion in Central Arizona. U.S. Geol. Surv. Water-Supply Paper 1860, 21 pp. + pl.

Bettenay, E., Blackmore, A. V., and Hingston, F. J. (1964). Aspects of the hydrologic cycle and related salinity in the Belka Valley, Western Australia, *Austral. J. Soil Res.* **2,** 187–210.

Chapman, T. G. (1964). Effects of ground-water storage and flow on the water flow. *In* "Water Resources Use and Management." pp. 290–301. Melbourne Univ. Press, Melbourne, Australia.

Clark, C. (1967). "The Economics of Irrigation." Pergamon, Oxford, 116 pp.
Denevan, W. M. (1970). Aboriginal drained field cultivation in the Americas, *Science* **169,** 647–654.
Denevan, W. M., and Turner, B. L. II (1974). Forms, functions and associations of raised fields in the Old World tropics, *J. Trop. Geog.* **39,** 24–33.
Dijk, D. C. van (1969). Relict salt, a major cause of recent land damage in the Yass Valley, southern tablelands, N.S.W., *Austral. Geog.* **11,** 13–21.
Downes, R. G. (1961). Soil salinity in non-irrigated arable and pastoral land as a result of unbalance of the hydrologic cycle, *Arid Zone Res.* **14,** 105–110.
Evenson, D. R., Orlob, G. T., and Lyons, T. C. (1974). Ground-water quality models: What they can and cannot do, *Ground Water* **12,** 97–101.
Feth, J. H., Roberson, C. E., and Polzer, W. L. (1964). Sources of Mineral Constituents in Water from Granitic Rocks, Sierra Nevada, California and Nevada. U.S. Geol. Surv. Water-Supply Paper 1535-I, Washington D.C., 70 pp.
Fleetwood, A., and Larsson, I. (1968). Precipitation, soil moisture content and ground water storage in a sandy soil in southern Sweden, *Oikos* **19,** 234–241.
Gerasimov, I. P. (1968). Basic problems of the transformation of nature in central Asia, *Sov. Geog.* **9,** 444–458.
Graves, J. (1971). Texas: "You ain't seen nothing yet." *In* "The Water Hustlers" (R. H. Boyle, J. Graves, and T. H. Watkins, eds.), pp. 15–129. Sierra Club, San Francisco.
Gustafsson, Y. (1968). The influence of topography on ground water formation. *In* "Ground Water Problems" (E. Eriksson *et al.*, eds.), pp. 3–21. Pergamon, Oxford.
Halvorson, A. D., and Black, A. L. (1974). Saline-seep development in dryland soils of northeastern Montana, *J. Soil Water Conserv.* **29,** 77–81.
Harshbarger, J. W., Lewis, D. D., Skibitzke, H. E., Heckler, W. L., Kister, L. R., and Baldwin, H. L. (1966). Arizona Water. U.S. Geol. Surv. Water-Supply Paper 1648, 85 pp.
Horton, J. S. (1972). Management problems in phreatophyte and riparian zones, *J. Soil Water Cons.* **27**(No. 2)**,** 57–61.
Horton, J. S., and Campbell, C. J. (1974). Management of phreatophyte and riparian vegetation for maximum multiple use values, U.S. Forest Serv. Res. Paper RM-117, 23 pp.
Jacobson, T., and Adams, R. M. (1958). Salt and silt in ancient Mesopotamian agriculture, *Science* **128,** 1251–1258.
Kayane, I. (1971). Hydrological regions in monsoon Asia. *In* "Water Balance of Monsoon Asia" (M. M. Yōshino, ed.), pp. 287–300. Univ. Tokyo Press, Tokyo.
Kelso, M. M., Martin, W. E., and Mack, L. E. (1973). "Water Supplies and Economic Growth in an Arid Environment: An Arizona Case Study." Univ. Arizona Press, Tucson, 327 pp.
Kimmel, G. E., and Braids, O. C. (1974). Leachate plumes in a highly permeable aquifer, *Ground Water* **12,** 388–392.
Kriz, G. J., and Skaggs, R. W. (1973). Water management using subsurface drains, *J. Soil Water Cons.* **28,** 216–218.
Leeper, G. W. (1970). Soils. *In* "The Australian Environment" (G. W. Leeper, ed.), 4th ed., pp. 21–31. CSIRO and Melbourne Univ. Press, Melbourne, Australia 163 pp.
Lindgren, W. (1897). Geological Atlas of the United States, Truckee Folio. U.S. Geol. Surv. Folio 39.
Mann, D. E. (1963). "The Politics of Water in Arizona." Univ. Arizona Press, Tucson, 317 pp.

Martin, W. E., and Archer, T. (1971). Cost of pumping irrigation water in Arizona: 1891 to 1967, *Water Resources Res.* **7,** 23–31.

McGuinness, C. L. (1963). The Role of Ground Water in the National Water Situation. U.S. Geol. Surv. Water-Supply Paper 1800, 1120 pp. + pl.

McMahon, T. (1964). Hydrologic Features of the Hunter Valley, N.S.W. Hunter Valley Research Foundation Monogr. 20, Newcastle, 158 pp.

Nace, R. L. (1969). Human use of ground water. *In* "Water, Earth, and Man" (R. J. Chorley, ed.), pp. 285–294 Methuen, London.

National Research Council, Committee on Water. G. F. White, chairman (1968). Water and Choice in the Colorado Basin; An Example of Alternatives in Water Management. U.S. Nat. Acad. Sci. Publ. 1689, 107 pp.

Ogrosky, H. O., and Mockus, V. (1964). Hydrology of agricultural lands. *In* "Handbook of Applied Hydrology" (V. T. Chow, ed.), Chapter 21. McGraw-Hill, New York, 97 pp.

Peck, A. J., and Hurle, D. H. (1973). Chloride balance of some farmed and forested catchments in southwestern Australia, *Water Resources Res.* **9,** 648–657.

Pereira, H. C. (1973). "Land Use and Water Resources in Temperate and Tropical Climates." Cambridge Univ. Press, London and New York, 246 pp.

Peters, H. J. (1972). Criteria for groundwater level data networks for hydrologic and modeling purposes, *Water Resources Res.* **8,** 194–200.

Schoeller, H. (1959). Arid Zone Hydrology; Recent Developments. Arid Zone Res. No. 12, Unesco, Paris, 125 pp.

Skibitzke, H. E., Bennett, R. R., da Costa, J. A., Lewis, D. D., and Maddock, T. Jr. (1961). Symposium on history of development of water supply in an arid area in southwestern United States Salt-River Valley, Arizona, *Internat. Asso. Sci. Hydrol. Publ. 57*, 706–742.

Southeastern Wisconsin Regional Planning Commission (1970). A Comprehensive Plan for the Milwaukee River Watershed., Vol. I. Inventory Findings and Forecasts. SEWRPC, Waukesha, Wisconsin, 514 pp.

Stallman, R. W. (1964). Multiphase Fluids in Porous Media—A Review of Theories Pertinent to Hydrologic Studies. U.S. Geol. Surv. Prof. Paper 411-E, 51 pp.

Thomas, H. E., and Leopold, L. B. (1964). Ground water in North America, *Science* **143,** 1001–1006.

Todd, D. K. (1959). "Ground Water Hydrology." Wiley, New York, 336 pp.

Tóth, J. (1971). Groundwater discharge: a common generator of diverse geologic and morphologic phenomena, *Int. Assoc. Hydrol. Sci. Bull.* **16,** 7–24.

United States Water Resources Council (1968). The Nation's Water Resources. The First National Assessment of the Water Resources Council. Washington, D.C., Parts 1–7 in 1 vol.

Visher, F. N., and Mink, J. F. (1964). Groundwater Resources in Southern Oahu, Hawaii. U.S. Geol. Surv. Water-Supply Paper 1778, 133 pp.

Walker, W. H. (1974). Our buried treasure—May it rest in peace, *Ground Water* **12,** 262–264.

Ward, R. C. (1967). "Principles of Hydrology." McGraw-Hill, New York, 403 pp.

Wentworth, C. K. (1947). Progress in the estimating of ground-water supplies in Hawaii, *Trans. Am. Geophys. Un.* **28,** 266–268.

Wiens, H. (1967). Regional and seasonal water supply in the Tarim Basin and its relation to cultivated land potentials, *Ann. Assoc. Am. Geog.* **57,** 350–366.

Wilken, G. C. (1969). Drained-field agriculture: an intensive farming system in Tlaxcala, Mexico, *Geog. Rev.* **59,** 215–241.

Wilken, G. C. (1975). Personal communication.

Chapter XVI

SURFACE TRANSPORTS FROM ECOSYSTEMS

In preceding chapters we have examined the on-site processes involving water: the delivery and impaction of raindrops and snow-flakes on the foliage of an ecosystem, the mostly downward progress of water from foliage to litter, and from litter to and into the soil. Some lateral movement occurs in the soil but the dominant line remains vertical, as water is pulled up through plant stems to be transpired or percolates downward into the ground water.

Water also moves horizontally, away from the ecosystem to which the atmospheric systems delivered it. Up to this point these horizontal movements have been mentioned only briefly; we now will look at them more closely.

All these off-site movements can be called "yields." We speak of the water yield of a site, including off-site movement, whether at or near the land surface or deep in the ground. The term "sediment yield" is in common use; and we also can think of yields of nutrients or other solutes deriving from an ecosystem, as well as biological products. All yields are ecosystem outputs. To the degree they are associated with the water fluxes, we will consider them in this chapter and the next.

MOVEMENT OF SNOW

Drifting

In unforested country, whether plains or mountains, strong winds reach the ground and can detach snowflakes that did not become cemented at once to snow already on the ground. The detached particles are carried away from their initial lighting place, often in great volume in blizzards that may continue for days after the end of actual snowfall. This kind of off-site transport is large from exposed lands, e.g., large wheat fields or pastures on the plains and above timberline in the mountains, where no woody vegetation sticks up high enough to brake the wind speed.

Maps of rates of snow drifting in the U.S.S.R. (Mikhel' and Rudneva, 1971) show that at the 0.05 level of probability, the amounts during winter along the coasts of the Arctic Ocean are as great as 1000 m^3 per running meter.* Inland, winter transport is about half as large except on uplands, "where the wind speed is higher, the frequency, duration and intensity of blowing snow is greater." In forested regions, drifting is as little as 100 m^3 per winter.

Measurements on the North American Great Plains showed that a shelterbelt received about 50 kg of snow in a winter per square meter of belt area (Sander, 1970). This extra water income is useful in summer when transpiration from the trees is increased by heat advection from the adjoining fields (the "clothesline" effect) by about 35 kg m^{-2}. In dry regions such winter transport of snow and subsequent buildup of soil moisture within tree belts can be vital to the survival of the trees. The same mesoscale transfer of water operates in the parkland belt of the Prairie Provinces of Canada. "Trees and shrubs improve their sites for future growth, by concentrating snowmelt supplies" (Laycock, 1972).

Loss of snow from dry-farmed lands where meltwater makes up a critical part of the yearly recharge of the soil-moisture reservoir needs to be prevented. Holding the snow on the fields gives the plants a cushion of safety if summer rains turn out poorly. Benching and stubble treatments of land in the northern Great Plains, intended to hold rainwater for infiltration, also hold snow on the fields.

Soviet land-management methods in similar environments include the use of grain stubble, rows of plant stalks (e.g., corn or sunflowers), shelterbelts, and other devices to hold snow on the site where the atmosphere delivered it. The drifting of snow is clearly an important mode of off-site movement of water; L'vovich (1973, p. 174) estimates the loss in Soviet forest-steppe and steppe zones as 16 km^3 yr^{-1}, and advocates cutting it in half by shelterbelts and other measures. The 8 km^3 so gained would be a substantial part of the additional water this region needs to increase its agricultural yield.

* Since we are more used to visualizing the movement of water substance in channels, it might be helpful to think of a small drainage basin, say of 10-km^2 area. A flow rate of 0.2 m^3 sec^{-1} in the outlet stream represents a yearly movement of 620 kg of water m^{-2} of basin area (or 620-mm depth). If this mass were transported out of the basin across a 5-km-long side, the rate of flow would be 1250 m^3 m^{-1} of border length, a number that can be compared with the values cited above.

Mountain ridges suffer the full force of the wind and are frequently swept bare of snow. Such "schneearm" or "snow-poor" ridges in the Austrian Alps average 3–4°C colder, over the whole year even at depths of a half meter (Aulitzky, 1961), and in winter as much as 15°C colder in the top few centimeters compared with normal sites at the same altitudes. They form a harsh environment for vegetation, having little snow either to keep the soil warm or to protect the aerial parts of the plants. As a result, these sites present a particular challenge to foresters trying to restore a vegetation cover as one means of holding snow at levels from which it would otherwise avalanche (Aulitzky, 1965).

Avalanching

Snow avalanches so dramatically transport snow from its initial lighting place on high slopes that their water-budget aspects might be overlooked; yet many glaciers in mountain valleys are nourished by avalanching snow, which prolongs the storage period of the mountain water. At lower altitudes, valley-floor ecosystems show the evidence of avalanche impact, as well as the massive input of water into their annual budget. From heat-budget considerations during summer ablation, Arai and Sekine (1973) estimate the inputs of snow necessary to hold snow patches through the warm season at various altitudes in Honshu. At 2 km altitude, an input of 8 tons of snow m^{-2}, mostly from avalanching, will maintain a snow patch; at 1 km the input is 14 tons m^{-2}.

In the high mountains of central Asia this vertical redistribution of mass amounts to an off-site movement of about 0.1 of the snow stored at high elevations (1–1.5 tons m^{-2}), and a larger fraction of that in the low-altitude reception zone (Abal'yan *et al.*, 1971).

In contrast with the transport of snow by drifting, which is powered by the kinetic energy of the wind, transport by avalanching is powered by the potential energy of precipitation delivered from the atmosphere to high-altitude slopes. As gravity is always in force, forecasting avalanche descents requires data on the strength of the snow mantle and observations are made to locate layers that are weak in shearing strength or poorly bonded to the ground. Although avalanches might occur at any time, they are especially likely to come near the end of snowstorms that have loaded to failure the weak layers holding the snow cover on a slope.

A classification of avalanches (Int. Assoc. Hydrol. Sci., 1973) points up the off-site transport of mass that is our theme in this chapter. In the zone of origin, for example, an avalanche may start from a point ("loose snow avalanche") or from a line ("slab avalanche"). The surface upon which sliding occurs may be at the ground or within the snow mantle. The path may follow an open slope ("unconfined avalanche") or a gully ("channeled avalanche"); snow may move along this path as a cloud of snow dust ("powder avalanche"), or along the fracture plane, or along the ground surface. Finally, in the deposition zone the snow may be clean, or contaminated by rock debris or tree branches. Parallels between the debris-carrying capacity of avalanches and that of the off-site movement of liquid water in sheets or channels are obvious, and we can see other parallels between aspects of this movement of snow, violent though it is, and the less dramatic off-site flow of liquid water, both representing conversions of potential energy bestowed by the ascent of condensing water vapor into the middle troposphere.

GRAVITY-POWERED MOVEMENT OF LIQUID WATER

Gravity-powered transport of water in liquid state does not require the steep gradients necessary in avalanche movement, but takes place over gentle slopes if they are smooth enough that their surface friction does not impede the flow. This phenomenon is not frequent; rain reaches the earth's surface during only a few percent of the hours of a year, and usually at intensities less than the capacity of the soil to take water in. Only at times are rainfall intensities great enough to exceed the infiltration rate and to build up on the soil surface a layer of detained water that will generate downslope flow. More frequently cases occur in which the upper soil layers are saturated and horizontal flow takes place near but just below the actual ground surface. The fraction of total off-site flow that moves on top of the ground surface depends on porosity, slope, and other factors. In such porous soils as those of forests it is almost never seen.

There is some advantage, nevertheless, in beginning with a rare case, the storm Camille in 1969 in central Virginia, for which rainfall conditions were discussed earlier. An eyewitness account follows:

> . . . The lightning was brilliant and almost constant, it seemed, accompanied by the sharpest claps of thunder I ever heard. But in between these there was another sound—a roaring like forty jet airplanes were stationary overhead. "It must be that hurricane picked up speed again," my wife said. "It must be blowing dreadful outside."

> Finally, at two o'clock, I got up to look out the window. The inside window was up, but we'd closed the storm window to keep the rain out. An inch of solid water was running down the pane, so I couldn't see out. I opened the window, and I got three surprises. They were so sharp they were practically shocks. First, the lightning was almost constant and very low, seeming to be practically parallel to the ground. Second, the lawn outside, which slopes down from the little swimming pool we have, was covered with a solid sheet of clear water inches deep, rushing by the house toward the creek. Third—and this was the biggest surprise of all, though all of them were pretty big, let me tell you—was that I had expected to find a screeching wind outside and it was absolutely still. Not a breath. In the lightning flashes the tree limbs hung down, the lower ones almost touching the ground with the weight of the water, and they weren't moving at all, looking frozen in the eerie light. The roaring sound we heard was the sound of water moving over the land (Kinkead, 1971).

Factors in Overland Off-Site Flow

This event was extraordinary. In rainstorms nearly every year, though, it is possible to see detention films that build in the grass and give rise to off-site flow on the ground surface. Since these events are usually brief and field observations are few, experiments are done with prepared plots subjected to simulated rainfall at high intensities.

Computer programs also can be developed to calculate the rapidly changing fluxes that make up this seemingly simple phenomenon. For example, the time required for an equilibrium flow rate from a slope under uniform rain to become established has been studied and found to be directly proportional to the length and roughness of the slope, and inversely proportional to rainfall intensity (Morgali and Linsley, 1965). We shall return later to a more detailed consideration of these factors, which are related to the depth of the film of water on the land surface.

The discussion of detention storage in an earlier chapter shows that not until the storage of water on the soil surface reached 5 mm did off-site movement of water begin. This did not come until 17 min after the rain started. During the 100 min of rain, off-site movement from this ecosystem occurred during only 12 min.

The detention layer built up from excess rainwater provides a partitioning opportunity. From it water can move one way off-site, or the other way down into the soil. Infiltration into the soil enjoys the senior priority; off-site movement gets what is left. It is residual.

There is controversy over the degree to which the transformation of rainfall to flood flow in a natural drainage basin, with its complicated geometry and soil conditions, can be understood in a deterministic way. Some hydrologists feel that present knowledge is too small to let us analyze the transformation deterministically, and that we must resort to stochastic methods. Others are more confident, though recognizing that the relations are seldom simple, being nonlinear and also varying with time. The problem of water flow on slopes and in channels, and of conceptualizing the processes by which water is concentrated spatially, are questions of hydraulics more than of the dynamics of water in soil–vegetation systems, however.

The rate of off-site flow of water q over the surface (per unit width) depends on the depth of the detention film h, and on the impelling force and the friction it must overcome. This force is the gradient of potential energy down the slope, and is opposed by the friction due to vegetation α. One formulation of this relation is

$$q = \alpha h^n$$

in which n is an exponent approximately equal to 2, and α a parameter expressing the gradient and roughness of the slope (Wooding, 1965). Increasing the depth of the film h, which reduces the frictional drag per unit mass of water, has a marked effect in increasing surface runoff q.

The retarding effect of vegetation increases the time of concentration of flow. The longer time period gives the rain a chance to slack off, and reduces the average delivery rate of water q at the bottom of the slope.

Overland Flow in a New Zealand Case Table I, from a study in New Zealand, represents the way infiltration capacity, rainfall intensity, and the detention storage of water impounded by vegetation friction all combine to generate a characteristic rate of off-site flow of water. The table shows runoff from a slope of 60-m length that results from rainfall supplied at the intensity equal to probability 1.0 at Wellington. (This intensity i decreases as the duration of the experiment increases.) The two kinds of grass in Table I illustrate different degrees of friction in the film of moving water. The taller grass holds a greater depth of water ponded in detention storage; its slope takes a longer time to reach steady flow T_c. Two representative values of infiltration capacity, 1 and 10 mm hr^{-1}, represent differences in soil condition. Four conditions of the vegetation–soil ecosystem are thus illustrated.

Because the mean intensity of rainfall i diminishes with the duration

TABLE I

Dominant Discharge from Small Areas with Three Types of Cover and Two Capacities of Infiltration[a]

Vegetation	Infiltration capacity f (mm hr^{-1})	Time to reach steady flow at downslope end of 60-m strip T_c (hr)	Probable rainfall intensity i during period T_c (mm hr^{-1})	Buildup of detention storage and surface runoff $i - f$ (mm hr^{-1})	Runoff from 2-ha area (m^3 sec^{-1})
Bermuda grass, dense uniform cover 10–20-cm ht	1	0.7	19	18	0.11
	10	1.4	13	3	0.02
Very short grass	1	0.24	30	29	0.16
	10	0.28	28	18	0.10

[a] Source: Campbell (1967).

of the period, the slopes where movement of water is slow need a long time to reach equilibrium flow, and over this period receive a relatively low mean intensity of rainfall. Slopes from which the full concentration of runoff occurs in a short time receive a higher mean intensity. Water making its way slowly through the dense sod of Bermuda grass takes 0.7 hr to get from the top of the slope to the bottom and establish steady flow; water coming down the short-grass slope takes 0.24 hr. These periods are called times of concentration T_c.

At 0.7 hr, the entire area of the Bermuda grass slope is contributing water that reaches the downslope end. This duration is long enough to allow the rainfall intensity to slacken a bit. This aspect of duration, which might not be obvious at first glance, derives from the tendency of rainfall intensity to decrease as its duration increases, which we considered in Chapter III. The mean intensity of the once-a-year rain at Wellington during the 0.7-hr concentration time associated with the Bermuda grass slope, is 19 mm hr^{-1}. This is a much lower rate than the mean intensity, 30 mm hr^{-1}, in 0.24 hr, the time relevant to the short-grass slope. A bare slope or a parking lot, which would have a still shorter concentration time, would have to handle a still higher intensity rainfall.

The other variable illustrated in Table I is infiltration capacity f, shown for each slope for two selected soil conditions, 1 and 10 mm hr^{-1}. In determining the rate at which detention storage builds up, the respective infiltration rate f is subtracted from the rainfall rate i. This excess water $i - f$ goes first to build up detention storage (including filling microdepressions), then into surface runoff. In the low-infiltration case of the Bermuda grass slope, the water excess is $19 - 1 = 18$ mm hr^{-1}. This is six times as much as on the same slope under high-infiltration conditions.

Runoff rates are high if vegetation is short and offers little frictional resistance to overland flow, and also if infiltration capacity is low. Runoff from the four soil–vegetation systems in this once-a-year rain ranges from 0.02 to 0.16 m^3 sec^{-1}, that is, over nearly one order of magnitude.

Changes in Soil–Vegetation Systems: New Zealand A later study in New Zealand identifies how the critical parameters of infiltration capacity and detention storage are changed as a result of practices that are transforming the agriculture of the country. Higher applications of phosphate fertilizers, often made by airplane to hill-country farms, have increased the growth and density of grass over millions of hectares (see Plate 34).

Plate 34. Hill-country agriculture on the North Island of New Zealand, with aerial fertilization (Te Kuiti, February 1966).

In one experimental area, where the grass production rate was tripled (Toebes *et al.*, 1968), the hydrologic results of the greater plant vitality were evident as increases in infiltration capacity and surface detention. A greater density of plant stems and leaves slows the flow of surface water; greater plant activity dried out the root zone, increasing its capacity to take in water from the next rain. These changes result in less off-site flow. Moderate rains that formerly caused off-site flow on untreated slopes were completely accommodated on the treated slopes, and produced no surface runoff.

Changes in Soil–Vegetation Systems: Coshocton Experimental Site Because deeper-rooted grass keeps the soil in better condition and dries it out more rapidly, water from each rainstorm finds greater opportunity to infiltrate, i.e., less water piles up on the surface to move off-site. Table II, surface runoff from summer rains at Coshocton, Ohio demonstrates that more vigorous plant growth means less off-site movement of water. Grass yield increases 5 times, water yield decreases to $\frac{1}{16}$.

Conversely, infiltration into an already moist soil is slower than into a dry one for the reasons discussed in an earlier chapter. In a protracted rainy period infiltration is low even at the beginning of each burst of rainfall, and surface runoff from the burst is correspondingly increased. Surface runoff was measured over thousands of plot

TABLE II

Surface Runoff in Summer from Two Kinds of Grass in Lysimeters at Coshocton, Ohio[a]

Parameter	Poverty grass (shallow-rooted)	Birdsfoot trefoil (deep-rooted)
Annual biomass yield (g m^{-2})	110	570
Summer runoff		
1959	2	1
1960	10	1
1961	8	tr
1962	11	tr
1963	10	0
Mean of five summers	8	½

[a] Units: mm. Source: Dreibelbis and Amerman (1964).

years at 35 soil erosion experiment stations in the United States and found correlated to rainfall energy and 30-min intensity. It was also found to be correlated with rainfall in the preceding 24 hr, which influences the level of soil moisture and hence the rate of infiltration (Nelson, 1958) and runoff.

Surface Runoff from Some of the Ecosystems in a Drainage Basin

Soil bodies already saturated at the beginning of a rain will not take in any water by molecular attraction, and detention storage will almost immediately start to form on the surface. Off-site runoff soon follows. It might be a flow on top of the saturated soil or a pressure wave forcing water that has filled the large pores of the soil to start moving out the downslope end of the soil body.

Saturated soil bodies are common near streams, in shallow soils, and at the foot of slopes. Their rapid response to a burst of rain that elsewhere in a drainage basin might infiltrate into the soil often supports the initial rise in the nearby stream channel. The conversion of rainfall to quick runoff is virtually complete, and the stream rise can be substantial even if no more than 0.1 of the area of the basin is contributing (Kirkby, 1969). Horton (1937) identified and studied "partial-area rises" in rivers, and presents a case study of the Muskingum River in June 1929. In later generalizations of the excess-rainfall concept for application to large areas, it was commonly assumed that all parts of a drainage area contribute equally, but more recently the

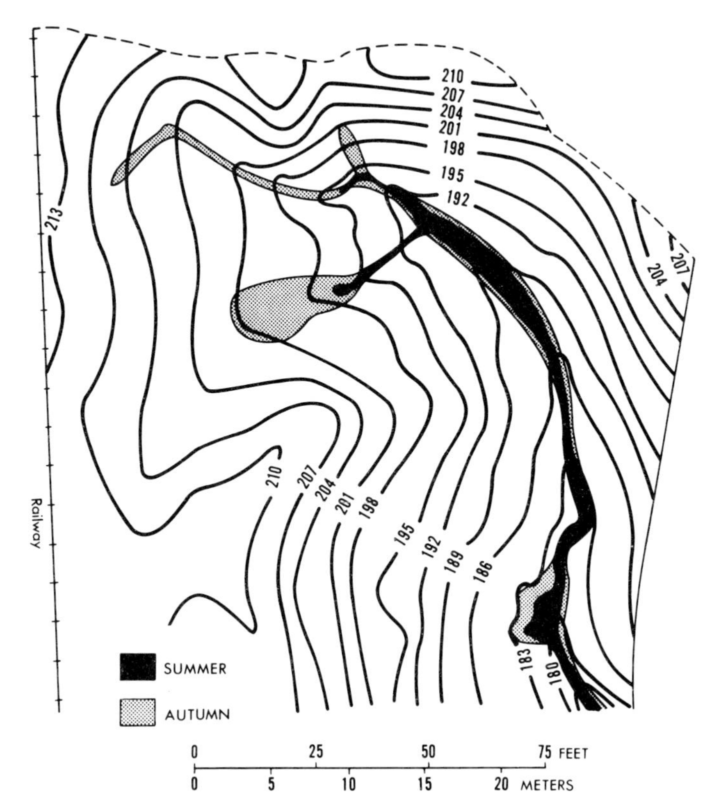

Fig. XVI-1. Areas of near-stream land that contribute storm flow in intense summer storms (black) and in autumn storms under generally wet conditions (from Dunne and Black, 1970).

partial-area idea has had a revival in explaining the time distribution of storm flow* as a spatially integrated flux from many ecosystems.

As a storm continues, the number of ecosystems with full storages of interception, detention, and soil-moisture capacity increases. Soon all are contributing storm runoff, and channel flow builds up to a peak.

A seasonal variation also occurs in the size of the contributing area. Storms in fall and winter deliver rain to a basin in which a large area is wet and ready to convert rain to storm flow; in summer the area ready to make this conversion is much smaller (Fig. XVI-1) (Dunne and Black, 1970).

* "Storm flow" refers to immediate off-site flow, whether it travels entirely on the ground surface, in the top layers of the soil, or both.

The idea that all the ecosystems to which water is delivered by a rainstorm are not necessarily equally ready to take it in expresses the diversity of nature, which is manifested in the contrasts among ecosystems with respect to soil depth and texture, vegetation density, and other properties of hydrologic significance. "The partial area concept describes, qualitatively, what one can observe in the field. Quantitatively, one cannot say as much. To date, no successful quantitative procedure has been developed to predict watershed runoff on this basis" (Engman, 1974, p. 514).

Flow Near the Surface of the Ground

In contrast with off-site flow across the land surface of experimental plots or thin bodies of saturated soils, some ecosystems almost never experience overland runoff. This behavior reflects ecosystem characteristics or situation.

In Forest Ecosystems The effect of different soil–vegetation systems on storm-flow events is shown in Table III from measurements in the dissected land of southwestern Wisconsin. The virtual absence of storm flow on forested slopes, even when they are very steep, indicates the hydrologic asset that the forest floor represents. Its open structure can accommodate large inputs of water. Much of this moves off-site within the soil, but more slowly than if it moved over the surface, and does not produce an initial steep rise of the flood wave.* This near-surface flow makes the main contribution to the flood wave itself, however. In the mountains of the northeastern United States, "the principal source of flood runoff delivered to the channels in forested watersheds is subsurface flow that moves rapidly downslope through the permeable forest floor and topsoil" (Lull and Reinhart, 1972, p. 83).

A good deal of folklore on forest effects on floods has accumulated during the past century, some of it exaggerating their undoubted benefits to runoff. Sartz (1969) discusses some of these exaggerations, many of them based on old research and crude measuring methods.

Relation to Slope Porous soil and high infiltration capacity of forested slopes permit little rainwater to move over the surface so there is little action of water on the surface soil. As a result, slopes remain uneroded, and unreduced in steepness. The coincidence of

* This condition exists even in the heavy rainfall of the southern Appalachians. Surface flow has never been seen to cross some slopes of the Coweeta Experimental Forest in more than 20 years of observations (Helvey *et al.*, 1972).

TABLE III

Surface Runoff from Ecosystems in the Coulee Experimental Forest, Wisconsin[a]

Vegetation	Slope	Yearly number of events in which storm flow exceeded 5 mm
Tilled land	0.15	3
Meadow	0.15	1
Abandoned field	0.15	1
Open pasture	0.25	1.5
Forest	0.35	0

[a] Period of study: 8 yr. Source: Sartz (1970).

steep lands with forest has been explained in many ways, including the shorter runs of wild fires in rough country thus weakening their tendency to favor grass over trees. Another explanation is that the forest preserves steep slopes by keeping their infiltration capacity high and minimizing erosion. Wood (1958) reports that in a sample of erosional terrain mean slope of forested lands was 0.14 and of grasslands only 0.07. On areas varying in size from 0.5 to 400 km^2 "the average relief of the forest samples is from two to three times greater than that of the grasslands."

At the same time, high infiltration generates a large flow of water *through* the soil. This water emerges at the toe of the slope, near the channel, and has a sapping effect on the slope. It enhances the stream's ability to carry away debris brought down by mass movement; by cutting back, it keeps the slope steep and thereby provides the gradient necessary for mass movement. This sapping effect is plainest where convex slopes lie above incised channels, and it is most likely that subsurface flow makes a substantial contribution to storm flow in the channels (Freeze, 1972).

Characteristics of Near-Surface Off-Site Flow When water fills all the large pores in the soil overlying a less permeable layer that restricts continued movement downward, off-site movement of water near the surface may bulk very large. While it encounters more friction than does runoff on the surface, it occupies a thicker cross section; slower speed is compensated by greater cross-sectional area of flow.

This depth of flow may be quite large in heavy storms, and tends to accelerate the volume of water moving subsurface. The lag of such flow was found in northern Japan to decrease in large storms. In the experimental basin of Iwate University in northern Japan, this lag was found to be 25 hr in storms of 10-mm flow, but only 5 hr in storms of 40 mm (Takeda and Ishii, 1968).

Subsurface runoff is common in snow-covered drainage basins, where the snow cover impedes flow on top of the ground, even though the rate of generation of meltwater seldom exceeds 5 mm hr^{-1}, a rate much less than the rate of infiltration. For example, in the basin of Castle Creek in California water runs on the soil surface for only a few meters downslope from a melting snow bank, then enters the soil, and continues as subsurface flow to the stream channel, a distance of 200–300 m (Chapter VII).

The speed of this subsurface movement allows it to cover this distance in 2–3 hr. In contrast, hydrograph analysis of a rainstorm heavy enough to generate extensive overland flow indicates much faster movement. Although slowed by microdams formed of pine needles, water reached the channels in less than an hour.

The large volume of near-surface movement of water in melt situations is indicated by measurements in the coarse, shallow granitic soil of the Idaho batholith. Overland flow is rarely seen in this soil, and near-surface flow is common. By interception of near-surface flow at a road cut, it was determined that the volume of water during one melting season was equivalent to 100-mm depth over the upslope contributing area (Megahan, 1972). Another 250 mm of water percolated deeper into the weathered granite. Part was retained in the nondraining small pores as a recharge to soil moisture, and part emerged into streams at lower elevations as an outflow from true groundwater.

Conditions of soil and surficial geology are of obvious importance to near-surface transport of water. A study of small drainage basins in the eastern United States shows that their responses to rainstorms are controlled "chiefly by porous mantle factors" (Woodruff and Hewlett, 1970). Averaging over 90 basins, the amounts of quick flow from all rainstorms in a year total about 100 mm. The size of the standard deviation among the basins, 75 mm, indicates large differences in geologic and soil factors that affect runoff.

Stephenson and Freeze (1974) found in volcanic terrain in Idaho, similar to that in the Sierra where we seldom observed surface runoff, that "transient saturated–unsaturated subsurface flow can deliver snowmelt infiltration through high-permeability low-porosity formations fast enough to be the sole generating mechanism of runoff from an upstream source area." They go on to suggest, however, that the peak flow might come from snow cover nearest the stream. An alternative might be that meltwater entering a subsurface reservoir transfers its effect faster than the speed of the moving water particles themselves, as suggested by use of a tritium balance (Martinec, 1975). This process is most likely to occur in the low-level ecosystems near the channel.

Such considerations of partial area contributions and subsurface flow are valuable in considering a specific ecosystem as a site for applying chemicals (fertilizers, biocides) or urban wastewater (Engman, 1974). Ecosystems near streams are not likely sites for such applications; in fact, when searching for the sources of pollutants found in streams such systems might well be examined first, as was indicated from a study of phosphate flow of a basin in Pennsylvania (Gburek and Heald, 1974).

Induced Surface Runoff

Compared with a forest floor with its high infiltration and large near-surface flow of water are systems in which overland flow is maximized. Certain land treatments isolate rainwater from the soil body entirely by reducing infiltration. If the on-site use of water by plants is discouraged, this also increases the off-site yield. The increase in surface runoff may or may not be beneficial.

The Negev Runoff Farms Such treatments to induce additional surface runoff for beneficial use on irrigated plots downslope are illustrated in southern Israel (Hillel and Rawitz, 1968; Tadmor and Shanan, 1969). Surface runoff was induced both by removing vegetation and by treating the soil surface by mechanical and chemical means. Both practices were effective in shifting the partition of surface-detention water away from infiltration and toward surface runoff. Analyses of daily data showed that the threshold amount of rainfall necessary to produce surface runoff (Tadmor and Shanan, 1969) is 2–4 mm day^{-1}.

The volume of surface runoff from bare plots was several times as great as from those covered by vegetation. When the relations found for daily off-site flows in the period of experimentation were extrapolated to an average 15-yr period the results in Fig. XVI-2 were obtained. In an average year the bare plot produces off-site yield (64 mm) equal to more than three times as much as the plot with vegetation cover (19 mm).

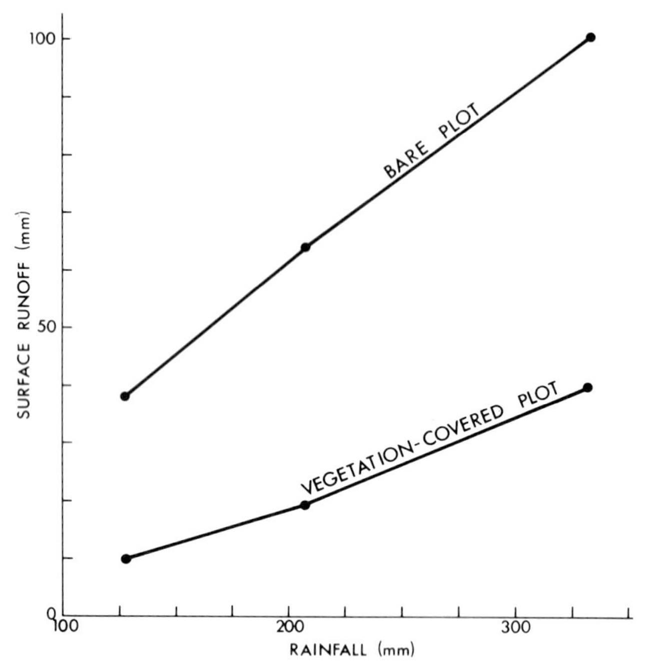

Fig. XVI-2. Relation of surface runoff from a plot kept bare to induce runoff, and from a vegetation-covered plot, for average annual conditions (from Tadmor and Shanan, 1969).

Restoration of an induced runoff farm built in Byzantine times required about a thousand man-days' labor.* This farm consisted of catchments of 12 ha of slopes with desert pavement, which supplied runoff to 0.8 ha of cultivated land. The ratio of catchment to receiving area, 15–1, depends on how much induced runoff can be obtained at a given level of probability from a year's rainstorms (Evanari *et al.*, 1968). "Catchment basin and terraced fields together constitute an indivisible entity: the runoff-farm unit" (Evanari *et al.*, 1971, p. 104) and provide an interesting case of the human linking of spatially separated ecosystems.

Water for domestic and livestock use also was obtained by induced runoff. Enough to supply ten families through a 300-m^3 cistern comes

* From these investigators' data, we can calculate the man-day energy costs of acquiring extra water. An increase of 25 m^3 ha^{-1} produced by clearing stones and allowing loess slopes to crust over is obtained at a cost of about one man-year. At 10% amortization, 1 m^3 of water so obtained annually costs about one man-day! For comparison, shallow-well irrigation described in Chapter XV delivers 1 m^3 to crop plants at an energy cost of 0.15 man-day.

from an area of 10 ha drained by 1200 m of ditches. "This cistern was certainly filled every year" (Evanari *et al.*, 1971, p. 157). The man–land ratio for water collection is 1 ha per family.

Australian Water Harvesting Similar practices of encouraging overland runoff and collecting it for stock water (and sometimes to irrigate a small field of some fodder crop in drought times) are common in Australia. Runoff is collected by long dikes that divert water across slopes and into a place where a small dam has been thrown across a gully. Infiltration, soil moisture, and biological production on the slopes are sacrificed in order to obtain stock water to be stored against unpredictable periods of drought.

The homemade nature of these storages, which in fact often fail because of poor design in either their hydrologic or soil-mechanics aspects, should not lead one to overlook their real economies. Capital costs of construction are about 3¢ m^{-3}, half the costs of the large engineering works of the Snowy Mountains Authority (Costin, 1971). Annual costs, even considering the frequent dam failures (two million dollars worth per year in New South Wales, for example), which could be minimized by better design, are only 0.3¢ m^{-3} of water stored.*

Surface Runoff Induced by Structures Where maximum appropriation of rainwater is desired, a small catchment area can be asphalted or covered by plastic or metal sheets. In the Kona district of Hawaii porous lava allows almost all the rainwater to move to a groundwater body hundreds of meters below the surface. Domestic water can be gotten only from expensive wells or by induced runoff from asphalted catchments or roofs.

A steel-covered catchment developed for the Great Plains has a minimal maintenance expense and produced water for livestock at a cost of about 80¢ m^{-3} (Fairbourn *et al.*, 1972). The same idea is seen in the general use of galvanized iron roofs to catch rain in places where groundwater is hard, saline, or scarce.

Urban Runoff

Induced Overland Flow The conversion of a rural land surface to an urbanized one produces an alteration in the surface hydrology that resembles the induced runoff treatments just described. Impervious pavements and roofs, which have zero infiltration and only slight

* Where water is scarce it is worthwhile covering such ponds with monomolecular films to reduce evaporation. They cost about 1¢ m^{-3} of water saved.

detention storage, now cover a half or more of the land.* They augment the volume and speed the flow of water, especially where roof drains have been allowed to feed into the network of sanitary sewers. In cities that permit (or wink at) these connections but neglect to consider the increased contributing area in the design of sewer lines or treatment-plant capacity, the result is much the same as if a flood hydrologist ignored a large number of ecosystems lying hydraulically higher than the dam he was designing. For example, a moderate rain (58 mm) in Milwaukee caused failure of a new treatment plant with damage of $1.5 million and continued the flow of massive pollution into Lake Michigan.

In flood hydrology, the need to know the capabilities of the upstream ecosystems to produce off-site flow of water brings about, as we saw in Chapter VIII for infiltration and Chapter XII for evaporation, special attention to ecosystem processes. A corresponding need exists in urban hydrology.

In figuring design flows for a network of storm sewers, common practice is to estimate that 1–2 mm of water are held in depression storage (in nondraining puddles in the pavement) and about the same amount in the film of detention storage. Both storages fill up promptly in a heavy storm and soon begin to spill. If a 3-hr duration is assumed for a design storm (Tholin and Keifer, 1960), the impervious areas are neither storing nor infiltrating water during most of the storm but are shedding it all.

Relative volumes of water moving off sites under different degrees of urbanization (as expressed by area of roofs and streets) are shown in Fig. XVI-3. Full urbanization approximately doubles the volume of off-site transport of water.

A runoff model of peak storm flow from 17 drainage basins in Baltimore, which had an average impervious area of 0.52, was found to give satisfactory reproduction of actual runoff if the average volume of stored water was taken as approximately 4 mm (about the thickness noted earlier). Except in very heavy rains, it could also be assumed that only the impervious areas of these small basins contributed to immediate off-site flow (Viessman *et al.*, 1970). In most rainstorms the other parts of the basins generate little or no off-site flow until volumes exceed 40–50 mm.

Because the storage capacity—4 mm in the Baltimore case—on impervious areas fills rapidly in a rainstorm that might deliver 25–50

* With full urbanization at high density, the fraction of impervious area is about 0.8. Even in low-density residential districts, there is still an appreciable change from rural conditions.

Fig. XVI-3. Relative volumes of water moving in off-site flow in areas of different degrees of urbanization (from Leopold, 1968).

mm of water, the time of peak flow is advanced. With urbanization, flows are not only more frequent but also earlier. The owners of downstream property find that a given flood level is being reached more times in a year, and with less warning time.

One result of the urban sprawl of recent years has been the creation of a new kind of water body; flooded residential districts not necessarily in or even near a stream channel receive overland runoff from adjacent higher areas that formerly rarely generated off-site flow. In many areas a modest rain of only 50–75 mm is now so efficiently converted into off-site runoff that houses are swamped. A loud cry arises for federal designation as a "disaster" area. Here the augmentation of off-site flow by urbanization, especially urbanization sprawling ahead of the storm-sewer network, not only produces more frequent channel floods but also has created a new feature of the landscape—an *urban wetland*. Problems such as this and the pollution in urban off-site flow are spurring the laggard field of urban hydrology to adopt scientific methods now common in such more advanced fields as snow hydrology, riverflood hydrology, crop evaporation, and wildland hydrology.

Augmented Flow from Cities Cities also generate another kind of off-site flow: the discharge of water they have brought in to operate their domestic and industrial processes. We think of the sewage discharge in terms of its load of dissolved and suspended materials, but most of the flow is water—99% or more. Its volume is indicated by a normal flow through Milwaukee's main sewage-treatment plant of 200 million gal day^{-1} (including much brewery waste water), which on an areal basis is equivalent to the runoff from a basin in a wet mountain range, 300–400 mm yr^{-1}. In the case of Milwaukee, this volume is about twice as large as the natural water surplus before urbanization, and about three times as large as the fraction of the preurbanization off-site flow that moved over and near the ground surface.

Cities gather the natural water surplus of a large area of catchment land (or lake surface in Milwaukee's case), move it at a relatively steady pace through the pipes undergirding the urban interface, and then discharge it in a highly concentrated off-site flow that may overwhelm the natural flow in some receiving streams in dry seasons of the year. The daily rate of throughput in Milwaukee is a little more than 1 kg m^{-2}.

This uniform flow, seemingly easily managed, does not always remain confined in pipes, however. In some communities the network sized to carry off this steady flow is also called upon to accept storm flow, which is typically several times greater in volume. The result might be expected—the combined sewer spills into the nearest body of water. The Milwaukee River and Lake Michigan are polluted this way about 50 times per year; lake beaches are usually closed to swimming after a rain of 6 mm. Moreover, the production of Milorganite fertilizer at the old sewage treatment plant loses nutrients that would otherwise be recycled—a conservation measure that began years ago and occasions considerable local pride.

OTHER FORMS OF MASS TRANSPORT ASSOCIATED WITH THE FLOW OF WATER AT AND NEAR THE SURFACE

Movement of water across the ground surface not only detaches soil particles, perhaps already loosened by rain impact, but also carries them away. Generally speaking, the factors in sheet erosion are the same as those in overland runoff: infiltration capacity of the soil, intensity of short-period rain, friction of the slope, and slope length and steepness. To these would be added some index of erodibility, to bring in the factor of cohesion or agglomeration of soil particles.

The magnitude of this off-site movement of soil particles is suggested by the volume of colluvium at the foot of hill slopes and alluvium in the valleys in landscapes subjected to long undisturbed weathering. In the Southern Tablelands of New South Wales, colluvial material covers very extensive areas, for instance one-third of the upper basin of the Shoalhaven River (Australia, 1969, p. 84). Alluvium, i.e., eroded soil particles transported still farther from their source areas, make up a further 0.03 of this basin.

The impoverishing of the source areas is shown by the statement that "the line that divides the colluviated and non-colluviated portion of the landscape has profound significance for soils, vegetation, and land use. Above this line . . . cultivation is impossible" (Australia, 1969).

Eroded sediment deposited in 43 reservoirs in California was related by Anderson (1974) to geologic and cultural (roads, fires) characteristics of the contributing lands, flow variability, and altitude. The altitude determines the frequency of snowfall rather than rain; a reservoir at an altitude of 2.8 km received only 0.06 the sediment of one in similar lands at a lower altitude. Of the total variance in sediment deposits, 0.78 was explained by Anderson's variables, of which 0.23 was due to altitude, 0.21 lithologic differences, 0.08 fire frequency, and 0.07 road density in the eroding ecosystems.

Exposed soil, without ecosystem foliage or litter to protect it from beating rain, is likely to generate a large off-site movement of loosened soil particles. In an experiment on deep loess soils (mean slope steepness = 0.09), "soil movement off the plots was primarily transport by raindrop splash to a rill system and then transport down the slope by runoff in the rill system" (Young and Mutchler, 1969). Raindrops are the first agent, bulk water in channels the second. Unlike the flow of water, however, that of soil is not always continuous: "A breakdown of the rill system occurred on the bottom of the concave slopes because of decreasing local steepness, resulting in sheet flow and sediment deposition" (Young and Mutchler, 1969).

Bare soil is common in dry climate ecosystems, often kept so by biological mechanisms that space out individual plants to exploit soil moisture efficiently. In a Colorado study, the off-site flow of water from 17 small basins was well correlated with the fraction of bare soil in each (Branson and Owen, 1970); the flow of sediment was additionally influenced by such shape factors as slope or density of drainage channels.

For several decades, radioactive strontium has been delivered to ecosystems of the United States, both wildland and cropland, which absorbed part of it near the surface and percolated part deeper. Two off-site fluxes can be distinguished: (a) the direct runoff of a fraction (up to 0.02) of the ^{90}Sr just delivered in rain, and (b) removal of accumulated strontium during sheet erosion episodes (Menzel, 1974). The direct-runoff flux is largest in winter and where rain falls on frozen or saturated soils. The other flux is closely correlated with sediment yield.

Dozens of examples of surface runoff and erosion in different ecosystems might be cited. We will look at a few that illustrate land management practices, and one illustrating a grosser impact of technology, i.e., suburbanization.

Effects of Wildland Management Practices

Grazing and Fire The loss of soil resulting from degraded groundcover is shown by Costin's work with fragile vegetation types in the Australian Alps (Table IV). A decrease of infiltration into the soil and concomitant increase in loss of soil materials, including nutrients that had slowly accumulated in its upper layers, are the results of fires. These were set to encourage new grass, but in the process killed the trees.

Sheep and cattle ate back the sprouts sent up by surviving root crowns of trees, as well as the palatable plants that formerly covered the bare areas between tussocks of snow grass. This combination of fire and grazing exposed the soil to "the erosive action of frost, wind and water" (Costin *et al.*, 1959). (See Plate 19, p. 244.)

A heavy rain occurred during the experiment, with effects shown in Table IV. The depleted land surfaces could not take in water at a rate of 50 mm hr^{-1}, and in a short time lost a large amount of the soil so slowly formed in this cold environment. Costin (1971) found that the mass of vegetation cover necessary to protect the soil surface is about 10 tons dry weight ha^{-1}, or 1 kg m^{-2}.

TABLE IV

Surface Runoff and Erosion from Plots of Vegetation in the Snowy Mountains, Australia[a]

Vegetation type	No. of plots	Fraction of ground covered by vegetation	In all rains		In a rain of 25 mm in 30 min	
			Runoff[b] (%)	Soil loss[c]	Runoff (mm)	Soil loss[c]
Forest						
Natural	5	0.96	0.3	0	0.8	0
Regrowth	4	0.90	1.7+	4	5.1	trace
Depleted	4	0.50	3.7+	54	9.7	2
Snow gum–snow grass						
Natural (snow grass)	14	0.99	0.6	0	1.8	0
Regrowth shrubs	6	0.97	0.6	1	2.1	0.4
Depleted	7	0.68	3.3	460	7.7	270+
Snow grass						
Depleted	4	0.30	8.1	5700+	—	—

[a] Source: Costin *et al.* (1960) (Catchment Hydrology II).
[b] Units: Percent of rainfall.
[c] Units: g m^{-2}

Some changes in infiltration ceased when grazing stress on the vegetation ended. In other areas, however, deterioration may have gone beyond the point of recovery, unless costly engineering structures were built to allow the soil–vegetation complex to renew itself. The effect of fire in a nearby area was to increase peak flow by five times and sediment flow by 100 (Pereira, 1973, p. 76).

Expensive means of regeneration might be justified on slopes from which eroded soil would quickly move into reservoirs of a hydroelectric project. In general, however, it is better economics to avoid the initial deterioration of the structure of the soil–vegetation community. In this case, the grazing value of the land could never touch the value of the annual yield of water for power—\$540 ha^{-1} of mountain land—or for irrigation—\$100–150 ha^{-1} (Costin, 1959).

Logging When forest land is logged, erosion occurs on the parts most disturbed—the roads, log landings, and skid trails. This consequence is well known, and can be minimized by careful operation. Off-site transport of material in dissolved form also occurs, however. This is seen in mass-budget measurements in Hubbard Brook experimental area in New Hampshire. Vegetation in one of the subbasins was cut in 1965, lopped, and left on the site, occasioning changes in the mass budgets of several ions over subsequent years. Results for two ions, Ca and NO_3, are shown in Fig. XVI-4.

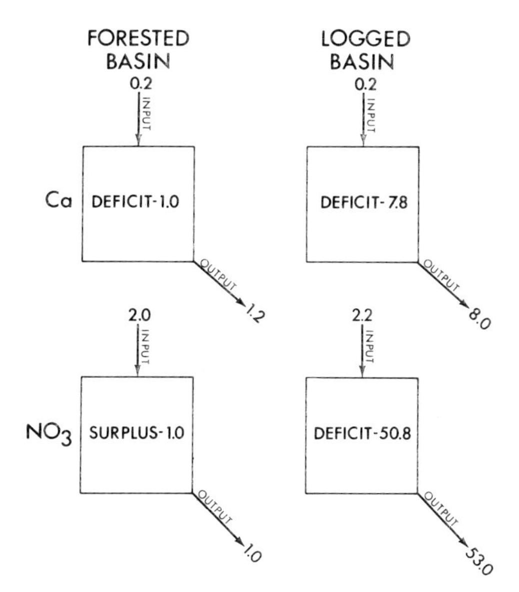

Fig. XVI-4. Effect of logging on the mass budgets of two ions in basins of Hubbard Brook, New Hampshire. Units: g m^{-2} of basin area (data from Pierce *et al.*, 1970).

Cutting results in a major mobilization of nitrogen, which, released from its biological bonds to the site, was flushed off in the off-site flow of water. Cation outflows like calcium increased by three to twenty times. Particulate-matter outflow increased by nine times, due to erosion of now unprotected stream banks and to an increase in off-site transport of soil particles.

A consequence of the partial area concept of overland flow described earlier is that pollutants near a stream channel are more likely to be carried into the channel than those farther away. For this reason guidelines for logging operations usually protect trees within a certain distance of a stream and require that roads be built far enough from streams that eroded fill particles will settle out before they reach the stream.

Effects of Cropland Management Practices

Plow agriculture can disrupt the soil system and increase its vulnerability to erosive attack if management practices do not pay attention to soil loss. For example, small basins of cultivated land in southwestern Wisconsin generate three storm-flow episodes per year with an average peak flow of 37 mm hr^{-1}, whereas meadow and abandoned fields, also on slopes of 0.15 gradient, generated only one flow episode, which peaked at 15 mm hr^{-1} (Sartz, 1970).

Erosion-Mitigating Practices One means of reducing erosion is to strip-crop meadow and grain crops alternately across a slope: the short-distance transport of soil particles can thus be both (a) reduced and (b) contained.

(a) Movement of soil from the strips of grain, on which soil is exposed to rain through part of the year, is *reduced* because the length of slope is effectively shortened. On short slopes the film of moving water does not get thick enough to generate excess energy that would be expended on the soil. Relative indexes of soil loss from a slope of 0.06 steepness are:

From a slope 20 m long	0.5
From a slope 50 m long	0.9
From a slope 100 m long	1.2

(FAO, 1965, p. 146).

(b) Soil eroded from each bare strip is *contained* in the next lower strip of grass, and not carried off the field. The combination of both processes acts to reduce total soil loss to about one-fourth that from up-and-down tillage (FAO, 1965, p. 152) (see Plate 35).

Plate 35. Just-plowed strips, with a deep furrow bordering the grass strips, near New Glarus, Wisconsin, May 1973.

Many methods of altering soil–vegetation systems to cover all the soil surface, strengthen soil structure, increase infiltration capacity, and reduce surface runoff and the associated soil detachment and transport, have been developed in the past forty years, building on older ideas of sound farming practice. The goal is to reduce yearly soil loss to 250 g m^{-2}, an objective that "is far from achieved" after a third of a century (Pereira, 1973, p. 174). Practices described by Colman (1953), chiefly for wildland areas, and by Kohnke and Bertrand (1959) for cultivated lands, include mulching the soil surface, strip-cropping slopes, grassing waterways, plowing on the contour, shaping the land into terraces, and constructing roads carefully. Reducing the period of bare soil, and such indirect measures as fencing from cattle or protection from fire, are good means of fostering soil-protecting vegetation and improving hydrologic characteristics of the soil–vegetation system.

The selection of the correct methods for each site, especially among the more massive land-shaping techniques like terrace building, depends on the characteristics of the ecosystems involved, and on the probable intensities of rainfall. Is it more important to hold water several days to infiltrate, or to get it off the slope, gently but as fast as possible, before more rain comes? The choice reflects the expected duration of heavy rain and the likely spacing between storms, which affects the general soil-moisture budget. Where rains are scarce, the aim is to maximize total infiltration; in wetter regions, the aim is to ease excess water into off-site flow at nondamaging intensities.

Unfortunately some techniques to control water on the land do not seem to be compatible with large cultivation machines, such as "the trend toward using 4-, 6-, and 8-row equipment." These practices are not conducive "to using the erosion control practices of the past. Many fields with terrace systems of the 1930s are now farmed along field boundaries with eventual destruction of the terrace system" (Soil Science Society of America, 1966).

In noting "new and bold concepts needed to replace the conventionally designed terrace system," this panel of soil scientists suggests "topographic modification," i.e., massive reshaping of slopes. This practice, however, creates "problems of soil fertility and of water intake, transmission, and storage" (Soil Science Society of America, 1966). The question of partitioning the water delivered to the land surface is still with us.

A study on managing the Wisconsin environment for quality notes that the change to larger farm machines, especially in systems producing cash crops rather than animal forage, has effects that have not yet been costed out. "Traditional conservation practices such as grass waterways, terraces and strip cropping are no longer looked upon with favor by large farm operators. This has created a serious problem from the wind and water erosion aspect in central and southeastern Wisconsin" (Wisconsin, 1971, p. 13).

Reduction of off-site transport of soil is usually not an all-or-none alteration but poses questions of optimum cost. Calculations for corn land in Illinois, which would yield sediment to a proposed recreational lake, shortening its life, indicate that substantial reductions in sediment flux can be obtained by changing conventional to conservation tillage practices. Further reductions, however, can be "achieved only by a shift to less profitable crops" that would lower farm income (Narayanan and Swanson, 1972).

Chemicalized Agriculture The increasing tonnages of fertilizers and biocides applied to agricultural lands supply new forms of matter that take part in off-site mass movements associated with surface runoff and erosion. Some of these new materials are soluble in water and are easily carried as it moves off-site; nitrates are an example of such a substance that is generally unwelcome in receiving waters.

Other materials, like DDT, not being water soluble, were initially

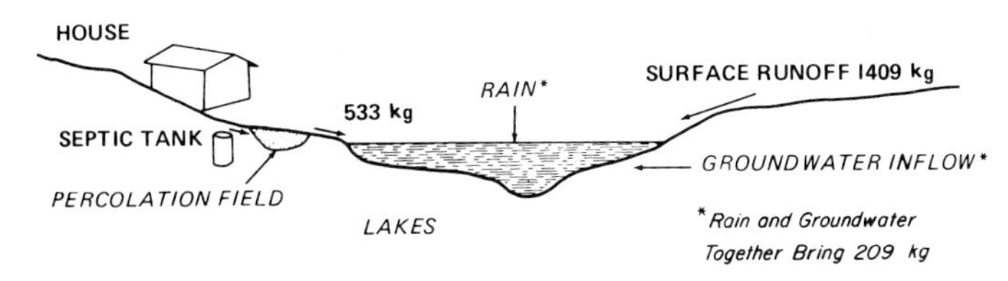

Fig. XVI-5. Annual mass budget of phosphorus in 16 small lakes in the Milwaukee River basin. Area of the lakes is 11.7 km^2 and total annual input of phosphorus is 2151 kg (data from Southeastern Wisconsin Regional Plg. Comm., 1971, pp. 315–331).

thought to remain where the farm operator put them. They surprised the experts, however, by appearing in high concentrations in receiving waters and continuing their injurious effects on the environment. Some, like the phosphates, are apparently transported on the surface of soil particles carried off in erosion. Cultivated lands under chemicalized farming have become areal sources of water pollutants that rival the more obvious point sources exemplified by factories and city sewer outfalls.

Fertilizers or manure applied to unterraced, frozen fields in southeastern Wisconsin do not enter the soil where they might attach themselves to soil particles, but are in part carried away in surface runoff. In the small lakes where this off-site transport ends, phosphorus becomes one cause of accelerated eutrophication, a process which threatens the amenity value, if not the life, of these valuable lakes (Fig. XVI-5). Table V shows the costs of combating effects of phosphorus, about \$50,000 yr^{-1} for each lake. The transfer of mass among components of this lake-dotted landscape is accompanied by a transfer of income among the land managers.

TABLE V

Annual Treatment Costs of 16 Small Lakes in the Basin of the Milwaukee River[a]

Annual costs of recovering lake quality	\$ yr^{-1}
Bench terracing of agricultural land	3 ha^{-1} of catchment area
Sewering of resident and seasonal population	100 per capita
Removing phosphorus from lake by algae control and weed harvesting	13 ha^{-1} of lake area
Total cost in terms of lake area	810 ha^{-1}

[a] Source: Southeastern Wisconsin Regional Planning Commission, (1971, 315–331).

The ubiquity of water-borne transports of material from lands managed by what has been an energy-rich, chemical-rich economy is evident in the nationwide concern about pollution of rivers and lakes. "Eutrophication, whether natural or cultural, results when nutrients are imported into the lake from *outside* the lake—that is, from the watershed. This is equivalent to adding nutrients to the laboratory microecosystem or fertilizing a field; the system is pushed back, in successional terms, to a younger or 'bloom' state" (Odum, 1969). Each body of receiving waters is dependent on the uplands where water and mass transports originate and cannot be understood without considering the "entire landscape catchment unit." Off-site flows provide an obvious linking function in these landscapes.

Biocides may play a different role where they end up than was intended where they were applied. In view of their powerful role in biological systems, this mass flux can be regarded also as a flow of information like the gene patterns carried in pollen (although less constructive).

A comprehensive model of pesticide transport, "piggy-backed on existing hydrologic and soil-loss models" (Bailey *et al.*, 1974) makes effective use of the association of pesticide transport with off-site transports of water and soil. Its analysis unit is the single rainstorm, which we have seen to be the principal forcing function in ecosystems hydrodynamics. In the spirit of the "Era of Evaporation" (Chapter XII) the model also analyzes processes at the surface between storms. It utilizes the natural time units for analysis—the rainstorm and the interstorm period.

Although this kind of off-site transport needs to be studied, it would also be helpful if the source strength were reduced. One factor here is the increasing cost of chemicals. Another is the enforcement of laws to protect the quality of water bodies receiving effluent from agricultural lands. The Federal Act of 1972, for example, gives special attention to the problem of pesticides (Section 104(1)), because of their effects on human health, and to research on area sources of agricultural pollutants (Section 104(p)).

Effects of Urbanization

Places where the soil is torn up more than even the worst farmer would allow are to be seen around many cities. Converting rural land to suburban sprawl seems to evoke a passion to eliminate vegetation, and large areas of land are bared, usually for several months. Around

Baltimore, the exposed area at any one time averages 19 km^2 (Wolman and Schick, 1967). Bulldozer grading tears up the soil more deeply than any plow and exposes to rain impact large areas of erodible subsoil.

In such "development" around Baltimore, the off-site movement of soil particles reached rates of several kilograms per square meter of land, a hundred times as large as from most eroding farmland. For each person added to the suburban population, about one ton of soil was detached and set moving downslope. Not all erosion can be blamed on hill farmers of Appalachia or tenant farmers of the southeast.

Such urban areas do not cease generating off-site movement of potentially harmful substances. Even after the new householders have sodded down their yards and erodible earth has been covered with concrete, transport of different kinds of material continues. It used to be thought that in comparison with the flow in sanitary sewers, the runoff from streets, roofs, and yards was clean, but closer examination has revealed that it is not. Often it is salty, carries oil and grease from cars (and often lead), and is organically polluted by the excrement of a large dog population and by runoff from over-fertilized lawns. Urban runoff is a major cause of polluted water bodies according to the U.S. Council on Environmental Quality (1972).

Cities also harbor many waste-producing systems that for a century have made use of water as a vehicle to carry off these wastes. These operations include not only industries, but also homes. The waste-laden water is collected by sewer systems and kept separated (in theory) from the environment until after it has been treated. Legislation to protect the quality of water bodies into which the effluent of these treatment plants might be deposited tends to encourage on-land disposal (to make use of the biological action of soil biota) or to limit the destructive potential of the effluent on river or lake quality. Thus the monthly average content of suspended solid particles in the effluent may not exceed 30 g m^{-3} * (Izaak Walton League, 1973, p. 24), nor its biological demand for oxygen 30 g m^{-3}.

The effluent has a zero-discharge limit for high-level radioactive wastes, and low if not zero limits for such toxic substances as mercury. Such limits may be tightened further if necessary "to protect public

* This limit would imply, in the case of Milwaukee, with an off-site flow of water of 100 kg m^{-2} yr^{-1}, a mass flux of 3 g m^{-2}. From a hectare of city land, this would amount to 30 kg of material permissible in the treatment-plant effluent into the lake.

water supplies" and other uses of water bodies (U.S. Environmental Protection Agency, 1973, p. 10). The results of effluent standards are intended to be seen in terms of undegraded lakes and rivers. Furthermore, the Act of 1972 (Public Law 92-500) provided for citizen action, if necessary to ensure adherence to these standards, backed up by an extensive program of effluent monitoring (U.S. Congress, 1972). It looks as though Americans may begin to have information never before available about off-site flows.

Cities, like other ecosystems, cannot be understood solely with reference to individual flows of water, other forms of matter, and energy. The residuals of urban systems come in many shapes—degraded thermal energy, long-wave radiation, carbon dioxide, heavy metals, and thousands of kinds of molecules. Many of these residuals have a special affinity for water, and move within or away from the city via water, that all-purpose vehicle. These interactions require us to monitor and study human settlements as integrated systems (Stearns and Montag, 1974, p. 124) in which human intervention, for instance, alters the form of a residual or recycles it. Cities are both working systems (though they ought to work better) and also suppliers of habitat. The intake, flow, and output of water in a city participate in these two roles.

TIME VARIATIONS IN OFF-SITE FLOW

Episodic Waves of Off-Site Flow

Off-site flow of water, which takes place in amounts that might total 100 or 200 mm yr^{-1}, is the sum of a number of individual events. In each event, overland and near-surface flow from an ecosystem begins as a film that gets deeper as it moves downhill. The energy economies inherent in deeper flow (less friction) give this concentration the ability to preempt water in its vicinity. As the concentrated flow gains volume, it benefits from larger energy economies and begins to convert energy in excess of what is dissipated in friction with the surface and vegetation; it is able to begin carving its own rill. This flow rises rapidly, peaks, and then recedes in a characteristic pattern.

Channel Cutting and Maintenance As a rill, it acquires greater depth, reduces the friction cost still more, gains more economies of scale and both lengthens and deepens the channel it is cutting in the earth's surface. Horton (1945)* spoke of a zone of overland flow (or a belt of "no erosion"), downslope from which flowing water collected itself into channels. This concept can also be expressed as an index of channel maintenance (Schumm, 1956), a number that represents the contributing area required to generate enough off-site flow to cut and maintain one unit length of channel. For example, in soft, impermeable clay, which generates a large volume of surface runoff, about 3 m^2 of land can maintain 1 m of channel. At the other extreme, up to 1000 m^2 of lands with high infiltration capacity may be needed to maintain 1 m of channel.

In the Sierra Nevada, daily pulses of meltwater having a volume of 20–30 kg m^{-2} of contributing area move along layers in the snow cover, on the actual ground surface, and just under it, for distances of 100–200 m before collecting into topographically observable channels. The index of channel maintenance in the basin of Castle Creek averages 165 m^2 m^{-1} of channel; such a contributing area daily generates between $3\frac{1}{2}$ and 5 tons of water for each meter of channel (Miller, 1948).

The reciprocal of the channel-maintenance index is called the "drainage density." In the 10-km^2 basin of Castle Creek, the total length of stream channels is 60 km; the drainage density is 60 divided by 10, or 6 km km^{-2}. This number indicates one way in which the off-site flow from the ecosystems in a drainage basin produces the drainage pattern that carries it away.

Atmospheric factors of rainfall intensity or rate of snow melting combine with such parameters of the earth's surface as infiltration capacity and erodibility to produce a specific type of dissection of that surface. In some places the scale of dissection is very fine, with a high drainage density; in others it might be quite coarse, with channels a long distance apart.

This organization of a landscape does not come about entirely as a matter of chance, although chance elements enter into it, for example, the random-walk concept. Basically, however, the off-site movement of matter and energy from systems within a drainage basin takes place

* This last major publication by Robert Horton, which formulated a new theory of the spatial and physical organization of stream channels, was as seminal in geomorphology as were his earlier works in hydrology. It was one of the beginnings of a rich period of investigations of land forms which, although outside our area of ecosystem dynamics, should be noted for inaugurating quantitative techniques in one of the older fields of geophysics.

in such a way as to bring about certain features of dissection. As Strahler (1964, p. 71) says,

> Conditions for a steady state within a drainage basin are such that, for a given Horton number (i.e., for a given intensity of erosion process), values of local relief, slope, and drainage density reach a time-independent steady state in which basin geometry is so adjusted as to transmit through the system just that quantity of runoff and debris characteristically produced under the controlling climatic regime.

In other words, the off-site yields of water and sediment are characteristic functions of the system and are associated with the geometrical form it assumes under conditions of equilibrium. The equilibrium is "maintained by mutual interaction of channel characteristics such as gradient, cross-sectional form, roughness, and channel pattern. It is a self-regulatory system" (Morisawa, 1968, p. 125). Channel conditions change to accommodate changes in the inputs the channel receives, as water or sediment, from the ecosystems that cover the slopes of the basin.

Interaction of Channels and Systems That Contribute Off-Site Flow The effect that water-generating systems have on channels is, however, not one-way, because the upland slopes themselves have been excavated by the off-site movement of soil into and through streams. Erosion results in the differentiation of an initially more uniform surface into a patchwork of terrain facets. These facets differ in exposure to rainstorms, and the ecosystems on them differ in capacity to take in and store water; different storage capacities produce differences in evapotranspiration and depletion of soil moisture. The resulting differences in storage capacity that is empty at the time a rain arrives produce differences in the way each ecosystem and slope facet responds to the storm, and these responses produce differences in the generation of off-site flows of water and soil. The whole system contains important feedbacks.

"Variations in rainfall, coupled with variation in evapotranspiration loss by variations in slope and aspect of the ground surface, make it appear that large heterogeneous catchments must be analysed in component parts" (Boughton, 1970, p. 57), if runoff is to be correctly determined from data on rainfall. This situation was found true in analyzing snowmelt runoff from Castle Creek basin, which was subdivided into 20 topographic facets, each of uniform slope and exposure, and 300 ecosystems, each responding differently to its inputs of energy by generating different outflows of meltwater. The very dissection of a landscape increases the local contrast among the parts of that landscape with respect to hydrologic processes and consequent yields of water and associated forms of matter. Terrain analyses, such as those of the Castle Creek basin, define the environment of the runoff generation process and aid in the transfer of findings to other basins.

Flood Flows in Channels Flood flows in a channel result from the off-site flows of water from hundreds of ecosystems in its drainage basin during rapid snow melting or a summer rainstorm. Although these outflows are different from different contributing systems, and reach the channel in different times, it still is possible to gain a large amount of usable information about the flow regime in a channel if we hypothesize a rainstorm that produces a uniform depth of "excess" (or uninfiltrated) water over all the contributing ecosystems. It integrates the rise, peak, and decline characteristics of the flows from each of many ecosystems.

This excess water (say 25-mm depth) collects in the channel in a characteristic time-distribution, which is called a "unit hydrograph." This empirical generalization of the behavior of water in the ecosystem across the collecting slopes and in the channels of a drainage basin (Chow, 1964) works well in basins of moderate size and relatively homogeneous composition, that is, where the same mix of forest, pasture, row crops, marsh, and so forth, prevails over the whole basin.

The wave of flood runoff, or the hydrograph, typically rises abruptly; it peaks, and then falls at a declining rate as the last increments of water arrive from distant parts of the drainage basin. Its three major characteristics are (a) the lag of its peak after the time when rain was falling in excess of the rate of infiltration; (b) the rate of streamflow at the peak; and (c) how fast the streamflow recedes following the peak.

(a) To the dweller on a flood plain, the lag represents the time in which a flood-warning system must operate. It ranges from a few hours to a day or so. To a geophysicist, the lag indexes the water-collecting and transporting efficiency of the net of stream channels and the speed of off-site flow over the slopes that move water to them from the runoff-generating ecosystems.

(b) The peak represents the highest stage of water over the channel bank and is highly correlated with damage to contents of houses and stores on the floodplain. Usually the stream flows fastest when it is deepest; this high energy increases its power to cause damage.

(c) The recession period is of particular concern in situations where several subbasins contribute peaks at different times to one channel, where flood water is being stored in a reservoir, or where water on the flood plain must be drained back into the channel.

An example of a flood hydrograph is shown in Fig. XVI-6, the flow of water in Castle Creek in the Central Sierra Snow Laboratory (U.S. Corps of Engineers and Weather Bureau, 1952). In a storm of 43-hr duration in mid-October 1947, rain amounting to approximately 90 mm fell on moist ground unprotected by a snow cover. The initial rain, falling at relatively low intensities of 1–2 mm hr^{-1}, produced little immediate effect on the creek, although the level of the shallow groundwater displayed a response. Heaviest rain, falling at rates of about 9 mm hr^{-1}, came at about 0600 on 16 October; the stream peaked about 1200 on the same day. Peak flow reached about 3 m^3 sec^{-1}. The lag was about 6 hr. The flow receded rapidly from the peak for 4 or 5 hr, then entered a more regular recession, in which each hour's flow was 0.9 of that of the preceding hour.*

Expressed in terms of basin area, the peak flow was equivalent to 1 mm hr^{-1}, considerably less than the peak rate of input of rainwater to the basin. Obviously many of the ecosystems were contributing little or no quick runoff. Over the whole period of the stream rise, total outflow of water was about 12 mm, again considerably smaller than the total rain (90 mm). Much of the water brought by the storm remained in the basin as an increment to soil moisture or to groundwater that would feed Castle Creek in later months.

Factors in Peak Flows Annual peak flows in streams are often summarized in terms of a 0.43 probability figure, otherwise called the "mean annual flood," with a slightly less than 50–50 chance of occurring in any given year. Over the United States, annual peak flow occurs at a rate of 3–4 mm hr^{-1}, indicating that the October flood just described was relatively small. In fact, it would often be exceeded during the spring melting season.

The peak flow clearly is a function of basin area, but not a linear one; the relation seems to be approximately $A^{0.7}$ (Leopold *et al.*, 1964, p. 251). This exponent reflects the greater opportunities for storage of water in the channels of large basins. It also recognizes the fact that many intense storms are restricted in area and would not deliver water

* The downstream progression of this flood wave belongs to the field of stream hydraulics, which, like the related field of fluvial geomorphology, lies outside the scope of this book.

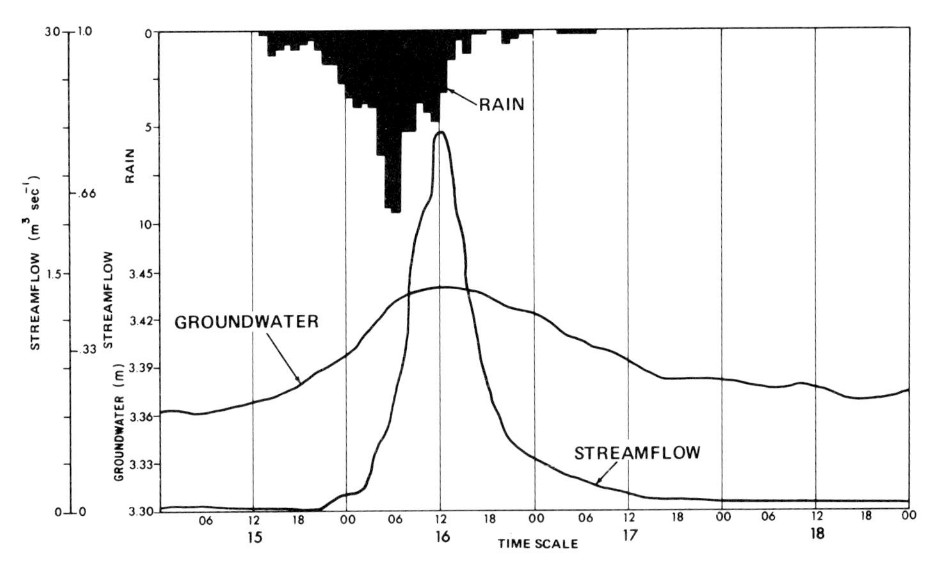

Fig. XVI-6. Hydrograph of Castle Creek, Central Sierra Snow Laboratory, California, in a rainstorm of 15–16 October 1947. Also shown is the groundwater hydrograph and the block diagram of rainfall intensities. Rates of mm hr^{-1} are shown for both rainfall and streamflow (from U.S. Corps Engineers and Weather Bureau, 1952, p. 13).

at high intensities to all of a large basin, a subject discussed in an earlier chapter under depth–area–duration analysis of hydrologic storms. Basin shape also affects peak flow, as well as lag and recession in Appalachian streams (Snyder, 1938).

Another drainage-basin characteristic that is related to peak flow is the density of vegetation cover. Dense, vigorous ecosystems generate relatively smaller volumes of off-site flow and hold it longer. Vegetation density can be expressed as a fraction of bare area, or the number of years since the last major fire. In a study of the floods in southern California in March 1938, peak flows from 40 basins were found related to rainfall, drainage-area size, years since the last fire, and a combination of Horton's indexes of drainage-basin dimensions (Anderson and Trobitz, 1949).

A later southern California study (Rice, 1970) of 35 rainstorms on 14 basins (mean area 3 km^2), grouped landscape properties by principal component analysis. These components seem to express concepts of basin area, shape, bedrock faulting, slope steepness, soil stability, and development of the drainage net.

In addition to these relatively stable properties of the basins studied are such transient conditions as the fire history (that is, age of vegetation, canopy density, perhaps canopy depth), soil moisture just before the storm, and several parameters of the storm itself. Rice concluded that the analysis increased the understanding of the "internal structure" of the basin, which might be taken to mean the manner in which its ecosystems and channels are linked by off-site flow.

Uncertainties in Peak Flows While heavy rains might strike a basin several times a year, few come at the precise time when soil–vegetation systems are ready to maximize the conversion of an input of rainwater into a large output of surface and near-surface runoff. The conversion of rainfall into overland runoff is most effective in seasons of high soil moisture, such as late winter. When the soil moisture is depleted, conversion of rainfall to off-site flow is reduced or entirely absent, because infiltration capacity is large (Chapter VIII).

Our knowledge of the frequency of flood waves is incomplete, due to the shortness of flood records. Our instrumental record of flood flows goes back half a century or so; our record of a few extreme floods is about as long as the period of white settlement. Floods of frequencies larger than 0.02 can be determined with a fair degree of reliability in channels where the mean annual flood is determined, but it is risky to extend these relations to rarer, bigger floods. The pattern of flood frequency is, of course, changed with a change in the ecosystems of the drainage area. Urbanization has a particularly large influence (Espy and Winslow, 1974) especially due to the increase in impervious systems and also the loss of interception and detention storages.

The uncertainties in the delivery of heavy rain to the earth's surface described in the first chapters are far greater for large floods. In floods, the element of indeterminacy in rainstorms is multiplied by the indeterminacy in the variable reception of rainwater at the surface, this depending on transient conditions of soil moisture and state of soil–vegetation systems.

Just as no rainfall station has yet experienced the heaviest rain that eventually will break over it, so some future flood in any channel will necessarily be higher than any flood in the recorded history. Moreover, this future flood might well exceed the record flood by a substantial amount, being perhaps double or triple its size. The excess above the past record is much larger for floods than for rainfall events.

Because the extreme flood that could destroy a dam is difficult to determine from short stream-flow records—and many streams have no flow record at all—it is usual for certain design purposes to calculate a "probable maximum" flood. This hypothetical event is the hydrologic follow-up, with basin saturated and channels full of water, of the probable maximum rainstorm described in the first part of this book, or the probable maximum snow-melting event. Developing such an event, whether it begins with a probable maximum rainstorm or transposes a historical storm (such as the Elba storm Snyder uses in the illustration in Chapter II), requires a conceptual framework that is physically coherent. In this framework one can reconstruct the sequence of delivery, interception, detention, and infiltration that will result in a logical derivation of off-site flow hour by hour through the flood.

For less critical conditions in which no loss of life would be caused by failure of the dam, a smaller flood called the "standard project flood" is commonly used. It is based on "the most severe combination of meteorologic and hydrologic conditions that are considered reasonably characteristic of the geographical region involved, excluding extremely rare combinations" (Dalrymple, 1964, pp. 25–26). Standard project floods commonly range from 0.4 to 0.6 of the probable maximum flood.

Flow Regimes through the Year

Off-site flow of water at or near the earth's surface, occurring in short episodes in some but not all of the rainstorms and snow-melting events happening in an ecosystem, is sporadic in occurrence. Except when a series of storms is so closely spaced that the later ones drop rain on a saturated basin and produce a succession of flood waves, they are isolated in time.

A difference between seasons of the year is found in many basins, in keeping with seasonal variations in rainfall intensity and basin condition. These seasonal differences influence not only the number of episodes of off-site flow but also the volume of flood water, which can be summed up in monthly periods.

The regimes of flow through the year differ in different climatic regions of the U.S. Figure XVI-7 presents the runoff regimes in small basins, in which stream flow represents chiefly the contributions of surface or near-surface off-site flow.

The quiescence of small streams in California during more than half the year contrasts with a high concentration of surface-runoff activity in late winter. In this season, soils have been thoroughly wetted by earlier rains, infiltration capacity is small, and heavy rains come in groups that last several days. Percolation, as in the example in the Sierra foothills (Chapter XIV), and off-site flow occur in this season.

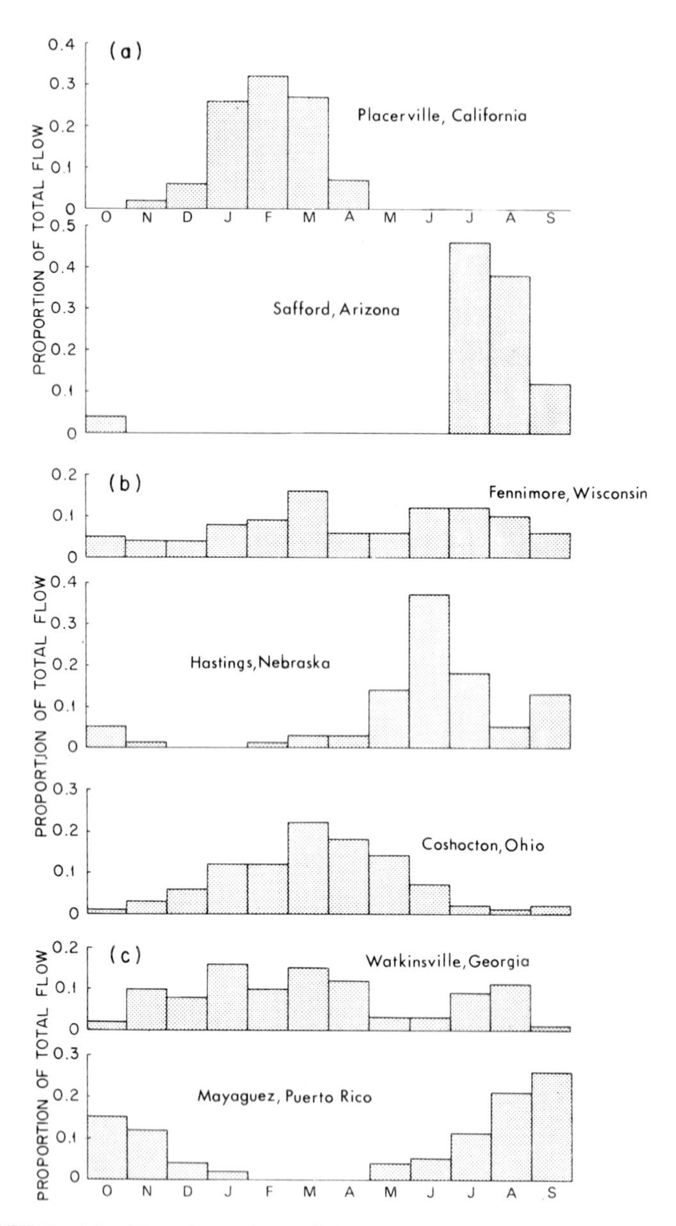

Fig. XVI-7. Monthly values of runoff from small basins, expressed as a fraction of total annual runoff: (a) western; (b) midcontinent; (c) southeastern (data from Ogrosky and Mockus, 1964, Table 21-20, from U. S. Soil Conservation Service).

A more extreme concentration of runoff is found in Arizona, but it comes in late summer. Winter storms normally are light, producing little runoff from rain intensities that are less than the same storm would have produced in California. In late summer, very humid air from the Gulf of Mexico or Gulf of California brings heavy convectional rains. While these are smaller in area than the winter storms, this dimension is unimportant in small drainage basins that can be completely blanketed by a good thunderstorm.

Surface runoff comes early in the summer in Nebraska, reflecting a rainfall maximum with peak intensities in this period in conjunction with wet ground carried over from spring. In Wisconsin, Ohio, and Georgia, the probability of off-site runoff is more evenly spread through the year than at the western stations. The probability is greatest in late winter and spring, a situation typical of most of eastern North America and indicating the initial presence of water as wet or frozen soil, producing low infiltration capacity. Surface runoff from the plot in Puerto Rico is markedly concentrated in the autumn. This indicates the probability of heavy rain from hurricanes, and like other spatial patterns of surface runoff suggests the role, described in Chapter V, of the large-scale spatial organization of atmospheric precipitation systems.

The graphs of flow cumulated from 1 October show that by the end of April, California systems have yielded all their annual off-site flow of water. Those in the eastern United States have yielded between 0.5 and 0.6 of their annual totals by April, while those in Arizona and Puerto Rico have hardly gotten started (Fig. XVI-8).

The mean of monthly values at all seven sites shows an interesting double-maxima curve. One peak comes in March, one in July–August, reflecting different combinations of favorable circumstances of regional water-vapor inflow, rain-producing mechanisms, and surface receptiveness. Total off-site flow as storm runoff averages about 200 mm yr^{-1} over North America; this is about the same as the worldwide average (L'vovich, 1972). In contrast, the mean for Australia is only 40 mm yr^{-1}.

Variations in Off-Site Sediment Flow

There is also a seasonality in sediment yield from drainage basins, although it is not necessarily parallel to the seasonality in surface runoff. "The sediment supply from a watershed is usually not a function of the stream discharge. At the beginning of a storm the

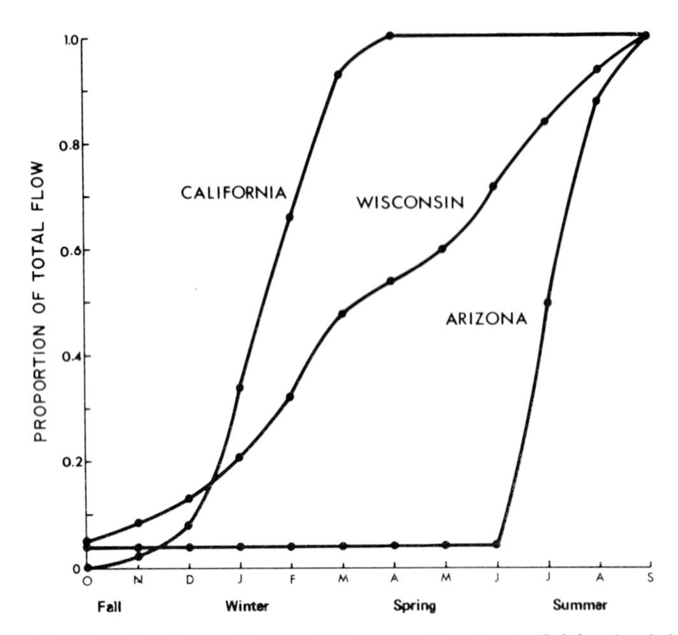

Fig. XVI-8. Cumulated monthly runoff from small basins in California, Arizona, and Wisconsin (see Fig. XVI-7) beginning 1 October.

water may find much more loose material ready to move than at the end of the storm. A winter storm may find entirely different watershed conditions from those of a summer storm" (Einstein, 1964). In West Coast basins, debris that has moved downslope during the dry season by creep and other forms of dry mass movement is ready and waiting at the stream channel when the first big storm comes. In the East Fork Trinity River of California, for instance, monthly mean values of runoff and debris flow in winter months show, as in the accompanying tabulation, the flushing out of the channels by the big

Month	Runoff ($kg\ m^{-2}$)	Suspended sediment ($g\ m^{-2}$)
November	30	10
December	80	600
January	100	280
February	130	80
March	110	35

storms of December (Source: Wilson, 1972). Clearly much of the mass flux takes place in a few days in December. Leopold *et al.* (1964, p. 73) note that the Cheyenne River in South Dakota, a stream of highly variable flow that drains badlands, transports 0.28 of its annual volume of suspended sediment in the ten highest days of the year. In contrast, the Niobrara River not far away in Nebraska, transports only 0.07 of its annual sediment volume in the ten highest days. There is a tendency for movement of the dissolved load (salts) to be less concentrated in time than the flow of water, and for the movement of the suspended load to be more concentrated (Leopold *et al.*, 1964).

Similar divergences in regime are seen in other water pollutants, many of which can be put into two classes: (a) those displaying a dilution trend, i.e., smaller concentrations at times of higher flows of water; and (b) those displaying a runoff trend, increasing in concentration at times of high flow (U.S., Council Environmental Quality, 1972, p. 13). The second class includes substances detached from the soil surface during storms or floated into channels by surface runoff, like the lead deposits on streets that are mobilized in rainstorms.

Off-site movement of suspended solids in the sewage flow from a city is likely to vary relatively little, due to the generally steady rate of operation of urban systems. The weekly average figure may not exceed 1.5 times the monthly average (which is 30 g m^{-3} of effluent) (Izaak Walton League, 1973, p. 24), a legal range predicated on a small natural variation.

The idea that the ecosystems near the drainage channels often contribute most of the storm flow, discussed earlier, also explains sometimes puzzling fluctuations in the off-site flux of certain kinds of suspended materials, like phosphates. A short transit *over* the soil surface affords less opportunity for phosphates to be adsorbed onto soil particles than does a longer transit through the soil in laminar flow, or a still slower journey via percolation to groundwater (Ryden *et al.*, 1973). The large off-site movement of phosphate when manure is spread on frozen ground is a well-known example of the role played by the exact path of the off-site flow of water in understanding the off-site flow of other substances, and hence in understanding their variations over time, and in applying data from plot experiments to large areas.

Flood Plains

Areally small but highly important components of fluvial landscapes, flood plains, result from another kind of interaction between the off-site flow of water and that of soil particles. Many things to many people, flood plains here can be examined as places where soil or water are temporarily stored in groups of well-watered ecosystems that show up clearly on aerial photographs (e.g., Plate 20 in Chapter XI).

Flood Plains as Sediment Storages The movement of debris from upland ecosystems might not necessarily go all the way to the ocean before stopping. Under most conditions a given particle goes only a few kilometers before being deposited in a sand bar or some other temporary resting place. In conditions of unsteady state, such as characterize most of the agricultural lands where cultivation began in the last two centuries, off-site flow is far greater than the volume of sediment carried all the way to the sea. Even considering the dissolving or the reduction in size of sediment particles during transport, large differences are measured between the sediment outputs of large and small basins as shown in the accompanying tabulation (Source: Gottschalk, 1964). It is clear that a large amount of the debris moving out of small upland basins is trapped downstream, before it issues from larger river basins. "Wide flood plains on streams draining large watersheds provide an environment for deposition which does not exist on steeper channels of smaller headwater tributaries" (Gottschalk, 1964). The benches and narrow alluvial strips along small creeks grow into wide meadows and swamps downstream.

Basin area (km^2)	Equivalent depth removed (mm)	$m^3\ km^{-2}$
<25	1.9	1900
25–250	0.8	800
250–2500	0.5	500
>2500	0.2	200

The flatness and continuity of flood plains makes them attractive sites for transportation lines and, near cities, for industrial and residential uses. All these functions follow from the way in which the sediments forming the flood plains are laid down by water.

Since flood plains are composed of soil particles from many parts of a drainage basin, often quite diverse geologically, they usually have soils that provide a good mix of plant nutrients. Especially where uplands lack one or another critical element and are all but sterile as far as cultivation is concerned, as in many parts of eastern Australia, the flood plains are the only cultivable parts of the drainage basins. In addition, flood-plain alluvium usually has good soil-moisture holding characteristics and is a good environment of groundwater at a convenient depth—further attractions for farming.

Flood Plains as Storage for Water Geomorphologists have observed that the general level of the principal flood plain in many channels conforms to the level of the flood wave that occurs, on the average, about every two years. In fact, it would be strange if there were not some kind of correspondence between a landform created by water-borne sediment and the floods in which much of the sediment is carried.

In floods higher than those influential in establishing the height of the main flood plain, the plain becomes a site where water is stored and which enlarges the cross section of the river. In both ways it accommodates itself to the fact that the ecosystems on the uplands of its drainage system experience spasms from time to time in which they produce great volumes of off-site flow. The width and elevation of the flood plain reflect the long-term frequency of occurrence of these events; those of medium rarity apparently have the greatest influence.

The 0.01-frequency flood is commonly taken as the criterion in establishing flood-plain zoning, and in many places the endangered area can be delineated on air and satellite photos by the contrast between its distinctive ecosystems and those on adjacent uplands. The flood-plain ecosystems include many species that tolerate flooding, sometimes for long periods, and have developed special soil series.

Flood waters are stored in the flood-plain soils and sediments as bank storage, and on the surface as overbank storage. This storage capacity, very large in some flood plains, reduces peak flows in the channel and prolongs the period of recession.

Flood Plains as Arenas of Conflict In this condition of equilibrium between flood-producing capabilities of the systems that shed waves of off-site flow into one stream channel, and the size of the channel receiving these waves, it seems foolish to interfere, considering the enormous tonnages of water involved.* Yet near many cities man confidently encroaches on flood plains that are only partially protected (and levees almost never afford complete protection). Such encroachment is most common near small streams, which might never have been gaged and whose flooding history and proclivities are unknown. For this reason and others, the $300–500 million annually invested, year after year, in flood-protection works in the United States have done little to decrease the average annual damage caused by floods, which, in any given year, make up two-thirds or more of the officially designated disaster areas.

* The median annual flood in the Milwaukee River—certainly not one of the nation's, or even the state's large rivers—brings down 12 million tons of water in a day.

This paradox has been investigated by White (1961, 1964, 1966) and other geographers and hydrologists. In addition to studying the flood hydrologies of many river basins, they examined the agricultural and urban occupance of flood plains and the perceptions that flood-plain residents have about the behavior of the river in whose domain they choose to live.

Flood damages will continue to grow without methods for meeting the occasional times when upslope ecosystems produce off-site flows of water in such volume as to require most of the valley a river has made for its own use. This growing risk puts an unfair burden on the engineers designing dams and levees, as well as on the taxpayers supporting their work. The studies by White led to new procedures for federal investment, in order to ensure that in each site the whole inventory of methods of alleviating flood hazard will be considered before funds are invested. For example, the underwriting of flood insurance is now tied in with such local actions as flood-plain zoning. The studies also improved our understanding of how we and our ecosystems can coexist on productive flood-plain lands with these rare dramatic cases of mass transport.

Of more than a thousand arenas where conflict has taken place between the occupation of flood plains by off-site flow from the uplands and its occupation by human activities, White's group gave special attention to a few, such as Darlington, Wisconsin. This market town in the southwestern part of the state, where commercial activity has extended from the courthouse down into a flood plain that accommodates flood waves of the Pecatonica drainage basin (area 690 km^2) that come relatively often: about 2.4 yr^{-1}. This high frequency is fortunate, because it keeps the possibility of flooding in the citizens' awareness. Measures to reduce damage are not allowed to rust away from disuse.

Some years ago the city decided not to sponsor a storage dam in the basin because it would take farm land out of cultivation (White, 1964, p. 33), and because the flood-warning system works well. The lag time of the flood wave, 6–8 hr, seems to be sufficient for an efficient warning procedure to operate. Some works constructed in the channel reduced the frequency of small floods and the resulting nuisance of frequently moving merchandise above their levels. New structures are built with elevated floor levels. Other actions suggest that a working adjustment has been reached between the central-place activities that Darlington supplies for the surrounding farm land and the regime of surface and near-surface runoff that is generated during heavy rain or rapid snow melting. This adjustment derives, probably, from the general awareness about the behavior of the off-site runoff.

That satisfactory accommodations of off-site flows and man can be reached is also attested in the humid low latitudes, in hundreds of flood plains and the deltas into which they merge. While most upland soils are leached and infertile, vulnerable to erosion, and difficult to bring into an annual cropping system, the soils of flood plains and deltas, in contrast, are of good texture, high fertility, and relatively easy to irrigate. It clearly is of advantage to use them.

Man has adapted a plant—rice—for intensive cropping of these sometimes ill-drained lands, and over the centuries has developed complex systems of managing and utilizing the flood waters that come from the uplands (Plate 36). One water-management district in Taiwan where rice is a principal crop was studied by VanderMeer (1968, 1971) over a period of change, when better knowledge of the water requirements of rice made possible a more rational basis for allocating water. The result was a decrease in the man-day costs of transporting water and increased levels of production.

Plate 36. Where volcanic upland soils are fertile, as in Indonesia, rice is grown on slopes as well as in deltas and flood plains, under water-retention management. These rice terraces are north of Bandung (photo by P. Irwin).

OFF-SITE FLOWS FROM ECOSYSTEMS

We have here discussed, not the storages and movements of water *in* ecosystems, but rather its movement *from* them when it travels over the surface or through porous soil layers close to the surface. Many factors influence the speed and volume of this movement, including such atmospheric variables as rain intensity, and such surface conditions as infiltration capacity that to a degree can be affected by man's practices in agriculture and urbanization.

Associated with the off-site flows of storm or meltwater at and near the surface are corollary but not always parallel flows of other forms of matter: soil particles, nutrients, rural and urban contaminants. These are also delivered to the channel networks that off-site flows of water have carved in the earth's surface and incorporated in flood plains. Not only the shaping but also the chemical makeup of downslope ecosystems reflect the off-site flows from ecosystems lying upslope from them. After the Eras of Infiltration and Evaporation and the emphasis on the volume of water yield, we are beginning to think about product quality. How welcome are these yields in the aptly named "receiving waters"?

The trend toward product quality in hydrology is particularly important for surface runoff. Demand for a cleaner environment, diminishing resources under population pressure, and consumer awareness have met a situation of increasingly complex contamination of water, as well as a broadening of area pollution sources from chemicals and energy being applied to rural ecosystems, at times, for example, in feed lots and strip mines, to an industrial level of intensity. If downstream users think water is getting scarce, they at least want it usable. We can anticipate more interest in the ecosystem processes that are most significant in generating a clean product. How should an ecosystem be protected or managed so that its off-site fluxes are acceptable in quality? How can an abused or malfunctioning system be renovated so that it will produce a marketable product?

The episodes of flow events and their seasonal regimes reflect basic atmospheric factors operating on the geology of a region through the intermediary layer of soil-vegetation systems that rocks and air, water and energy have brought into being. These atmospheric and geologic factors also produce a quieter form of off-site movement, which will be described next, followed by an evaluation of all the yields of water and other substances from ecosystems.

REFERENCES

Abal'yan, T. S., Mazurova, L. J., and Nikonorova, S. N. (1971). Use of aerial photographs for snow cover study in the mountain basin of the Varsob River, *Sov. Hydrol.* 151–160.

Anderson, H. W. (1974). Sediment deposition in reservoirs associated with rural roads, forest fires, and catchment attributes. *Int. Assoc. Hydrol. Sci. Publ.* **113,** 87–95.

Anderson, H. W., and Trobitz, H. K. (1949). Influence of some watershed variables on a major flood, *J. Forestry* **47,** 347–356.

Arai, T., and Sekine, K. (1973). Study on the formation of perennial snow patches in Japan, *Geog. Rev. Japan* **46,** 569–582 [*Engl. res.*].

Aulitzky, H. (ed.). (1961). Ökologische Untersuchungen in der subalpinen Stufe zum Zwecke der Hochlagenaufforstung, *Mitt. Forstl. Bundes-Versuchsanstalt Mariabrunn* **59,** Bd.1, 431 pp.

Aulitzky, H. (1965). Waldbau auf bioklimatischer Grundlage in der subalpinen Stufe der Innenalpen, *Cbl. Ges. Forstwesen* **82,** 217–245.

Australia (1969). Commonwealth Scientific and Industrial Research Organization. Land Research Division, Lands of Queanbeyan–Shoalhaven Area, A.C.T. and N.S.W. Land Res. Ser. 24, 164 pp. maps, Melbourne.

Bailey, G. W., Swank, R. R. Jr., and Nicholson, H. P. (1974). Predicting pesticide runoff from agricultural land: a conceptual model, *J. Envir. Qual.* **3,** 95–102.

Boughton, W. C. (1970). Effects of Land Management on Quantity and Quality of Available Water. Aust. Water Resour. Council., Univ. N.S.W. Water Res. Lab. Rep. 120, Sydney, 330 pp.

Branson, F. A., and Owen, J. B. (1970). Plant cover, runoff, and sediment yield relationships on Mancos shale in western Colorado, *Water Resources Res.* **6,** 783–790.

Campbell, A. P. (1967). Forecasting downstream effects from dominant discharge and other interpretive data. *In* "Forest Hydrology" (W. E. Sopper and H. W. Lull, eds.), pp. 673–686. Pergamon, Oxford.

Chow, V. T. (1964). Runoff. *In* "Handbook of Applied Hydrology" (V. T. Chow, ed.), Chapter 14. McGraw-Hill, New York, 54 pp.

Colman, E. A. (1953). "Vegetation and Watershed Management: An Appraisal of Vegetation Management in Relation to Water Supply, Flood Control and Soil Erosion." Ronald Press, New York, 412 pp.

Costin, A. B. (1959). Vegetation of high mountains in Australia in relation to land use. *In* "Biogeography and Ecology in Australia" (A. Keast, R. L. Crocker, and C. S. Christian, eds.). The Hague: Junk. *Monogr. Biolog.* **8,** 427–451.

Costin, A. B. (1971). Water. *In* "Conservation" (A. B. Costin and H. J. Frith, eds.), pp. 71–103. Penguin Books, New York.

Costin, A. B., Wimbush, D. J., Kerr, D., and Gay, L. W. (1959). Studies in Catchment Hydrology in the Australian Alps. I. Trends in Soils and Vegetation. Aust. CSIRO, Div. Plant Ind., Tech Paper 13, 36 pp.

Costin, A. B., Wimbush, D. J., and Kerr, D. (1960). Studies in Catchment Hydrology in the Australian Alps. II. Surface Runoff and Soil Loss, Australia. CSIRO, Div. Plant Ind., Tech. Paper 14, 23 pp.

Dalrymple, T. (1964). Flood characteristics and flow determination. *In* "Handbook of Applied Hydrology" (V. T. Chow, ed.), Chapter 25-I. McGraw-Hill, New York.

Dreibelbis, F. R., and Amerman, C. R. (1964). Land use, soil type and practice effects on the water budget, *J. Geophys. Res.* **69,** 3387–3393.

Dunne, T., and Black, R. D. (1970). Partial area contributions to storm runoff in a small New England watershed, *Water Resources Res.* **6,** 1291–1311.
Einstein, H. A. (1964). River sedimentation. *In* "Handbook of Applied Hydrology" (V. T. Chow, ed.). Chapter 17, pp. 35–67. McGraw-Hill, New York.
Engman, E. T. (1974). Partial area hydrology and its application to water resources, *Water Resources Bull.* **10,** 512–521.
Espy, W. H. Jr., and Winslow, D. E. (1974). Urban flood frequency characteristics, *Am. Soc. Civil Eng. Proc. J. Hydraul. Div.* **100,** Hy-2, 279–294.
Evanari, M., Shanan, L., and Tadmor, N. H. (1968). "Runoff farming" in the desert. I. Experimental layout, *Agron. J.* **60,** 29–32.
Evanari, M., Shanan, L., and Tadmor, N. H. (1971). "The Negev. The Challenge of a Desert." Harvard Univ. Press, Cambridge, Massachusetts, 345 pp.
FAO (1965). Soil Erosion by Water: Some Measures for Its Control on Cultivated Lands. FAO *Agric. Dev.* Paper 81, Rome, 284 pp.
Fairbourn, M. L., Rauzi, F., and Gardner, H. F. (1972). Harvesting precipitation for a dependable, economical water supply, *J. Soil Water Conserv.* **27,** 23–26.
Freeze, R. A. (1972). Role of subsurface flow in generating surface runoff. 2. Upstream source areas, *Water Resources Res.* **8,** 1272–1283.
Gburek, W. J., and Heald, W. R. (1974). Soluble phosphate output of an agricultural watershed in Pennsylvania, *Water Resources Res.* **10,** 113–118.
Gottschalk, L. C. (1964). Reservoir sedimentation. *In* "Handbook of Applied Hydrology," (V. T. Chow, ed.), Chapter 17-I. McGraw-Hill, New York. 34 pp.
Helvey, J. D., Hewlett, J. D., and Douglass, J. E. (1972). Predicting soil moisture in the southern Appalachians, *Proc. Soil Sci. Soc. Am.* **36,** 954–959.
Hillel, D., and Rawitz, E. (1968). A preliminary field study of surface treatments for runoff inducement in the Negev of Israel, *Int. Congr. Soil Sci., 9th Trans.,* **1,** 303–311.
Horton, R. E. (1937). Determination of infiltration-capacity for large drainage-basins, *Trans. Am. Geophys. Un.* **20,** 371–385.
Horton, R. E. (1945). Erosional development of streams and their drainage basins; hydrophysical approach to quantitative morphology, *Bull. Geol. Soc. Am.* **56,** 275–370.
International Association of Hydrological Sciences, International Commission on Snow and Ice. Working Group on Avalanche Classification (1973). Avalanche classification, *Hydrol. Sci. Bull.* **18,** 391–402.
Izaak Walton League (1973). A Citizen's Guide to Clean Water. Izaak Walton League, Washington, D.C., 94 pp.
Kinkead, E., (1971). Big rain, *The New Yorker,* 31 July 1971, pp. 66–74.
Kirkby, M. J. (1969). Infiltration, throughflow, and overland flow. *In* "Water, Earth, and Man" (R. J. Chorley, ed.), pp. 215–227. Methuen, London.
Kohnke, H., and Bertrand, A. R. (1959). "Soil Conservation." McGraw-Hill, New York, 298 pp.
Laycock, A. H. (1972). The diversity of the physical landscape. *In* "The Prairie Provinces" (P. J. Smith, ed.), pp. 1–32. Univ. Toronto Press, Toronto.
Leopold, L. B. (1968). Hydrology for Urban Land Planning—a Guidebook on the Hydrologic Effects of Urban Land Use. U. S. Geol. Surv. Circ. 554, 18 pp.
Leopold, L. B., Wolman, M. G., and Miller, J. P. (1964). "Fluvial Processes in Geomorphology." Freeman, San Francisco, California, 522 pp.
Lull, H. W., and Reinhart, K. G. (1972). Forests and Floods in the Eastern United States. U. S. Forest Serv. Res. Paper NE-226, 94 pp.
L'vovich, M. I. (1972). Vodnyi balans materikov zemnogo shara i balansovaia otsenka mirovykh resursov presnykh vod, *Izv. Akad. Nauk Ser. Geog.* **5,** 5–20.

L'vovich, M. I. (1973). "The World's Water: Today and Tomorrow." Mir Publ., Moscow, 213 pp. (*English transl.* by L. Stoklitsky of Vodnye Resursy Budushchego. Izdat. Proveshchenie, Moskva, (1972).
Martinec, J. (1975). Subsurface flow from snowmelt traced by tritium, *Water Resources Res.* **11,** 496–498.
Megahan, W. F. (1972). Subsurface flow interception by a logging road in mountains of central Idaho. *In* Watersheds in Transition, pp. 350–356. Am. Water Res. Assoc. and Colorado State Univ.
Menzel, R. G. (1974). Land surface erosion and rainfall as sources of strontium-90 in streams, *J. Envir. Qual.* **3,** 219–223.
Mikhel', V. M., and Rudneva, A. V. (1971). Description of snow transport and snow deposition in the European USSR, *Sov. Hydrol.* (4), 342–348.
Miller, D. H. (1948). A terrain sample of the Sierra crest region, *Yrbk. Assoc. Pacific Coast Geog.* **10,** 46–47.
Morgali, J. R., and Linsley, R. K. (1965). Computer analysis of overland flow, *Am. Soc. Civil Eng. Proc. J. Hydrol. Div.* **91,** Hy-3, 81–100.
Morisawa, M. (1968). "Streams: Their Dynamics and Morphology." McGraw-Hill, New York: 175 pp.
Narayanan, A. V. S., and Swanson, E. R. (1972). Estimating trade-offs between sedimentation and farm income, *J. Soil Water Conserv.* **27,** 262–264.
Nelson, L. B. (1958). Building sounder conservation and water management research programs for the future, *Proc. Soil Sci. Soc. Am.* **22,** 355–358.
Odum, E. P. (1969). The strategy of ecosystem development, *Science* **164,** 262–270.
Ogrosky, H. O., and Mockus, V. (1964). Hydrology of agricultural lands. *In* "Handbook of Applied Hydrology" (V. T. Chow, ed.), Chapter 21. McGraw-Hill, New York, 97 pp.
Pereira, H. C. (1973). "Land Use and Water Resources in Temperate and Tropical Climates." Cambridge Univ. Press, London and New York, 246 pp.
Pierce, R. S., Hornbeck, J. W., Likens, G. E., and Bormann, F. H. (1970). Effect of elimination of vegetation on stream water quantity and quality, *Int. Assoc. Sci. Hydrol. Publ.* **96,** 311–328.
Rice, R. M. (1970). Factor analysis for the interpretation of basin topography, *Int. Assoc. Sci. Hydrol. Publ.* **96,** 253–268.
Ryden, J. C., Syers, J. K., and Harris, R. F. (1973). Phosphorus in runoff and streams, *Adv. Agron.* **25,** 1–45.
Sander, D. H. (1970). Soil water and tree growth in a Great Plains windbreak, *Soil Sci.* **110,** 128–135.
Sartz, R. S. (1969). Folklore and bromides in watershed management, *J. Forestry* **67,** 366–371.
Sartz, R. S. (1970). Effect of land use on the hydrology of small watersheds in southwestern Wisconsin, *Int. Assoc. Hydrol. Sci. Publ.* **96,** 286–295 (Symp. Represent. Exptl. Basins).
Schumm, S. A. (1956). Evolution of drainage systems and slopes in badlands at Perth Amboy, N. J., *Bull. Geol. Soc. Am.* **67,** 597–646.
Snyder, F. F. (1938). Synthetic unit-graphs, *Trans. Am. Geophys. Un.* **19,** 447–454.
Soil Science Society of America. Water Resources Committee, C. H. Wadleigh, chairman (1966). Soil science in relation to water resources development. I. Watershed protection and flood abatement, *Proc. Soil Sci. Soc. Am.* **30,** 421–424.
Southeastern Wisconsin Regional Planning Commission (1971). A Comprehensive Plan for the Milwaukee River Watershed, Vol. 2, Alternative Plans and Recommended Plan. SEWRPC, Waukesha, 625 pp.

Stephenson, G. R., and Freeze, R. A. (1974). Mathematical simulation of subsurface flow contributions to snowmelt runoff, Reynolds Creek watershed, Idaho, *Water Resources Res.* **10,** 284–294.

Stearns, F., and Montag, J., eds. (1974). "The Urban Ecosystem: A Holistic Approach." Dowden, Hutchinson & Ross, Stroudsburg, Pennsylvania, 217 pp.

Strahler, A. N. (1964). Quantitative geomorphology of drainage basins and channel networks, *In* "Handbook of Applied Hydrology" (V. T. Chow, ed.), Chapter 4, pp. 39–76. McGraw-Hill, New York.

Tadmor, N. H., and Shanan, L. (1969). Runoff inducement in an arid region by removal of vegetation, *Proc. Soil Sci. Soc. Am.* **33,** 790–794.

Takeda, S., and Ishii, M. (1968). Movement of sub-surface flows from forest area and mechanism that affects soil erosion. Iwate Univ., Mtn. Land Use Res. Lab. Bull. 1, 15 pp.

Tholin, A. L., and Keifer, C. J. (1960). The hydrology of urban runoff, *Trans. Am. Soc. Civil Eng.* **125,** 1308–1379.

Toebes, C., Scarf, F., and Yates, M. E. (1968). Effects of cultural changes on Makara experimental basin. Hydrological and agricultural production effects of improving intensively grazed small catchments, *Int. Assoc. Sci. Hydrol. Bull.* **13,** No. 3, 95–122.

U. S. Congress, 92nd. (1972). Public Law 92-500, Federal Water Pollution Control Act Amendments of 1972. 86 Stat. 816.

U. S. Corps of Engineers and Weather Bureau. Cooperative Snow Investigations (1952). Hydrometeorological Log of the Central Sierra Snow Laboratory 1947–1948 Water Year. Corps of Engineers, San Francisco, California, 220 pp. + app.

U. S. Corps of Engineers and Weather Bureau. Cooperative Snow Investigations (1952). Hydrometeorological Log of the Central Snow Laboratory 1950–1951 Water Year. So. Pac. Div., Corps of Engineers, San Francisco, California, 212 pp. 6 app.

U. S. Council on Environmental Quality (1972). Environmental Quality. 3rd Annu. Rep. Washington, D.C., 450 pp.

U. S. Environmental Protection Agency (1973). Action for Environmental Quality: Standards and Enforcement for Air and Water Pollution Control. Envir. Prot. Agency., Washington, D.C., 21 pp.

VanderMeer, C. (1968). Changing water control in a Taiwanese rice field irrigation system, *Ann. Assoc. Am. Geog.* **58,** 720–747.

VanderMeer, C. (1971). Water thievery in a rice irrigation system in Taiwan, *Ann. Assoc. Am. Geog.* **61,** 156–179.

Viessman, W., Jr., Keating, W. R., and Srinivasa, K. N. (1970). Urban storm runoff relations, *Water Resources Res.* **6,** 275–279.

White, G. F. (1961). Comments. *In* "Economics of Watershed Planning" (G. S. Tolley and F. E. Riggs, eds.), pp. 329–330. Iowa State Univ. Press, Ames.

White, G. F. (1964). Choice of Adjustment to Floods. Univ. Chicago Dept. Geog. Res. Paper 93, 164 pp.

White, G. F. (1966). Optimal flood damage management: retrospect and prospect. *In* "Water Research" (A. V. Kneese and S. C. Smith, eds.), pp. 251–269. Johns Hopkins Press, Baltimore, Maryland.

Wilson, L. (1972). Seasonal sediment yield patterns of U.S. rivers, *Water Resources Res.* **8,** 1470–1479.

Wisconsin Natural Resources Council of State Agencies (1971). Quality Management for Wisconsin: Preserving and Improving the Quality of the Air, Land and Water Resources. Madison, 97 pp.

Wolman, M. G., and Schick, P. A. (1967). Effects of construction on fluvial sediment, urban and suburban areas of Maryland, *Water Resources Res.* **3,** 451–462.
Wood, W. F. (1958). Flat grasslands and hilly forest. Why? [abstr.], *Ann. Assoc. Am. Geog.* **48,** 298.
Wooding, R. A. (1965). A hydraulic model for the catchment–stream problem. I. Kinematic-wave theory, *J. Hydrol. (Amst.)* **3,** 254–267.
Woodruff, J. F., and Hewlett, J. D. (1970). Predicting and mapping the average hydrologic response for the eastern United States, *Water Resources Res.* **6,** 1312–1326.
Young, R. A., and Mutchler, C. K. (1969). Soil movement on irregular slopes, *Water Resources Res.* **5,** 1084–1089.

Chapter XVII

OFF-SITE YIELD OF ECOSYSTEMS

Off-site movement of water over or near the surface of the earth is something we directly experience from time to time. We are less aware of the steady yet invisible movement of water out of groundwater storage below a site, yet this outflow makes up a large fraction of the total yield of water from that site. In this chapter we illustrate this underground component of the off-site flow of water, and discuss the total yield of water and its associations with crop yield that recall our theme of ecosystem hydrodynamics.*

OUTFLOWS FROM GROUNDWATER STORAGE

This flow of mass is, in general, beneficent: It seldom threatens us with floods; it has a counterseasonal trend of temperature—warm in winter, cold in summer; it is clear and clean; and, most important of all, it is stable in time. Called "base flow" when it is identified in the regime of a river, this outflow makes up about 0.3 of the total off-site yield of all terrestrial ecosystems. Its value as a resource is even larger than the fraction suggests by reason of its dependable rate of flow, temperature, and cleanness. Its occurrence indicates the healthy functioning of ecosystems, and, at a larger scale, the regional geology.

A groundwater body is usually of considerable size, and its steady outflow reflects the environment of its downslope end, notably the permeability of the aquifer as a medium and the gradient of the water table toward the outlet (see Plate 37). Because considerable friction attends movement of water through the ground, this gradient often

* The special kind of yield as vapor rises out of ecosystems was described in Chapter XII.

Plate 37. Watercress farm located in the overflowing groundwater of the basal aquifer of Oahu, near Pearl Harbor. This crop adds much to the pleasures of Hawaii's oriental cuisine (April 1970).

must be steeper, and hence more easily measured, than the gradient in a stream channel. It can be mapped from well or piezometer readings, and such maps show the direction of flow. With auxiliary data on aquifer permeability, the maps can be used to calculate the rate of flow, a different method than that needed to keep track of the episodic pulses of water that travel over and near the ground surface.

If water comes out of the rock in a concentrated flow and reaches the channels of a drainage network, much of it escapes being evaporated locally as described in an earlier chapter, but rather is conveyed entirely out of the locality, entitling us to include it in the idea of "yield." In the basin of the Central Sierra Snow Laboratory, for example, much groundwater from the springs along the contact between the volcanic and granitic rocks comes out while the landscape is snow covered or still wet with meltwater, and reaches one of the 60 km of channels that carry water in this season.

Mingled with the direct meltwater flow, this flow is conveyed out of the basin into the South Yuba River. Even in early summer, springs support a large enough flow in the larger channels to keep a continuous thread of water moving out of the basin. Only in August do these larger channels go dry so that further outflows from groundwater bodies cannot escape the basin. The total volume of groundwater that flows out of the basin is about 40 mm.

Base Flow in Networks of Natural Channels

The largest outflow from groundwater in humid lands occurs where fluvial erosion has incised valleys down into the volume of rock occupied by a groundwater body. Water reaches these streams through headwater springs and undefined seepages through channel banks along their courses, and, by reason of its virtually continuous nature, is called "base flow."

In small channels in the headwaters of a drainage basin there is some indication that flow is not entirely drawn from the zone of saturation, i.e., the groundwater body itself, but that some comes from the layer of unsaturated soil. This contribution becomes minor as one goes downstream to channels of higher order; the more deeply incised channels tap strata that may contain large volumes of groundwater. Large channels therefore can tap more groundwater outflow than small ones, and carry base flow more continuously and in larger volume. Their regime is less flashy than that of small streams; they carry less sediment and more dissolved matter.

The increased base flow in large channels represents water that bypassed the small headwater tributaries; thus drainage-basin area becomes an important parameter in base-flow determination. This relation is shown in Fig. XVII-1, in which the open circles depict

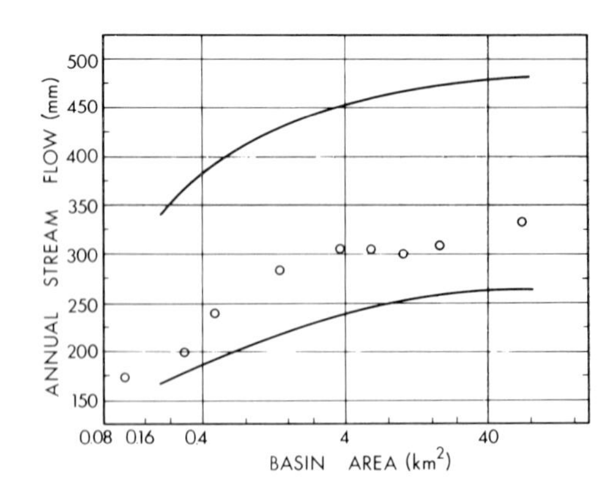

Fig. XVII-1. Yield of water in base flow plus storm flow in streams near Coshocton, Ohio in relation to area of their drainage basins (from McGuinness *et al.*, 1962).

annual flow volume in streams draining areas of different size. Streams that drain basins of up to about 30 ha in area carry about 200 mm of runoff, most of which would be off-site flow moving at or near the ground surface. Streams that drain basins of 500 ha or more area carry about 300 mm of runoff. The excess over the 200 mm of surface flow represents additional base flow.

The same relation is found in the wettest and the driest of the years, as shown by the solid lines. Streams that are shallowly incised into the rock layers of the Appalachian plateau are bypassed by a third of the off-site movement of water.

The excess 100 mm or more of water has percolated, fed the groundwater, and issued into the deeper stream valleys, where it helps them cut their channels still deeper. Also, by favoring erosional sapping of the toe of the side slopes, base flow helps keep the slopes steep. The association of high infiltration and percolation with steep slopes, which was noted earlier, is a part of the same relationship.

A similar increase in base flow with area is found in drainage basins in Michigan (Hudson and Hazen, 1964). The 7-day low-flow rates, which represent pure groundwater contributions to the streams, when converted to a unit-area basis of millimeter outflow, increase with one-fourth power of the increase in basin area.

Volume of Base Flow The variation with time displayed by base flow fits the general theory of an outflow rate dependent on the volume of water remaining in storage. As storage diminishes, so does the rate; the decline is logarithmic, i.e., proportional to the amount in storage at any given time.

In this respect it probably does not differ from other outflows from storages in ecosystems, such as the outflow from storage of intercepted water on foliage, or of detention water at the ground surface. However, concentrated in stream channels, its systematic changes are visible and are in contrast with water fluxes like rainfall that seldom display a pattern of steady change.

The base flow fraction of total yield, which varies with the geological environment of the underground water and the volume of percolation, averages 0.23 over the Soviet Union. In different economic regions the fraction varies as low as 0.12 in Moldavia to as much as 0.43 in White Russia (Dreyer, 1969). Insofar as excessive surface flow suggests that an ecosystem is not functioning well, is out of equilibrium with its environment, and may be generating mass fluxes that can cause damage downslope and downstream, a higher fraction of base flow in total yield indicates a healthier long-term situation, geological factors being equal.

In the historical sequence of land-use changes in the central chernozem region of European Russia, discussed earlier with respect to infiltration, the following long-term changes (from Grin, 1965) are estimated to have occurred in the fraction of yield contributed by groundwater:

Ninth century	0.64
End of nineteenth century	0.28
1925–1950	0.31
Early 1960s	0.36
Near future	0.46
Distant future	0.53

This valuable fraction of the total yield of water has slowly recovered.

Over the whole world, the average volume of base flow is 90 mm yr^{-1} (L'vovich, 1972). This amounts to 0.31 of the total volume of river flow, and is augmented by artificial storage in reservoirs, which, at great expense, adds about 14 mm. The large investments in such reservoirs indicate the value we attach to stability of river flow.

Base flow is less than 5 mm yr^{-1} in semiarid lands like the Great Plains of North America. It increases eastward to about 50 mm in Wisconsin. In the Milwaukee River it is 0.26 of 195 mm, or about 50 mm again. Still larger amounts of base flow are found in the rivers of the southeastern United States.

Geological Factors Specific characteristics of rock texture and structure that affect the yield of base flow can be illustrated from the Hunter Valley in Australia. A set of streams that head in tertiary basalt capping uplands north of the Valley behaves quite differently in low-flow periods than streams that head 10–20 km downslope from the

TABLE I

Low-Flow Characteristics of Two Groups of Streams in the Hunter Valley, Australia[a]

Stream group	Western Barrington	Central Barrington
Monthly flow exceeded 0.9 of the time (mm $month^{-1}$)	1.2	0.5–0.2
Flow over 36-months duration at frequency 0.02 yr^{-1} (mm)	150	19
Catchment deficiency (month)	20	40–50

[a] Source: McMahon (1968).

edge of the basalt cap (Table I). Low flow at the 0.9 level in the western Barrington group of streams is 1.2 mm $month^{-1}$, which is much greater than low flow of the same frequency in the central Barrington group, which lacks the benefit of the basalt storage. Catchment deficiency, defined to index the "inability of a catchment to yield water during low rainfall periods" (McMahon, 1968), is shown in the table, indicating the amount of storage capacity that must be supplied by dam construction in order to attain a desired level of reliable yield. This index is much larger for the streams lacking the basalt headwater geology.

Large-scale differences in the geology of shields, platforms, and folded mountains produce differences in groundwater yield, as shown by Popov (1968, p. 201) in the accompanying tabulation. Much of the greater turnover of groundwater in the Alpine structure zones results from their greater degree of dissection, as well as high precipitation.

Geologic factors	Area of U.S.S.R.	Groundwater yield	Mean annual amount of groundwater yield (kg m^{-2})
Platforms	0.56	0.44	44
Shields	0.05	0.04	45
Mountains	0.39	0.52	76

Geologic conditions are naturally important in determining low flow. They determine the amount of water in underground storage, and its rate of discharge into stream channels. Figure XVII-2 (Hutchinson, 1970, p. 32) presents flow-duration curves for the Root River, draining a rather tight basin, and the Fox, draining an area with much coarse glacial outwash, just to the west of the Root basin. At the 0.5 frequency, the flow rates in both streams differ in proportion to the difference in their drainage areas, but at the 0.9 frequency the Fox draws proportionally much more base flow from its sandy basin than the Root.

Figure XVII-2 is compiled from all days of record, thrown into the pot without regard to possible clustering in time of high or low values. Low values, however, do tend to cluster, a reflection of the large-scale and hence long-period phenomenon of the depletion of water in a large aquifer.* Clustering is particularly important because it strains

* Such depletion is related to the typical seasonality of recharge in low-energy periods (described in Chapter XIV) and also to atmospheric circulation patterns that bring few rain-generating systems (Chapter V).

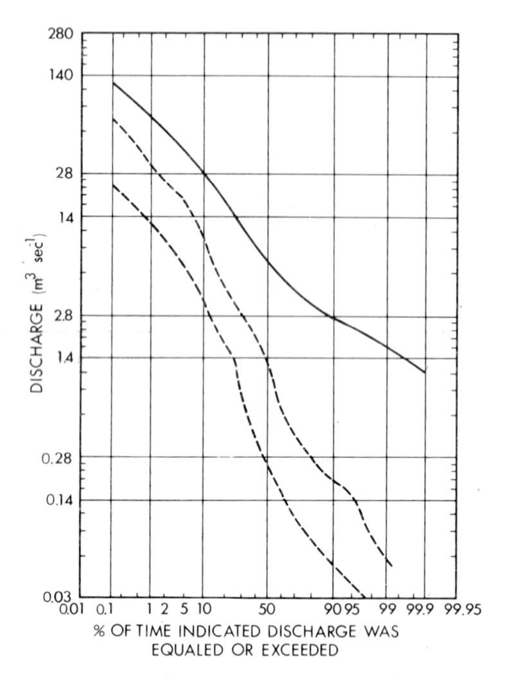

Fig. XVII-2. Flow-duration curves of two rivers in southeastern Wisconsin, the Fox River at Wilmot (drainage area of sandy soil, 2200 km^2) (solid line), and the Root River at Racine (470 km^2 clay and silt) (dashed line). At the 50% ordinate, the flow in the Fox River, which comes from five times as large a drainage basin as that in the Root, is five times as large as flow in the Root. At the 90% ordinate, however, the Fox River, which drains a more permeable terrain, carries 14 times the flow in the Root. The bottom dashed line portrays flow in the Root River Canal, draining 145 km^2 of tight silt and clay (Hutchinson, 1970).

the capacity of storage reservoirs in small city water-supply systems. It is customary, therefore, to calculate low-flow probabilities in terms of the lowest 30 consecutive days, the lowest 7 consecutive days, and so on (Fig. XVII-3). The 7-day flow is often used to index the base-flow behavior of a river; for the Fox River it is 2.4 m^3 sec^{-1} at an every-other-year probability, but only 1.6 at a frequency of one in ten. This last number, the so-called 7-day Q_{10}, is taken as the critical value, below which the Wisconsin standards for water quality do not apply (Wisconsin Statutes, 1973).

Urbanization The process of urbanization tends to choke off the volume of water supporting base flow. Streets and roofs, as noted in the preceding chapter, shift the partition of water at the earth's surface toward fast overland runoff and away from the downward stream of infiltration and percolation. The resulting reduction in groundwater recharge is seen in the reduction of base flow in the local stream channels.

Many urban areas, however, make a compensating contribution to low flow in the form of effluent from sewage-treatment plants. Since the throughput of domestic water is uniform throughout the year, effluent holds up well during the dry season, when it might contribute half or even more of the total volume of low flow.

Significant Characteristics of Base Flow Off-site movement of water from deep bodies of groundwater is characterized by steadiness. The individual inputs of percolate from separate storms slow down and are smoothed out as they travel through the deep storage-outflow sequence. The longer the underground period, the steadier the outflow, as was described earlier. As it feeds into stream channels it sustains them long after the wave of storm flow has passed, hence its name. This sustained thread of water in a stream channel provides a stable biological habitat that is quite different from all other habitats in the landscape.

Unfortunately our inadequate knowledge of the groundwater environment sometimes limits our understanding of base flow. For instance, it is hard to separate the rapid outflow from shallow groundwater from the slow response of deeper groundwater. One reviewer (Hall, 1968) concludes that basic concepts and mathematical expressions of the recession of base flow have long been at hand, yet the practical art is not well advanced. This gap seems to be the consequence of barriers between scientists in different countries (and of different languages), and in different disciplines (Hall, 1968). The same barriers of monolingual, monodisciplinary training that hindered the study of snowfall interception and wasted much research effort (Chapter VI) appear to be a problem in the study of base-flow recession.

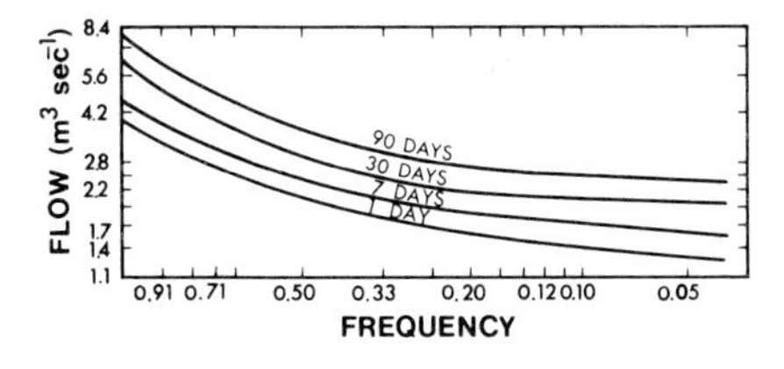

Fig. XVII-3. Mean rate of low flow over 1, 7, 30, and 90 consecutive days, at frequencies from 0.95 to 0.04 in the Fox River at Wilmot, Wisconsin (drainage area 2200 km^2) (Hutchinson, 1970).

We can think of base flow as somewhat analogous to memory of past rainstorms. One model defines base flow as "propagating antecedent or historical effects, carry-over signals from the past, now stored and in transit. . ." (Appleby, 1970). Depending on local geologic conditions that affect size of the groundwater body and the speed of movement of water through it, base flow contributes varying amounts of water to the total amount of off-site yield.

Steadiness of flow is one of the desirable characteristics of base flow in a river. Another is its counterseasonal temperature contrast with the surface environment. Cold-water fish like trout usually are found only in streams that carry a large fraction of groundwater outflow, as in the sand plain of central Wisconsin and certain dissected regions near the Mississippi River.

The relative warmth of groundwater outflow in the spring is utilized in the Schwarzwald of Germany to speed up the warming of the soil (Plate 38). Irrigating mountain pastures by groundwater outflow, then substantially warmer than the air or top soil, raises the soil temperature and improves the climate of plant roots.

A third valuable characteristic of base flow is its relative freedom from turbidity and suspended sediment.

Distant Outflows from Groundwater

Water travels slowly through the rocks of deep aquifers but occupies a very large cross-sectional area so that the total rate of transport can be significantly large. Some of the sandstone formations that crop out at the edge of the central Rocky Mountains in an area favoring high percolation convey water far out into the Great Plains, where it is tapped by deep wells. Such groundwater bodies function as pipelines, although movement is so slow that changes in the conditions in the intake areas might not register in their outflows for a long time.

Water in a pervious layer cut off from the earth's surface by upper impervious rocks travels downslope and comes under increasing pressure. Artesian flow may result when this pressurized water body is tapped by wells drilled through the overlying cap rock. This happens in the Great Artesian basin of eastern Australia, which slopes inland from the wet mountains near the Queensland coast. It is drilled into and water rises in the bores to the surface a thousand kilometers inland. Although warm and saline it can be utilized for watering sheep.

After a few decades of excessive release of the pressurized water and much wastage of both the water and the pressure resource, the basin

Plate 38. Water-management ditches in the German Schwarzwald later will convey relatively warm groundwater to speed up the spring warm-up of the soil (December 1971).

has now more or less been brought under control (Costin, 1971). The regulated outflow averages, on an areal basis, about 3 mm yr^{-1}, small in volume but vital for the population of animals that is the best means of harvesting the biological production that utilizes local soil moisture. We obtain "protein from the wasteland" (Macfarlane, 1968).

Mass Fluxes Associated with Base Flow

While little suspended sediment is carried by the outflow from bodies of groundwater, dissolved material is ubiquitous and organic pollutants also are being found. The long time of intimate contact between water and underground rocks permits solution to do its work, and most outflows from groundwater are mineralized.

In limestone country like southeastern Wisconsin, both the groundwater and the base flow in the Milwaukee River is hard, with high concentrations of calcium, magnesium, and bicarbonate ions (Southeastern Wisconsin Regional Planning Commission, 1970, p. 280). All three aquifers contain hard water; the content of dissolved solids being 380 in the uppermost one, the sand and gravel aquifer, which probably supplies most of the base flow in the river, 470 in the middle or dolomite aquifer, and 900 in the deep sandstone aquifer (Southeastern Wisconsin Regional Planning Commission, 1970, p. 270).

Base flow in the Colorado River as it leaves its upper basin, which amounts to about 650,000 tons day^{-1}, carries 100 tons of salts (Iorns *et al.*, 1965, p. 31), some of which represent outflow from deep groundwater bodies, some deriving from accelerated solution as irrigation water percolated through the soil, into shallow groundwater, and into the upper river. This salt burden presents a problem to farmers in the lower basin, e.g., in the Imperial and Coachella Valleys, and to city users. It is indeed possible to remove this salt loading from the water, just as any load can be removed from a vehicle; the question, however, is whether this should be done at the expense of the downstream users or the upstream polluters.

Although, as we saw in Chapter XIV, percolation fluctuates from year to year, the beginning of irrigation produces a still more radical change, in that the volume of percolate is much increased and is held at this high level year after year. The water-budget consequences may appear in stirred-up saline ground water; the mass-budget consequences appear in accelerated weathering of soil minerals (Rhoades *et al.*, 1974). Over-irrigation "can result in the discharge of more salt from the soil than was applied in the irrigation water because of mineral weathering and dissolution phenomena." This finding applies to the western United States, but we noted earlier the effect of cane irrigation on water quality in the basal aquifer of Oahu.

The enrichment of groundwater by fertilizer nitrates in many basins is evident in the base flow in streams draining this farm area. Such off-site transport of nitrates from the San Joaquin Valley of California has reached such a size as to occasion proposals to construct a drainage network that would completely bypass the San Joaquin River itself.

Mass fluxes from agricultural ecosystems are carried in different proportions by surface runoff and percolate; a study in New England found that "nitrates but not phosphates will move both vertically and laterally through the soil to subsurface drains, [and] that surface runoff contains few nitrates but significant concentrations of phosphates" (Benoit, 1973). Since much off-site transport of nutrients is associated with the transport of sediment particles, percolate and base flow are a desirable form of water yield for their relative cleanness as well as their stable regimes. In a study of more intensive farming in Iowa, base flow carried only 0.04 of the nitrogen and 0.05 of the phosphorus from a contour-planted corn-cropped basin, which shed great quantities of both nutrients in association with sediment (Burwell *et al.*, 1974). Reducing sediment transport by instituting level terracing correspondingly reduced transport of sediment-associated nutrients; the level-terraced basin also produced more base flow (96 versus 63 mm), which moreover carried off less phosphorus (4 mg m^{-2} versus 5 from the contour corn basin) and less nitrogen (100 mg m^{-2} versus 146).

Studies of land cleared for wheat cultivation in Western Australia (Peck and Hurle, 1973) indicate that when percolation of water under cleared land increased by 30–60 mm yr^{-1} and under entire drainage basins by 20 mm yr^{-1}, the outflow from groundwater into the streams increased. Unfortunately, this augmented base flow is of high salinity. The salt outflow from forested basins in the region is about the same as the input of salt in rains and by dry fall; from basins even partly cleared the salt flow is far greater, being increased by about 30 g yr^{-1} m^{-2} of land area. This degradation of base flow was set in motion by the increased input of percolate water to the groundwater body, as we saw earlier in the swamping process in the Belka case study.

One of the costs associated with mining is the loading of groundwater outflows with acids or other pollutants. When the pyrites commonly associated with coal beds are exposed to the air, they oxidize and the sulfur is mobilized. Acidity depends on the underground environment, limestone counteracting it, but often is a damaging characteristic of mine-drainage water. The Water Quality Act of 1972 (U.S. Congress, 1972), in Section 107(a) provides for demonstration projects on "approaches to the elimination or control of acid or other mine water pollution resulting from active or abandoned mining operations." Many of these are found in the Appalachians, often representing a legacy from mining companies long gone out of business.

Base flow has value as a reliable source for withdrawals of water for domestic, industrial, or irrigation needs, and also for the in-channel uses of water: as habitat, for navigation, and as amenity. Aquatic life can survive dry seasons if base flow does not drop too low or get too warm or become too loaded with oxygen-demanding organic pollutants. The stream is most vulnerable to man's actions at these periods of low flow, and is most needed then.

WATER YIELD AS ASSOCIATED WITH BIOLOGICAL YIELD

On-site transpiration goes with on-site photosynthetic production. This relation is not causal, but expresses the fact that stomata that are open to let CO_2 diffuse inward to the wet walls of the leaf cells are also open to let water vapor diffuse outward into the free air. Increased biological yield therefore means less water available to move off-site.

This association is, in general, a useful generalization, but a few departures should be mentioned. For example, evaporation from the bare soil, say in the early growth stages of an annual crop before it covers the ground, is "unproductive." The diffusion of CO_2 into a leaf is not controlled by quite the same resistances that control the diffusion of water vapor.

Rauner's (1973) index η_E* shows relative yields of biomass and water in different conditions

$$\eta_E = 4000\bar{m}/600E_T,$$

where $\bar{m}$ is production of dry matter, and E_T evapotranspiration, hence the complement of water surplus or water yield.

For a given air temperature, η_E increases with rising leaf temperature, i.e., in strong sunshine, and indicates that the biomass yield tends to outweigh the yield of water. For a given leaf temperature, η_E is larger in cool air. In unstable conditions of the lower air, η_E increases with wind speed, representing accelerated input of CO_2 to the vegetation canopy. On north and south slopes of a steppe landscape in central Russia, η_E is larger on the north slopes than south; on the uplands η_E is the smallest, indicating that it plays an important role in water yield.

Shawcroft *et al.* (1974) demonstrate the use of a soil–plant–atmosphere model to estimate the implications for water use of possible changes that geneticists might make in leaf area or angle. At ordinary leaf angles, for instance, a leaf-area index of 4 seems to result in the highest ratio of biomass production to transpiration.

If the desired plant product is a fruit that represents an incomplete conversion of the primary productivity of the plant, transpiration departs from the energy yield, as illustrated in Table II. Oranges are not eaten for their energy content, and yield only half as many joules per cubic meter of applied irrigation water as does rice. Sugar cane is well known for its efficient utilization of both solar energy and water, but produces only "empty" calories from a nutritional standpoint. Foods of animal origin incur the $\frac{9}{10}$ loss attendant on the plant-to-animal conversion; when dairy or beef cattle graze, irrigated pastures

* This index is also called water-use efficiency, although the appropriateness of "efficiency" in reference to a process that physically cannot be expected to come anywhere near 100% might be questioned.

TABLE II

Food Values of Irrigated Crops per Cubic Meter of Water[a]

Crop	Energy values (kJ)	Protein (g)
Oranges	3670	15
Rice	8600	33
Sugar cane	13,550	0
Beef	460	6
Milk	1100	14

[a] Source: Costin (1971).

that might have primary productivity of 12,000–15,000 kJ m^{-3} of water evaporated,* they yield far less energy—and even less protein—than crops raised directly for food.

By and large, however, the association of evapotranspiration rates with the rates of primary productivity is a useful one. In periods of a high level of plant activity, water flows through the vegetation upward and away into the free atmosphere and is not available to percolate to groundwater and later to support the base-flow component of off-site yield.

Soil that has been evacuated of water by evapotranspiration absorbs rain avidly, and the increased capacity for infiltration reduces the volume of storm flow to come from the succeeding rain. Both components of water yield, base flow and storm flow, are reduced by an increase in biological productivity. Let us look at some data on the association of biological and water yields.

We do not ordinarily think of a farming region as producing much besides food, but it usually also produces a water surplus. Even irrigation districts generate sizable flows of drainage water, although not always of good quality. Alternative ecosystems can thrive in a given region, and the choice of what to foster should be made with water yield as well as the more conventional kinds of harvest in mind. Such decisions, however, are hampered by the fact, as Colman (1953, p. 188) puts it, that there is "much still to be learned about the control of water as a product of the land."

* Vaporization of 1 m^3 of water represents, of course, the conversion into latent heat of 2.5×10^6 kJ of radiant and sensible energy. The difference between these two energy–water ratios is about two orders of magnitude, consonant with experimental data on photosynthesis as an energy-converting process.

Contrasts within the Yearly Cycle

Surplus water from farming regions may be valuable downstream as a source for city or industrial supply, or it may become a threat. The great floods of April and May 1973 in the Mississippi River represented for the most part a massive off-site movement of water from farm lands of the Midwest. These lands were sodden, too soft to plow, and inundated in low spots. There was "water running out of the base of the hills," to cite the comment of a Wisconsin farmer. Ditches were full, and creeks and rivers were up onto their flood plains. The land was certainly incapable of storing any more water. Each rain was transformed nearly completely into off-site flow with very little delay. Evapotranspiration was small and biological production low where there was grass or alfalfa, zero in the bare, muddy fields, which often remained unplowed.

In contrast, late summer in the Midwest is often dry. Soil moisture has been used up, plants wait for rain, and creeks shrink into trickles. Rain from a moderate storm is immediately absorbed by the soil. The life processes involved in survival claim all available water and leave only what is delivered at extreme intensities for off-site flow (Chap. IX).

The eastern part of North America experiences this changeover almost every year in response to its annual energy cycle. Figure XVII-4 shows this change by use of areal averages (Lettau, 1969; Rasmusson, 1968) over the land from the Rockies to the Atlantic. In winter (DJF on the figure), energy is low, about half the mean over the whole year; precipitation, much of it as snow, averages 1.5 mm day^{-1} and is 0.9 mm day^{-1} greater than evapotranspiration. This difference, 0.9, represents the generation of a large yield of water. At the same time the rate of evapotranspiration, 0.6 mm day^{-1}, represents, even including the perennial ecosystems of the southeastern United States, a relatively low rate of biological production. In spring (MA on the figure), when the energy input is at about the yearly mean level and precipitation has increased to 1.8 mm day^{-1}, the surge in biological activity increases the rate of evapotranspiration to 1.0 mm day^{-1}. Water generation is less than in winter, biomass production is greater. In May and June (MJ) this trend continues. Rainfall, at 2.7 mm day^{-1}, is half again as great as in spring (MA), the energy input has grown still faster, more than doubling the March–April rate of supply. As a result, generation of water is down and biological yield is up. By midsummer (JA) the energy input has not changed much; a slight increase in rainfall does not compensate for the activity of the fully developed ecosystems. Evapotranspiration exceeds rainfall; no water yield is

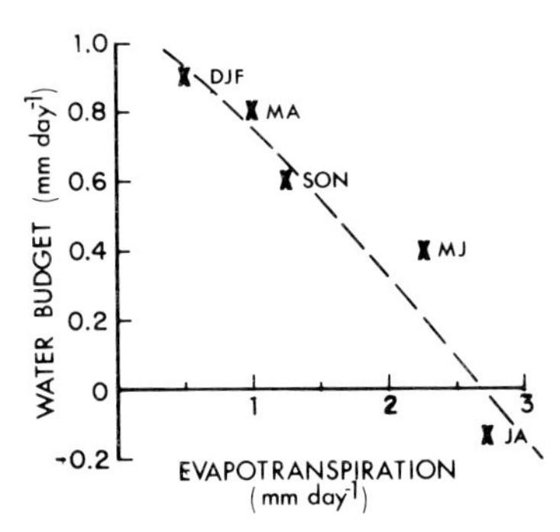

Fig. XVII-4. Areal means of seasonal evapotranspiration and water surplus from ecosystems of eastern North America (mm day^{-1}) (data from Rasmusson, 1968, and Lettau, 1969).

Months	Evapotranspiration	Water yield (precipitation − evapotranspiration)
DJF	0.6	0.9
MA	1.0	0.8
MJ	2.3	0.4
JA	2.7	−0.1
SON	1.3	0.6

generated but biological yield is high. Fall (SON) brings the familiar tapering off. While rainfall is less than in summer, evapotranspiration is cut in half and a small difference (0.6 mm day^{-1}) forms the first water surplus anew.

Soil and aquifer storages lag in converting the monthly differences of rainfall and evaporation into off-site yield, so the ordinates on Fig. XVII-4 do not portray streamflow or groundwater levels. While they show only the amounts of water available to build up soil-moisture storage, groundwater storage, or off-site yield, they index the relation between ecosystem demands for water and nature's supplying of it.

Contrasts in Land-Management Practices

A long set of measurements of water fluxes and crop yields at the Northern Appalachian Experiment Station, Coshocton, Ohio, shows some trade-offs between water yield and crop yield. These are especially plain in a contrast of two practices of managing the corn–meadow rotation typical in the region. "Prevailing practice" in Table III represents customary methods in eastern Ohio in the 1930s, when the research began; "improved practice" includes erosion-control

TABLE III

Vapor Flux (Evapotranspiration) and Crop Yield under Two Management Practices in Coshocton[a]

Crop	Prevailing	Improved	Fractional increase
Corn			
Summer evapotranspiration	455	505	+0.11
Yield	0.82	1.19	+0.44
Meadow			
Summer evapotranspiration	365	485	+0.33
Yield	0.59	1.05	+0.78

[a] Units: kg m^{-2}. Source: Data from Dreibelbis (1963).

cultivation and some fertilization, largely to increase the soil-holding properties of the vegetation, but which in the process resulted in greater vigor and growth, and greater transpiration.

These improved practices of cultivation resulted in greater detention of water on the soil, providing longer opportunities for infiltration and hence promoting a higher level of soil moisture. Vegetation responds to this increased soil moisture by growing more vigorously and transpiring more water, continuing such activity longer into periods between storms at times when the crops farmed under customary practices were beginning to experience moisture stress.

Table III shows that the improved practices caused evapotranspiration to increase by 0.11 from corn and by 0.33 from grass. The off-site yield of water therefore decreased by about 50 mm from cornfields and 120 mm from grassland. The new practices, however, produced an increase in crop yields (0.44 for corn, 0.78 for grass) that was proportionately greater than the loss in yield of water.

While not all of the crop increase can be ascribed to the reduced stress for soil moisture, because tilth and nutrient levels also were improved, a large fraction of it is clearly associated with the greater activity of the vegetation. The agricultural practices prevailing in the 1930s allowed about 200 mm of water to run off and to percolate annually. The improvement of practices reduced these forms of off-site yield by 50–100 mm, and also reduced the off-site transport of soil, increasing the tonnages of hay and grain trucked away by the farmer.

An associated off-site yield is that of sediment, which in the general Coshocton area is about 100 g m^{-2} yearly (Gottschalk, 1964) and can be reduced.

Shelter Reducing wind speed and turbulence in crop ecosystems in the shelter of lines of tall vegetation is a common practice in the windy coastal lands of the eastern North Atlantic (Denmark, northwestern Germany, Brittany) and in the dry interiors of North America and the Soviet Union. Shelterbelts in the Russian steppes in particular have been studied by the energy–water-budget method (Konstantinov and Struzer, 1965; see also publications of the Main Geophysical Observatory) in order to design belts of optimum width, permeability, and spacing for reduction in evapotranspiration and increase in grain yield. Especially where excessive evaporation is caused by sensible-heat advection, as on the Great Plains, a reduction in turbulence can improve the ratio of biological production to water use (Hagen and Skidmore, 1974); further improvement might be obtained if crop ecosystems were tailored to gain the greatest benefit from shelter.

Drained Croplands Drainage of a soil–vegetation system appears to be a means of increasing the yields of both crops and water. In reality, however, once the water table has been lowered the off-site flow of water is perhaps only a little greater than it was before drainage, and might be less, if increasing the depth of the aerated root zone improves the vigor of plant growth and its rate of transpiration. An increase in the capacity of the root zone increases the soil's capacity to take in water by infiltration; less water moves off the site as surface runoff.

The slow movement of water through the soil requires that drainage lines be spaced closer than is true of the stream channels that remove surface runoff. An idea of the required density is illustrated in the Imperial Valley in California, which is so plagued by the accumulation of salts brought in the irrigation water from the Colorado River that rapid disposal of salty ground water is vital. Here an area of 1000 km^2 is drained by 19,000 km of tile lines (Houston, 1967, p. 14), giving a drainage density of 19 km km^{-2}, which is much larger than the density of 6 for stream channels in the basin of Castle Creek mentioned earlier.*

* The index of channel maintenance, i.e., the reciprocal of 19 km km^{-2}, is 52 m^2 of cropland m^{-1} of tile line. Houston indicates a capital cost of \$3 m^{-1} of tile, which is equivalent to 6¢ m^{-2} of field area. To put this in terms of costs per ton of water transported, let us assume that 300 mm percolates and drains away. The capital cost of the tile drains is 19¢ for capacity of 1 ton yr^{-1}. Assuming rough depreciation and interest rates, and adding for the cost of the larger ditches into which the tile lines empty, we arrive at an annual cost of about 5¢ ton^{-1} of water, which would be comparable with the cost of getting the irrigation water to the Valley.

Wildlands Many wildland ecosystems, or the assemblies of ecosystems in drainage basins, are found under single ownership, often public, a fact that facilitates thinking of multiple yields from them—both biological and water products. Land administered by the U.S. Forest Service, for example, is managed under a multiple-use principle that dates from the establishment of this agency and was recently affirmed in legislation. While we tend to think of timber as the primary yield, it was never the intent to give timber production overriding priority over other forms of biological yield—forage for cattle and sheep, habitat for wild game, vegetation providing an environment for recreation and amenity—or over the yield of high-quality water with a minimal yield of suspended sediment and dissolved materials.

Cooper (1969) describes the systems approach as a practical way to bring into harmony the sometimes conflicting practices associated with each form of yield; e.g., the proposal to scalp off trees from a drainage basin in order to maximize the volume of water yield. He notes several occasionally incompatible goals for water itself (maximum volume, reduced flood peaks, augmented low flows, maximum quality), and continues: "When desires for maximum yields of high quality forest products, wildlife, livestock, and esthetic and recreational opportunities are added to this list, a complex of objectives appears that clearly cannot be fully satisfied." The best set of tradeoffs must be determined, and this can best be done by systems analysis and constructing a model in which different combinations of practices can be tried. Such an analysis highlights gaps in our knowledge about the relations of water to biological processes, and might help guide research directions.

TOTAL OFF-SITE MOVEMENT OF WATER

Effect of Physical State of Water

The water delivered to a section of the earth's surface is converted into a variety of outputs. Before looking at the case of the present day, let us consider the formerly more common case of off-site movement in the form of ice.

We can draw an interesting comparison in the ways water moves off the site to which it was delivered from the clouds. The different ways it moves affect the land surface and leave their marks on it. It can move, as we have seen, as drifting or avalanching snow, as ice, as liquid flow on the surface, within the upper porous layers of the soil, and in deeper bodies of groundwater. Ice flow occurred recently enough that the marks are very plain in many landscapes.

Off-site movement of water in the form of snow may be so great as to remove all the precipitation delivered to a site. This condition is seen, for example, where a poorly sited high-rise building conveys the kinetic energy of the strong winds at its top down to the pavement below; no snow remains on the pavement. It is removed as fast as it would be from a bare ridge in the Rockies.

The high slopes of the Snowy Mountains of Australia hold relatively little snow where they are covered by low ecosystems (Plate 39). In contrast, the liquid-water yields from snow-gum (*Eucalyptus niphophora)* sites are 1.2 or more than that of low vegetation (Costin, 1971).

Snow that remains on a site and is not melted during the following summer metamorphoses in firn and then into ice. Eventually a depth accumulates that provides enough weight to force the lower ice layer to move. Potential energy at the ice–air interface is transmitted into motion, just as the gradient of a water–air interface puts liquid water into motion. With greater viscosity, a greater potential energy gradient is necessary with ice.

Plate 39. Lines of snow gums across the smooth slopes of the Snowy Mountains above Guthega Dam in late winter. These lines often follow ridges and increase the interaction between the earth's surface and the fast-moving atmosphere of winter storm systems (August 1966).

The contrasts lie less in the way off-site movement of water is powered than in the speed of response. Ice requires a large cross-sectional area to transport the local (plus upstream) surplus in the water budget. This large cross-sectional area means that the whole site must lie under deep ice.

The gravity potential is supplied by the slope of the upper surface of the ice, which is relatively independent of irregularities in the rock–ice interface. The ice bed therefore takes on forms that are more related to streamlining than to the task of supplying gravity potential to force the water to move. An ice-covered surface does not display an organized, downslope graded arrangement of valleys.

Movement off-site of liquid water in a fluvially dissected landscape requires only a small cross-sectional area in the flow channel, because the velocity of movement is high. The grade of the water surface is therefore closely followed by the channels themselves. The result is a graded network of channels, each with optimum depth to move water, sediment, and solutes efficiently down the potential gradient and off the site of origin. Organization of off-site fluxes of liquid water thus introduces order into the forms of the landscape, as quantitative geomorphology (Chapter XVI) has shown.

Movement off-site of water within the ground is slow, as with glacial ice, but, like ice, it occupies a large cross-sectional area, perhaps 1000 m thick. As we have seen, the total volume of water moving in this fashion and ending in the bigger streams as base flow is of the order of 100 mm yr^{-1}.

At the present time, the usual situation is that much of the delivered water leaves the site in the form of vapor, which, in its high-energy, mobile state, moves much more easily than water or ice. The amounts are large, averaging 400 mm yr^{-1} from terrestrial ecosystems. The rest leaves the site in liquid form, as surface or near-surface flow and outflow from groundwater.

For large areas of the earth's surface we can combine the forms of surface, near-surface, and groundwater flow and speak simply of liquid-water yield, expressing it as a mean depth of water over the source area in millimeters.* The global total of yield from land areas is 38,830 km^3, in unit-area terms 295 kg m^{-2}, or 295 mm (L'vovich, 1972). Although the off-site fluxes of water are as different as night and day, there is logic in discussing their combined form, because it is the surplus item in an ecosystem's water budget. Moreover, to a degree it lies within our power to alter the way this surplus is partitioned. Poor

* 1 mm yr^{-1} = 1 kg m^{-2} yr^{-1} = 0.032 liter sec^{-1} km^{-2}.

land management, for example, can shift the division away from base flow toward surface flow, a kind of yield generally lower in quality than base flow but without, perhaps, greatly changing total yield. Let us look at some of the seasonal and other time variations in yield, and then at its areal distribution.

Variations over Time

The Spring Flood of 1948 Comparing seasons of the year, we find that maximum yields in spring are most common. They result from either the melting of winter snow cover and frozen ground, or from spring rains on wet or frozen soil, or from a combination of both. Both conditions, i.e., the buildup of water as snow cover or as soil moisture, indicate the low-energy situation of winter rather than any unusual amount of precipitation.

Figure XVII-5 illustrates this spring maximum by showing how the timing of yield from the ecosystems of a drainage basin (this one being Skyland Creek in the northern Rockies) is shifted from the timing of precipitation. During the water year 1947–1948 (from September 1947 to August 1948), water was supplied abundantly and steadily for ten months, September–June, at an average rate of 4.6 mm day^{-1}. Energy was supplied at rates capable of vaporizing such quantities of water only through the summer months; it was supplied

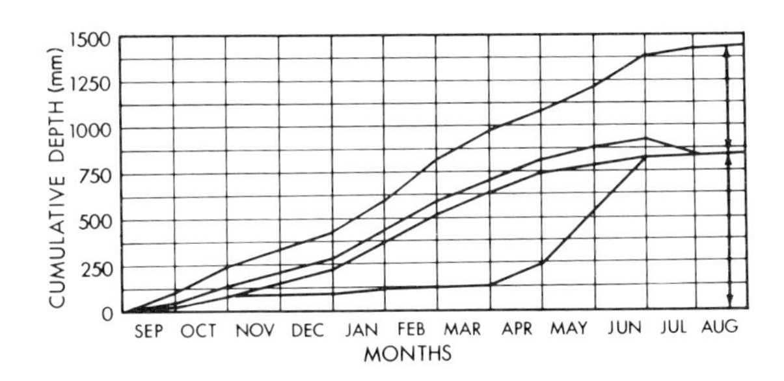

Fig. XVII-5. Water fluxes into and out of the basin of Skyland Creek, Montana (Upper Columbia Snow Laboratory) in the water year 1947–1948. Values accumulated from 1 September 1947 to the end of each month are plotted for precipitation (top line), precipitation less evapotranspiration (second line), and stream flow (lowest line). The space between the second and third lines from the top represents storage as soil moisture, depleted by mid-summer; the space between the third and fourth lines represents storage in the snow cover. Storage of water in soil and snow mantle builds up and then declines in spring and summer. Note the unusually large inputs of rain and snow during the winter, and into May and June, when heavy snow melting contributed to the massive flood of 1948 in the Columbia River (U. S. Corps Engineers, 1956).

only weakly in winter. Water therefore accumulated in and on the ground in large volumes. Soil-moisture storage built up to approximately 100 mm, and snow-cover storage to 525 mm at its maximum in early April. This is a large amount for the interior location of the northern Rockies, and presaged heavy runoff in the Columbia River. We see that it turned out worse than expected. April was cold and snowy and melting went slowly, but the later the snow cover stays on, the greater the likelihood of its coming into the relatively high-energy weather of late spring. This is what happened in 1948. The upper circulation shifted in the middle of May (U.S. Geological Survey, 1949) and melting began. Ecosystems in Skyland Creek basin yielded nearly 300 mm of water, which rapidly moved down into the rivers of the upper Columbia basin, on the heels of long surges of meltwater from hundreds of thousands of square kilometers of high-yield land at lower altitudes. To see a giant river like the Clark Fork or Kootenai in its rise—an irresistible massive movement of 15,000 tons of water every second, deep eddies slowly turning on its surface and revealing the power and menace of the water lapping at the railroad—makes a sight that is never forgotten.

The 1948 flood in the lower Columbia River and the destruction of Oregon's then third largest city, Vanport (never rebuilt), were record consequences of this late spring liquidation of snow-mantle storage. Figure XVII-5 shows how the snow accumulated from November on, with neither winter evapotranspiration nor winter stream flow removing much water from the basin. The total duration of the snow cover was about 230 days. Its presence was chiefly responsible for a delay in yield from the median date of precipitation, about 15 February, to the median date of yield, about 20 May, or nearly 100 days. Of this, no more than 10 can be attributed to delays in movement of water through the saturated soil of the basin, although a smaller storage, about one-fifth of that in the snow mantle, is represented by soil-moisture buildup and decline. Soil-moisture storage began in late September, held at 60–80 mm through the winter, and rose to 100 mm in the melting season, but its depletion was slow until July, so the total duration of storage was about 250 days, or $3\pi/2$ of the annual cycle.

The snow cover produces approximately a lag of 90 days, or $\pi/2$, in a large volume of water and provides an example of storage of water that is delivered to ecosystems at a season of low energy. The water was in a physical state such that when the energy input did arrive, it was converted not to vaporizing water but to liquefying it; the consequences are not in on-site biological production* but rather in off-site yield reaching tonnages of 200,000 or more day^{-1} from this small experimental area.

Melting occurred so fast that the author's measurements of liquid-water content of the upper layers of snow at the Columbia Laboratory reached levels higher than 0.1. These thick films of liquid water on the snow grains and in the pores were cold, however, and evaporated only in small part. Most of the meltwater drained down into the saturated soil and soon reached Skyland Creek, then Bear Creek, then the Middle Flathead River, and so on down to the mainstem Columbia. The average regime of the Columbia River shows the existence of late spring floods in most years (Fig. XVII-6a); that of 1948 was a little later and much bigger.

Figure 6c shows another late spring maximum, that of the Missouri River near its mouth—not surprisingly as this water is from melting snow in the Northern Rockies not far from the slopes that feed into the upper Columbia River. The peak on the Missouri follows an earlier pulse of water from rain on the lower part of the basin. Early spring maxima also characterize the Ohio (Fig. XVII-6b).

Peak Yield in Other Seasons Winter yield is produced where winter precipitation is heavier than summer (the Pacific regimen) and comes as rain, not snow. It typifies California hill-country rivers and those of the Coast Ranges and lower Cascades on to the north (see p. 460).

The coastal streams of the Pacific Northwest display a winter maximum yield generated by rain from the same storms that deliver snow to the inland mountains that feed the upper Columbia River. Ecosystems of the snowy region differ from those of the coast not only in their structure and biomass but also in their later growing seasons and later yield.

The juxtaposition of these contrasting yield regimes in the Pacific Northwest makes possible a means of forecasting the later yield from the interior basins (Rockwood and Jencks, 1961). It also provides for a convenient exchange of hydroelectric power between the two regions, power being transferred eastward in fall and winter and westward in spring and summer. In winter the short, steep rivers of the Coast Ranges and Cascades are flowing full and those of the interior are low; in spring and summer the rivers rising in the northern Rocky Mountains are high.

* A large part of the evaporation that occurred was that of water on tree foliage, hence biologically unproductive.

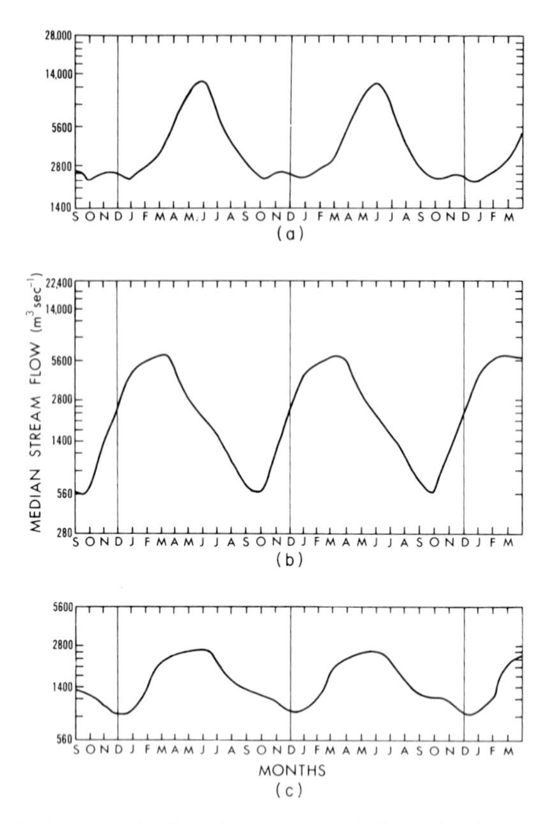

Fig. XVII-6. Spring rises in three large rivers, (a) the Columbia River at The Dalles (drainage area 595,000 km²), (b) the Ohio River at Louisville (drainage area 230,000 km²), and (c) the Missouri River at Hermann, Missouri, near its mouth (drainage area 1,320,000 km²). Median monthly flows are shown (U. S. Geological Survey, Water Resources Review, 1973).

A similar hydropower exchange is made between the North Island of New Zealand, where there is a winter yield maximum from rains, and the South Island, where heavy snow melting in the Alps brings a maximum in late spring. This arrangement takes place because of the fortuitous proximity of different yield regimes, which here are associated with the fact that the North Island apparently represents an underthrusting oceanic plate with attendant volcanism* but only low to medium altitude mountains. The South Island, in contrast, repre-

* This generates an upward flux of water vapor as high-pressure steam, utilized in geothermal power stations with attendant river pollution problems from sulfur and mercury in the steam—a special kind of mass flux associated with water.

sents the meeting of pieces of two continental plates with upthrusting to great altitudes of mountains that catch heavy snowfall. The same storms cross both islands, but the receiving surfaces are of medium altitude in the North Island and high in the South Island.

It is possible to have a maximum in water yield when inputs of water and energy to ecosystems are in phase in summer, but this occurs only where rainfall is so great that it can satisfy evapotranspiration and still force a large volume of water over or through the ground. A summer maximum in rainfall in Arizona and New Mexico elicits little response in streamflow because "at this time of year space is available in the soil for storing water" (Colman, 1953, p. 19) except insofar as high rain intensity produces brief surface runoff, as noted in Chapter XVI. Yield maxima in fall also suggest large rainfall maxima, often due to the occurrence of hurricanes in many years with their attendant heavy and long rainfalls.

In order to establish stream flow regimes, the volumes of surplus water determined each month by the soil-moisture bookkeeping procedure can be routed downstream at a rate depending on the degree to which groundwater and other forms of basin storage of water slow the outflow. For example, it might be assumed "that only about 50 percent of the available surplus water will run off in a given month" in which it was generated (Mather, 1974, p. 139), with the rest emerging in later months.

Converting unit-area water surpluses, expressed as kg m^{-2}, into volumes of stream flow (in m^3) assumes that we know fairly precisely the area that contributes surface runoff, subsurface flow, and groundwater outflow to a stream network above a given point on a river, yet in fact we often do not have this information.*

Yield Variations among Years Changes in yield from year to year are larger than changes in rainfall, because yield is a residue after ecosystems have taken their bite out of the water budget. The dispersion of yield values from year to year in the Hunter Valley of Australia, as indexed by standard deviation, varies from 0.55 of the mean in one of the wetter drainage basins to 1.55 of the mean in the driest (McMahon, 1964). As in most places, the reliability of yield is usually small where the yield itself is small.

* It is interesting that in selecting experimental drainage basins of the Snow Investigations for their hydrologic qualities, it was necessary to accept small stretches of undefined boundary lines in all three basins. Boundaries of underground flow are, of course, difficult to establish in rough or impermeable terrain where the water table is difficult to identify or generalize over a wide area.

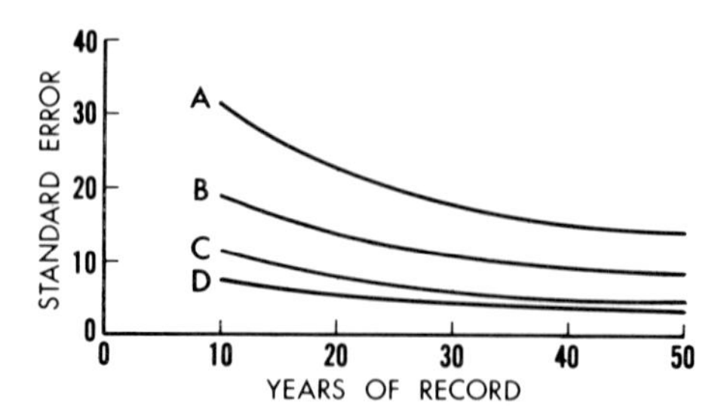

Fig. XVII-7. Standard error of stream flow (averaged over varying periods of record) in ecological zones of the Soviet Union, expressed as a percentage of the mean flow. Large values indicate the unreliability of calculated averages. A—semiarid, B—steppe, C—forest steppe, D—forest (from Sokolovskii, 1959, p. 81).

A different expression of the variation from year to year is the standard error of the mean, which expresses the reliability of the figure computed from the record available. In the forest zone of the Soviet Union, for example (Fig. XVII-7), a fifty-year record locates the computed mean within 3.5% of the true mean. In drier areas the computed mean is less reliable.

Relying on nothing but a record of yield or streamflow, even one that seems comfortably long in comparison with the usually brief records of this expensively measured variable, often leads to problems. This is illustrated by the well-known controversy over the size of the yield of water from the basin of the upper Colorado River, one of the major water-exporting regions of the country. Political difficulties have arisen from the fact that the interstate compact for dividing the waters of the Colorado River took as premise a yield estimate that was not borne out in subsequent years. (The record was about 25 years long at the time of the compact. Figure XVII-7 suggests that, for a semiarid region, the standard error of estimate exceeds 20% of the mean, which is a large margin of error.)

Nearly 40 years after the compact was made the Special Master in *Arizona vs. California et al.* in the Supreme Court said that "It is impossible to predict, with any useful degree of accuracy, future supply in the Lower Basin" (i.e., yield of water from the upper Colorado basin) (quoted in National Research Council Committee on Water, 1968, p. 33). Such are the consequences when variability in yield is not properly taken into account.

It has proven practicable to estimate variation in yield by determining water surplus in individual years from calculations of the budget of soil moisture. This procedure is best applied in terrain where errors in measuring groundwater contributing area, streamflow, precipitation, and evaporation energy inputs are not excessive. Muller (1972) calculated the water surplus in winter and spring in Louisiana in each year from 1941 to 1970. From his results the 30-yr mean can be compared with a wet winter, 1953; both patterns show a large surplus in central Louisiana but the wet winter generated 1100 mm, double the average winter yield. In the coastal swamp zone, on the other hand, 1953 experienced about the same freshwater surplus (350 mm) as the average, with presumably minor changes in the brackish water environments of this zone's aquatic ecosystems and no interruption in their production of shrimp, oysters, and other New Orleans staples.

Applying these water-budget methods to a small river in Delaware produced a good reconstitution of annual flows, the correlation coefficient being +0.94 (Mather, 1974, p. 139), over a 15-yr period in which individual years ranged from 225 to 650 mm annual yield.

Longer Term Variations Due to Changes in Ecosystems Ecosystem water yields are likely to change following any change in ecosystem characteristics that affect rainfall delivery, interception, infiltration, energy inputs that support evaporation, and so on, as described in several earlier chapters. Evaluating these effects, however, is not easy because it will be remembered that some of these hydrodynamic processes are measured unreliably or not at all; even the geometric outlining of the area contributing runoff and base flow to a given stream channel is sometimes uncertain. In an approach to the question of evaluating the effects of a change in land use, the paired basins approach is sometimes used, in which the yields of two nearby drainage basins are measured together until an acceptably high correlation is attained—if this happens at all. If a high correlation should be found, the problems of undefined contributing area and leakage through deep ground water flow that bypasses the stream gage are minimized, because both can with some justification be assumed to be reasonably stable over time and perhaps can be thought to afflict both basins equally.

At this point the ecosystems in one of the pair of basins are abruptly modified—forest is clear-cut or patch-cut, riparian vegetation is killed, shrubs converted to grass, gyppo or modern logging carried out, logging roads built, and so on. Thereafter the actual yield from the treated basin is compared with a yield estimate made by use of the correlation with yield from the untreated basin.

A problem that remains is that atmospheric conditions are different in any two basins, even adjoining ones. Even so simple a factor as the absorption of solar radiation is not likely to be the same. Exposure to storm winds and to atmospheric ventilation during evaporation periods is also dissimilar in the two basins. We know so little about mesoscale patterns in wind fields that we cannot adjust for terrain effects with much confidence, especially if the friction on the lower boundary is changed by cutting or whatever other ecosystem modifications were carried out in the experiment.

A combination of the paired basin approach with a physical one, in which the best possible atmospheric observations are made to help in estimating precipitation and evaporation, would appear the optimum solution. Sampling difficulties, however, are serious in mountainous country for all atmospheric variables, and tend to obscure real differences among basins. They also hinder the transfer of empirical results from the experimental basins to those of operational concern—the aim of the whole effort.

Another basic difficulty need only be mentioned, since phases of it have been discussed in many preceding chapters. This is the "black-box" approach, so named as an allusion to a computer that converts a given input into an output by inner workings that are invisible and mysterious. Rainfall-runoff studies ignore the complicated chain of hydrologic and hydraulic processes that intervene between the time rain is delivered to a basin and the time the last of the yield goes down the river. Yet, as we have tried to show, these water fluxes and storages in ecosystems are not always well understood either in theory or in the field.

Any modification of the ecosystems of a drainage basin affects one or several of these processes, described in individual chapters. We need to know *which* processes are affected and to what extent if we are to determine quantitatively the changes of yield from drainage basins with different terrain and ecosystems. Such information is needed in hydrologic modeling, especially if the model is meant to reproduce peaks and recessions in stream flow throughout the year, not simply total annual yield of water as a lumped figure.

Lee (1970) feels that for these and other reasons, including the large degree of uncertainty in the basic data on stream flow, basin area, leakage, precipitation, and evaporation estimates* (no direct measurements of this flux have yet been made over tall ecosystems or in

* Random errors in determining potential evapotranspiration were found to have less significance in basin-hydrology modeling than bias errors in it (Parmele, 1972); but for this flux we have even less to go on than we do for rainfall in evaluating the degree of bias. At least for rainfall we can be fairly certain that our measurements are too low; for the energy flux represented in potential evapotranspiration we cannot be certain whether we are too low or too high!

mountains), a large uncertainty is inherent in any estimate of yield. In wet areas with a high yield, this uncertainty is about 0.2; that is, if the actual difference between the yields of the treated and the control basins is less than 0.2, then the calculated values are "not helpful either in confirming or challenging the accuracy of water balance measurements. . . ." In drier basins where yield is a third or less of the precipitation, this uncertainty index is larger.

In special circumstances, however, estimates of the effect of treating ecosystems in a particular basin do seem large enough to overcome the uncertainty effect Lee derived, if the treatment is radical and if the yield is large. This was true in an experiment at Coweeta Experimental Forest in the wet country of the southern Appalachians. While the year-to-year changes in yield resulting from differences in volume or timing of water or energy inputs remain large, those resulting from cutting a dense forest also are large enough to be discriminated. Cutting produced a flush of yield that gradually diminished as the forest reestablished itself over two decades, smoothing out the random year-to-year changes associated with atmospheric circulation changes (Hibbert, 1967).

Forest clearing in the Hubbard Brook basin in the northern Appalachians resulted in a mean initial increase in yield of about 280 mm, which is expected to diminish as the ecosystems recover and again drain the soil reservoir. The effect of the forest cutting is to reduce the steady draft on this reservoir; it is more likely to be nearly full at the time a rain comes, and therefore more likely to percolate water to deeper layers and to reject some water at the surface and force it into surface and near-surface flow, or "storm flow." The relative effects of a wetter soil on these two modes of water flow are best shown in the June–September period, when the drying effect of the forest would have been strongest if it had not been removed. In this period the components of off-site yield increased as follows after cutting (Hornbeck, 1973):

Base flow from percolated water	160 mm yr^{-1} = 0.72
Storm flow from surface and near-surface runoff	64 mm yr^{-1} = 0.28
Total change in yield	224 mm yr^{-1} = 1.00

The augmentation of percolation is a desirable outcome of the ecosystem conversion and indicates the openness of soil bodies created by forest systems. The question is how long the forest soil can survive the demise of the forest itself.

TABLE IV

Water-Budget Fluxes in the Don River Basin[a]

Parameter	Past (before about 1950)	Present	Long-term expectation	Change from Present
Precipitation	167	167	170	+3.0
Infiltration	144.3	148.5	156.4	+7.9
Fate of infiltrated water				
Evaporation	137.5	141.3	148.8	+7.5
Percolation	6.8	7.2	7.6	+0.4
Yield components				
Base flow	6.8	7.2	7.6	+0.4
Surface runoff	22.7	18.5	13.6	−4.9
Total yield	29.5	25.7	21.2	−4.5

[a] Units: km^3. Source: L'vovich (1969, p. 142).

The yield of solutes also increased. Nutrient outflows rose many fold (Likens *et al.*, 1969), indicating a breakdown of the nutrient-conserving mechanisms of the ecosystems. This increased yield also might result in an undesired enrichment of downstream bodies of water.

In connection with schemes for increasing yield of water, especially by manipulating ecosystems in a drainage basin, reference to two recommendations by the National Water Commission (U.S. National Water Commission, 1973, p. 138) is relevant:

> (1) The Congress and the President should direct Federal agencies having land management responsibilities to give adequate consideration to water yield as an objective of multiobjective land management plans.
>
> (2) Local non-Federal water management agencies, whose constituents would benefit from an increase in water supplies derived from land management practices, or public and private landowners who would benefit, should finance the additional cost of those management practices which are attributable to the water supply objective.

The first recommendation should promote sound research on whether a real yield increase can be achieved. The second says that if it can, the beneficiaries should pay the costs for what they get; these might be in foregone biological yield, habitat, or amenity value of the ecosystems that were manipulated.

Still longer term changes are implicit in changed practices of land management in the chernozem region of the Soviet Union. More shelterbelts, fall plowing, erosion control, and so on, all favor on-site conversion of water by transpiration and discourage off-site yield. Based on studies at Alekhin National Park near Kursk of several individual hydrologic processes in different ecosystems typical of the zone, L'vovich developed present and possible future water budgets (Table IV). The result of changes in infiltration, evapotranspiration opportunity, and percolation in the crop ecosystems of this region, which is 0.7 in cultivated land, over a long period of time shows in the yield of water in the rivers, vital for urban and industrial uses. Unregulated quick runoff is expected to decrease and base flow to increase, but not so much. As a result, total yield will decrease. The areal patterns of yield are shown in Fig. XVII-8. Industries and cities depending on stream flow will benefit from a more reliable supply but will have to use it more conservatively, recycling much of what they withdraw.

Areal Variation in Off-Site Flow

Mountains versus Lowlands The areal pattern of yield displays as much variability as its time pattern. Mountains, for example, often yield several times as much water per square kilometer as foothills or plains. They usually receive more rain and snow, experience lower rates of evapotranspiration over shorter periods of the year, and with shallow soil they store less water that plants can retrieve and transpire.

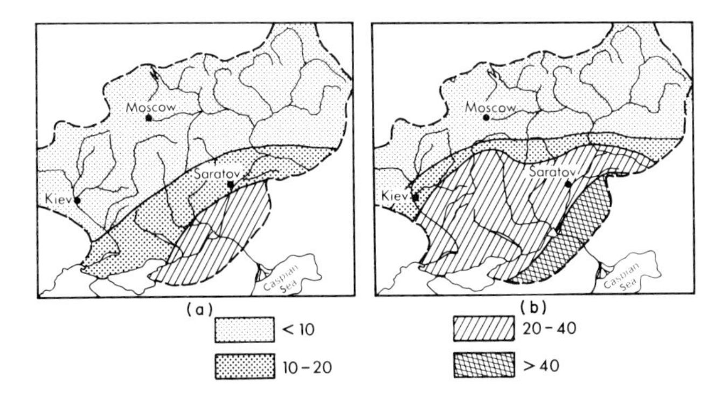

Fig. XVII-8. Changes in yield experienced in (a) the early 1960s from the 1950 level, and (b) expected reductions in the future as a result of changes in agricultural practices on crop ecosystems of southern European U.S.S.R. (L'vovich, 1969, p. 140). In percent.

A large fraction of mountain precipitation goes promptly into streams or into storage in snow cover, or groundwater where rocks are shattered. In all these forms of storage, it is relatively secure from vaporization and is likely to become liquid-water yield.

The largest values of specific yield cited by Parde' (1964) are those of seaward slopes of mountain ranges facing stormy latitudes of the ocean, 3000 mm or more in Norway, the Chilean Andes, and the South Island of New Zealand. Glacial regions in the Alps have yields around 2500 mm. Large basins in the Alps and Himalayas average 1000–1500 mm, three times the yields in adjacent lowlands.

A small mountainous area in a drainage basin may produce a major portion of the basin yield. The small area of the Barrington Tops, which makes up 0.05 of the basin in Australia's Hunter Valley, generates nearly a third of the yield (Fig. XVII-9). Its precipitation is more than 1100 mm yr^{-1} and the yield several hundred millimeters. The driest half of the basin, in contrast, generates only 10% of the yield (Table V).

A similar distribution characterizes the western United States, where the islands of high yield stand out individually. Most of them are confined to the high altitudes of the central and northern Rockies and the Sierra Nevada. The vast stretches of intervening lowlands and low mountains generate little excess water, as is true generally where ecosystems receive less than 200 mm of rain and snow. This pattern shows up well on a national yield map (Fig. XVII-10).

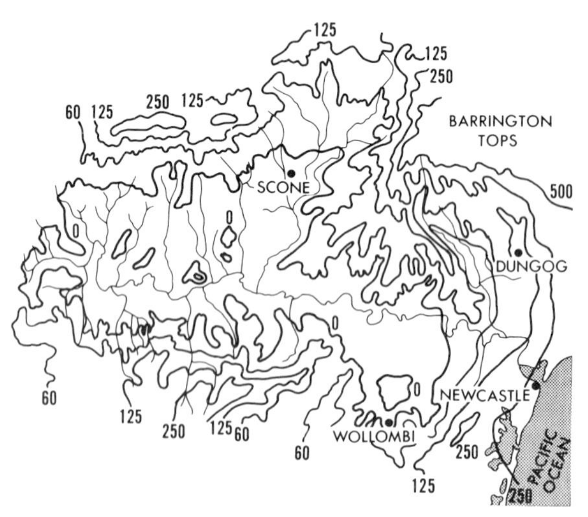

Fig. XVII-9. Annual surplus of water in different parts of the Hunter Valley, Australia (mm yr^{-1}) (Tweedie, 1963).

TABLE V

Mean Annual Yield Generated by Cumulated Areas of the Hunter Valley, Australia (Beginning with the Most Productive)[a]

Area (% of basin)	Yield (% of total volume)	Precipitation class for given percentage of area (mm)
5	29	>1100
10	43	>1050
24	77	>850
54	90	>675
75	97	>600
100	100	>400

[a] Source: McMahon (1964).

Runoff Regions of the United States The high cost of transporting water in sufficient bulk for economic use means that few large countries as such have a truly nationwide water problem, as far as yield is concerned, but their component regions often do. Such water resources regions as the upper Colorado basin (roughly that drainage area within the states of Utah, Colorado, and Wyoming), the Lakes states, and so on, are described in a national survey (U. S. Water

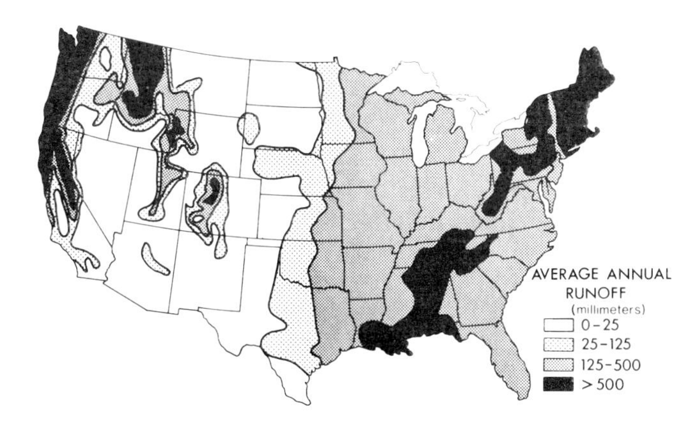

Fig. XVII-10. Average yield (mm yr^{-1}) in the conterminous United States. The high-yield islands of the western half of the nation rise above a general lowland yield less than 25 mm yr^{-1}, and are especially small in the southwest. Most of the agricultural ecosystems of the country (except spring wheat) also yield 100–400 mm of water annually (from U. S. Water Resources Council, 1968, p. 3-2-5).

TABLE VI

Off-Site Water Yield and Population of Selected Water-Resource Regions of the United States[a]

Region	Total annual water yield (1965) Mean	0.9 Freq.	Population (1960) ($\times 10^6$)	Per capita water yield Mean[b]	0.9 Freq.
	(10^6 tons day^{-1})			(tons day^{-1})	
North Atlantic	610	470	43.9	14	11
Great Lakes	236	174	25.5	9	7
Upper Colorado	49	33	0.3	160	110
Lower Colorado	12	4	1.5	8	3
California	245	126	15.6	16	8
Hawaii	50	—	0.6	83	—

[a] Source: Tables 3-1-2 and 3-2-1, U.S. Water Resources Council (1968) and U. S. Census (1960).

[b] For comparison, the following mean per capita amounts of stream flow are given by L'vovich (1973, p. 88) (tons day^{-1}): North America as a whole, 52; South America, 154; Australia (continent only), 66; Asia, 18; Europe, 13.

Resources Council, 1968). All regions have water problems, but each has a different mixture of them.

From these data it is possible to work out the annual water output in per capita terms. The data and the ratios of water output to population of each region are shown in Table VI for regions of special interest: the North Atlantic coast (Megalopolis), the Lakes region, three regions in the southwest, and Hawaii. Approximate mean conditions and those at a 0.9 frequency are presented. The spread between the mean and the 0.9 frequency columns in the table is an indication of reliability of yield. Reliability is greatest in the northeast, least in the southwest. Per capita yields are similar (approximately 8 tons per capita per day) in the Great Lakes and Lower Colorado water regions, and in California and Megalopolis (where they are slightly higher). These are small when compared with Hawaii and the Upper Colorado regions. The favorable situation of these two regions is, perhaps, an argument in favor of the restrained population growth they have had.

The report also assesses the seriousness of water-related problems of each water region of the country. Those having to do with adequacy of volume of yield, without reference to quality, are as follows (U. S. Water Resources Council, 1968, pp. 1–29):

(A) Adequacy is only a minor problem in some parts of these

regions: Those in the southeastern United States; Hawaii, Alaska, Puerto Rico.

(B) A moderate problem in some areas, or a minor problem in many areas: Regions of the northeastern United States (Ohio basin and north; west to include the Upper Mississippi); Columbia Basin—North Pacific Coast.

(C) A major problem in some parts of these regions, or a moderate problem in many parts: The Great Plains, from Canada to Texas; California.

(D) A severe problem in some parts, or a major problem in many parts of these regions: Intermountain country and the Rio Grande basin.

The geographic picture is a simple one, with few outliers. California, for instance, falls in class (C) with the states of the Great Plains rather than with the intermountain country (Fig. XVII-11).

The concept of "adequacy" refers to the level or intensity at which water is used in the economies of these regions at the present time, rather than to any uniform standard. In the intermountain regions, for instance, much water is used at low intensities, producing low-value products. Water used at higher intensities would make the matter of adequate volume less pressing.

Patterns in Other Countries The availability of liquid water for industrial purposes is vital in any modern economy. It has been

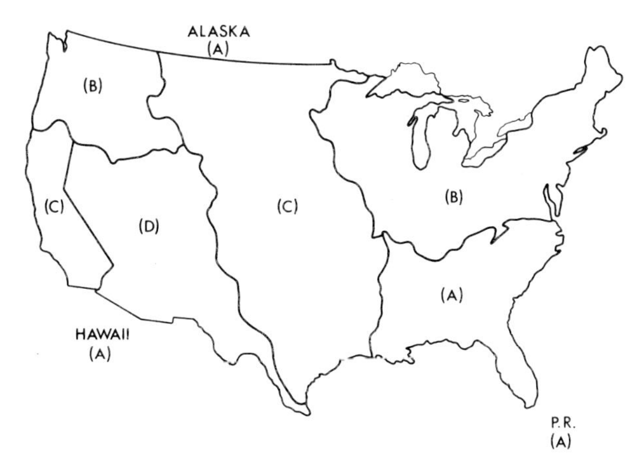

Fig. XVII-11. Adequacy of yield of water in different regions of the United States. Legend: (A) adequacy no more than a minor problem in a few areas; (B) a moderate problem; (C) a major problem; (D) a severe problem in some areas or major throughout (data from U. S. Water Resources Council, 1968).

calculated for the economic regions of the Soviet Union (Dreyer, 1969). The following is a selection of some regions of high and low yield (mm):

Caucasian republics	382
Northwestern region of the RSFSR	318
Central region RSFSR (near Moscow)	198
Central Chernozem region, RSFSR	96
Ukrainian Republic	86
Middle Asian republics (mountains and deserts)	92
Kazakh Republic	21

The distribution of yield is very uneven in Australia. A narrow strip in the low latitudes produces half the total of the continent. Most of the rest also comes from lands within 200 km of the coastline; the interior yields less than 10 mm yr^{-1}. Over the whole continent the average is 47 mm (L'vovich, 1972).

In Australia and in the other continents except South America areas near the coast produce large yields of water, while areas of low yield occupy much of the interiors. Yields less than 300 mm yr^{-1} characterize the interiors of Asia, North America, and much of Africa.

Equatorial South America, however, receives heavy rain not in the mountains but in the lowlands.* The basin of the Amazon, an area of 5 million km^2, yields, according to new measurements, 1060 mm yr^{-1}. Most of this comes from the lowland parts of the basin, but the Andes contribute most of the sediment in the characteristic "white waters" that contrast so markedly with the "black" and "clear" waters from other parts of the basin of this gigantic river† (Douglas, 1972).

Patterns of Quality Not only the adequacy of yield, but also its quality, are significant characteristics. An adequate yield is compromised by sediment, urban wastes, salt, thermal pollution, and so on. If we combine the official ratings of each of eight kinds of water problems (U.S. Water Resources Council, 1968) in a specific region, assigning arbitrary weights ("minor in some areas" of a region = 1, "moderate in some areas" = 2, "major in some areas" = 4, and "severe in some areas" = 8), we arrive at an aggregate number for that region. For the Great Lakes region, this number is 28 (Miller, 1972); for

* In these latitudes, the rain-generating systems of mountains and lowlands are often separate and independent (Chapter II).

† As an old saying goes, there are only two kinds of rivers: the Amazon, and all the others.

the Lower Mississippi region it is about the same, but represents a different mix of problems.

The serious problems in the Lakes region are thermal and urban pollution in receiving water bodies, while in the Lower Mississippi region they are flood damages, erosion of river banks, and excessive sediment in streams—and over wider areas, erosion of soil and threats to wetland ecosystems. The river problems result to a large degree from malfunctioning of the region's present ecosystems under the impact of heavy rainfall. The map in Fig. XVII-12 shows the distribution of this composite problems index over the country. Two severe problems are volumetric in the Rio Grande basin, inadequate yield and overdrafts on groundwater bodies; others represent ecosystem problems, the large yields of salt and sediment. The region's index is 46, much greater than in the regions of the Lakes and lower Mississippi. The problems index is the lowest (15 or 16) in one region which has only recently felt the impact of man and technology, Alaska, and in two where water management has been a matter of public policy and good stewardship over the years (Hawaii and the Tennessee Valley).

Yield from the Continents

The continental yields of water in base flow and in surface plus near-surface runoff are shown in Table VII. The outflow from ground-

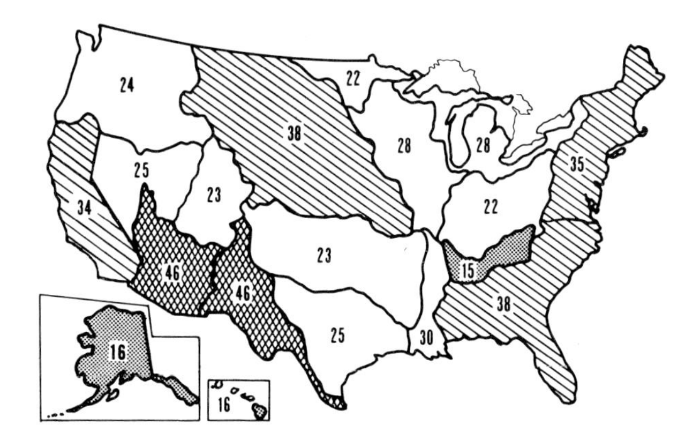

Fig. XVII-12. Weighted scores assigned to water-resources problems of different classes of severity. Note the generally favorable conditions in three widely separated regions (Alaska, Hawaii, and the Tennessee Valley), and difficult problems in the interior southwest (Miller, 1972, from data in U.S. Water Resources Council, 1968, p. 1-29).

TABLE VII

Annual Off-Site Yields of Water and Sediment by Continents[a]

Continent	Base flow	Surface runoff regulated by lakes and reservoirs	Unregulated surface and near-surface runoff	Total	Suspended sediment
Eurasia	83	15	201	298	0.47
North America	84	26	177	287	0.10
South America	210	5	368	583	0.06
Africa	48	15	76	139	0.03
Australia (alone)	7	2	38	47	0.05
Lands (except polar)	90	16	188	294	0.15
Fractional	0.30	0.06	0.64	1.00	

[a] Units: kg m^{-2}. Sources: L'vovich (1972); Holeman (1968).

water and the flow regulated by lakes and reservoirs can be grouped as "stable" flow, which, on a worldwide basis, is about 106 mm yr^{-1}. This amounts to more than a third of the total yield, 294 mm yr^{-1}. The northern continents are near the world average in total yield. South America with 583 mm yr^{-1}, has far more, thanks to the Amazon. Africa and Australia produce the smallest yields of liquid water from their soil and vegetation. All except 2×10^{-4} of the water of the planet lies almost motionless in the rocks or ocean basins (average residence times 3 millennia) or in the eternal ice (mean residence time 8.6 millennia) (Sokolov, 1974). It is hard to visualize water as taking part in any of the periodic processes we call cycles that are defined by a regular coming and going of the same molecules.

Associated Mass Fluxes Table VII also shows the suspended sediment carried off each continent by streams. As we saw earlier, this flow is only a fraction of the total sediment yield contributed by terrestrial ecosystems; much of it remains in temporary storage in flood plains or is filling lake basins and has not yet reached the sea. Here again the northern continents differ from the others, which all together contribute only a tenth of the total continent-to-ocean transport of 20×10^9 tons yr^{-1}.

The great rivers that rise in the mountains of interior Asia deliver a major part of the total: the Yellow and Yangtze in China deliver 2.7 billion, and the Ganges, Brahmaputra, and Indus in south Asia deliver 2.9 billion tons. Yields of sediment from their basins range from 0.5 to 3 kg m^{-2} yr^{-1}; other Asian rivers drain unstable mountain areas of similar large sediment yields.

The global mean flux into the oceans is 0.15 kg m^{-2} as suspended sediment. The flux of dissolved solids is of the same order of magnitude. From the United States it averages about 0.03 kg m^{-2}, the Mississippi basin 0.05 kg m^{-2} (Leifeste, 1974).

It begins to look as though the unbridled power of technologists to cause large off-site yields of forms of matter like mercury or cadmium, elements brought to our surface environment from long isolation in the depths of the rocks, or forms created *de novo* like DDT and vinyl chloride, is being slowed. The comfortable dream that mercury could be dumped into the nearest stream and sink to the bottom where it would lie until judgment day without affecting the environment, was rudely shattered on the biological realities. In the summer of 1970 a Canadian student showed that mercury did not lie eternally in the river bed but was being brought into active circulation in aquatic ecosystems. (Montague, K. and Montague, P., 1971, p. 73). When we looked, we found it in Wisconsin rivers too. Consternation ensued, and the rivers were closed to fishing.

Similar revelations had occurred earlier with DDT, to take the most blatant example, and later with asbestos and various polymers. The Water Quality Act of 1972 came none too soon to slow down the poisoning of man's habitat by off-site transports from urban and industrial (including industrialized farming) systems. Monitoring the fluxes of these potent substances and assessing the precise kind of danger they present to us should give a future writer on water problems much better factual material than we now have for describing these dangerous yields.

Many of the substances, like a fraction of the volume of eroded soil particles, end in the oceans. The tonnages tell the story less effectively than does the increase in the trace fractions of such elements as lead, which are damaging in even minute concentrations. The problem is particularly serious for substances that cannot be volatilized from the ocean, and therefore accumulate there. On the geologic time scale, some of these are incorporated into oceanic sediments and again locked away, but many are not. Remaining in the ocean water, they preserve a record of the yields of matter associated with off-site flow of water from terrestrial ecosystems.

TOTAL YIELD

The manner in which water moves away from the ecosystems on the earth's surface to which it was delivered by atmospheric processes depends on the delivery—usually sporadic and highly unpredictable—and on the receptiveness, storage capability, transmission, and energy availability characterizing these ecosystems. Delivery of rain at a high rate to an unreceptive surface produces rapid off-site runoff and flooding of other systems. Slower rain may not overtax the absorptive and storage capacities of the land, so that water is held on the site and eventually evaporated if time is long enough for the energy supply to power this slow phase-change process. Where precipitation comes as snow it remains in the solid state until sufficient energy is received to melt or evaporate it.

The yield of an ecosystem depends not only on deliveries of water and energy but also, particularly as regards the distribution of yield in time, on buffers in the system. Episodic delivery of atmospheric water confronts an ecosystem with a problem of uncertainty, which is dealt with by a soil-moisture storage that has a capacity ten times the size of a typical rain. This buffer is essential to the life of the system. The outputs from the system are even more episodic than the inputs. A healthy system, however, tends to minimize flashy surface runoff and to maximize percolation, which enters a large buffering storage, the groundwater. Base flow is yield of the steadiest kind.

Transmission capabilities of the land are important if water is delivered as rain and is mostly still in the liquid state when it leaves the site. What counts is the ability of the ecosystem and underlying rock to take in, store, percolate, and discharge water. If this capacity is large, outflow takes place as the steady movement of groundwater into streams of stable regime. If it is small, outflow expresses the episodic nature of the atmospheric processes; streams experience flash floods of silt-laden water, followed by long dry spells. The effects of the partitioning of water in ecosystems are seen as far as the rivers reach, far from the site from which the water came.

REFERENCES

Appleby, F. V. (1970). Recession and the baseflow problem, *Water Resources Res.* **6,** 1398–1403.

Benoit, G. R. (1973). Effect of agricultural management of wet sloping soil on nitrate and phosphorus in surface and subsurface water, *Water Resources Res.* **9,** 1296–1303.

Burwell, R. E., Schuman, G. E., Piest, R. F., Spomer, R. G., and McCalla, T. M. (1974). Quality of water discharged from two agricultural watersheds in southwestern Iowa, *Water Resources Res.* **10,** 359–365.

Colman, E. A. (1953). "Vegetation and Watershed Management: An Appraisal of Vegetation Management in Relation to Water Supply, Flood Control, and Soil Erosion." Ronald Press, New York, 412 pp.

Cooper, C. F. (1969). Ecosystem models in watershed management. *In* "The Ecosystem Concept in Natural Resource Management" (G. M. van Dyne, ed.), pp. 309–324. Academic Press, New York.

Costin, A. B. (1971). Water. *In* "Conservation" (A. B, Costin and H. J. Frith, eds.). pp. 71–103. Penguin Books, New York.

Douglas, I. (1972). The geographical interpretation of river water quality data, *Prog. Geog.* **4,** 1–81.

Dreibelbis, F. R. (1963). Land use and soil type effects on the soil moisture regimen in lysimeters and small watersheds, *Soil Sci. Soc. Am. Proc.* **27,** 455–460.

Dreyer, N. N. (1969). Water resources of the RSFSR and the other Union republics, *Sov. Geog.* **10,** 137–144.

Gottschalk, L. C. (1964). Reservoir sedimentation. *In* "Handbook of Applied Hydrology" (V.T. Chow, ed.), Chapter 17-I. McGraw-Hill, New York, 34 pp.

Grin, A. M. (1965). "Dinamika Vodnogo Balansa Tsentral'no-Chernozemnogo Raiona." Nauka, Moscow, 147 pp.

Hagen, L. J., and Skidmore, E. L. (1974). Reducing turbulent transfer to increase water-use efficiency, *Agric. Meteorol.* **14,** 153–168.

Hall, F. R. (1968). Base-flow recessions—a review, *Water Resources Res.* **4,** 973–983.

Hibbert, A. R. (1967). Forest treatment effects on water yield. *In* "Forest Hydrology" (W. E. Sopper and H. W. Lull, eds.). pp. 527–543. Pergamon, Oxford.

Holeman, J. N. (1968). The sediment yield of the major rivers of the world, *Water Resources Res.* **4,** 737–747.

Hornbeck, J. W. (1973). Storm flow from hardwood-forested and cleared watersheds in New Hamsphire, *Water Resources Res.* **9,** 346–354.

Houston, C. E. (1967). Drainage of Irrigated Land. Univ. Calif. Agric. Expt. Sta. Circ. 504, 40 pp.

Hudson, H. E. Jr., and Hazen, R. (1964). Droughts and low streamflow. *In* "Handbook of Applied Hydrology" (V. T. Chow, ed.), Chapter 18, pp. 1–26. McGraw-Hill, New York.

Hutchinson, R. D. (1970). Water Resources of Racine and Kenosha Counties, Southeastern Wisconsin. U. S. Geol. Surv. Water-Supply Paper 1878, 63 pp + pls.

Iorns, W. V., Hembree, C. H., and Oakland, G. L. (1965). Water Resources of the Upper Colorado River Basin—Technical Report. U. S. Geol. Surv. Prof. Paper 441. 370 pp. + pl.

Konstantinov, A. R., and Struzer, A. R. (1965). "Lesnye Polosy i Urozhai," Gidrometeorol. Izdat., Leningrad, 176 pp.

Lee, R. (1970). Theoretical estimates versus forest water yield, *Water Resources Res.* **6,** 1327–1334.

Leifeste, D. K. (1974). Dissolved-solids Discharge to the Oceans from the Conterminous United States. U. S. Geol. Surv. Circ. 685, 8 pp.

Lettau, H. (1969). Evapotranspiration climatomy. I. A new approach to numerical prediction of monthly evapotranspiration, runoff, and soil moisture storage. *Mon. Weather Rev.* **97,** 691–699.

Likens, G. E., Bormann, F. H., and Johnson, N. M. (1969). Nitrofication: Importance to nutrient losses from a cutover forested ecosystem, *Science* **163,** 1205–1206.

L'vovich, M. I. (1969). "Vodnye Resursy Budushchego." Izdat. Prosveshchenie, Moskva, 174 pp.

L'vovich, M. I. (1972). Vodnyi balans materikov zemnogo shara i balansovaia otsenka mirovykh resursov presnykh vod, *Izv. Akad. Nauk Ser. Geog.* **5,** 5–20.

L'vovich, M. I. (1973). "The World's Water: Today and Tomorrow." Mir Publ., Moscow, 213 pp.

Macfarlane, W. V. (1968). Protein from the wasteland. Water and the physiological ecology of ruminants, *Austral. J. Sci.* **31,** 20–30.

Mather, J. R. (1974). "Climatology: Fundamentals and Applications." McGraw-Hill, New York, 412 pp.

McGuinness, J. L., Harrold, L. L., and Amerman, C. R. (1962). Hydrogeologic nature of streamflow on small watersheds, *Am. Soc. Civil Eng. Trans.* **127,** Pt. I, 763–775.

McMahon, T. (1964). Hydrologic Features of the Hunter Valley, N.S.W. Hunter Valley Res. Foundation Monog. 20, Newcastle, 158 pp.

McMahon, T. A. (1968). Geographical interpretation of hydrologic characteristics in the Hunter Valley, *Austral. Geog.* **10,** 404–407.

Miller, D. H. (1972). Some aspects of [the] regional hydrology of the United States. Freiburger geog. Hefte 12 (Runoff Regimen and Water Balance II, Ber. 2, IGU—Comm. on IHD, R. Keller, ed.), pp. 51–87.

Montague, K., and Montague, P. (1971). "Mercury." Sierra Club, San Francisco and New York, 158 pp.

Muller, R. A. (1972). Application of Thornthwaite water balance components for regional environmental inventory, *Publ. Climatol.* **25** (2), 28–33 (Thornthwaite Memorial Volume I).

National Research Council, Committee on Water (1968). Water and Choice in the Colorado Basin, An Example of Alternatives in Water Management (G. F. White, Chairman). U. S. Nat. Acad. Sci. Publ. 1689, 107pp.

Pardé, Maurice (1964). "Fleuves et Rivières," 4th ed. Lib. A. Colin, Paris, 224 pp illus.

Parmele, L. H. (1972). Errors in output of hydrologic models due to errors in input potential evapotranspiration, *Water Resources Res.* **8,** 348–355.

Peck, A. J., and Hurle, D. H. (1973). Chloride balance of some farmed and forested catchments in southwestern Australia, *Water Resources Res.* **9,** 648–657.

Popov, O. V. (1968). "Podzemnoe Pitanie Rek." Gidrometeor. Izdat., Leningrad, 291 pp.

Rasmusson, E. M. (1968). Atmospheric water vapor transport and the water balance of North America. II Large-scale balance investigations, *Mon. Weather Rev.* **96,** 720–734.

Rauner, Iu. L. (1973). Energeticheskaia effektivnost' produktsionnogo soobshchestv, *Izv. Akad. Nauk Ser. Geog.* **6,** 17–28.

Rhoades, J. D., Oster, J. D., Ingvalson, R. D., Tucker, J. M., and Clark, M. (1974). Minimizing the salt burdens of irrigation drainage waters, *J. Envir. Qual.* **3,** 311–316.

Rockwood, D. M., and Jencks, C. E. (1961). Forecasting river runoff by coastal flow index, *Am. Soc. Civil Eng., J. Hydrol. Div.* **87** (Hy-2), 121–148.

Shawcroft, R. W., Lemon, E. R., Allen, L. H. Jr., Stewart, D. H., and Jensen, S. E. (1974). The soil–plant–atmosphere model and some of its predictions, *Agric. Meteorol.* **14,** 287–307.

Sokolov, A. A. (1974). Mezhdunarodnoe sotrudnichestvo po probleme mirovogo vodnogo balansa. *In* "Vlagooborot v Prirode i ego Rol' v Formirovanii Resursov Presnykh Vod" (G. P. Kalinin, ed.), pp. 14–26. Stroiizdat, Moscow.

Sokolovskii, D. L. (1959). "Rechnoi Stok. Osnovy Teorii i Praktiki Raschetov," 2nd ed. Gidrometeorol. Izdat, Leningrad, 527 pp.

Southeastern Wisconsin Regional Planning Commission (1970). A Comprehensive Plan for the Milwaukee River Watershed, Vol. I, Inventory Findings and Forecasts. SEWRPC, Waukesha, 514 pp.

Tweedie, A. D. (1963). Climate of the Hunter Valley, CSIRO, Melbourne, *Land Res. Ser.* No. 8, 62–80.

U. S. Census (1960).

U. S. Congress, 92nd. (1972). Public Law 92-500, Federal Water Pollution Control Act Amendments of 1972. 86 Stat. 816.

U. S. Corps of Engineers. Snow Investigations (1956). Snow Hydrology. Corps of Engineers, Portland, Oregon, 437 pp.

U. S. Geological Survey (1949). Floods of May–June 1948 in Columbia River basin, with a section on magnitude and frequency of floods by S. E. Rantz and H. C. Riggs. U.S. Geol. Surv. Water-Supply Paper 1080, Washington, D.C., 476 pp.

U. S. Geological Survey (1973). Water Resources Review, May 1973, p. 9.

U. S. National Water Commission (1973). New Directions in U. S. Water Policy. Summary, Conclusions and Recommendations from the Final Report of the National Water Commission. GPO, Washington, D.C., 197 pp.

United States Water Resources Council (1968). The Nation's Water Resources. The First National Assessment of the Water Resources Council, Washington, D.C., Parts 1–7 in 1 vol.

Wisconsin Statutes (1973). Statutes Ch. NR 102.03(3), Register Sept. 1973, No. 213, p. 13.

Chapter XVIII

WATER IN ECOSYSTEMS

In the preceding chapters we have followed the paths of water through ecosystems at the earth's surface. The centering of our thoughts on the dynamics of water has led us also to see that water occupies different environments in ecosystems, and that its moves from one to another and its exchanges with the local rocks and local air can be quantitatively expressed in the framework of water budgets, as also can be the various other kinds of matter that it carries.

The budget framework puts a spotlight on gaps in our knowledge and errors in what we think we are measuring. It also provides a way to identify spatial and time variations. We can see microscale patterns, mesoscale differences among ecosystems due to terrain and geology, and macroscale distribution over the surface of the continents of the globe. We can follow the regular changes with day and year, the sequence of events after an episodic input of atmospheric water when a rainstorm or snowstorm passes, and the uncertainties of their timing, as well as the longer changes that result from the slower atmospheric changes and human impact. In short, the budget of water in a given environment helps us see what happens between input and output and sort out some of the everchanging complexity of nature.

We have usually expressed our information about water, either at rest when in storage, or in motion, in terms of mass per unit area of the earth's surface—kilograms per square meter. Compared with the tonnages of water in the oceans or the great ice plateaus, millions of kilograms of water per square meter, the mass of water moving through terrestrial ecosystems and the associated local air and rocks is miniscule:

In local air	up to 10 kg m^{-2}
In air in plant layer—forest	0.5
—grassland	0.005

On leaves in rain	1–5
In plant tissue—forest	50
—grass	5
In soil	100–500
In accessible groundwater	5000

The fluxes of water are, in contrast, more impressive. In hourly rates, rain and snow are delivered at 1–100 kg m^{-2}. Snow melts at rates up to 5 kg m^{-2}. Water is pulled into dry soil at 50 kg m^{-2}, and pulled out again by transpiring plants at 1 kg m^{-2}.

Water percolating downward from ecosystems might average 0.1 kg m^{-2} hr^{-1}, groundwater bodies support base flow in streams at 0.01, and total off-site yields of water from ecosystems range up to 0.1–0.2 kg m^{-2} hr^{-1}. These units are comparable with the way fluxes of chloride and other forms of matter are expressed, as well as the biological harvests from ecosystems, which average around 0.1 g m^{-2} hr^{-1}.

ENVIRONMENTS OF WATER IN ECOSYSTEMS

Terrestrial ecosystems are complicated in structure, and the way they harbor water requires us to distinguish different environments that water occupies in them. We described raindrops, fog droplets, and snowflakes flung about in the boundary layer of a storm atmosphere as they approach the earth, and saw them impacted on leaves. A brief storage as intercepted water on leaves, described in largely qualitative terms because we have few measurements in the crown environment, displays a balancing between input and outgo.

The water storage in the foliage feeds down into transient storages on top of the ground, water detained in a short-lived liquid film or a longer lived mantle of snow. From this quiet environment, gravity outflows take the water into off-site runoff or capillary forces pull it into the soil, and so on through a succession of environments, each one offering its particular kind of surfaces to hold water and its particular mix of forces that bring water in and move it out. These forces constitute its energy environment.

Water in each level or environment of a living ecosystem is at the same time also a habitat for life. Humidity in the local air affects many ecosystem functions. Fungus spores take up residence in dew or rainwater on leaves, as in potato blight, and snow cover shelters fir reproduction. Moisture is a basic aspect of the soil climate that surrounds plant roots and the abundant life of the soil. Local groundwater outflows support lush wetlands and aquatic ecosystems, and base flow provides continuity of habitat in stream channels.

Storage of Water: Budget Analysis

Water can come to rest in each environment of an ecosystem, creating a storage. This might be no more than a film of dampness on leaves or litter, or it might be a snow cover as large as 1000 kg mass m^{-2}. What role is played by these storages?

A common feature of all such storages is a temporary lack of the energy necessary to move the water farther. The water remains at rest. The storage builds as long as inflow continues and outflow remains small, a simple restatement of the budget idea. In some environments, like ice caps or deep groundwater, the energy supply is small and water remains in storage a long time. Ecosystems, however, live at the earth's surface in a zone of high energy inflow, and so the water storages associated with them are usually brief. Sooner or later the energy supply increases and the stored water is again set in motion.

Another feature that characterizes water storages in ecosystems is that they afford an opportunity for more than one outflow of water, that is, for branching of the initial inflow into two or more avenues of outflow. Snow on pine crowns might blow downwind into a clearing; it might fall under the tree that intercepted it, it might melt and drip off, or it might even, although rarely, evaporate. Interception storage affords opportunities for several kinds of outflows to occur, depending on how much energy in which form is supplied to the tree crown and on the structure of the crown environment.

Liquid water detained on the surface of the ground affords a branching opportunity between overland runoff, near-surface runoff, and infiltration, depending on energy conditions and the features of this particular environment of water. The way this branching goes, especially as between the ecosystem and some off-site destination, is important to the future water status of the system and hence the object of many agricultural practices. It also determines the quality of system yield.

Setting up the water budget of one of these storages helps us evaluate the branching process and determine the relative importance of the different outflows. The budget says that the sum of the outflows equals the inflow corrected for any change in the amount of storage itself, and thus provides a check list for measurement programs. Moreover, some outflows are accelerated by a rise in storage. When a burst of rain thickens the film of water detained at the ground surface, there is only a minor response on the part of the outflow via infiltration but a large response in surface runoff—a budget concept basic to flood hydrology. In the case of a burst of snow falling on a snow cover, however, the thickening of the storage has little current effect on outflows from the snow cover, which wait for a rise in energy supply to put them in movement. Here the budget concept is expressed in terms of the growth of storage. A later budget will be cast in the spring to determine the outputs from this storage at that time, and the way water is partitioned into off-site flow, evaporation, and infiltration into the soil of the site.

Fluxes among Environments: Budget Analysis

Water moves from storage in one environment of an ecosystem to another; the outflow from one storage zone becomes the inflow to another. This dynamic aspect of water is in part a cycling or recirculation, but in larger part a throughput of a substance that also is a vehicle for other forms of matter, nutrients or pollutants, as the case might be, that are dissolved or suspended in it.

To move any form of matter requires the expenditure of energy. Water being no exception, each water flux in an ecosystem is associated with a conversion of energy from one form to another, and we must look to see where the energy comes from. Solar radiation, a direct measure of energy supply, was found more closely related to the daily regime of evapotranspiration from well-watered meadow than is air temperature, which is merely an indirect indicator of the presence of energy of one particular kind. We noted that dew is deposited in strict accordance with the rate at which the heat of condensation can be removed, and that a cloud passing overhead radiates enough heat to the ecosystem to upset the chilling requisite to dew formation. Snow melting, especially in the West, is highly dependent on the input of radiant energy to it and becomes mobile with the increase of this input in spring. While the subject of energy in ecosystems is deferred, it is clear that we cannot talk about water fluxes without some mention of energy. The intimate connection between water fluxes and those of energy and of other forms of matter is a fortunate circumstance that offers us additional avenues for investigating specific situations.

Alternating Storage and Fluxes

The alternation of water fluxes and water storages in a succession of environments in an ecosystem lends itself to the casting of water budgets. In each environment—the local air, foliage, litter, ground surface, soil, local rock—we search out the inputs, storages, and outputs of water. Investigating individual environments in terms of inflows and outflows affords the methodological advantage of showing us where measurements are most needed and where existing information, perhaps hallowed by long practice though it may be, might be based on wrong physical principles and be misleading in our research. It also brings out some of the connections of the different parts of an ecosystem as a system.

The succession of environments is tied together by moving water and the flows of energy that make it move, more or less in the same order as the chapters in this book. In some systems the above-ground parts are adapted to collect the maximum amount of rainwater and get it into the soil as fast as possible. Even where no functional adaptation is present, a connection exists where a water flux exists. This connectivity role of water is enhanced by the action of water as a solvent and a vehicle, easily loaded up and efficiently carrying its freight as a mass flux. Unloading is best done in nature's own time, if human actions have not overloaded the vehicle; suspended sediments settle out, solutes are left behind at sites where water evaporates, organic compounds oxidize or are decomposed. The transport of metabolites washed off leaves into the soil represents, for instance, a nutrient cycling important to the whole ecosystem though apparently small in mass-budget numbers. Both this cycling and the availability of water for transpiration are factors in biological productivity of a system.

UNKNOWNS AND UNCERTAINTIES IN WATER BUDGETS

The discipline of balancing outflows from an environment against inflows makes us aware of fluxes otherwise easy to overlook. Perhaps they are not visible, being in vapor form, or have only a brief, spasmodic existence, or have been slighted in the folklore that still is found in geophysical thinking. On the other hand, some fluxes are given more credit than they are due, like evaporation of intercepted snow.

Whatever the reason for ignoring or wrongly evaluating a water flux, we are faced with reality when we have to place it in an inflow–outflow budget and make the budget balance. If we also cast the associated energy budget, we soon see whether a certain flux can actually move in the volume we think it does. Then we come to question the folklore about all the snow in the trees evaporating. Similarly, the slow downward flux of moisture in unsaturated soil came to be recognized when soil-moisture budgets cast in field research on evapotranspiration produced inconsistent results.

As a consequence of casting budgets, we have even come to question the accuracy with which we measure the premier flux in ecosystem, basin, and regional hydrologies for two hundred years—precipitation; and it is about time we did so!

Inconsistent findings in the budgets of adjacent ecosystems also alert us to an unpleasant fact of life, sampling variability. This problem has come to rival those due to instrument malfunction or exposure, and has occasioned the resort to methods of statistics that have also been valuable in dealing with real variations over time or space.

Measurement of all Fluxes

Budget framework methods require accurate measurements of *all* the fluxes in a budget. This becomes a stringent condition, given the state of the art in measurement. For example, if we elect to use a water-budget framework to evaluate evapotranspiration from a corn field, we must measure the following:

(1) infiltration of water into the field, or, alternatively, rainfall delivered to the vegetation cover less intercepted water evaporated, and less surface and subsurface runoff from the field;

(2) percolation of water to layers below the soil occupied by plant roots, and upward movement of water under capillary forces into the root zone, if the groundwater table is high;

(3) change in the amount of moisture stored in the soil between the beginning and end dates of the period in question.

The sum of all these quantities estimates the vapor flux from the corn field over the whole period, but not necessarily over parts of it. For shorter periods it is necessary to take soil-moisture readings more frequently; the accuracy of available instruments (see Chapter IX) sets a time interval below which real differences cannot be detected and in which, therefore, evaporation cannot be determined. These periods are several days to a week long.

The water-budget framework, faultless in concept, presents difficulties in practice if we want to evaluate a residual unmeasured component of it, because measurements of the other fluxes in the budget are not accurate. In many terrains precipitation is hard to measure, as we saw earlier. In some lithologic provinces the rate of deep percolation is unknown, a difficulty found even in carefully selected experimental drainage basins. Delineation of the boundaries of a groundwater basin is usually imprecise. Movement of water in unsaturated soil is not well understood; and even surface runoff is not usually measured with accuracies better than ±10%.

Each difficulty represents an area of ignorance about moisture movement in the media overlying and underlying the drainage-basin surface—the lower air, the vegetation cover, the soil, and the upper layers of rock. Each is magnified in dissected landscapes, where the shape of the earth's surface influences airflow and exposes a complicated pattern of rock outcrops and soils. Attempts to estimate evapotranspiration from water-budget considerations in rough terrain are commonly made in the absence of other procedures, but the estimators must expect a substantial margin of error.

Rugged drainage basins often are forest covered. The correlation of steep slopes and forest cover is high as a result of the selective action of such agents as fire, agricultural land clearing, soil-fertility patterns, and also by the relative invulnerability of forested slopes to being lowered by surface runoff, which is diminished by the high porosity of the forest floor. Even lowland forests present aerodynamic problems in the determination of water delivery and interception.

Precipitation and the other water transports are not well measured in mountains, and as a result the water budgets of forested mountains are inaccurate even over intervals as long as a year. Evapotranspiration estimates, calculated from precipitation and runoff, are likely to be in error by a quarter or a third.

The Era of Product Quality Uncertainties and unknowns in ecosystem dynamics are increased as ecosystems have come to be recognized as potential area sources of pollutants. Yield quality affects downslope and downstream systems of all kinds—wetlands and swamps as aquatic ecosystems, rivers that produce fish and offer recreation values, urban water supplies, industries in need of clean water, the Great Lakes and the ocean. Attention once focused only on volumetric analyses but has turned to include chemical analyses of nutrients, salts, organisms and toxic substances that water, especially off-site flow, carries in great variety and in an environment that maximizes the opportunity for further chemical interaction.

The Era of Infiltration led hydrologists to look beyond the raingages of the period when gross rainfall-runoff indexes represented sophistication in hydrology (not so long ago in some applications of urban hydrology), and to study the characteristics of ecosystem soils that affect infiltration and soil storage capacities. The Era of Evapotranspiration led them to look at ecosystems between storms, the water relations of plants, and feedbacks in plant–soil–atmosphere water and energy relations. Both these shifts of emphasis tended to supplant ordinary input–output relations by including the storages, fluxes, and chemical recombinations that take place in the environments of ecosystems.

They correspondingly complicated the task of measurement. From a time when a few raingages were read once a day at minor cost, we went to infiltration theory with its recording gages, depth–area–duration analyses of storms, and hard-to-interpret plot experiments; all these brought problems in data acquisition. Snow hydrology and evapotranspiration hydrology introduced the need to determine energy fluxes, and instrument and sampling problems multiplied. Now the Era of Product Quality demands chemical and biological analyses of water yields, especially from malfunctioning or disturbed ecosystems that threaten to produce polluted runoff. Improving yield quality creates a need to understand processes within ecosystems in which mass fluxes are carried by water fluxes, yet the difficulties in representative time sampling of a host of solutes and suspensions are formidable. The unknowns and uncertainties have increased.

PATTERNS OF DISTRIBUTION

Spatial Differentiation in Water Budgets

The budgets of water, along with those of associated fluxes of energy and other forms of matter, provide a means of comparing not only different levels in an ecosystem but also one ecosystem with another. Each ecosystem has its special coupling with the boundary layer in a storm; the amount of rain or snow that it collects is likely to be different from what its neighbor receives. From there on, its individuality grows. Adjacent ecosystems intercept and pass on rain and snow differently; they shelter different depths of snow; they infiltrate greatly different amounts of water into soil storage, as is evident in the large variation in sampling soil moisture. Differences in slope friction and infiltration associated with grassland systems in New Zealand, for example, alter the duration of effective rain, hence its intensity, and shed vastly different volumes of surface runoff. Evaporation from different ecosystems and their total yields of surface runoff and percolation differ widely, often in ways fostered by land management.

Since the presence of particular ecosystems often reflects a special mutual interaction with soil and a dependence on terrain shape, the water fluxes exhibit spatial variation on a mesoscale. This is evident when we try to determine the sources of water in a flood wave coming out of a mountain drainage basin; some slopes contribute little or nothing, while others are supplying most of the streamflow. On this scale, spatial variations in the local air (shelter and exposure, warmth, moisture content, and so on) and in the local rocks (porosity, permeability, chemical composition, and so on) become apparent in the water budgets of ecosystems. Since the basin water budgets needed for water management are composited from the mix of ecosystem budgets, this mesoscale variation is important in determining timing and volumes. The interpretation of water processes occurring at different spatial scales provides for much of the scientific attraction of the budget. We have the long controversy between plot experiments and the often oversimplified studies of operational drainage basins, suggesting that short-cutting the budget method has been too prevalent.

Regional water budgets are again a mixture of those of many small drainage basins and do not make sense if the smaller scale budgets do not accurately balance. When properly drawn, however, even if they are skeletonized into precipitation, yield, and evaporation, they portray the regional environments of crop and wildland ecosystems across the nation. On a still greater scale there are water budgets of continents and the globe.

In Fig. XVIII-1 inputs and yields are shown for large areas. Net yield is shown by the departures from the 45° line, there being none from the landlocked interiors of several of the continents. Intensity of exchange increases toward the upper right. It can be seen that water turnover is most rapid over the oceans and only moderate over terrestrial ecosystems. Terrestrial ecosystems bear the following relations to the whole:

(a) they encompass 0.30 of the global area;
(b) they receive 0.21 of the rain and snow; and
(c) they energize 0.12 of its total evaporation.

This minor place is, however, significant because of the yield of terrestrial ecosystems, and is intellectually interesting because of the complexity of the environments they provide for water storages and fluxes.

Variations over Time

Rhythmic The framework of the water budget with its auxiliary energy and other mass budgets has time as an inherent element. Water fluxes are measured in kilograms per square meter in each hour

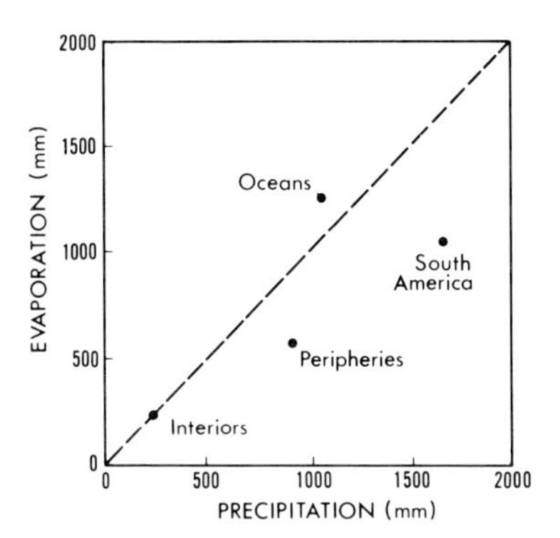

Fig. XVIII-1. Mean annual precipitation and evaporation from the oceans, continental interiors, and peripheral regions, and South America (mm) (data from L'vovich, 1973; Karasik, 1974).

or day, and the filling or emptying of a storage takes time, sometimes little, as in the wetting of a leaf canopy in a heavy shower, sometimes a great deal, as in the building up of the Sierra snow cover or the draining out of a body of groundwater. Residence times of a water molecule in a particular environment vary from a few hours for intercepted rain, a few days in the local air and 10–12 days in the free atmosphere, to months or a year in the soil-moisture reservoir, to times of geologic scale in water-bearing rocks.

Variations of a given flux or storage over time result from variations in the initial input of water or energy. Thus the regular cycles in the solar energy absorbed by an ecosystem force regular diurnal and annual cycles in snow melting and particularly in evapotranspiration, and these cycles result in regular variations in their competitor fluxes, like the percolation pulses each winter and spring from the lysimeters at Coshocton.

Just as water fluxes are moderated by the presence of storage capacity for water, so energy fluxes are moderated by storage capacity for heat. The large capacity of Lake Ontario to store heat transmutes the June maximum of solar input into a regime of energy available to power evaporation that is out of phase by nearly half the cycle. The circadian and annual regimes of all the water and energy fluxes tell a fascinating narrative; each one is differently damped down and lagged by the movement of water in or out of storage in different zones.

For example, the regime of actual evapotranspiration from a corn field is influenced by the regimes of solar energy, soil temperature, and air temperature, of snow melting, rainshowers, and soil moisture, and also by the ecosystem's own rhythm in extending its roots, covering the soil with its leaves, and shifting from vegetative growth to grain development. As we saw, all these affect transpiration, as well as such other water fluxes in the corn ecosystem as infiltration and percolation.

Episodic On a time scale intermediate between the day and year occurs another, but irregular variation in ecosystem water budgets. These episodic time variations are significant both for land managers and river engineers, and have occasioned much research in American hydrology. This time scale is the length of the rundown or exhaustion period after a pulse of rain has been fed into the system. Since these pulses often are sporadic in appearing, the length of post-storm periods is never certain.

Many of the water fluxes in an ecosystem spring up during the rain, like the drip from wet leaves, infiltration into the soil, or surface runoff, then decline and die out soon afterward, depending on the size of the storage that nourishes them. The pulse of surface and near-surface runoff gets into stream channels and is on its way down the river in a few hours. We try to follow this pulse by means of the rainfall–infiltration–surface flow budget. In fact, infiltration analysis is a sequential application of the water budget at short intervals of time.

Other fluxes start up after a rainstorm in a more deliberate manner, for example, the renewed extraction of moisture as plant roots experience the rewetting of the soil body. The length of the period of transpiration after soil wetting depends, as we saw, on the energy input, the storage capacity of the soil for water at low capillary tensions, and rooting behavior. It is one of the major factors in the biological activity of ecosystems and their development, growth, and yield. We try to follow it by computing the day-to-day budgets of soil moisture, going through one post-storm exhaustion period after another over a growing season.

The importance of these two forms of post-storm ecosystem yield (storm flow into a drainage network, and biological production resulting from replenished soil moisture) gives point to a fundamental characteristic of the bursts of rain that set these and their associated fluxes into motion—their uncertainty. Unlike the predictable time regimes of day and year, the arrival of the initial water inputs that generate storm runoff and replenish soil moisture can only be described as episodic. This essential unpredictability, which was emphasized in the early chapters of this book, is bound to persist through all the subsequent fluxes of water in an ecosystem. They also can be described as episodic, although episodes of the more distant fluxes like percolation might be more frequent in one season than another. The outflows from a groundwater body come from a storage large enough to buffer single rain episodes, but still may show a response to a month of heavy rains, for instance, or to several months of little rain. Because water budgets account for all the water every instant, successive casting of budgets can help follow these pulses as they die down after the storm.

The episodic and unpredictable nature of atmospheric rainstorms and their aftermath has been a theme running through every chapter of this book. Ecosystems have to live with unpredictability, and usually manage to do so, though at a cost. Our attempts to understand the water fluxes and storages in ecosystems also encounter the uncertainty arising from this atmospheric process; some fluxes operate only briefly, rising abruptly to high rates of transport. As a result, measurement is difficult and often inaccurate, as we have seen. Casting successive budgets can, of course, provide a running check on the accuracy of measurements and assess the role of such built-in ecosystem buffers as soil-moisture storage.

The effects of purposeful or inadvertent changes brought about in ecosystems by man are sometimes visible only during the short episodes of infiltration or surface runoff, when measurement is difficult. Trying to understand these effects by analyzing only monthly or longer period totals of rainfall and runoff, for instance, has generally been futile because it leaves the investigator in ignorance about what actually happened at the critical environments in the ecosystem. Horton's frequently quoted recommendation to walk over a drainage basin during a rainstorm in order to see what is going on is good, needing to be supplemented only by an injunction to measure as well as to see these fluxes of water, and to verify the measurements by plugging them into a total water budget. Of course there is still the question of how one applies the observations of a storm of 10 kg m^{-2} to estimate what might happen in a deluge of 500 kg m^{-2}! And who knows when such a deluge might occur? Or how many ecosystems it will cover?

Long Range On a time scale longer than the year, there are other sources of uncertainty and unpredictability in the hydrodynamics of ecosystems. These are the changes from year to year and over longer periods of time, including those brought about by technology.

One year is different from the next with respect to the water budget of an ecosystem, particularly as a consequence of a different timing or frequency of rainstorms. Spring 1973 in Wisconsin had frequent rains following on the snow-melting time, then after a severe storm on 16 June only small, far apart rains occurred for several months. Most of the fluxes set in motion by rains died out; while base flow from groundwater held up well, the store of accessible moisture in the soil was exhausted. The streams ran high while the pastures dried up. Transpiration declined, and the total yield of the basic crops dropped substantially, as did the milk production dependent on these crops. The drop in yield from the preceding year was a result primarily of timing of the rain inputs into cropland ecosystems and hence of the subsequent water fluxes in the budgets of these systems, especially the extraction of soil moisture for transpiration.

Changes in the frequency or spacing of rainstorms can persist over still longer periods of time, like the series of dry years over the Plains in the 1930s and the 1950s. These catastrophes can be meaningfully described by their diminished water budgets, soil moisture in particular. Such data are then used to delineate their shifting shapes and to compare one year or decade with another.

Human Impact on Water in Ecosystems

The impact of man on ecosystems can be expressed in several ways, including the destruction of amenity or recreational environment; another way is by determining changes in the components of the water budget. In spite of the difficulties in measuring changes in some of these fluxes, e.g., evapotranspiration or percolation, the budget still presents a framework to guide analysis.

If interception of snowfall should be reduced by a logging program, what changes in other water fluxes will follow, and what change in ecosystem or basin yields? If deep plowing is done every fall, or if shelterbelts are planted, how are the water fluxes changed, and how are transpiration and off-site yield changed?

If weather modification could, perhaps, increase the volume of rainstorms but not their spacing, what ecological effects would follow and will off-site yield be increased? When irrigation water is applied to dry-land ecosystems and soils, how long before swamping and salting show up and the project asks Congress for a bail-out?

When Illinois farmers flood their cornfields with cheap ammonia, what nitrate concentration will be found in the local body of groundwater? When salt covers the roads, how much of it is carried off-site and how much of it lodges in the groundwater? As creeping urbanization extends impermeable surfaces over rural land with or without a sewer network, what changes occur in each water flux in the new suburban tract ecosystem and in off-site flooding and groundwater contamination?

All these interesting modifications in ecosystem structure or function, with their potentials for benefit or for damage, which is perhaps irreversible, can be analyzed by means of the water budget. Not all the consequences of man's impact on them can be determined from such analysis, but many things are associated with water in one way or another. Changing a system's water fluxes affects its operation and yields, and can be understood in the framework provided by the water budget.

WATER IS EVERYWHERE

Water exists not only inside the organisms that comprise ecosystems at the surface of the earth, but from time to time occupies their environments at different levels. It exists in storages over varying lengths of time when input of water to an environment exceeds outflows, and it moves from one environment to another. The outflow from one storage becomes the inflow to the next, often serving as the vehicle for fluxes of other forms of matter. Its presence and movements can be observed and measured within the framework of the water budget.

Budget analysis identifies areas in which our knowledge about water dynamics in an ecosystem is lacking or dubious. Often in practice, if combined with budgets of energy and of other forms of matter, it is a useful guide to field measurements, a check on folklore about water, and a key to a better understanding of one aspect of the natural world.

Determination of the fluxes and storages of water in an ecosystem becomes a quantitative means of differentiating it from other ecosystems. Such mesoscale differences in water budgets can be aggregated into the water budgets of drainage basins and regions, to depict the diversity of the surface of the earth.

Water fluxes and storages in an ecosystem change over time, and the water budget provides a means of following these sequences. Some changes are rhythmic, set in operation by diurnal or annual pulses of solar energy. Others are episodic, being the chain of fluxes and storages initiated when a passing storm cell in the atmosphere impacts a volume of water on the canopy of an ecosystem. These episodic disturbances slowly run down, as the water is eventually dissipated in vapor outflow or off-site yield during the interstorm period.

Longer range variations in the water budget of an ecosystem result in years when rains are more frequent than usual, or less frequent, or come in a different season depending on soil-moisture and other storages available. They also result from changes in ecosystems structure at the hands of man and machines. Modifications in ecosystems can thus be interpreted in terms, for instance, of an altered opportunity for water to be infiltrated, a change in its interception, an increased drainage of local groundwater, a lengthened period of transpiration, a reduction in surface runoff or in infiltration rate or in percolation, or degradation of yield. The altered budget of the fluxes and storages of water measures the direction and intensity of the modifications made by man, as it does other changes over time or in space in ecosystems at the earth's surface.

REFERENCES

Karasik, G. Ia. (1974). "Vodnyi Balans Iuzhnoi Ameriki." Akad. Nauk SSSR, Mezhduved. Geofiz. Kom., Rezul'taty Issledovanii po Mezhdunarodnym Geofizicheskim Proektam. Moskva: Sovetskoe Radio, 1974, 111 pp. Engl. summ.

L'vovich, M. (1973). "The World's Water: Today and Tomorrow." Moscow: Mir Publishers, 213 pp. (Transl. by L. Stoklitsky of Vodnye Resursy Budushchego. Moskva: Izdat. Prosveshchenie, 1969.)

INDEX

In this index an effort has been made to bring out themes that integrate water dynamics in ecosystems. These include the variability, uncertainty, and episodic occurrence of most water fluxes, the sequential alternation of fluxes and storages and their environments and energy sources, and the spatial contrasts and linkages of systems at the surface of the earth.

B

C

D

F

G

H

I

M

Q

R

T

U

International Geophysics Series

EDITED BY

J. VAN MIEGHEM
(July 1959–July 1976)

ANTON L. HALES
(January 1972–December 1979)

WILLIAM L. DONN
Lamont-Doherty Geological Observatory
Columbia University
Palisades, New York

Volume 1 BENO GUTENBERG. Physics of the Earth's Interior. 1959

Volume 2 JOSEPH W. CHAMBERLAIN. Physics of the Aurora and Airglow. 1961

Volume 3 S. K. RUNCORN (ed.). Continental Drift. 1962

Volume 4 C. E. JUNGE. Air Chemistry and Radioactivity. 1963

Volume 5 ROBERT G. FLEAGLE AND JOOST A. BUSINGER. An Introduction to Atmospheric Physics. 1963

Volume 6 L. DUFOUR AND R. DEFAY. Thermodynamics of Clouds. 1963

Volume 7 H. U. ROLL. Physics of the Marine Atmosphere. 1965

Volume 8 RICHARD A. CRAIG. The Upper Atmosphere: Meteorology and Physics. 1965

Volume 9 WILLIS L. WEBB. Structure of the Stratosphere and Mesosphere. 1966

Volume 10 MICHELE CAPUTO. The Gravity Field of the Earth from Classical and Modern Methods. 1967

Volume 11 S. MATSUSHITA AND WALLACE H. CAMPBELL (eds.). Physics of Geomagnetic Phenomena. (In two volumes.) 1967

Volume 12 K. YA. KONDRATYEV. Radiation in the Atmosphere. 1969

Volume 13 E. PALMEN AND C. W. NEWTON. Atmospheric Circulation Systems: Their Structure and Physical Interpretation. 1969

Volume 14 HENRY RISHBETH AND OWEN K. GARRIOTT. Introduction to Ionospheric Physics. 1969

Volume 15 C. S. RAMAGE. Monsoon Meteorology. 1971

Volume 16 JAMES R. HOLTON. An Introduction to Dynamic Meteorology. 1972

Volume 17 K. C. YEH AND C. H. LIU. Theory of Ionospheric Waves. 1972

Volume 18 M. I. BUDYKO. Climate and Life. 1974

Volume 19 MELVIN E. STERN. Ocean Circulation Physics. 1975

Volume 20 J. A. JACOBS. The Earth's Core. 1975

Volume 21 DAVID H. MILLER. Water at the Surface of the Earth: An Introduction to Ecosystem Hydrodynamics. 1977

Volume 22 JOSEPH W. CHAMBERLAIN. Theory of Planetary Atmospheres: An Introduction to Their Physics and Chemistry. 1978

Volume 23 JAMES R. HOLTON. Introduction to Dynamic Meteorology, Second Edition. 1979

Volume 24 ARNETT S. DENNIS. Weather Modification by Cloud Seeding. 1980

Volume 25 ROBERT G. FLEAGLE AND JOOST A. BUSINGER. An Introduction to Atmospheric Physics. Second Edition. 1980

Volume 26 KUO-NAN LIOU. An Introduction to Atmospheric Radiation. 1980

Volume 27 DAVID H. MILLER. Energy at the Surface of the Earth. 1981

Volume 28 HELMUT E. LANDSBERG. The Urban Climate. 1981

Volume 29 M. I. BUDYKO. The Earth's Climate: Past and Future. 1982

In preparation

Volume 30 ADRIAN E. GILL. Atmosphere–Ocean Dynamics